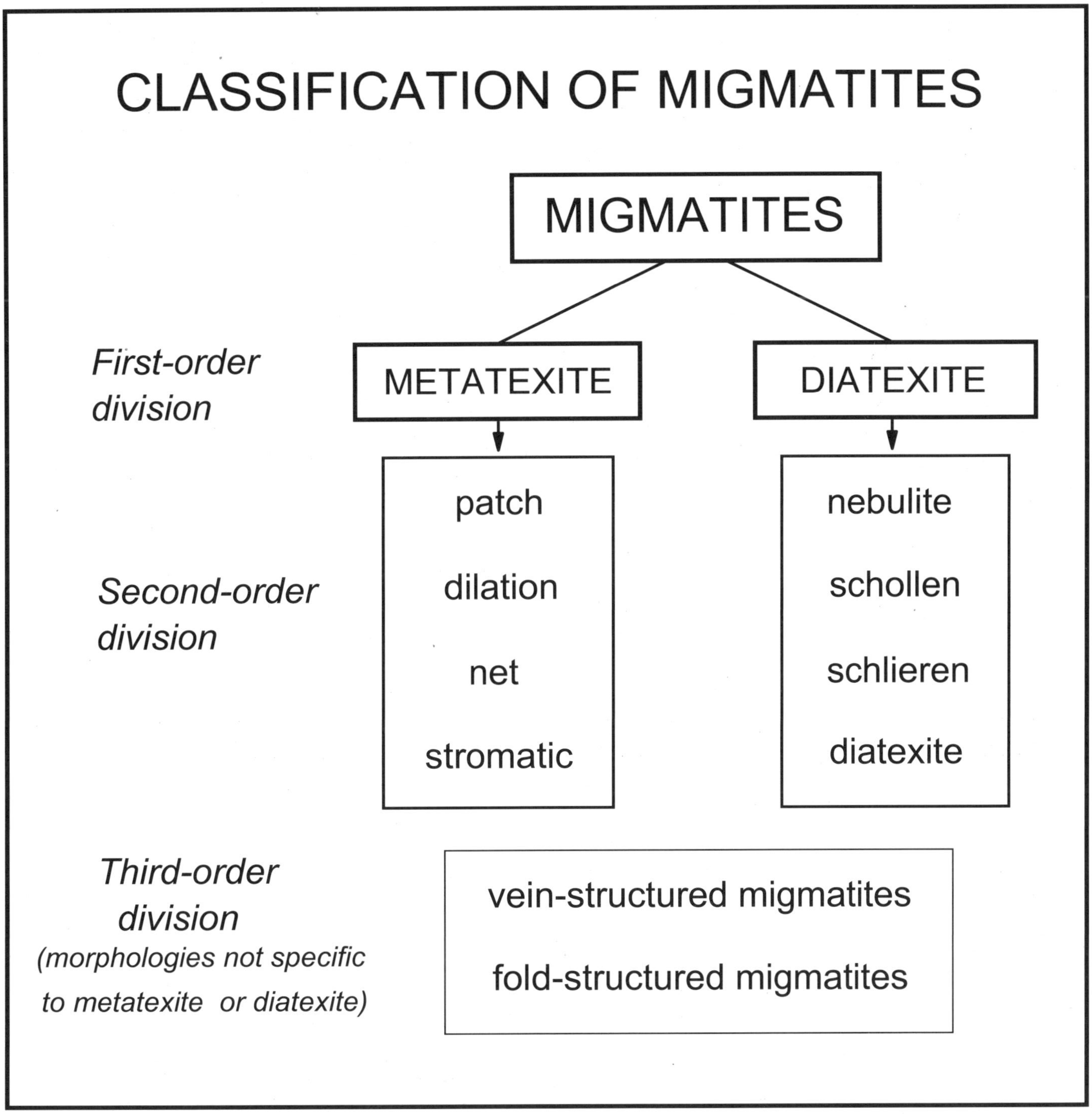
CLASSIFICATION OF MIGMATITES
MIGMATITES
First-order division
METATEXITE
DIATEXITE
Second-order division
patch
dilation
net
stromatic
nebulite
schollen
schlieren
diatexite
Third-order division (morphologies not specific to metatexite or diatexite)
vein-structured migmatites
fold-structured migmatites

Atlas *of* Migmatites

About the cover:

Front cover
Metatexite migmatite developed from a foliated granodiorite showing leucosome in dilation structures. Laag Mountain area, British Columbia. Photograph by Paul McNeill.

Rear cover
Diatexite migmatite with schlieren, developed from a pelitic protolith. Quetico Subprovince, Ontario.

The Mineralogical Association of Canada gratefully acknowledges the financial contribution of the following organizations:

Canadian Geological Foundation

Natural Resources Canada

Atlas *of* Migmatites

Edward W. Sawyer

The Canadian Mineralogist
Special Publication 9

Mineralogical Association of Canada
Association minéralogique du Canada

NRC·CNRC
NRC Research Press

Printed in Canada on acid-free paper. ♾

ISBN 978-0-660-19787-6
ISSN 1717-6387
NRC No. 46331

Library and Archives Canada Cataloguing in Publication

Sawyer, Edward William, 1951–
Atlas of migmatites/ E.W. Sawyer

"NRC Research Press"
Issued by: National Research Council Canada
Co-published by Mineralogical Association of Canada
Includes bibliographical references
ISBN 978-0-660-19787-6

1. Migmatite. 2. Migmatite — Pictorial works. 3. Petrology.
I. National Research Council Canada II. Mineralogical Association of Canada
III. Title.

QE475.M5S38 2008 552'.4 C2008-980022-2

NRC Monograph Publishing Program

Inquiries: Monograph Publishing Program, NRC Research Press, National Research Council of Canada, Ottawa, Ontario K1A 0R6, Canada.
Web site: http://pubs.nrc-cnrc.gc.ca

Correct citation for this publication: Sawyer, E.W. 2008. *Atlas of Migmatites*.
The Canadian Mineralogist, Special Publication 9. NRC Research Press, Ottawa, Ontario, Canada. 371 p.

Also in this series

Encyclopedia of Mineral Names
W.H. Blackburn & W.H. Dennen
Special Publication 1 (1997)

Glossary of Mineral Synonyms
J. de Fourestier
Special Publication 2 (1999)

Atlas of Micromorphology of Mineral Alteration and Weathering
J. Delvigne
Special Publication 3 (1998)

New Minerals 1995-1999 *(2001)*
J.A. Mandarino
Special Publication 4 (2001)

The Health Effects of Chrysotile Asbestos: Contribution of Science to Risk-Management Decisions
R.P. Nolan, A.M. Langer, M. Ross, F.J. Wicks & R.F. Martin, eds.
Special Publication 5 (2001)

Mineral Species Discovered in Canada and Species Named after Canadians
László Horváth
Special Publication 6 (2003)

Mineral Species First Described from Greenland
Ole V. Petersen and Ole Johnsen
Special Publication 8 (2005)

Available from:

Mineralogical Association of Canada
490, rue de la Couronne
Québec, QC G1K 9A9
Canada

www.mineralogicalassociation.ca

Editor, The Canadian Mineralogist
Robert F. Martin

Table of Contents

Preamble

It is with pleasure that I introduce the latest addition to the list of Special Publications produced by the Mineralogical Association of Canada. This ***Atlas of Migmatites*** contributes directly to the educational mandate of our Association. The book deals with migmatites, a very widespread group of rocks, especially common in the Archean crust that makes up a major proportion of our country. Everyone has learnt a bit about migmatites, but by and large, the subject matter "falls between the cracks" in our university curricula. Typically, an upper-level undergraduate course in metamorphic petrology does set the stage for an understanding of anatectic reactions in terranes that have undergone metamorphism at conditions of upper amphibolite or granulite facies. Students in such a course do learn about the concept of anatexis and dehydration-induced melting, but really not much about the extraordinarily complex array of products of such anatexis, the importance of fractional crystallization of the anatectic liquid, and the fate of assemblages of residual minerals. Furthermore, contact-related anatexis usually is not discussed. On the other hand, students in an upper-level course in igneous petrology do deal with products of partial melting in the crust, and can speak at length about the physical properties of a silicate magma and the results of its fractional crystallization. But they deal with ready-made plutons. Such students do not develop a good understanding of the steps that precede the formation of a pluton, where small batches of leucosome coalesce and rise through a deforming mass of neosome and resister lithologies.

As is made clear below, Professor Sawyer has remedied the situation by writing the first authoritative treatise about the petrology of migmatites since the work of Karl Richard Mehnert (1913–1996), of Berlin, published 40 years ago. He has abandoned the purely descriptive approach of Mehnert in favor of an openly genetic approach: a migmatite is a rock that is the product of partial melting. If melting can be shown not to have taken place, the rock is simply not a migmatite. Furthermore, he has recognized the importance of anatectic reactions not only in amphibolite-facies rocks, but also in the realm of the granulite facies. He has examined key occurrences of migmatites throughout the world, and presents here in a systematic way the results of his scrutiny on all scales. Thus, the reader is shown photographs of key exposures of migmatites in the field and petrographic details in 273 photographs, each with a detailed caption, supplied by the author in 68% of the cases, the remainder being contributed by respected colleagues also active in the characterization of migmatitic assemblages. Field examples are presented from 12 countries, with the author personally involved in field studies in six of those. In this atlas, the author emphasizes the latest contributions in the area of migmatite-related research, either in a regional setting or in a contact aureole.

Society-run, not-for-profit organizations like the Mineralogical Association of Canada find it very challenging to undertake such a publication, owing to the high costs of production. In this instance, the Association is most fortunate to have coproduced the volume with NRC Research Press. On behalf of the MAC, I thank Suzanne Kettley, Mark Boileau, and Diane Candler for their valuable input and their key roles in creating this book. In addition, I acknowledge the involvement of Pierrette Tremblay, Director of Publications of the Association, in ensuring close communications with NRC Research Press throughout the preparation of this book. Also, Pierrette applied for and obtained funds from the Canadian Geological Foundation and Natural Resources Canada, part of which paid for the printing of a poster to publicize this book. I thank Vicki Loschiavo, who entered the editorial corrections on the master files of text and captions as I progressed through the book. I am grateful to have had this opportunity to help bring this major contribution in the area of migmatite research to fruition.

Robert F. Martin
Editor, ***The Canadian Mineralogist***

Preface

Migmatites are some of the most confusing-looking, yet aesthetically pleasing rocks. The most striking consist of light-colored quartzofeldspathic segregations in a darker host, with the light segregations showing a diverse and often spectacular range of appearances, in some cases in swirls, in some others cross-cutting, and in yet others discrete, or diffuse, or folded, or containing high-grade metamorphic minerals like garnet and pyroxene. Migmatites justify their age-old name; they are "mixed rocks."

Migmatites commonly look "squishy." This is no coincidence, for they are interpreted to be rocks frozen in the act of partially melting. A mixture of melt, disaggregated minerals, and unmelted rock is mechanically highly heterogeneous. It is no wonder that migmatites show such a wide range of textures and structures.

Migmatites are of crucial importance in understanding the genesis of the huge volumes of granitic magma that are found in batholiths and that make up much of the continental crust. Migmatites are widely considered to be examples of assemblages where such magma was caught in the act of being generated and escaping. Migmatites thus furnish a unique perspective on a fundamental process in the evolution of the Earth.

The arrival of this ***Atlas of Migmatites*** is very timely. It is the first book devoted to migmatites since Mehnert's iconic 1968 text. In the intervening 40 years, dramatic strides have been made in the understanding of these complex rocks.

Ed Sawyer is without question the world's leading expert on migmatitic textures and structures. In this atlas, he provides a wide range of superb field and thin-section photographs of migmatites from around the world, both his own and those of others, each carefully described and interpreted. The photographic part of the atlas is preceded by a substantial introductory section in which he clarifies the dizzying range of descriptive terms for migmatites, both new and old.

Although migmatites will remain among the most complex of rocks to understand, this atlas goes a long way to making them more understandable.

David R.M. Pattison
University of Calgary

Acknowledgements

I would like to start by expressing my gratitude to Mike Brown and Richard White for their encouragement and their reviews of the text. They pointed out my errors, omissions, and tracts of muddled writing; the changes that they suggested have resulted in a far better text. The remaining faults are entirely mine. Many people took the time to send me their photographs of migmatites for this atlas. The choice of which ones to use was far from easy, but each photograph, whether included or not, has had an influence. I thank everyone who sent me a photograph or electronic image for his/her contribution.

My perception of migmatites and what happened within them has been influenced by seeing and discussing migmatites with many colleagues from around the world. To have seen such a wide range of migmatites has been invaluable, and I am grateful for the hospitality and generosity of all those who have shown me their field areas. In particular, and more or less in chronological order: Ned Chown (Grenville Front), Ron Vernon and Bill Collins (Wuluma and Mount Hay), Daniel Lamothe and Alain Leclair (Ashuanipi Subprovince), Mark Severson and Steve Hauck (aureole of the Duluth Igneous Complex), Mike Brown (Saint-Malo), Gary Solar (Maine), Abdelali Moukhsil (Eastmain area), Geoff Clarke and Richard White (Mount Stafford, Broken Hill and Wuluma), Fernando Bea (Peña Negra), Olivier Vanderhaeghe (Massif Central), and Dirk van Reenen (Limpopo Belt). There has always been an easy exchange of thoughts and views within the migmatite community, and next I would like to thank Marian Holness, Tracy Rushmer, Bernardo Cesare, Roger Powell, Alfons Berger, and Claudio Rosenberg for illuminating some of the connections between what one can see in migmatites and the various processes that have happened to them.

In most endeavors, progress comes through sustained and focused effort, together with some serendipity, of course. My work on migmatites, and ultimately this atlas, have been no different and would not have been possible without continuity of funding. Therefore, I am grateful for the philosophy behind the discovery grant program of the Natural Sciences and Engineering Research Council of Canada, which funds long-term, curiosity-driven research programs.

Finally, I thank Pierrette Tremblay and Robert F. Martin for the invitation to write a book on migmatites and for their encouragement throughout, and to Robert for all his hard work in reading and editing.

Edward W. Sawyer
Université du Québec à Chicoutimi

Introduction

1.

THE SCOPE OF THIS ATLAS

Migmatites are spectacular, complex-looking rocks that can inspire, fascinate, or confuse geologists. All migmatites viewed in an outcrop represent the sum of a series of processes that acted in parallel, or sequentially, and the influence of various local factors (see the photographs in section A). In order to begin to understand the complexity and seemingly endless variety in migmatites, and to provide the reader context for the subject of this book, some of the key factors and processes that make individual migmatites the way they are should be outlined at the outset.

First, there is the petrological process of partial melting. If the temperature becomes sufficiently high (above about 650°C), rocks may begin to partially melt. Some rocks have compositions that will produce more melt than others at a given temperature, a property of the rock called fertility. Thus, one can immediately see that complexity begins at this stage, because some rocks in a sequence will make more melt than others; some will not melt. The highest temperature attained also has an effect. If the temperature only just surpasses the solidus, the migmatite will contain a few small patches of melt scattered about in the most fertile rock. If the maximum temperature was, say, another 250° higher, then melting in the fertile layers might be pervasive, and well advanced in other less fertile rocks. Thus, the two migmatites would look completely different, even if generated from exactly the same sequence of rocks.

Second, the nature of the protolith has an influence too. For example, the migmatite produced from a sequence of rocks that contains a single thin fertile layer (say a pelite) enclosed in a thick sequence of infertile rocks (say quartzite) will look completely different from one generated from a sequence that contains a thin quartzite in a thick sequence of pelites. Similarly, a migmatite generated from an isotropic rock (e.g., a granite) will not look the same as one generated from a sequence of thinly bedded shales and sandstones.

Third, deformation during partial melting has a major effect on the morphology of migmatites. Only rarely (e.g., in some contact aureoles) do migmatites form under conditions where there is essentially no deformation. Generally, migmatites are deformed while they contain melt. Deformation of heterogeneous rocks produces variations in the differential stress from place to place, and this sets up pressure gradients that drive the movement of material. There are likely to be variations in viscosity or competence from rock to rock in the protolith, but once partial melting of the juxtaposed rocks starts, much larger variations are created. The partial melt has a viscosity that is much lower than that of nonmelted rocks; therefore, during deformation, the melt in the rock moves more rapidly down pressure gradients than the solid fraction, and collects in low-pressure sites. Deformation thus creates petrological diversity, because the melt is separated, or segregated, from the solids, which enhances the mechanical extremes in the migmatite. Deformation becomes concentrated, or partitioned, into the weakest places; these are located where there is melt. Consequently, strain becomes more heterogeneous, and the migmatites develop more and more complex and deformed geometries.

Finally, the period of time that a migmatite has had to form also exerts an influence. The heating and cooling cycle in contact aureoles is rapid (days to tens of thousands of years), such that melt may quench to glass or crystallize to granophyre, and primary microstructures, such as the shapes of the pores that the melt occupied, are preserved because recrystallization did not reset microstructures. Strains are small, and deformation is restricted to small cracks and shear zones, so that the initial geometrical relations in the migmatite are generally preserved. In contrast, the slow heating and cooling (millions to tens of millions of years) of regional metamorphic terranes mean that melts crystallize slowly, and subsequent recrystallization extensively modifies, or even eliminates, the primary microstructures. Moreover, large strains can accumulate over long periods of time, so that the original geometry of the migmatite is extensively folded, boudinaged, and transposed.

This book contains two parts. One is a summary of the advances in the understanding of migmatites and related rocks, including partially melted xenoliths of crustal rocks, that have been made since Mehnert's (1968) influential book. The other consists of seven series of photographs, each with a substantial explanatory caption, that illustrate the aspects of migmatites covered in the first part. The two parts are complementary, and in the text, I refer to the illustrations throughout, but the two parts could also be read separately, with the text part as a generic discussion of migmatites, and the illustrated part as a series of detailed interpretations applied to individual migmatites.

Section 2, on this page, begins with the terminology associated with migmatites; the various constituent parts of migmatites are dealt with first, then the diversity in outcrop appearance as a function of deformation, melt fraction, and parent rock-type are described. The terms necessary for the study and description of migmatites are defined; text is shown in bold type to introduce a term where it is first defined, whereas text in italics is used to indicate a new definition. Next, the processes that control the morphological diversity of migmatites are considered, so as to derive a classification scheme that is useful as a mapping tool, and as a means for understanding why migmatites have a particular morphology. A two-tiered scheme of classification is proposed. All parts of migmatites and types of migmatites mentioned section 2 are systematically illustrated in sections B, C, D, and E. The weight given to the importance of metasomatism and the ingress of H_2O-rich fluids in the formation of migmatites has changed considerably since migmatites were first defined; current results and thinking are summarized in section 4. The interpretation of microstructures has always been central to metamorphic petrology and to the study of migmatites in particular. In recent years, the repertoire of microstructures that can be used to identify a specific process has grown considerably, and includes some microstructures derived from experiments as well as from natural rocks; both are discussed in section 5 and shown in section F. Because the acquisition of chemical compositions of whole rocks and minerals is now a routine part of regional mapping, a brief discussion of what to sample, and the use of geochemical data in identifying the processes contributing to migmatites, are given in section 6; however, this a guide and should not be regarded as an exhaustive survey of what can be done with compositional data on rocks and minerals. Some rock types that are not formed by partial melting can resemble migmatites; thus, in section 7, I discuss rocks that are commonly mistaken for migmatites; examples of these are shown in section G. Finally, sections 8 and 9 deal with the problems of mapping of migmatites; topics range from what to show on maps to the tricky matter of the interpretation of "generations" of leucosome.

2.

GENERAL TERMINOLOGY AND DEFINITIONS NEEDED FOR WORK ON MIGMATITES

2.1 The heritage of migmatite terminology

The debate over what migmatites represent and the processes involved in their formation has been closely connected with the prevailing views of how high-grade regional metamorphism occurs, and how granites and granitic magmas are formed. Hutton (1795) believed that partial melting changed pelitic sediments and crystalline schists into gneisses. Lyell (1855) went a step further and considered that granites form where the partial-melting process was stronger and more complete. However, the present terminology did not start to develop until later. One of the first terms specific to crustal melting is **diatexis**, introduced by Gürich (1905) to describe cases where "partial" melting is complete, or where it has occurred throughout a rock, to lead to the formation of granite. Sederholm (1907) made the obvious comment that in most cases where a term is required to explain the process of making granite, fusion was not complete; he introduced the term **anatexis**, meaning melting, or remelting, of rocks, to cover the entire range from incipient to complete melting.

In the late nineteenth and early twentieth centuries, mapping in the Precambrian shields of Fennoscandia and North America revealed many places where metasedimentary rocks appeared to pass into granite. These transitional zones attracted a great deal of interest. Not surprisingly, there appeared in the literature many opinions as to what processes were involved; each hypothesis reflected an individual's own field experience.

One school of thought, which saw intrusions of granitic magma as the source of heat for the transformations, developed from work done at the edges of granite plutons by Barrois (1884), Lacroix (1898, 1900), Michel-Lévy (1893), Greenly (1903), and Sederholm (1907, 1923, 1926). Sederholm (1907) described the manner in which an older, foliated granite underwent "refusion" and was "reborn" into intrusive masses by the injection of the Hango (Finland) granite, which imparted what he called a "new eruptivity" to it. Because he believed that the rocks in the Hango area were produced by a combination of processes and not simply by partial melting, Sederholm (1907) introduced the term **migmatite** to describe them. He defined migmatite as "the mixture of two genetically different constituents . . . one is intrusive relative to the other . . . To this group belongs the gneissic granitic rocks which show 'net structure' characteristic of incipient melting, frequently 'blind-ending veins', breccia-like granites with innumerable fragments of more-or-less completely dissolved older rocks, and finally some striped granites in which only the still preserved parallel structure indicates a faint remnant of the original properties of the rock." Sederholm (1907) called the migmatite-forming process **palingenesis**, and although it specifically included partial melting and dissolution, he also regarded magma injection and its associated veined and brecciated rocks as fundamental to the process. The veined rocks were later defined by Sederholm (1923) as **arterite**, and the brecciated rocks as **agmatite**. In subsequent work, Sederholm (1926) placed more emphasis on the roles of assimilation and the actions of fluids, for which he introduced the term **ichor**, in the formation of migmatites.

An alternative view of migmatites was developed by geologists working in regional metamorphic terranes. One of them, Holmquist (1916), found high-grade gneisses that contain many small patches and veins of granitic material. As there were no granites nearby, he interpreted the patches and veins to be the collection sites for partial melt exuded from the mica-rich parts of the host gneiss. Holmquist gave these migmatites the name **venite** to emphasize their internal origin (Holmquist 1916, 1920, 1921) and to distinguish them from Sederholm's arterites, with their veins of injected material. The absence of nearby granites indicated to him that the heat source for the partial melting lay deeper in the Earth, like that inferred for regional metamorphism in general. Thus, Holmquist (1916) introduced the term **ultrametamorphism** in order to indicate the higher-than-normal degree of metamorphism required for partial melting (i.e., anatexis).

The term **metatexis** was introduced later by Scheumann (1936) to mean the general process of partial melting in the continental crust, and he intended that it replace Sederholm's terms anatexis and migmatite. However, the existing term "anatexis" was perfectly adequate, and both metatexis and diatexis disappeared from use, leaving anatexis to mean partial melting of the continental crust.

The dispute over the origin of granite (see the discussions by Read 1957) also influenced thought on migmatites. Bowen's (1928) demonstration that a granitic magma is the product of the extreme fractional crystallization of basaltic magma, and the subsequent finding that there is far more granite than could possibly be made by this mechanism, stimulated new interest in partial melting of the continental crust. However, the lack of a structural model to explain how large volumes of granitic magma move in the crust led to the so-called "space problem" and to the idea that bodies of granite form *in situ*. Debate then turned to whether or not granite has a magmatic origin, and among the nonmagmatic protagonists, to the "wet" versus "dry" granitization dispute. Since migmatites were commonly seen in the field in an intermediate position between regional metamorphic rocks and granite, opinion as to their origin also diverged. The legacy of that controversy was the development of a nongenetic descriptive terminology for migmatites, first by Dietrich and Mehnert (1960), and finally by Mehnert (1968).

Since the publication of Mehnert's book in 1968, a very large majority of work done points to the fundamental role of partial melting in the formation of migmatites. There are few studies of subsolidus processes in migmatites, but they show that the leucocratic subsolidus segregations present in the migmatites had formed prior to the onset of partial melting (e.g., Blom 1988). Other authors have shown that metasomatism is not a significant factor in the formation of migmatites, although the influx of aqueous fluids into rocks that are close to their solidus temperatures may provoke partial melting and the formation of migmatites (e.g., Pattison 1991, Harris et al. 2003, Johnson et al. 2003).

Thus, the modern view of migmatites corresponds closely to Holmquist's concept of ultrametamorphism, and to Sederholm's concept of anatexis, but is far from the concept of palingenesis, or the various metasomatic and subsolidus processes proposed during the granitization debate. In modern usage, migmatite is synonymous with anatexite. Consequently, it is now highly debatable whether a general definition of migmatite needs to be nongenetic; in this book, I use a genetic, partial-melting-based definition and descriptive terminology for migmatites. Any definition of migmatite should certainly include reference to basic characteristics in the field, the most important of which are that (1) migmatites occur only in the high-grade parts of regional metamorphic terranes (upper amphibolite and granulite facies) and contact aureoles (pyroxene and sanidinite hornfels facies), and (2) because migmatites contain parts where melting occurred, parts where melt was extracted, parts through which melt migrated, parts where melt collected, and parts that did not melt at all, they are morphologically complex rocks at almost every scale of observation, from the microscopic to the macroscopic.

2.2 A definition of migmatite

Mehnert (1968) defined a migmatite as ". . . a megascopically composite rock consisting of two or more petrographically different parts, one of which is the country rock generally in a more or less metamorphic stage, the other is of pegmatitic, aplitic, granitic or generally plutonic appearance." For a nongenetic definition, the use of the terms "pegmatitic", "aplitic", and "granitic" was unfortunate, as the three carry unmistakable genetic implications. A second problem is the imprecise reference to metamorphic grade. Ashworth (1985) rectified both of these problems in his definition of migmatite as "a rock found in medium-grade to high-grade metamorphic areas, that is pervasively inhomogeneous on a macroscopic scale, one part being pale-coloured and consistently of quartzofeldspathic or feldspathic composition." Unfortunately, this nongenetic definition does not exclude Sederholm's agmatites, rocks that Brown (1973) had already argued are intrusion breccias and not migmatites. There is further advantage to be gained by excluding agmatites. The phrase ". . . two genetically different constituents . . ." in Sederholm's (1907) definition of migmatite becomes irrelevant if agmatites are excluded because for partial melting (and indeed subsolidus segregation and metasomatism), the newly generated parts of a migmatite ought to be petrogenetically related. Hence, the following revised definition is proposed.

Migmatite: *a rock found in medium- and high-grade metamorphic areas that can be heterogeneous at the microscopic to macroscopic scale and that consists of two, or more, petrographically different parts. One of these parts must have formed by partial melting and contain rocks that are petrogenetically related to each other (called the neosome) and to their protolith through partial melting or segregation of the melt from the solid fraction. The partially melted part typically contains pale-colored rocks that are quartzofeldspathic, or feldspathic, in composition, and dark-colored rocks that are enriched in ferromagnesian minerals. However, the partially melted part may simply have changed mineralogy, microstructure, and grain size without developing separate light or dark parts.*

Note that the proportions are not specified and that some, but not necessarily all, of the light-colored parts have to be petrogenetically related to the other parts. There are four basic parts to a migmatite, although not all may be found in any particular migmatite, especially if the outcrop examined is small (less than several m^2); scale is an important factor in the study of migmatites. There are parts where partial melting has occurred, parts from which the melt fraction has been removed, parts where the melt fraction has collected, accumulated, or been injected, and of course, there are parts that did not melt at all. Each of these parts has its own wide range of morphologies (or structure), mineral assemblage, bulk composition, and microstructure (also called texture); consequently, there is a terminology to define and describe each. In the next section, I deal with the definitions for the constituent parts of a migmatite; these are illustrated in section B of this book (Figs. B1–B54). First, the terminology used to differentiate the old (nonpartially melted) from the newly created (i.e., partially melted) parts in migmatites is discussed, and then the terms that are used to describe the features and variations within these broad divisions at a smaller scale are outlined. This level of terminology is, therefore, applicable to the fine-scale, detailed description of migmatites in a small outcrop (less than several m^2), or large hand-samples.

2.3 Descriptive terms and definitions for the principal parts of a migmatite

Imagine that one is able to compare a large outcrop (say 400 m^2) of migmatite with the rocks that existed in the same outcrop before partial melting occurred. Some rocks, because they have suitable bulk compositions, will have been affected by the partial melting, whereas others, of unsuitable compositions, will not. Those rocks newly formed by partial melting are called neosome, meaning "new rock"; the term is defined as follows.

Neosome: *the parts of a migmatite newly formed by, or reconstituted by, partial melting. The neosome may, or may not, have undergone segregation in which the melt and solid fractions are separated.*

A general characteristic of neosome is that it has a coarser grain-size (e.g., Figs. B1–B4) than the rest of the migmatite. Furthermore, the structure (e.g., foliation, folds, layering) and microstructure (shape, size, and orientation of grains) that existed prior to partial melting are progressively degraded as the degree and extent of partial melting increase, and eventually replaced by a new microstructure created by the neosome-forming processes. A neosome displays a wide range of morphology; as the examples in section B of this book show, the melt and solid fractions have segregated in some (Figs. B5–B14), but not in others (Figs. B15–B25). The neosome in many migmatites is *in situ* (e.g., the spatially focused neosome described by White et al. 2004), but in migmatites where the melt fraction was high, the neosome may have moved. If the portions (e.g., beds or compositional layers) of the rock from which the neosome formed can be identified in the pre-migmatite state, then these parts are called the **protolith** (Johannes 1985), or the **parent rock** (Ashworth 1985). This means, of course, that the protolith cannot be present in a migmatite; it will have been converted to neosome.

Naming of the non-neosome parts of a migmatite has been confused by the inconsistent use of terms, and by the use of inappropriate terms. Mehnert (1968) used the term "paleosome," meaning "old rock" throughout the text of his book for the non-neosome part of a migmatite, but this usage is inconsistent with the definition ("parent rock of a migmatite") given in the appendix of his book. To Ashworth (1985), such a definition means that paleosome cannot be present in a migmatite; it must all have been converted to neosome by definition. Some authors (e.g., Johannes and Gupta 1982) have used the term "mesosome" for the non-neosome part of a migmatite, but that is problematical too, as the non-neosome part need not be mesocratic, and moreover, in many cases, the neosome has mesocratic parts. The root of the problem is that there is no clear opinion on which "old rock" is the relevant one. If it is the "old rock" of the pre-partial-melting state, then paleosome refers to the lithologies that will become neosome, but this usage is redundant, as adequate terms, i.e., protolith or parent rock, already exist for that part.

The alternative, that "old rock" refers to that part of the migmatite not affected by partial melting and which, therefore, contains only structures that pre-date the partial melting, is far more useful. Hence, the following definition is proposed.

Paleosome: *the non-neosome part of a migmatite that was not affected by partial melting, and in which structures (such as foliations, folds, layering) older than the partial melting are preserved. The microstructure (size, form, and orientation of grains) is either unchanged, or only slightly coarsened, compared to that in similar rocks just outside the region affected by anatexis.*

Paleosome exists because its composition was such that it did not partially melt and did not become neosome (Olsen 1985); however, as some of the photographs in the atlas show, it is not always easy to decide what has melted and what has not. Further subdivision of the paleosome may be possible, and even useful, in mapping some migmatites. If, in comparing the pre- and post-partial-melting states, certain lithologies persist unchanged into the highest-grade parts of the migmatite, then these paleosome lithologies can be called **resisters** (Read 1957) or **refractory**. Layers of calc-silicate, quartzite, and metamafic rocks are common resister lithologies in migmatite terranes. The neosome and paleosome parts of migmatites are shown specifically in Figs. B1–B4 of section B, but as the other figures in section B show, there is no "typical" or "classic" morphology to either the neosome or the paleosome part of migmatites; their morphology is highly varied.

Terms specific to the neosome

The nongenetic terminology of Mehnert (1968) is based on changes in the relative proportions of light- to dark-colored minerals that occur as a migmatite forms. Because the development of leucocratic and melanocratic rocks in a neosome is nearly always the result of well-defined petrological processes (e.g., partial melting, segregation of the melt from the solid fraction, and fractional crystallization) that have combined to create migmatites, this nomenclature has proven to be very useful and is easily transferred to a genetic scheme. The onset of partial melting changes a one-phase (solid) protolith to a two-phase (melt + solid) neosome. The melt fraction has a lower viscosity and density than the residual solid, and consequently, the two parts of a neosome can become separated, or segregated. The part of a migmatite from which the melt fraction has been removed can be defined as follows.

Residuum: *the part of the neosome that is predominantly the solid fraction left after partial melting and the extraction of some, or all, of the melt fraction. Microstructures indicating partial melting may be present.*

Residuum is a general term; there is no particular reference to rock color or to mineral assemblage. For some bulk compositions, the residuum may be dominated by light-colored minerals, such as feldspar or quartz; typically, because these minerals were so abundant in the protolith, they dominate the residuum as excess phases with respect to the melt-producing reaction. However, partial melting of the common crustal rock-types typically generates residua in which ferromagnesian minerals are major constituents. Consequently, residua are most commonly melanocratic, and these are given the special term **melanosome**.

Melanosome: *the darker-colored part of the neosome in a migmatite that is rich in dark minerals such as biotite, garnet, cordierite, orthopyroxene, hornblende, clinopyroxene, and even olivine. The melanosome is the solid, residual fraction (i.e.,* ***residuum****) left after some, or all, of the melt fraction has been extracted. Microstructures indicating partial melting may be present.*

The complement to residuum is, of course, derived from the anatectic melt, and this part of a neosome is called the **leucosome**.

Leucosome: *the lighter-colored part of the neosome in a migmatite, consisting dominantly of feldspar and quartz. The leucosome is that part of the migmatite derived from segregated partial melt; it may contain microstructures that indicate crystallization from a melt, or a magma. Leucosome may not necessarily have the composition of an anatectic melt; fractional crystallization and separation of the fractionated melt may have occurred.*

The size, physical form, orientation, and grain size are not factors in determining what is residuum, leucosome, or melanosome, and the photographs in section B show some of the range of diversity present in migmatites. However, these parameters should all be recorded, as they may be important in fully describing all the types of leucosome and of melanosome (or residua) in a migmatite, and could be of significance in determining exactly how each rock in a migmatite formed.

The solid, residual fraction of a neosome can, in most circumstances, be regarded as being in place (i.e., *in situ*), but the melt phase is potentially mobile, a fact recognized in the older literature by use of the term "mobilizate" for the melt-derived parts of a migmatite. Therefore, a set of terms that convey how far the fraction of anatectic melt has moved from where it formed to where it crystallized are useful in making a more complete field-based description of the individual products of melting in a migmatite, and for the subsequent understanding of the petrogenetic relationships between the melt products and the rocks around them. The following terms are proposed and adopted throughout the book.

***In situ* leucosome**: *the product of crystallization of an anatectic melt, or part of an anatectic melt, that has segregated from its residuum, but has remained at the site where the melt formed.*

In-source leucosome: *the product of crystallization of an anatectic melt, or part of an anatectic melt, that has migrated away from the place where it formed, but is still within the confines of its source layer.*

Leucocratic vein or dike: *the product of crystallization of an anatectic melt, or part of an anatectic melt, that has migrated out of its source layer and has been injected into another rock, which maybe nearby, or farther away, but is still in the region affected by the anatectic event.*

Granite (or granodiorite, tonalite, etc.) dike or sill: *the product of crystallization of a felsic melt that has migrated out of its source region completely, and is injected into host rocks of lower metamorphic grade or into nonmetamorphosed rocks.*

Each of the four can be recognized by a combination of morphological and geochemical characteristics, and by the petrogenetic relationship between the melt-derived part and its host. *In situ* leucosome is in contact with its own residuum, which typically forms a melanosome around it (see, for example, Figs. B5 and B11). Contacts between the leucosome and melanosome are generally diffuse on a millimeter to centimeter scale. The composition of an *in situ* leucosome corresponds to an initial anatectic liquid or, if some melt was lost after crystallization had started, to a cumulate composition derived from an initial melt. The petrogenetic relationship between the leucosome and host is quite specific; the leucosome is derived from the anatectic melt with respect to which the melanosome is the residuum. However, if some loss of melt has occurred, an excess of residuum relative to leucosome can be expected.

An in-source leucosome is derived from melt that has moved and, therefore, may be discordant and have sharp borders with its host (e.g., Fig. B27); many occurrences, however, are stromatic (i.e., layered; see below), and have some diffuse and some sharp contacts. An in-source leucosome may have a composition corresponding to an initial anatectic melt, a cumulate or a fractionated anatectic melt. Because it is derived from a melt that has moved, an in-source leucosome is detached from its own residuum, but as it remains in its source layer, it is hosted by another, similar residuum (melanosome) that was formed from the same source layer. Thus, there is a petrogenetic connection between the leucosome and its host (e.g., similar *Mg-number*, mineral assemblage, or isotopic compositions), but they are not exact complements. The residual host may, for example, have formed from a slightly lower, or higher, degree of partial melting, or may have experienced a higher, or lower, degree of melt loss.

Leucocratic veins or dikes typically have sharp contacts, irrespective of whether they are discordant or concordant to the structure in their host (e.g., Fig. B28). There is no direct petrogenetic relationship between the leucocratic vein or dike and its host rock; the age of crystallization in the former should be consistent with the metamorphic age in the latter, however. Leucocratic veins or dikes generally have compositions that indicate derivation from a fractionated anatectic melt, but some have compositions of initial melts, and others, compositions of cumulates.

Granite dikes typically have sharp contacts with their host and, because they were injected into cool hosts, may have fine-grained border zones (chilled margins). There is no petrogenetic relationship between a granite dike and its country rock. The dike may be a product of crustal anatexis, or it may be derived from the crystallization of a felsic, intermediate, or even mafic magma. Granitic dikes commonly have a composition consistent with crystallization from a fractionated melt.

Well-defined bands of leucosome are very common in migmatites, but as the photographs in the book show, a leucosome may also be poorly defined and may form somewhat nebulous patches. Furthermore, it is quite common for several different forms of leucosome to be present in the same outcrop of migmatite (e.g., Oliver and Barr 1997). Most outcrops are relatively small, and on these, it may not be possible to trace leucocratic veins, or dikes, back to their source layers and to examine their relationship with *in situ* leucosome. However, the root zones of leucocratic veins and dikes can be found in some larger outcrops (see Figs. B43–B48), and these show a characteristic petrological continuity from *in situ* leucosome through in-source leucosome to the leucocratic dikes and, commonly, with a strong structural control (i.e., by shear zones or folds) on the initial location of the dikes.

The best-known form that the residual part of a migmatite can take is the melanosome that occurs along the margin of *in situ* leucosome (e.g., Figs. B5–B14). Far less attention has been paid to the other, and far more common, morphologies that the melanosome can adopt in migmatites; it may occur as patches, as continuous layers, and in some cases, as irregularly shaped bodies (e.g., Kenah and Hollister 1983, Sawyer 2001, White et al. 2004). Some residual rocks formed in migmatites in granulite-facies terranes are not particularly melanocratic, although they are very widespread (e.g., Guernina and Sawyer 2003).

The neosome in many migmatites has segregated into clearly defined leucosome and melanosome. However, if the melt and residual solid fractions do not separate, leucosome and melanosome do not form, or are far less evident. For these nonsegregated examples, the term "neosome", used alone, suffices. Examples of a neosome that has not developed leucosome and melanosome parts through segregation are shown in Figs. B15–B26.

The recognition of neosome and its leucosome and melanosome parts is, generally, a simple matter in migmatites developed from mesocratic or melanocratic protoliths (e.g., metagreywackes, metapelites, metadiorites, and metamafic rocks). It may be a much more difficult matter if the protolith is leucocratic (e.g., some granites, tonalites, trondhjemites, and metapsammites); the dilution of the mafic minerals by feldspar and quartz means that subsequent changes in modal proportions produce only subtle changes in color, which may be difficult to detect in the field. For such rocks, distinguishing the neosome from the paleosome in the field may best be accomplished by using changes in microstructure, such as fabric or grain size, rather than changes in mineral modes or color. Examples of this approach will be found in the atlas (e.g., Figs. B15 and B16).

Two further terms specific to the neosome remain to be defined.

Melt: *a silicate liquid without crystals.*

Magma: *a silicate liquid that contains crystals, which might have crystallized from the melt (liquidus phases), be the solid products of the melting reaction (also called peritectic products by some authors), or be minerals in excess.*

Terms for the other parts of a migmatite

In morphologically complex migmatites, there are, inevitably, some parts that are neither leucocratic nor melanocratic. Mesosome has been used as a descriptive term for these, but its usefulness in any genetic sense is very limited. In part, this is because of inconsistent usage in the past. For Henkes and Johannes (1981), mesosome corresponded to the protolith, whereas for Ashworth (1985), it was the paleosome. A far more serious problem is that mesocratic rocks can occur in either the neosome or the paleosome; hence, the petrogenetic significance of mesosome is inherently less precise than that attached to leucosome or melanosome, which are, by definition, parts of the neosome. For this reason, I suggest that mesosome should not be used. Because mesocratic rocks can occur in the neosome or paleosome (or both), it is necessary to record where the particular mesocratic rock is located in the migmatite (e.g., in the paleosome, or in the neosome).

If the melt and solid fractions do not separate in a neosome, it will not develop leucosome and melanosome, and overall, the neosome will be mesocratic. However, in such cases, a coarser grain-size and a more isotropic fabric will typically have replaced the microstructure and fabric that pre-dated anatexis, and will aid in identifying what is neosome.

As more work is done on migmatites, it is becoming clear that some example of *in situ* and in-source leucosome and many leucocratic dikes, or veins, in migmatite terranes are separated from their host rocks by a narrow rim that is compositionally, mineralogically, and microstructurally different from the host (see Figs. B49–B54). These compositional rims are not the residuum left after extraction of anatectic melt; therefore, the term used for them is **selvedge**. Field observations from many migmatite terranes indicate that selvedges are most commonly developed around leucocratic veins and dikes; the injected melt thus does not seem to have been in chemical equilibrium with its host. Consequently, three main mechanisms have been proposed for the formation of selvedges: (1) reaction between the host rock and an aqueous fluid exsolved as the melt that produced the leucosome or leucocratic dike crystallized, (2) reaction between minerals in the adjacent wallrock and the injected anatectic melt, and (3) diffusional exchange of components between the anatectic melt (or the crystallized leucosome or leucocratic dike) and its host in response to activity gradients.

Selvedges can be leucocratic, mesocratic, or melanocratic. However, a very common type consists of a thin (mm-wide) biotite-rich (or hornblende-rich in metamafic rocks) melanocratic rim adjacent to leucocratic dikes and some cases of in-source leucosome; these are called **mafic selvedges**. Mafic selvedges are also common at the margin of granitic dikes and sills intruded into low-grade country rocks; mafic selvedges are, therefore, not found only in migmatites.

3.

MIGMATITES: THE PROCESSES AND MORPHOLOGIES

The next step is to introduce the terms that describe the overall appearance of a migmatite at a scale larger than a small outcrop. Mehnert (1968) introduced 13 morphological terms (agmatitic, diktyonitic, schollen, stromatic, surreitic, folded, ptygmatic, veined, ophthalmitic, stictolithic, schlieren, nebulitic, and homophaneous) for this purpose. His scheme was not explained in terms of petrological processes, and so, overall, it has proved unsatisfactory for three main reasons. First, this scheme does not provide a way of understanding how one morphology of a migmatite is related to another. Second, it does not provide any insight into the petrological processes that contributed to the formation of migmatites. Consequently, the scheme has turned out to have little practical use in making a map of a migmatite terrane. Third, the terms were designed to be nongenetic, but in practice most of the morphologies are highly origin-specific, i.e., they occur exclusively as a result of partial melting.

The overall morphology of migmatites will be considered next, with the intention of determining which factors control their appearance at the scale of a large outcrop. This approach leads to a much clearer picture of how different morphologies of migmatite form and are related to one another, in addition to providing a workable basis for mapping migmatites.

3.1 The first-order morphological division of migmatites

Field studies from a great many anatectic terranes (Breaks et al. 1978, Brown 1979, Jamieson 1984, Weber et al. 1985, Sawyer and Barnes 1988, Wickham 1987*a*, Bea 1991, Obata et al. 1994, Sawyer 1998, Oliver et al. 1999, Solar and Brown 2001, Johnson et al. 2001*a*, White et al. 2005) reveal a similar change in migmatite morphology with increasing metamorphic grade. Examples from several migmatite terranes are shown in section C of this book. In the lower-grade parts of anatectic terranes, paleosome is dominant in the migmatite (i.e., the proportion of neosome to paleosome is low), and old, pre-partial-melting structures such as bedding, compositional layering, foliation, and folds are widely preserved in it. The neosome part is characterized by narrow bodies of leucosome of various orientations, bordered by melanosome, with its residual mineralogy and bulk composition. From a rheological aspect, the bulk behavior of these migmatites differs little from solid rocks that are not partially molten, although the rocks are weaker.

Toward the higher-grade parts of anatectic terranes, there is a change in migmatite morphology; neosome becomes the dominant feature. It is significant that in many of these higher-grade migmatites, leucosome is far more abundant than residual material. The latter typically occurs as schlieren of mafic minerals in the leucosome, together with schollen or rafts of paleosome and melanosome. Overall, paleosome is not abundant in higher-grade migmatites, and may even be absent. Characteristically, pre-partial-melting structures are absent (except where they are preserved as schollen or rafts); they were destroyed during neosome formation and replaced by syn-anatectic structures, typically a magmatic or submagmatic-foliation, or a flow banding. In terms of rheology, the neosome in those migmatites was magma-like.

The transition from one morphology to the other is gradual in some migmatite terranes. The progressive change in the morphology from the lower-grade to the higher-grade parts of migmatite terranes is systematically covered in section D of the book. However, the passage from one to the other is abrupt in many terranes, and commonly tectonic, as it coincides with domains of high shear strain (e.g., Brown and Solar 1998*a*, Solar and Brown 2001). The abrupt nature of the many contacts suggests that some type of threshold behavior may be controlling the changeover. Exactly the same change in the morphology of migmatites occurs between the outer and inner parts of contact aureoles affected by partial melting (Flood and Vernon 1978, Pattison and Harte 1988, Grant and Frost 1990, Hobson et al. 1998, Barnes et al. 2002, Johnson et al. 2003).

This two-fold morphological division of migmatites readily falls into the old scheme of metatexis and diatexis. The paleosome-dominated types were regarded as having formed from low degrees of partial melting, and were called **metatexites**, whereas the neosome-dominated ones were interpreted to be the result of nearly complete fusion, and called **diatexites** (e.g., Mehnert 1968, Brown 1973, Ashworth 1985). The reasons why only these two basic types of migmatites (usefully called metatexite and diatexite) form needs to be understood before the terms can be properly defined. To do this, it is necessary to examine the results from recent petrological studies of migmatites and from studies of the physical properties of partially molten rock.

3.2 Temperature, degree of partial melting, and fraction of melt

Estimates of the metamorphic temperature reached in diatexite migmatites (Brown 1979, 1995; Weber et al. 1985; Percival 1991; Guernina and Sawyer 2003; Sawyer 1998; Solar and Brown 2001; Obata et al. 1994; Oliver and Barr 1997; Johnson and Brown 2004) are typically between 750 and 900°C. These temperatures are commonly a little higher than in nearby metatexites (Grant and Frost 1990; Obata et al. 1994, fig. 11; Oliver et al. 1999, fig. 1). In some terranes, however, there is no significant difference in the metamorphic temperatures recorded in metatexite and diatexite migmatites. In some cases, the transition from metatexite to diatexite migmatite may be due to the influx of hydrous fluids, which promoted more melting, where the diatexite migmatites subsequently formed (e.g., White et al. 2005). The temperatures reached in diatexite migmatites are clearly much lower than required for complete fusion (e.g., 1150–1250°C, Vielzeuf and Holloway 1988). At complete fusion, the composition of the anatectic melt must, necessarily, be the same as that of its protolith. The composition of some diatexite migmatites does indeed coincide with that of the protolith, but this is invariably because the anatectic melt did not separate from its residuum (Sawyer and Barnes 1988; Sawyer 1996, 1998; Milord et al. 2001; Solar and Brown 2001), and not because of complete fusion.

Petrological and geochemical modeling indicates that diatexite migmatites can form with as little as 16% partial melting, but that more commonly, 30–60% partial melting occurred (Sawyer 1998, Milord et al. 2001, Johnson et al. 2001*b*). Similar studies on leucosome–melanosome pairs show that the degree of partial melting (*F*) in metatexite migmatites is generally less than 20–30% (Sawyer 1987, 1991; Barbey et al. 1990, 1996). Higher degrees of partial melting (60–70%) found in certain layers in some metatexite migmatites have been attributed to the local influx of aqueous fluids, or to "wet" melting (Weber et al. 1985).

Petrological (Brown 1979, Johnson et al. 2001*b*) and geochemical (Sawyer 1998, Milord et al. 2001, Solar and Brown 2001) studies show that there is a considerable variation in the fraction of melt (M_f) inferred to have been present within diatexite migmatites. Some parts contained a high fraction (up to 100 vol.%) of anatectic melt (or a melt derived from it by fractional crystallization), whereas other parts, typically those that are richer in residual minerals, contained very little melt (<5 vol.%); yet other parts of diatexite migmatites are rich in the minerals that crystallized from the anatectic melt, most commonly plagioclase. The proportion of leucosome in metatexite migmatites is typically less than 20 vol.% (Weber et al. 1985, Symmes and Ferry 1995, Barbey et al. 1996, Kohn et al. 1997, Johnson et al. 2001*a*, Solar and Brown 2001, Guernina and Sawyer 2003) and may indicate that metatexite migmatites contained a lower fraction of melt than diatexite migmatites. Caution must be exercised, as the leucosome commonly does not have the composition of an initial anatectic melt (see below, section 6.2). Outcrops of metatexite migmatite with greater than 20 vol.% leucosome have been described, and are attributed to veining by granitic melt (Pattison and Harte 1988, Symmes and Ferry 1995, Guernina and Sawyer 2003) or to the early crystallization of feldspar in places where large volumes of melt have passed through the crust (Brown 2004).

A large discrepancy between the degree of partial melting (*F*) and the fraction of melt (M_f) exists for some metatexite and diatexite migmatites, where $M_f < F$ indicates a net loss of melt, and where $M_f > F$ indicates a net gain of melt. Discrepancies indicate that melt was redistributed within migmatites during anatexis (Sawyer and Barnes 1988, Barbey et al. 1990). Because the degree of partial melting in diatexites is typically less than 60 vol.%, a redistribution of melt is the most likely mechanism by which the diatexites that are close to 100 vol.% melt, or mostly leucosome, may have formed (Weber et al. 1985; Sawyer 1996, 1998; Brown and Rushmer 1997).

3.3 The partial-melting process

Where the input of heat has caused the temperature to rise sufficiently that melting starts, small isolated pockets and tubes of melt form at the junctions between the reactant phases (Mehnert et al. 1973). The addition of more heat causes further melting, and where the tubes and pockets of melt have grown sufficiently to link and form an interconnected network, the rock becomes permeable (Waff and Bulau 1979). Melt can then move out of (i.e., segregation of the melt can begin), into or through, the solid framework. This important step was called the permeability threshold by Maaløe (1982), and the liquid percolation threshold (LPT) by Vigneresse et al. (1996). Less than 2 vol.% melt is needed to ensure that sufficient grain-boundaries contain melt, so that felsic systems become permeable (e.g., Dell'Angelo and Tullis 1988, Laporte et al. 1997, Lupulescu and Watson 1999); Rosenberg and Handy (2005) suggested that almost all of the grain boundaries will contain melt where the fraction of melt reaches 0.07. For the segregation of melt to actually occur, there has to be both a driving force and a dilatant site (sink) in which the melt can collect (Sawyer 1994).

The greater viscosity of felsic melts means that gravitational forces are far less effective at driving large-scale segregation of melt in the continental crust (Wickham 1987*b*) than they are in segregating basaltic melt from the mantle (McKenzie 1984). Most partial melting in the continental crust was synchronous with tectonic deformation; therefore, segregation and migration of the melt fraction in migmatites were probably driven by differential stress (McLellan 1988; Barbey et al. 1990; Brown 1994; Sawyer 1991, 1994; Brown et al. 1995; Rutter 1997; Brown and Rushmer 1997; Vanderhaeghe 1999; Marchildon and Brown 2001). Partial melting under conditions of lithostatic stress in which gravity is the only driving force for melt segregation is probably very rare in the continental crust. Nevertheless, it represents an end-member case to be considered first.

3.4 A special case: melting under lithostatic stress conditions (so-called "static melting")

At low melt fractions, near the onset of partial melting and under conditions of purely lithostatic stress, gravity is insufficient to drive the viscous granitic melt out of the framework of solids in which it formed over the period of a typical anatectic event, say 30–40 My (e.g., Rubatto et al. 2001). Consequently, the anatectic melt and residual solids remain together, the neosome is isotropic and without leucosome or melanosome. The distribution of neosome in the migmatite depends on how much melting occurred, and whether melting occurred at specific sites, or was pervasive. As more melting occurs, the fraction of melt eventually becomes sufficiently large that the contacts between all the solid grains disappear; this marks the onset of a magma-like (very weak) rheology. For a matrix initially consisting of uniform, rigid spheres, this occurs at about 26 vol.% melt (MacGregor and Wilson 1939, Arzi 1978). Under perfectly lithostatic conditions, little happens as the volume of melt reaches and then exceeds this value, which was referred to as the melt-escape threshold (MET) by Vigneresse et al. (1996) and the solid-to-liquid transition (SLT) by Rosenberg and Handy (2005); the melt and residual solids remain interspersed. Holness (1999) reported that in the absence of deviatoric stresses, the melt and solid fractions did not separate from each other in the shallow contact-aureole of the Rum Igneous Complex, even though the melt fraction locally reached 0.95. Typically, a new, coarser, isotropic fabric develops in the neosome once the melt fraction crystallizes; migmatites that display these characteristics occur in some deeper contact-aureoles, but very rarely in regional metamorphic terranes (see below, Figs. A1, B17, and B19). However, if gravity-induced separation of dense minerals or convection could occur (a matter of some contention; see Scaillet et al. 1998), then some separation and orientation of minerals may occur.

3.5 The general case: melting under differential stress (so-called "dynamic melting")

Crustal rocks are anisotropic; therefore, during tectonic processes, they deform heterogeneously, and dilatant structures form in the more competent layers. Furthermore, local differences in differential stress create pressure gradients (Robin 1979). Once sufficient melt has been produced that permeability is achieved in the matrix, any additional melt generated moves down the pressure gradients and collects in nearby low-pressure sites. Further deformation produces more dilation and increases the pressure gradients, and together these drive more movement of melt. Where the crust melts under differential stress, the fraction of melt at the site of melting should not normally exceed that required for permeability to be maintained, because any excess would move away to nearby low-pressure sites (Sawyer 1994). In many regionally developed migmatite terranes, the continuous segregation of melt from its residuum appears to be a relatively nondestructive process. For example, some granulite-facies migmatites that now consist largely of strongly melt-depleted rocks still contain well-preserved sedimentary structures (Waters and Whales 1984, Guernina and Sawyer 2003).

The movement of melt in migmatites can be divided into three stages (e.g., Petford 1995, fig. 5; Rutter 1997; Vanderhaeghe 2001). First, it may move a short distance, up to several tens of centimeters, by porous flow. The melt moves from where it formed on grain boundaries, through an interconnected network consisting of microscopic, melt-filled tubes located along grain edges and melt-filled grain boundaries and into small-scale (mm to perhaps cm) low-pressure sites nearby. These may be dilatant foliation and bedding planes, shear bands, interboudin partitions, or ductile fractures. The porous flow stage moves melt away from its immediate point of origin, and concentrates it in various small-scale sites still within the confines of the original source.

The second stage marks the onset of a regime of flow in clearly defined channels. In the initial phase of channeled flow, the melt moves tens of centimeters through a mesoscopic network of linked channels. Thus, the melt has moved farther away from its point of generation and has accumulated in larger (cm–m scale), more stable low-pressure sites, but still within its source. The sites in which the melt accumulates are commonly dilatant foliation or bedding planes and interboudin partitions. The channeled-flow network is made up of cm-scale linked segments consisting of various dilatant structures, the exact type and orientation of which depend on local circumstances, such as the distribution of rock types and the style of deformation. The melt remains trapped in these dilatant sites until it is forced to move again. Upon crystallization, this part of the channel network will become an array of in-source leucosome.

The last phase of the channeled-flow regime is called the transfer stage. It occurs once the melt that accumulated in the second-stage mesoscopic network migrates out of its source entirely. The flow of melt is through a tertiary network of a few commonly discordant, macroscopic channels (fractures, veins, or dikes) over distances of tens to hundreds of meters. This segment of the channeled-flow regime produces leucocratic veins or dikes, and if the transfer distances are sufficiently large, granitic dikes.

Thus, many of the bodies of leucosome seen in migmatites mark either the channels through which melt flowed, or the sites in which anatectic melt collected. Examples of flow-path geometries are given by Brown et al. (1999), Sawyer (2000, 2001), Guernina and Sawyer (2003), Marchildon and Brown (2003), and Brown (2004).

Although some melt-filled sinks and channels may be long-lived, others are transient. Simakin and Talbot (2001*a*, *b*) used a numerical model of an isotropic protolith to show that melt-rich "veins" can grow or atrophy, depending upon their orientation relative to the maximum and minimum principal stresses. Sawyer (2001) argued that melt channels would collapse and disappear if the supply of melt stopped before crystallization began. Similarly, the direction of melt flow may change. Simakin and Talbot (2001*a*, *b*) found that if a vein array is reoriented, or the orientation of the stress field changes, the direction of melt flow may reverse; change of melt-flow direction has also been demonstrated from field studies (Sawyer et al. 1999).

Robin (1979) considered the migration of melt during layer-parallel extension and found that low-pressure sites develop perpendicular to the extension direction in the most viscous (i.e., competent) layer, which attracts melt from the nearby less viscous layers. The most viscous layers are most commonly paleosome, but if the residuum develops to the extent that it contains a large modal proportion of strong minerals, such as garnet or pyroxene, then the residual layers (melanosome) could start to attract melt, rather than lose it. This model has been applied to migmatites (Pattison and Harte 1988, Brown et al. 1995) and relates principally to the segregation of melt into the space between boudins in competent layers and the growth of some stromatic migmatites.

From the discussion above, it is relatively easy to understand how metatexite migmatites form during melting of a heterogeneous continental crust under differential stress. In contrast, it is not clear how diatexite migmatites can form. Two important topics that are central to the formation of diatexite migmatites need to be considered next: (1) the replacement of pre-anatectic structures in migmatites by syn-anatectic flow-structures, and (2) the generation of large-domains that contain high fractions of melt in migmatites.

Rosenberg and Handy (2005) have re-examined data from deformation experiments conducted on partially melted crustal rocks. They found that the major (90%) decrease in strength occurs where the fraction of melt reaches about 0.07. Rosenberg and Handy called this drop in strength the "melt connectivity transition" (MCT) and suggested that it occurs at the point at which virtually all of the grain boundaries in the partially molten rock become melt-bearing. The deformation mechanisms that operate at the high strain-rates in experimental studies include inter- and intra-granular cracking, rigid rotation of grains, and sliding along pockets of melt (Rosenberg and Handy 2005). At the slower strain-rates of natural deformation, other mechanisms, such as grain-boundary sliding accommodated by diffusion, intracrystalline plasticity, and the dissolution of grain–grain contacts into the melt (e.g., Hirth and Kohlstedt 1995; Park and Means 1996, 1997; Rosenberg and Handy 2000) may operate. Although the flow rates achieved by shear stresses at low fractions of melt (ca. 0.07) may be lower than those attainable in a magma (suspension of crystals in a melt), the macroscopic effect is, nevertheless, similar. The pre-partial-melting structures in the rock are progressively destroyed as a new flow-induced structure develops, with the result that the neosome created is a diatexite migmatite. Sawyer (1996, 1998) has suggested that in some instances, a diatexite migmatite that resulted from low degrees of partial melting and contained low fractions of melt ($M_f < 0.2$) may have formed in this way.

Stevenson (1989) developed a model for the migration of melt during layer-parallel compression. In his model, decreases in pressure in the least competent layers (i.e., the layers containing most melt) create a pressure gradient that brings more melt in from the surrounding, more competent layers. Thus, a melt-rich layer grows by a simple feedback-type mechanism. Stevenson's model is particularly interesting because it is a way of concentrating melt in various layers to achieve melt fractions well above the liquid-percolation threshold (LPT) in the general case of melting under differential stress.

High fractions of melt may also be achieved in a migmatite if melt is produced at a rate greater than the rate at which it migrates away (Sawyer 1994). One mechanism for triggering such rapid melting could be the influx of an aqueous fluid (e.g., White et al. 2005). Whether the high fraction of melt in a migmatite undergoing synchronous melting and deformation arises from rapid melting, or from melt migration as outlined by Stevenson (1989), the effects on the morphology of the migmatite are essentially the same. Once melt has accumulated to the point that sufficient grain-to-grain contacts in the matrix have disappeared, the partially molten rock becomes a suspension of crystals in a melt, i.e., a magma, which will then flow in response to the shear stresses imposed on it by local or far-field tectonic forces. Where magma flow occurs, the pre-partial-melting structures in the matrix are destroyed and are replaced by magmatic flow structures, thus forming a diatexite migmatite. Many examples of diatexite migmatite, especially those that have bulk compositions indicating that they contained a high, or very high, fraction of melt, formed in this way (Sawyer 1998, Milord et al. 2001). Estimating the fraction of melt required to change from a rock-dominated to melt-dominated rheology has attracted a great deal of attention. This transition has been termed the rheological critical-melt percentage (RCMP) by Arzi (1978), the critical melt fraction (CMF) by van der Molen and Paterson (1979), and the melt-extraction threshold (MET) by Vigneresse et al. (1996). Recently, Rosenberg and Handy (2005) have called it the solid-to-liquid transition (SLT), but they and others (Takeda and Obata 2003) have suggested that this rheological threshold is much less important than previously considered. Estimates place this threshold at melt fractions between 0.26 and 0.4.

Studies of the microstructures in diatexite migmatites clearly indicate that many flowed as suspensions of crystals in a melt, and some acquired their microstructures in a submagmatic state as the melt crystallized. Nevertheless, some diatexite migmatites may have lost their pre-anatectic structures and developed flow foliation at far lower fractions of melt (<0.2) through the other mechanisms of deformation outlined above. Thus, from the combination of how a high fraction of melt was achieved and the possible mechanism of flow involved, there are five end-member ways of forming a diatexite migmatite. (1) In a closed system where *F*, and hence M_f, are well below the solid-to-liquid transition, by grain-boundary sliding, melt-enhanced diffusion and dissolution processes, leading to bulk flow (i.e., flow of a solid containing a small volume of melt). (2) In a closed system, by an increase in temperature (or possibly a drop in pressure), so that *F*, and hence M_f, rise above the solid-to-liquid transition, leading to the creation of a suspension of crystals in an anatectic melt, and hence magma flow. (3) In a locally open system (the system could be closed on a terrane scale, however), by the movement of anatectic melt within the migmatite, so that M_f is locally increased above the solid-to-liquid transition, and magma flow can occur. (4) In an open system, by the intrusion of an external granitic magma into the migmatite, which augments the locally derived anatectic fraction so that the net M_f rises above the solid-to-liquid transition and enables magma flow to occur. This external magma might be derived from deeper in the same source, or it might be derived from a completely different source-volume. (5) In an open system, by the addition of an aqueous fluid into rocks already at a high temperature, which enables rapid fluid-present partial melting to occur, so that *F* and M_f are locally increased to above the solid-to-liquid transition, and magma flow can occur. In outcrop, the diatexites may look very similar, but petrological (including microstructural) and geochemical studies generally allow a specific process to be determined. In the closed systems, M_f corresponds to *F* determined petrographically or geochemically from the rocks; these could be called **primary diatexite migmatites**. However, in the open systems, *F* determined from the rocks in the migmatites would not correspond to M_f; these could be called **secondary diatexite migmatites**.

3.6 Definitions of metatexite and diatexite

For the same protolith composition, metatexite migmatites generally form at lower temperatures than diatexite migmatites; therefore, the passage from metatexite to diatexite could be viewed as one of increasing degree of partial melting with temperature, but this is an incomplete view. Field, petrological, and geochemical results from migmatites, and the theoretical consideration of melt flow in partly molten rocks discussed above, suggest that movement of melt is also a significant mechanism in leading to elevated fractions of melt in migmatites where melting and tectonic deformation were synchronous. The fraction of melt (M_f) may be increased to 1, or decreased to the liquid-percolation threshold, by migration of the melt, even if the degree of partial melting (*F*) was relatively low, as Sederholm noted. Only in a closed system does $M_f = F$. Therefore, the first-order morphological division in migmatites is best considered in terms of fraction of melt rather than the degree of partial melting. This represents a change in emphasis, from the degree of partial melting to the fraction of melt present, and a divergence from the original reason that the terms "diatexite" and "metatexite" were introduced. However, both terms are too well established and too useful to be discarded. Accordingly, the definitions of diatexite and metatexite are changed slightly from those given by Mehnert (1968), Brown (1973), and Ashworth (1985), and go somewhat beyond the etymological roots of their original definition given by Gürich (1905) and Scheumann (1936).

Metatexite: *a migmatite that is heterogeneous at the outcrop scale, and in which coherent pre-partial-melting structures are widely preserved in the paleosome (where*

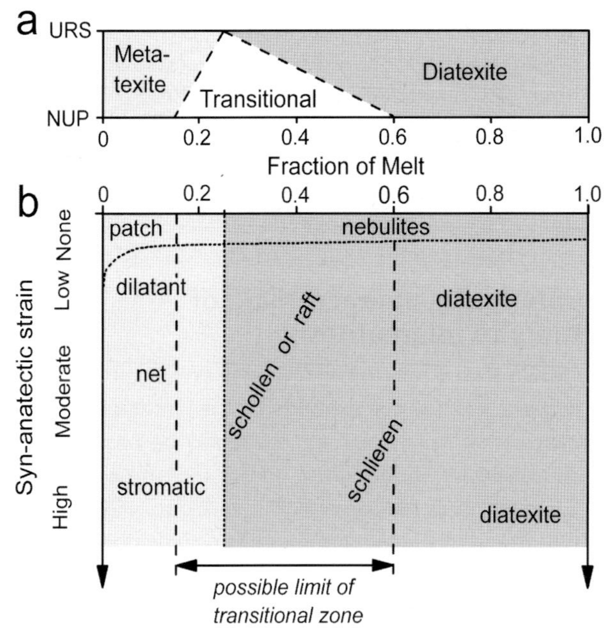

Fig. 1. Classification scheme for migmatites. (a) The first-order division of migmatites into metatexite and diatexite migmatites is considered a function of the fraction of melt and the properties of the solid grains in the partially melted rock. The passage from metatexite to diatexite migmatite occurs abruptly at a fraction of melt equal to 0.26 if the solid crystals are treated as uniform, rigid spheres (URS model). However, for a model in which the solid particles are nonuniform in size and shape (NUP model), there is a transition zone between metatexite and diatexite migmatites that may extend from 0.16 to 0.6 in terms of fraction of melt. (b) The second-order morphologies of metatexite and diatexite migmatites are shown on a plot of syn-anatectic strain versus fraction of melt. The diagram is shaded for the URS model, but the vertical dashed lines indicate where the boundaries are for the transitional zone in a NUP model. The terms shown, except diatexite, should be used as a prefix to either metatexite or diatexite as appropriate (e.g., patch metatexite migmatite or schollen diatexite migmatite).

the microstructure appears unchanged) and, possibly in the melanosome (residuum) part of the neosome, where the fraction of melt was low. The neosome part is generally segregated into leucosome and melanosome, but neosome in which melt and residuum did not segregate may also occur.

Diatexite: *a migmatite in which neosome is dominant, and melt was pervasively distributed throughout. Pre-partial-melting structures are absent from the neosome, and are commonly replaced by syn-anatectic-flow structures (e.g., magmatic or submagmatic foliations, schlieren), or by isotropic neosome. The neosome is variable in appearance, reflecting a large range in the fraction of melt, and it can range from predominantly leucocratic to predominantly mesocratic (e.g., unsegregated melt and residuum) to predominantly melanocratic. Paleosome occurs as rafts or schollen, but may be absent.*

The transition from metatexite to diatexite is abrupt in some terranes and transitional in others. The first-order morphological division of migmatites into metatexite and diatexite is represented on Fig. 1*a* as a function of the fraction of melt (abscissa) and the characteristics of the solid particles (ordinate). The upper part of the diagram shows the change from metatexite to diatexite at a fraction of melt of 0.26, based on the model that the solid particles in the rock undergoing partial melting are uniform, rigid spheres (URS). In rocks, the crystals are uniform neither in size nor aspect ratio; a transition therefore exists between the loss of solid–solid contacts at the local scale and the widespread loss of solid–solid contacts in the bulk sample. Thus, the lower part of the diagram is for the general case in which the rock undergoing partial melting contains a range of particle sizes and shapes. In such a nonuniform particle model (NUP), there is a transitional region, as suggested by Lejeune and Richet (1995) and Renner et al. (2000), extending in terms of fraction of melt from about 0.16 to approximately 0.6 between the metatexite and diatexite fields.

3.7 The second-order morphological divisions in migmatites

In this section, I describe the morphologies of migmatites at scales involving individual leucosome and melanosome on the one hand, and the overall scale at which metatexite and diatexite morphologies are determined on the other. All the morphologies discussed below are shown in section D of the book in the same order as here in the text. The morphology of a migmatite right at the onset of partial melting is likely to be controlled by the physicochemical factors that determine the exact place where the first melt and, therefore, *in situ* neosome are formed. This early morphology is probably best conserved in environments where differential stresses are least, e.g., contact aureoles. However, differential stresses acting as the *in situ* neosome grows will cause the melt fraction to migrate to dilatant structures in the migmatite. Consequently, further changes in the outcrop-scale morphology of migmatites are not solely due to the petrological process of partial melting; they result from the interplay of two factors. (1) The proportion of melt in different parts of the migmatite. Melt will be redistributed in response to differential stresses; thus, the amount of melt present may range from zero to 100 vol.%. (2) The way the rocks respond to differential stress while melt is present. The location of dilatant sites is controlled by the distribution of the competent layers and by the way in which strain occurred. Part of this response will be influenced by the initial, pre-anatectic distribution of competent and incompetent lithologies, but it will also be controlled by the structures that form during deformation and by the locus of melt collection. As the melt fraction increases, strain is increasingly partitioned into the melt-bearing parts (Vigneresse and Tikoff 2000) of the migmatite. Consequently, the morphology of migmatites may be very heterogeneous; a migmatite viewed on a 1-m^2 outcrop might not be described in the same way when seen on a 10 000-m^2 outcrop.

Figure 1*b* shows how the second-order morphologies, which will be described below, are related to the first-order metatexite–diatexite subdivision of migmatite, and to each other by variation in relative syn-anatectic "strain." Strain is not used in a quantitative manner; it is simply used qualitatively to differentiate relatively more deformed rocks from less deformed rocks.

Most of the morphological diversity in metatexite migmatites derives from the anisotropy inherent in the paleosome (and protolith), rather than the magnitude of the syn-anatectic strain or fraction of melt present. High strain generally results in simpler morphologies, because all parts of the migmatites are strongly attenuated or transposed, and paleosome, leucosome, and neosome tend to become parallel, and the overall morphology is stromatic, or layered. Thus, the criteria that are wholly strain-related, such as the lateral continuity of leucosome, do not provide a good basis on which to define metatexite migmatites, or on which to distinguish metatexite from diatexite migmatites. In contrast, the morphological variations that occur within diatexite migmatites are almost wholly a function of the fraction of melt present; more melt necessarily means less paleosome and more neosome. The effect of syn-anatectic strain on the morphology of diatexite migmatites is subtle. Because magmas are weak, flow structures and foliations are initiated in them at comparatively low shear stresses; higher shear strain simply results in better preferred orientations for the enclaves and minerals (Ildefonse et al. 1992, Arbaret et al. 1997), not in a significant change in the morphology. The second-order morphological terms should be used as a prefix to the first-order one, e.g., stromatic metatexite migmatite and schollen (see below) diatexite migmatite.

Figure 1*b* indicates that most of the second-order morphologies are characteristic of relatively low-"strain" rocks. However, recent results from numerical models (e.g., Beaumont et al. 2004, Jamieson et al. 2004) indicate that many migmatite terranes exhumed from the deep parts of orogens are likely to have experienced very high syn-anatectic strain. The weakening that crustal rocks undergo as they partially melt results in the partition of strain into, and the flow of material through, an orogen along discrete zones or channels located in the middle and lower part of the crust. The migmatites from regions where this type of crustal-scale channel flow has occurred are likely to have strongly layered morphologies, but there may still be small domains where other morphologies can be recognized (see below, Figs. A2, A3, D23, and D24).

Morphologies characteristic of metatexite migmatites

Very low fractions of melt and low-strain conditions, near the origin in Fig. 1*b*, corresponds to the onset of melting; migmatites that have lost melt could also be near the origin. Although the very low fraction of melt precludes any eye-catching morphologies from developing (e.g., Figs. D1–D4), the rocks formed at the onset of melting are very important, as they mark the lower-grade limit to areas of migmatite formation. The field evidence for the onset of partial melting is commonly overlooked because the first signs of partial melting in outcrops are subtle. Reactant minerals become slightly rounded where melting has begun, and a sugary texture can develop in lithologies not too rich in mica. If the melting reaction produces a distinctive solid product, that mineral might be recognizable. The first evidence of partial melting in some migmatites is the development of fine-grained, quartzofeldspathic films, which weather white, along some grain boundaries; the films represent crystallized melt (Holness and Clemens 1999, Sawyer 1999, Holness and Watt 2002). A pink coloration at the edge of muscovite grains was found by Holness and Watt (2002) to mark the onset of partial melting in pelitic rocks in a contact aureole in Scotland.

Patch migmatites are formed where slightly higher fractions of melt have been generated, and melting has occurred at discrete sites, so that small, scattered patches of nonfoliated *in situ* neosome develop (Fig. 1*b*). Examples of patch metatexite migmatites can be seen in Figs. D7–D10 and also in Figs. B1, B2, B15, and B30. Paleosome is dominant, and the patches of neosome are generally round or oval in shape. Such neosome was called "blind ending" by Sederholm (1907) and "blind patches" by Platten (1983), and is characteristic of the incipient stages of partial melting in their hosts (Weber and Barbey 1986, McLellan 1988, Grant and Frost 1990, Hobson et al. 1998, Sawyer 1991, Timmermann et al. 2002, Slagstad et al. 2005). Consequently, the first appearance of patchy neosome in migmatite terranes is useful in locating the "melt-in" isograd for the lithology in which it occurs. However, it may be difficult to identify the neosome if the paleosome is a coarse-grained, nonfoliated, felsic plutonic rock. Oriented, tabular bodies of neosome can arise if melting is confined to thin layers, or to certain planes. At higher degrees of partial melting, the neosome grows, or it becomes more plentiful; lobed shapes can develop by coalescence. If coalescence results in the creation of large patches of neosome that have diffuse borders, then the term **nebulite** can be used (see Figs. D11 and D12).

Typically, in patch migmatites, the melanocratic and leucocratic parts are uniformly distributed throughout the neosome because the melt and residuum have not separated; hence, overall the neosome is commonly mesocratic. However, variations result if the melt and residuum do segregate. In one case, the leucosome may be surrounded by a rim of melanosome, whereas in another case a leucocratic rim, or moat, may surround a melanocratic core, which is commonly a single crystal (Weber and Barbey 1986). This latter type of patch neosome (e.g., Figs. B30, D10) occurs in the granulite-facies parts of some migmatite terranes (Waters and Whales 1984, Stüwe and Powell 1989, Powell and Downes 1990, Sawyer et al. 1999, Waters 2001, White et al. 2004); mass-balance estimates for these patches show that many have lost a considerable volume of melt ($M_f < F$). White et al. (2004) described patches of neosome that occur in metapelitic rocks from the upper-amphibolite-facies to granulite-facies transition and that involve garnet produced by an incongruent melting reaction involving the breakdown of biotite and sillimanite. The inability of garnet to easily form nuclei in these rocks has meant that the location of the melting reaction itself and of the other products of reaction was determined by the locus of nucleation of the garnet. This situation results in a migmatite that contains patches of *in situ* neosome consisting of a large garnet porphyroblast surrounded by the other solid products of reaction (e.g., K-feldspar) and the melt.

Patch migmatites are best preserved where syn-anatectic strain is low. Such environments include incipient melting where rocks are stronger, and layers are competent (e.g., Slagstad et al. 2005). Hence, the patch migmatites extend to greater strain than nebulitic migmatites on Fig. 1*b*. The existence of even a small differential stress deforms patch migmatites into high-aspect-ratio shapes (Timmermann et al. 2002), and drives the segregation of melt. As the *in situ* patches of neosome in the migmatites described by White et al. (2004) become larger and more abundant, differential stresses cause the melt and solid fractions to separate as the melt moves into dilatant structures. Consequently, the morphology of the migmatites changes from patch to net-structured, or stromatic. Thus, even low syn-anatectic strain results in the loss of some, or most, of the melt fraction, and the *in situ* patches of neosome are dominated by residual minerals. At higher syn-anatectic strain, patch migmatites no longer exist, and net, or stromatic (also called layered) morphologies form instead.

Net, dilational, and stromatic describe three metatexite migmatite morphologies in which the distribution and form of the neosome (the leucosome in particular) reflect different ways in which the anisotropy due to the layering in the protolith responded to strain during partial melting; some examples are shown in Fig. 2. Since these morphologies are defined with reference to their pre-melting structures, they are divisions of metatexite migmatites. The order of the three types along the "strain" axis of Fig. 1*b* is somewhat arbitrary; all may form at low and intermediate strains. However, at high strains, the leucosome in dilational and net types tends to become parallel, in which case the term "stromatic" or "layered metatexite migmatite" is more appropriate.

Dilation-structured migmatites (surreitic structure of Mehnert 1968) are distinctive, as can be seen in Figs. D13–D22. The leucosome is located in dilatant (i.e., low-pressure) structural sites, such as the spaces between boudins, in pressure shadows, or in fractures (Fig. 2) in the more competent layers of the migmatite. Typically, the dilational sites are comparatively short and restricted to particular layers and do not form an outcrop-scale net structure. This morphology implies the existence of a competency contrast between the various layers in the migmatite while layer-parallel extension occurred. Typically, the competent layers are the less fertile (e.g., greywacke in a pelite), or the resister lithologies (e.g., calc-silicates, quartzite, metamafic rocks, or pegmatites), but they could also be newly formed residual layers rich in strong minerals, such as garnet or pyroxene. Dilational structures may also form owing to strain incompatibilities across the foliation planes (Platt and Vissers 1980), as in the migmatites of southern Britanny (Jones and Brown 1990, Brown and Dallmeyer 1996, Brown 2004) and northwestern Australia (Oliver and Barr 1997). Other types of dilational sites can develop in, and around, the hinges of folds developed in rocks with a strong planar anisotropy (e.g., Collins and Sawyer 1996).

Net-structured migmatites (diktyonitic structure of Mehnert 1968) are a very common type of metatexite migmatite, and Sederholm (1907) considered their development to be an indicator of partial melting; examples are shown in Figs. D23–D30. The essential feature of this morphology is that the leucosome occurs in two or more systematic sets, such that their intersection creates a net-like pattern outlining lozenge-shaped, or polygonal blocks of darker rock. At the early stages of melting, the leucosome is narrow with a high aspect-ratio and is bordered by melanosome; the centers of the lozenges are paleosome. The net-like pattern is very commonly the result of one or more sets of extensional shear bands in which the leucosome is located. It is identified by curvature of the foliation from the host to, or some way into, the leucosome. The

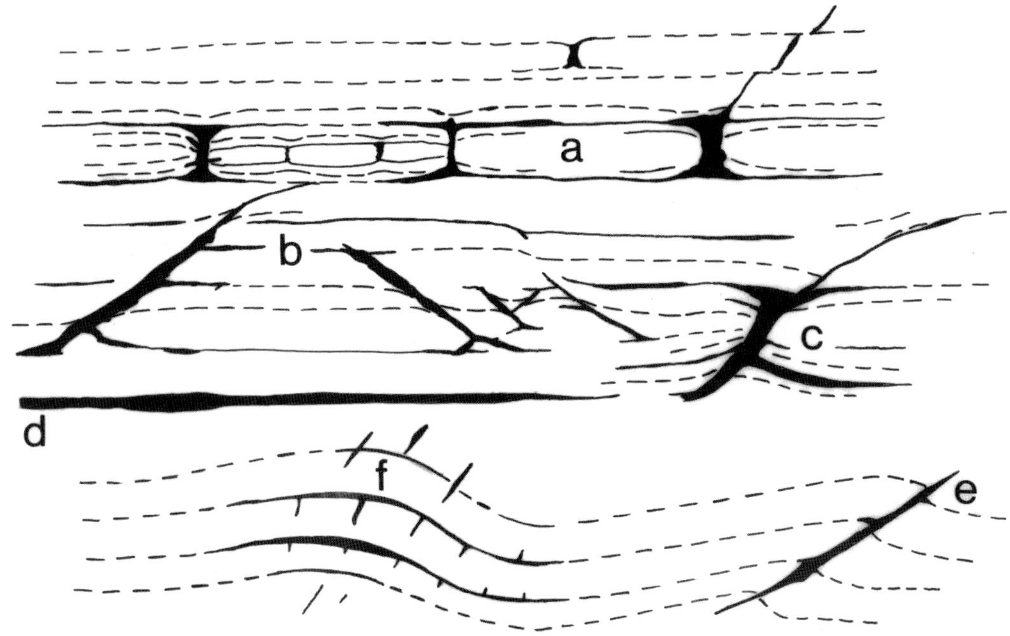

Fig. 2. Schematic representation of some of the structural sites in which the leucosome in dilational metatexite migmatites can be expected to occur. Solid areas and lines represent leucosome, whereas the dashed black lines are the traces of bedding or foliation. (a) Leucosome in interboudin partitions that develop in the competent layers of migmatites; these layers may be paleosome resisters, or even melanosome. Some boudins contain smaller internal boudins. (b) Leucosome located in extensional shear bands; both synthetic and antithetic examples are shown. (c) Leucosome located in an asymmetrical foliation boudin. (d) Stromatic leucosome oriented parallel to the principal plane of anisotropy, which may be either bedding or foliation. (e) Leucosome located in a reverse shear, cutting the short limb of an asymmetrical fold. (f) Leucosome associated with parallel folds; leucosome in the fold hinges is located in the space between dilated bedding planes; shorter domains of leucosome are located in extensional fractures developed on the outside of the folded competent layers and are not axial-planar, but tend to be radially disposed. Leucosome that is located parallel to a fold's axial plane occurs in the less competent layers.

anatectic melt thus seems to have migrated a short distance by porous flow from the host into the shear bands as the bulk rock underwent layer-parallel extension (e.g., McLellan 1988; Sawyer 1991; Brown 1994, 2004; Oliver et al. 1999). Deformation experiments on partially molten analogue materials confirm this (Rosenberg and Handy 2000); the onset of partial melting causes deformation to become concentrated into shear bands, and the melt then migrates to the shear bands. Migration of melt into shear bands, or expulsion from them, depends on whether the shear band is shortening or lengthening (Mancktelow 2002). A set of leucosome bands commonly is oriented parallel to the compositional layering or foliation. The orientation of all sets of leucosome bands, foliations, and shear zones together with their directions of movement, should be recorded in the field so as to determine the local kinematics of the syn-anatectic deformation.

With further partial melting, the paleosome is changed to a gneiss in which folia of residual minerals alternate with quartzofeldspathic layers, and the leucosome bands no longer have obvious melanosome around them (e.g., Oliver and Barr 1997). The rock in the lozenge between leucosome bands progressively becomes neosome, and is generally composed of melanocratic residuum. However, if melt is injected into these layers, then the lozenge may be mesocratic in color. With increased fraction of melt, net-structured migmatites pass, via a transitional stage, into raft or schollen diatexite migmatites (Fig. 1*b*).

Stromatic or layer-structured migmatites have numerous thin and laterally persistent bands of leucosome that are oriented parallel to the major plane of anisotropy in the paleosome. Examples are shown in Figs. D31–D40. A single layer is called a *stroma* (plural *stromata*). The planar anisotropy is, typically, the compositional layering (bedding or igneous layering), or a foliation. Generally, a band of leucosome has melanosome on both sides; in some cases, however, the melanosome is only on one side, and in yet others, there is no melanosome.

Because of the large number of leucosome domains and the regularity of their spacing, stromatic migmatites have attracted a lot of attention. Michel-Lévy (1893) suggested that they are the result of repeated injections of magma between the layers, or foliation planes; hence, he introduced the term *lit par lit* to describe this morphology. The finding in certain layered migmatites that (1) seemingly layer-parallel bands of leucosome actually cross-cut the layering locally, stepping from one plane of anisotropy to the next if traced out across large outcrops (e.g., Sawyer et al. 1999), and (2) adjacent melanocratic bands do not have appropriate residual compositions (Jung et al. 1999), support the melt-injection origin. Many investigators have proposed an *in situ* origin; Holmquist (1916) and Brown et al. (1995) have suggested that the segregations of melt are derived from adjacent rocks. In contrast, Johannes (1983*a*) and Johannes et al. (1995) have argued that the layered structure results from the partial melting of only those layers with suitable bulk compositions (i.e., the fertile layers); the separation of biotite into narrow zones between the layers created the melanosome. Petrographic and geochemical studies have demonstrated that in some layered migmatites, the melanocratic part is the residuum left after the extraction of a partial melt, which corresponds to the composition of the adjacent leucosome (Sawyer 1991, Oliver and Barr 1997), which supports an origin by *in situ* partial melting. The injection and *in situ* layered migmatites can form in the same portion of the fraction of melt – strain space (Fig. 1*b*) as net and dilatant types.

Transposition and the morphology of metatexite migmatites

Park (1983) considered that strain during partial melting could generate a layered structure (see Figs. D35–D40) in migmatites by transposition. Timmermann et al. (2002) and Slagstad et al. (2005) described a migmatite from the Grenville Province in which strain partitioning resulted in the preservation of small, equant bodies of leucosome in a competent host, but transposed them to form layered (high aspect-ratio) leucosome in adjacent, less competent rocks. The competent porphyroblasts of garnet with short (a few millimeters to centimeters in length), but elongate domains of leucosome in their pressure shadows (e.g., Williams et al. 1995, Brown et al. 1999) and hence growing parallel to the foliation, provide another example.

Some migmatites show a curvature of the structural fabrics (such as foliation or compositional layering) and low-aspect-ratio leucosome into a part that exhibits stromatic leucosome and a more regular, and finer-scale compositional banding than the adjacent nonlayered migmatites. The layered structure in these migmatites formed by the attenuation and transposition of originally more equant-shaped accumulations of melt, or partially crystallized leucosome, in outcrop-scale syn-anatectic shear zones. At a much larger scale, metatexite migmatites that develop within crustal-scale shear zones (including zones of channel flow in the continental crust) typically have a pronounced layered, or stromatic, morphology due to transposition and progressive deformation, most evident by abundant, parallel and laterally persistent bodies of leucosome (Figs. A2, A3). The layered structure may be so strong as to obscure any previous morphology that the metatexite migmatite may have had; in this case, the name "layered metatexite migmatite" should be sufficient. Although all pre-anatectic structures in these migmatites may have been destroyed (by transposition, for example), they should not be called diatexite migmatites (unless a high fraction of melt can be demonstrated to have existed in them); they should be viewed primarily as tectonites. However, leucosome generally is present, and more rarely neosome (i.e., leucosome + melanosome), which postdate the main layered structure, and these may have any of the morphologies typical of the lower-strain part of Fig. 1*b*. Evidently, a layered or stromatic morphology in metatexite migmatites can form in several different ways, and over a wide range of conditions of syn-anatectic strain.

Morphologies characteristic of diatexite migmatites

At low syn-melting strains, a fraction of melt above about 0.26 eliminates the contiguous crystal-based framework, and the whole migmatite becomes nonfoliated (i.e., isotropic) and unsegregated neosome; for this stage, the term "nebulitic diatexite migmatite" should be used (Fig. 1*b*). However, under rare circumstances, nebulitic migmatites may also form at much lower fractions of melt instead of a patch metatexite migmatite. This will occur if the partial melting occurred, or spread, throughout the whole volume of the protolith, rather than at a few discrete sites. Pervasive melting leads to the widespread development of neosome in which the residual and remaining solid minerals can undergo extensive recrystallization and grain growth because of the presence of melt along the grain boundaries.

Nebulitic migmatites are preserved only in places where the syn-anatectic strain was very low. A partially molten rock containing a melt fraction of more than 0.26 is very weak, and in general, differential stress creates shear stresses, and flow of the magma readily occurs. Consequently, the nebulitic migmatite develops a foliation due to flow, and then another of the morphological terms shown on Fig. 1*b* becomes appropriate.

Schollen or raft-structured migmatites are characterized by the raft-like form of the remnants of paleosome, resister lithologies, or melanosome displayed in the neosome. This type of migmatite is common at the transition from metatexite to diatexite migmatites (e.g., Solar and Brown 2001), as well as in the lower-melt-fraction part of diatexite migmatites (Fig. 1*b*). Examples of schollen diatexite migmatites from the transition can be seen in Figs. D41–D46. The proportion of paleosome, melanosome, and resister lithologies is highest at the start of the transition zone, where the lateral continuity of these starts to become disrupted. The layers and rafts (also called schollen) of paleosome are large and have high aspect-ratios, and some show rounding at the ends, but there is generally little rotation of the rafts. Farther into the transition zone, and toward the diatexite domain, there is a progressive decrease in the size, aspect-ratio, and number of rafts or schollen (Figs. D47–D52). Typically, the rafts are more rounded, rotated, and dispersed in leucocratic neosome. The leucocratic portions generally have a flow foliation defined by the orientation of platy minerals, such as the feldspars and micas. The migmatites called inhomogeneous diatexites (Mehnert 1968, Brown 1979), or heterogeneous diatexites (Solar and Brown 2001), have schollen structures and are typical of the transition from metatexite to diatexite.

Schlieric migmatites have well-developed flow-induced structures indicated by trains of platy or elongate minerals, most commonly biotite, but also sillimanite, plagioclase, orthopyroxene, and amphibole, that are called schlieren (the singular is *schliere*). The microstructure of schlieren is shown later, in section F, Figs. F97–F100. Rafts or schollen of paleosome, resister lithologies, and melanosome may be present, but they are far less abundant than in schollen migmatites, as can be seen in Figs. D53–D56. The passage from schollen migmatite to schlieric migmatite is achieved by an increase in the melt fraction, or neosome:paleosome ratio (Fig. 1*b*); strain has little effect.

Small outcrops of schollen and schlieric migmatites tend to show a parallelism between successive bands of leucosome, schlieren, and raft-rich layers. However, larger outcrops invariably show regions where the foliation and compositional banding or layering are truncated at a low angle by groups of similar-looking, although generally slightly more leucocratic layers; Mehnert (1968) referred to these as interpenetrating forms. These structures are evidence of syn-magmatic shear zones, or flow discontinuities in the partially molten rock, and are probably due to the partitioning of strain into the zones with the highest fraction of melt. Folds may develop in these shear zones, in which case they typically have a sheath-like geometry.

Diatexite migmatites, in Fig. 1*b*, occupy the high fraction-of-melt part of the fraction of melt versus strain plot; thus, they are dominated by neosome (Figs. D57–D62), and relics of paleosome are rare, or absent. Pre-partial-melting structures, such as foliation, folds, and bedding, occur only in the scattered schollen of paleosome. However, a foliation defined by the orientation of platy or tabular minerals, most commonly plagioclase and mica, acquired during flow in a magmatic or submagmatic state, is virtually ubiquitous in the neosome. A flow banding due to layers of different mineralogy, grain size, or microstructure also may be present, and schlieren may be present but are scarce. Diatexite migmatites are gradational from schollen and schlieric migmatites through an increase in melt fraction (or neosome: paleosome ratio), and from nebulites through an increase in syn-anatectic strain (Fig. 1*b*), which account for their characteristic flow-induced foliation and banding. As diatexite migmatites tend to show subdued, or subtle, variations in outcrop, the term "homogeneous diatexite" has been applied to them (Mehnert 1968, Brown 1979, Solar and Brown 2001). Unfortunately, it is not always the same property that is being described as homogeneous. For Mehnert, homogeneous meant lacking a foliation, and he coined to term "homophaneous" specifically for these diatexites. Brown (1979) also used homogeneous in a structural sense, whereas Solar and Brown (2001) used it in more of a morphological context to mean lacking schollen and schlieren. Other investigators have found significant petrological diversity within, and among, diatexite migmatites in individual anatectic terranes and have, accordingly, chosen to subdivide diatexite migmatites on a petrological basis, e.g., the melt-rich and residual diatexites of Sawyer (1998) and the leucocratic, mesocratic and melanocratic diatexite migmatites of Milord et al. (2001).

Almost all diatexite migmatites, including the schollen and schlieric varieties, contain leucocratic patches and veins, consisting of quartz, K-feldspar, and plagioclase that appear to postdate the magmatic foliation and flow structures in their host. In all the cases where the composition of these veins and patches has been investigated, the leucosome represents a fractionated melt (also called residual melt). Diatexite migmatites contain sufficient melt that as they cool and crystallize under applied differential stress, the crystals in them become aligned, giving the rock a foliation (see section 5.3). A framework of touching crystals develops where crystallization reduces the fraction of melt below about 0.4. If deformation continues, the framework then deforms heterogeneously, and this creates dilatant structures to which the evolved (fractionated) melt, located in the interstices of the enclosing framework of crystals, migrates; analogous structures are well known from crystallizing granites (e.g., Cuney et al. 1990, Pons et al. 1995, John and Stunitz 1997). Geometrically, therefore, these masses of leucosome resemble those in metatexite migmatites and those shown in Fig. 2, but they differ in morphology in that the melt-depleted halos around them are not generally melanocratic, because they are dominated by feldspars.

High strain and the morphology of diatexite migmatites

High shear strain tends to generate better alignments of minerals, rafts (or schollen) of paleosome material and of the schlieren in diatexite migmatites. Deformation during crystallization can lead to the segregation of the residual melt and to transposition (e.g., Figs. D63 and D64); the resulting alignments commonly produce well-developed parallel planes of foliation or compositional banding.

3.8 Migmatite morphologies outside the metatexite–diatexite division

Two of the migmatite structures defined by Mehnert (1968), vein and folded, are different from the other types described so far because they occur in both metatexite and diatexite migmatites. The principal control on the morphology of fold structures in migmatites is the mechanism by which the folds formed, and this is linked to the fraction of melt present, and to the form, or geometry, of the anisotropy present in the migmatite. In metatexite migmatites, where paleosome dominates, folds might form by processes such as buckling, but in diatexite migmatites, where neosome dominates, the folds may form because of flow instabilities (i.e., passive flow folds). Vein migmatites, however, are singular; their morphology is not controlled by melt fraction, but by the ability of their host to fracture, either by brittle or by ductile fracture, and by the timing of the veining process relative to the rest of the anatectic history.

Fold-structured migmatites, which develop at low fractions of melt, in the early stages of anatexis, generally have morphologies that are controlled by the relative difference in competency between the layers in the paleosome. In relatively incompetent rocks, the folds in the paleosome are similar types, but if there are more competent layers, buckle (parallel) folds may develop (see Figs. E1–E8). During folding, some of the melt in a migmatite migrates into dilatant sites that develop as the folds grow, resulting in leucosome in the fold hinges and parallel to and between the folded layering, in axial planar shears, or in shears developed on the fold limbs. If competent layers are present, then the melt will also occupy the array of radial fractures that may develop around the fold hinges in these layers (Collins and Sawyer 1996). At this stage, then, the control on the distribution of leucosome is similar to the dilatant metatexite migmatites discussed above. Several investigators (e.g., Allibone and Norris 1992, Collins and Sawyer 1996) have noted that there is generally more leucosome in the fold hinges than on the fold limbs; this finding could indicate that limbs are regions of net loss of melt, and the hinges, of net gain of melt, which is confirmed by the analogue experiments of Barraud et al. (2004).

Metatexite migmatites in which bodies of leucosome are oriented parallel to the axial planes of syn-anatectic folds appear to be very common (e.g., Edelman 1973, Hand and Dirks 1992, Brown 1994, Vernon and Paterson 2001), but they are quite diverse in appearance. In the majority of the examples that have been described, the bodies of leucosome are thin, have smooth margins and high aspect-ratios. However, in some cases, the leucosome accumulation is thicker and more equant, with irregular outlines; these are typically also coarser grained. Melanosome is evident around the leucosome in some cases, but absent in others. In some migmatites, the leucosome in the axial plane is K-feldspar-rich compared to the leucosome elsewhere (e.g., Jamieson 1984). Such K-rich leucosome presumably was derived from anatectic melts with evolved compositions owing to fractional crystallization. Why concentrations of leucosome should be located in axial surfaces has long been a controversial point amongst structural geologists working on migmatites; Brown and Rushmer (1997) and Vernon and Paterson (2001) discussed this matter. Skelton (1996) suggested that the flow of melt on the limbs of a fold is parallel to the layering and not across it; thus, the melt from both flanks necessarily accumulates at the nearest fold-hinge, where layer-parallel flow is no longer possible. Consequently, the melt either accumulates in the hinge, or else must escape across the layering, that is, in a direction necessarily subparallel to the axial plane of the fold. Where examined closely in the field, many bodies of leucosome described as axial planar are, in fact, not so; they are oriented so as to cross both fold limbs.

Some investigators (e.g., Edelman 1949, Kranck 1953, Wynne-Edwards 1963, McLellan 1984) have noted that where the fraction of melt reaches and exceeds the melt-escape threshold (0.26–0.4), there is a progression from planar, cylindrical folds to noncylindrical, nonplanar disharmonic and convoluted types of folds, with décollement surfaces between some layers. However, hinge directions commonly remain systematic, and subparallel to the flow lineation; sheath-fold geometries commonly develop.

It is important that a study of the microstructures be made to determine whether the folding occurred when the rocks contained a high fraction of melt, or whether it occurred later, at lower fractions of melt when the rocks were crystallizing, or whether folding occurred entirely in the solid state, after all the melt had crystallized. The latter, of course, are folded migmatites and not fold-structured migmatites.

Vein-structured migmatites (Figs. E9–E14) contain one or more generations of discordant, leucocratic veins (of granitic, granodioritic, or tonalitic composition), which are typically superimposed on an earlier metatexite or diatexite migmatite morphology. All of the sets of veins, or only one of them, may be deformed, but commonly the youngest set consists of undeformed planar fracture-fillings that postdate the peak of anatexis and the peak production of a melt fraction in their host. Interesting microstructural relationships have been reported from vein migmatites. Some veins contain trains of mafic minerals oriented parallel to their walls, whereas others have helicitically preserved structures from their wallrocks. Collectively, these microstructures may indicate injection into partially molten wallrocks and erosion of the vein walls as the melt flowed in the vein (Brown 1994, Sawyer et al. 1999, Solar and Brown 2001). Wide borders with residual compositions are rarely developed around discordant leucocratic veins. On the other hand, narrow mafic selvedges at the edges of the veins are very common features, and these may have formed as a result of a reaction between the wallrocks and the injected melt, or with an aqueous fluid derived from the melt. The lack of a mass balance between the volume of melt in the vein and the volume of material in the adjacent mafic selvedge (or residual border) is evidence that the melt was not derived from its immediate host, but was injected. It is important to determine that the melt in the veins belongs to the same anatectic event as the migmatites; if the veins belong to a younger event, then the term "vein-structured migmatite" does not apply; it is a veined migmatite.

3.9 Descriptive terms that should be abandoned

Bedded migmatite was used by Greenfield et al. (1996) and White et al. (2003) to describe migmatites in which the neosome, and the leucosome in particular, are confined to certain layers of a particular bulk composition, generally metapelite; examples are shown in Figs. A1, B17, and B18. The examples described in the literature display a considerable range in proportions and morphology of the neosome, and especially the leucosome. Some examples in which the proportion of neosome is relatively small have leucosome material located in dilatant structures (e.g., between boudins) or oriented parallel to the layering and foliation, and associated with prominent melanosome. However, in others the layer has been entirely converted to neosome in which the melt and the residuum, typically, have not segregated. Nevertheless, for both of these, an *in situ*, closed-system origin for the neosome is plausible in which melting occurred in only those layers that were fertile; these are, therefore, exactly analogous to the origin of stromatic metatexite migmatites described by Johannes and Gupta (1982), Johannes (1983*a*), and Johannes et al. (1995).

In contrast, some of the layers have a very high proportion of leucosome, which suggests that melt was added, i.e., open-system behavior, unless perhaps the bulk composition was optimal for a very high fraction of melt to form. For these examples with a high fraction of melt, the bed or layer may have acted as a dilatant sink into which melt from the outside has moved. At a sufficiently high fraction of melt, magmatic flow can occur in the layer if it is deformed (e.g., Sawyer 1996). At Mount Stafford, in Australia, shear stresses were low and neither flow nor segregation of the melt from the solid fraction occurred (Greenfield et al. 1996, White et al. 2003); consequently, the neosome has an isotropic appearance (e.g., Fig. B17).

The term "bedded migmatite" is inappropriate, as it does not refer to any process active during the formation of these migmatites. Furthermore, structures such as bedding or compositional layering that are older than the partial melting are still preserved in a coherent fashion; therefore, these are, necessarily, metatexite migmatites. Consequently, the term "bedded migmatite" ought to be abandoned.

Agmatite was defined by Sederholm (1923) as a rock with "fragments of older rock cemented by granite", and was regarded by him to be a type of migmatite. However, rocks matching this description can be found around igneous intrusive bodies in low-grade or unmetamorphosed country-rocks. Brown (1973) argued that agmatites are not migmatites, and should be called intrusion breccias. Consequently, the term "agmatite" ought to be abandoned.

Ptygmatic migmatites. The formation of ptygmatic folds reflects a particular combination of vein and host-rock properties and strain conditions that are not specific to migmatites. Moreover, it is common that not all the bands of leucosome in a particular outcrop display ptygmatic folding. Therefore, the term "ptygmatic" should be reserved to describe the style of folding displayed by some bands of leucosome, and not used to describe the overall morphology of a migmatite.

Ophthalmite migmatites, as defined by Mehnert (1968), are characterized by the presence of augen, or lenticular aggregates of newly formed minerals. Many of the augen would now be called porphyroclasts, and the lenticular aggregates viewed as disrupted, coarse-grained quartzofeldspathic layers. In addition, the rocks typically have a strong planar fabric and, in some cases, oblique subsidiary foliations, which, taken together, are geometrically equivalent to 'C' planes, 'S' planes, and shear bands, respectively. Thus, the structure of ophthalmite migmatites closely resembles that found in many coarsely crystalline rocks deformed to high strains under high, but subsolidus, temperatures and that have subsequently undergone some recrystallization. Therefore, the ophthalmite structure commonly reflects post-anatectic, not syn-anatectic, processes. Consequently, ophthalmites are not migmatites, although their protolith may have been a migmatite, and they are better called augen gneisses or blastomylonites.

4.

METASOMATISM AND MIGMATITES

In the first half of the twentieth century, three lines of reasoning contributed to the idea that fluids are fundamental to the generation of granites and migmatites. First, the opinion of Bowen (1915, 1928) that granitic magma could only be formed at the final stages of the fractional crystallization of a basaltic magma influenced many geologists to believe that the granitic rocks found abundantly in shield areas may not be magmatic. Second, observations from many terranes appeared to show gradational contacts between metamorphic rocks and granites, and were interpreted as indicating an *in situ*, nonmagmatic transformation. The view that granites were formed by an *in situ* transformation was strengthened by the finding of fragments and screens of metasedimentary rocks in granites that still retained the orientation of the rocks outside the granite (Misch 1949, Pitcher 1952). Third, whole-rock geochemical data appeared to show a systematic increase of granitic components with increasing depth (Lapadu-Hargues 1945) and the existence of compositional fronts in the metamorphic rocks and migmatites around granites; this was interpreted as evidence of the movement of granite-making material (and of granitization) in the crust. One school (e.g., Perrin and Roubault 1937, 1939) argued that the movement of material occurred by diffusion in the solid (dry) state, whereas the other (e.g., Sederholm 1926, Backlund 1946, Misch 1949) argued that aqueous fluids moved the material. More recently, Korzhinski (1971) called these two alternatives **diffusion metasomatism** and **infiltration metasomatism**, respectively.

Subsequent systematic study of the whole-rock compositions of metasedimentary rocks ranging from low grade to high grade (e.g., Shaw 1954, 1956; Haack et al. 1984; Sawyer 1986) has shown that crustal metamorphism is close to being isochemical, except for the loss of H_2O and CO_2. Consequently, neither diffusion nor infiltration metasomatism is now considered to be an important, ubiquitous, or necessary process in crustal anatexis, although there are other views (e.g., Litvonovsky and Podladchikov 1993). Nevertheless, there are some migmatites where infiltration metasomatism has played a significant role, and these fall into two broad groups: (1) the ingress of aqueous fluid into hot rocks and (2) metasomatism prior to anatexis that changed (improved) rock fertility.

4.1 Influx of aqueous fluid into hot rocks causing partial melting

Large-scale influx of fluid

The extensive development of migmatites with a high proportion of leucosome derived from leucocratic tonalitic, trondhjemitic, granodioritic or granitic protoliths is common in many Archean and Proterozoic shield areas. These migmatites are problematical. The absence of muscovite and the small amount of biotite or hornblende (or both) in these rocks preclude the generation of a large fraction by dehydration melting. Consider the hornblende-free migmatites first; five modal percent biotite yields about 8 vol.% granitic melt. Furthermore, the general absence of cordierite, garnet and orthopyroxene from these migmatites also precludes volatile-phase-absent incongruent melting of biotite. Consequently, the high fraction of melt in these migmatites is generally attributed to an influx of H_2O close to, or just above, the solidus temperature, which enabled H_2O-saturated (congruent) melting of quartz + sodic plagioclase + K-feldspar to occur (e.g., Kenah and Hollister 1983, McLellan 1988, Nédélec et al. 1993, Sawyer 1998). Many of the migmatites derived from felsic, plutonic protoliths contain hornblende in the melanosome; this could be a product of melting reactions involving the breakdown of biotite. Recent experimental studies (Gardien et al. 2000) indicate that the formation of product (residual) hornblende from the breakdown of biotite requires the addition of H_2O, and hence volatile-phase-present melting. In some cases, the abundance of the large-ion lithophile elements (LILE) is greater than can be attributed to the protolith, such that they may have been introduced with the fluid, generating potassium-rich migmatites from rather potassium-poor protoliths.

The source of the fluid, or how it enters the rock, is not understood, but the protoliths of most of these migmatite terranes were calc-alkaline rocks formed in plutonic arcs (although the anatexis may have occurred many millions of years later; e.g., Slagstad et al. 2005), and both the associated subducted material and the underlying metasomatized mantle wedge are possible sources for the fluid and the

LILE. A similar process seems to occur presently in active island arcs; H_2O entering the base of the arc fluxes partial melting of hot, mantle-derived mafic rocks there, and generates the more felsic magmas found at higher levels (e.g., Hansen et al. 2002). Partial melting fluxed by the large-scale influx of H_2O into hot rocks has been suggested for the generation of upper-amphibolite-facies migmatites (Yardley and Barber 1991, Johnson et al. 2001*b*, White et al. 2005) and for lower-granulite-facies metatexite and diatexite migmatites (Otamendi and Patiño Douce 2001) derived from pelites and aluminous metagreywackes, respectively. The aqueous fluids that fluxed melting were probably released by prograde metamorphic dehydration-type reactions occurring in adjacent rocks, which, because of their bulk compositions, had slightly higher solidus temperatures (Fornelli et al. 2002, White et al. 2005).

Castro (2004) has proposed that K-rich granodioritic magmas are generated from older tonalitic continental crust where H_2O-rich intermediate (e.g., andesitic, ferrodioritic or monzodioritic) magma generated from the hybridized parts of a mantle wedge, suitably modified by fluids released from subduction zones, are injected into the lower crust. The H_2O liberated from the crystallization of the intermediate magma comes into contact with the adjacent hot, dry tonalite and promotes fluid-present partial melting. The resulting anatectic melts are enriched in K and Rb (and possibly the other LILE) transported from the intermediate magmas in the hydrous fluid. The CaO content of the anatectic melt is increased to levels typical of granodiorite by the dissolution of microgranular enclaves consisting of the intermediate magma. Therefore, these migmatites show not only the usual paleosome–neosome relationships, but also evidence of the interaction between anatectic melt and the intermediate magma, such as millimeter- to centimeter-scale lobate interface structures and microgranular enclaves, and of the synchronous intrusion of dikes of intermediate composition into the protolith.

Small-scale influx of fluid

Results from studies on a hand-sample scale, using many different techniques including neosome–protolith mass balance (Olsen 1982, 1984, 1985; Olsen and Grant 1991; Pattison 1991), whole-rock compositions (Weber and Barbey 1986), fluid inclusions (Olsen 1987, Whitney and Irving 1994), reaction progress (Symmes and Ferry 1995), and consideration of melting reactions (Johnson et al. 2001*b*), have established that some migmatites formed when a local influx of H_2O enabled already hot rocks to partially melt. The whole-rock and mass-balance methods commonly show that the fluid also introduced K, Rb, and other elements into the rocks.

One common scenario is the intrusion of anatectic magma into the parts of a migmatite terrane where metamorphic temperatures were above the solidus, yet insufficient for significant dehydration melting of biotite to have occurred in the muscovite-free rocks. Aqueous fluids (and heat) released by the crystallization of the intrusive body infiltrates into the hot country-rocks and enables H_2O-present melting to occur, generating what Finger and Clemens (1995) have called **secondary migmatites**.

Several investigators (e.g., Symmes and Ferry 1995, Harris et al. 2003, Johnson et al. 2003) have shown that fluids generated by dehydration reactions in the outer part the contact aureoles have migrated inward, up the temperature gradient, and enabled H_2O-present melting to occur close to the intrusive body; H_2O-absent partial melting occurs still closer to the intrusive contact. In some instances, the fluids have transported trace elements, such as B, Ba, and Li, which serve to reduce the temperatures required for partial melting to start (e.g., White et al. 2003). The breakdown of tourmaline during upper-amphibolite-facies anatexis may also lead to a boron-rich melt and a boron-depleted, residual crust (Kawakami 2001, Kawakami and Ikeda 2003).

Instead of, or in addition to, local-scale fluid infiltration, Buick et al. (2004) suggested that more extensive partial melting may occur at relatively low temperatures in contact aureoles if the fluid from the low-grade protolith remains in the system once partial melting starts (i.e., the system is closed to loss of volatiles). This scenario is plausible during rapid heating in the innermost parts of high-temperature contact-aureoles and, in effect, renders the protolith more fertile than if heating had been slower and if the fluid had had sufficient time to migrate away. Büttner et al. (2005) proposed that a similar process may have occurred in regionally metamorphosed rocks in Argentina. There, fluids generated by prograde dehydration-type reactions in the greenschist and amphibolite facies remained in the rocks and fluxed higher degrees of partial melting in granulite-facies rocks.

4.2 Metasomatism and changes in the fertility of rocks

If the bulk composition, and hence the fertility, in parts of a metamorphic terrane are changed as a result of infiltration metasomatism that occurred before anatexis, then the distribution of anatectic migmatites will reflect this. Sawyer (1991) noted that the first sign of anatexis and migmatites in mafic protoliths was in strongly foliated metamafic rocks, and not in the massive equivalents. The foliated metamafic rocks represented an earlier, pre- to syn-partial-melting shear zone that had been hydrated by a fluid that introduced H_2O, K_2O, Rb, Ba, and Cs. Thus, the shear-zone rocks were made more fertile than the adjacent massive metamafic units, and so melted first.

4.3 Morphology of migmatites affected by infiltration metasomatism

Where the influx of fluid has caused partial melting, the resulting migmatites could have any of the morphologies characteristic of metatexite and diatexite migmatites. However, there are two additional factors to consider, which could affect the appearance of the migmatites that form in outcrop. (1) Melting occurred where the aqueous fluid was present, and hence the overall morphology of the migmatite should mirror the path taken by the infiltrated fluid. In many cases, the fluid infiltration was along foliation planes or lithological layering, and this is reflected by the stromatic (layered) or bedding-confined morphologies of the resulting migmatites. However, net geometries can also be generated (e.g., Pattison 1991) if fluid ingress was through a network of fractures (Figs. B31, B32). Dilatancy pumping in shear zones (Barr 1985, Johnson et al. 2001*b*) can move aqueous fluids into hot rocks, and the geometry of the resulting anatectic leucosome should be related to the pattern of dilatant structures that formed, and hence to shear-zone kinematics; thus the morphology of the migmatite could be quite varied. (2) The degree of partial melting is a function of how much fluid was available in a given volume of rock. Furthermore, permeability may differ from layer to layer. Thus, on a local scale, there can be greater variation in the degree of partial melting from place to place, and correspondingly in migmatite morphology, than is generally the case for dehydration melting. The layers that generated much melt could become entirely neosome, whereas adjacent layers may not have melted at all, either because of refractory bulk compositions, or because they were impermeable.

5.

MICROSTRUCTURES IN MIGMATITES

In recent years, there has been a considerable advance in understanding how microstructures form in solid and partially molten rocks. In this section, I identify some of the key microstructures and quantitative analytical techniques that may be used to identify the processes that have contributed to the formation of migmatites. Most of the microstructures discussed indicate the former presence of melt, and hence constitute petrographic criteria for distinguishing migmatites from the migmatite-like rocks formed by subsolidus segregation or other processes (see section 8). Photomicrographs that illustrate the microstructures discussed here are shown in section F of the book, and are arranged in the same order.

In order to fully understand the microstructures, the mineral parageneses have to be identified, and the sequence of overprinting parageneses must be determined. Once this has been accomplished, the petrogenesis may be interpreted by reference to quantitative petrogenetic grids. In recent years, the insight that can be gained from mineral parageneses has been greatly increased through equilibrium thermodynamic modeling of the specific bulk compositions of the samples being investigated. This has become possible because internally consistent thermodynamic datasets are now reasonably comprehensive (e.g., Holland and Powell 1998), and models for the activity–composition relationships for the minerals of interest in anatexis have been improved (e.g., Holland and Powell 2001, NCKASH; White et al. 2001, NCKFMASH; Tinkham et al. 2001, MnNCKFMASH). In addition, software such as Thermocalc (Powell and Holland 1988, Holland and Powell 1998) enables equilibrium phase-diagrams for the specific composition of the bulk rock, or equilibrium domain, of interest to be constructed as a so-called pseudosection or as a chemical potential diagram. This book does not contain any general phase-diagrams for partial melting for this reason.

5.1 Mineral paragenesis

The approach is the same as that used by metamorphic petrologists in general. All minerals are identified, and the way in which they are spatially related is documented (i.e., whether minerals only occur as inclusions in others, whether grains are corroded or not, whether all minerals are in mutual contact somewhere in the rock, etc.). This information is gathered systematically across the migmatite terrane to determine the sequence of mineral parageneses.

Generally, this information is arranged in order of increasing metamorphic grade, together with information on the bulk-rock compositions in which they occur. Then, a petrogenetic grid or an appropriate pseudosection is used to identify a sequence of metamorphic reactions that may account for the mineral parageneses found, and the measured modal abundances of the minerals.

In most cases, the greatest diversity of mineral parageneses and microstructures is to be found in the residual rocks and paleosome. Resister lithologies are commonly competent, and along with strain shadows, should not be ignored; they may contain minerals and microstructures that could reveal the earlier part of the *P–T–t* history destroyed in rocks where strain was higher. Some recent examples of the use of mineral parageneses and microstructures to reconstruct the sequence of melting reactions are provided by Hartel and Pattison (1996), Raith and Harley (1998), Greenfield et al. (1996), Fitzsimons (1996), Cesare (2000), White et al. (2003), and Johnson et al. (2003). These investigators show that the careful examination of the microstructural or textural relationships among minerals in migmatites remains an extremely important part in understanding their development.

In the past, the conditions of partial melting were either calculated using suitable geothermometers and geobarometers, or estimated from a petrogenetic grid (e.g., Grant 1985, Vielzeuf and Holloway 1988, Grant and Frost 1990, Spear et al. 1999). As more studies of partial melting have been done, the range of bulk compositions and reactions investigated has expanded and now includes metapelites (Vielzeuf and Holloway 1988, Le Breton and Thompson 1988, Patiño Douce and Johnston 1991, Carrington and Watt 1995, Patiño Douce and Harris 1998, Pickering and Johnston 1998, Koester et al. 2002, Spicer et al. 2004), metagreywackes (Conrad et al. 1988; Vielzeuf and Montel 1994; Gardien et al. 1995, 2000; Patiño Douce and Beard 1995, 1996; Stevens et al. 1997; Montel and Vielzeuf 1997; Castro et al. 1999; Nair and Chacko 2002; Grant 2004), amphibolites (Rushmer 1991; Rapp and Watson 1995; Wolf and Wyllie 1991, 1994; Springer and Seck 1997; Skjerlie and Patiño Douce 2002; Selbekk and Skjerlie 2002), tonalite (Rutter and Wyllie 1988; Skjerlie and Johnston 1992, 1996; Singh and Johannes 1996*a*, *b*; Patiño Douce 2005), trondhjemite (Johnston and Wyllie 1988), granite (Holtz and Johannes 1991, Castro et al. 2000, Litvinovsky et al. 2000, Acosta-Vigil et al. 2006), and felsic to intermediate volcanic rocks (Conrad et al. 1988). The conditions of melting in migmatites are now commonly deduced by direct comparison with the experimental results done on similar bulk compositions.

The development of a thermodynamic model for silicate melts (Holland and Powell 2001, White et al. 2001) has made it possible to calculate petrogenetic grids and pseudosections for individual rocks in the region of *P–T–X* space where partial melting occurs, and to adjust their bulk compositions for the loss of some, or of all, of the granitic melt (White and Powell 2002). This powerful new approach is being applied to partial melting; the consequences of melt extraction on mineral assemblages, mineral modes, and mineral compositions in pelitic, semipelitic, and psammitic rocks may now be investigated in a quantitative manner. Recent examples of this application are given by White and Powell (2002), White et al. (2003–2005), Johnson et al. (2003), and Johnson and Brown (2004).

5.2 Quantitative analysis

All methods of quantitative analysis have the potential problem that the microstructure preserved in the rock might not be that which developed when anatectic melt was present. The possibility that the microstructure has been partially, or wholly, modified subsequently must be carefully evaluated. Deformation tends to reduce the grain size and to produce changes in grain shape and grain orientation. Recrystallization tends to increase grain size, to eliminate the small grains, and to produce more equant grains (i.e., to reduce the total area of grain boundaries). In addition, retrograde metamorphism may introduce new minerals. Fortunately, the grain-scale petrographic features resulting from deformation, recovery, and recrystallization are well documented (Nicolas and Poirier 1976, White 1977, Vernon 1981), and are easily identified in thin section, so that later microstructural modification can be recognized in most cases. The papers by Berger and Kalt (1999) and Berger and Roselle (2001) offer introductions to this theme, as well as to quantitative analytical methods.

The grain-contact method

This method is based on the assumption that rocks that crystallized from a melt under conditions of hydrostatic stress should develop a random spatial distribution of the constituent mineral species. Rocks that crystallized in the solid state, or from flowing magma, or rocks subsequently deformed penetratively, all have mineral distributions that depart from randomness. Contacts between like minerals can be either more (aggregate distribution) or less (dispersed distribution) common than in the random case. The grain-contact method, described by Ashworth (1976), McLellan (1983), and Ashworth and McLellan (1985), involves determining the frequency of contacts between the various mineral species in a rock, and then conducting a statistical test on its randomness. Symmes and Ferry (1995) applied the technique to domains of leucosome in

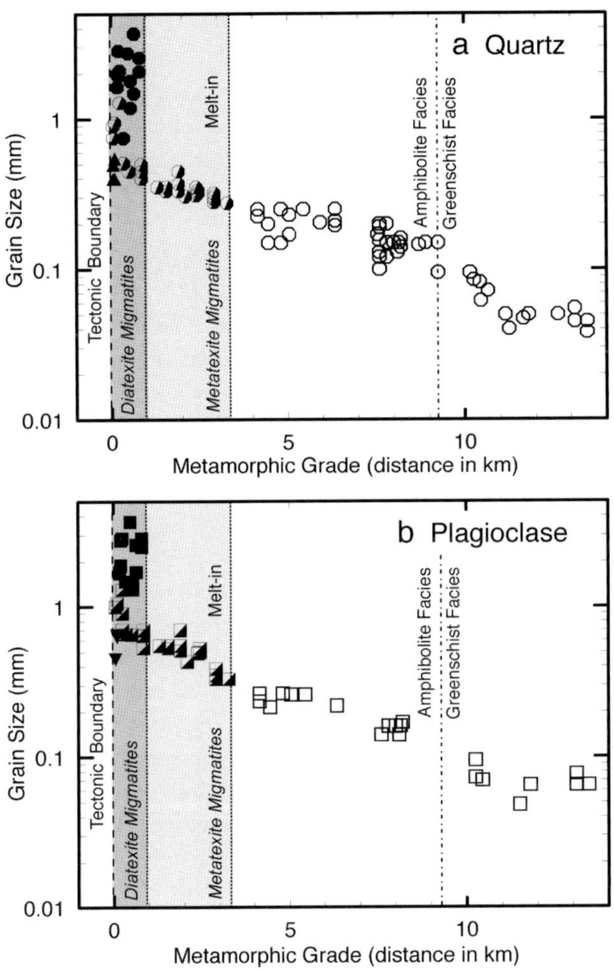

Fig. 3. Change in the grain size of metamorphic minerals versus grade during 3.5 kbar regional metamorphism in the Quetico Subprovince, Canada; (*a*) quartz and (*b*) plagioclase. There is a systematic increase in the grain size from the greenschist facies to the start of melting in the upper amphibolite facies. The grain size in the metatexite migmatites increases at about the same rate as in rocks below the "melt-in" isograd. A very marked increase in the grain size of both quartz and plagioclase coincides with the appearance of diatexite migmatites and affects both the residual rocks in metatexite migmatites and the diatexite migmatites. However, it does not affect the paleosome and resister lithologies (triangles), which did not melt, and which occur as schollen in the diatexite migmatite. The increase in grain size for these rocks follows the same subsolidus, grain-growth trend as that defined by the unmelted greenschist- and amphibolite-facies rocks. Data are provided for metagreywacke (plagioclase + biotite + quartz psammites) and metapelitic rocks. Symbols, circles and squares: open, rocks from the greenschist and amphibolite facies; half-filled, residual part of metatexite migmatites; filled, diatexite migmatites. Triangles, unmelted, massive psammitic, paleosome lithologies forming schollen in the diatexite migmatites.

the aureole of the Onawa Pluton in Maine. They found that leucosome from the sillimanite zone had dispersed distributions but that leucosome from the innermost zone had random distributions. The former were interpreted as of subsolidus origin, and the latter, as of anatectic origin.

Crystal-size distributions

The study of crystal-size distributions (CSD) has the potential to reveal information about the crystallization history (nucleation and grain growth) of rock-forming minerals. The technique is most commonly used to investigate rapidly cooled volcanic rocks, where subsolidus modifications to the grain size and shape are believed to be minimal; the resulting size-distribution patterns thus largely reflect crystallization from the magma. In slowly cooled rocks (e.g., plutons and metamorphic terranes), processes such as surface-energy-driven and strain-energy-driven grain-boundary migration tend to produce rather uniform grain-sizes, and the crystal-size distribution is peaked. More complex histories of simultaneous coarsening, nucleation, and growth lead to a reduction in the number of small grains and an increase in the number of larger grains (e.g., Higgins 1998). Berger and Roselle (2001) found a variety of crystal-size distributions in the Bayerische Wald migmatites. Euhedral cordierite from diatexite migmatites exhibits a linear crystal-size distribution consistent with crystallization in a melt. In contrast, samples of leucosome show more complex patterns. The K-feldspar from quartz–K-feldspar leucosome has a concave-upward pattern, indicating an overproduction of large grains, which could be due to the latent heat of crystallization, or to homogeneous nucleation. Plagioclase crystals from nearby areas of leucosome have linear crystal-size distributions, but with a deficiency of small grains, attributed to heterogeneous nucleation or grain coarsening. Berger and Roselle (2001) illustrated the potential range of problems that can be addressed in studies of crystal-size distribution applied to migmatites, and identified some of the problems in applying the technique to these rocks.

Studies of grain size, aspect ratio, and orientation

Most quantitative studies (e.g., Ashworth 1976, Johannes and Gupta 1982, McLellan 1983) have shown that grain size increases more abruptly after melting begins (see Fig. 3); this is attributed to the fact that the melt phase will aid diffusion, and hence Ostwald ripening. Furthermore, a systematically smaller grain-size from leucosome to melanosome to paleosome is typical, although grain size in each generally increases with metamorphic grade. However, regions where melt has been extracted can show a decrease in grain size (Dougan 1983). Berger and Kalt (1999) found that differences between the extent of lattice-preferred orientation and shape-preferred orientation of biotite, cordierite, and plagioclase are related to whether the rocks were melt-rich (leucosome), or melt-poor (melanosome and paleosome), and whether they have experienced melt segregation, and hence either submagmatic or solid-state deformation.

Milord and Sawyer (2003) compared residual grains of biotite in schlieren with those in the host diatexite and anatectic granite. They found that flakes of biotite in the schlieren are generally larger, have greater aspect-ratios, show a marked absence of the small grains, are better oriented, and display a slightly narrower range of compositions, with marginally higher *Mg number* and TiO_2 contents, than biotite from the host. They attributed these features to recrystallization of the biotite after it had been concentrated into the schlieren.

5.3 Diagnostic microstructures in migmatites

Because migmatites contain parts where melting occurred, parts from which melt was removed, parts through which melt has passed, parts where melt has collected, and indeed parts that did not melt, the type of microstructures (and also mineral assemblage and bulk composition) preserved in each may be quite different. The melanosome should be dominated by microstructures involving minerals that were present at the time of melting, either those in excess or the solid products of incongruent melting reactions. In contrast, the leucosome, and other melt-rich migmatites (e.g., diatexites), are dominated by minerals and microstructures that result from the crystallization of the anatectic melt or magma. The microstructure of the paleosome should reflect high-temperature solid-state metamorphism. The presence of glass is generally considered sufficient evidence for the former presence of a melt, or melts, in a rock. The problem in slowly cooled migmatites is that the melt would normally have crystallized, and not quenched to glass. Fortunately, there are microstructural criteria that can be used to infer the former presence of melt; in some cases, it is possible to determine whether the microstructures formed during melting, or later, as the melt fraction crystallized. The microstructures produced in quenched melting experiments (e.g., section F in the book, Figs. F1–F4) serve as a good starting point for an interpretation of the microstructures in migmatites, and will be outlined first.

Microstructures produced in partial-melting experiments

Anatectic melt first forms a thin film along grain boundaries and edges, and small triangular and tetrahedral, cuspate pockets at the corners among the reactant minerals (e.g., Mehnert et al. 1973). The reactant minerals become progressively more corroded, or embayed, as the fraction of melt increases, and commonly form small, rounded inclusions in the melt (e.g., Büsch et al. 1974, Holyoke and Rushmer 2002, Acosta-Vigil et al. 2006). The products from the incongruent melting of ferromagnesian minerals, e.g., orthopyroxene from the breakdown of biotite, can occur as needles in the melt close to the reactant mineral. The melt tends to form pools in many experiments; Jurewicz and Watson (1984) showed that pools of melt form during partial melting and that the pools have triangular, cuspate margins with concave edges because the reactant minerals adjacent to the pool of melt become rounded as they dissolve into the melt phase. In contrast, when a pool of melt crystallizes, the shape of its margin changes; it becomes more rectangular, or blocky, in outline because the crystals grow into the pool of melt and have crystal faces (Jurewicz and Watson 1985, Wolf and Wyllie 1991). In some cases, these crystals grew entirely from the melt, and in others they are simply overgrowths on former reactant phases (Acosta-Vigil et al. 2006). A common feature of the quenched partial-melting experiments is that the crystals that grew in the melt have crystal faces against the glass (e.g., Gardien et al. 1995). However, the acicular habits and skeletal morphology of some crystals may be due to the rapid rate of cooling imposed on the experiment.

Virtually all deformation-melting experiments were performed at confining pressure and strain-rate conditions such that intergranular and intragranular fractures and shear zones developed and accommodated most of the deformation. In these experiments, much of the melt that was generated is located in shear zones and fractures that develop in the sample (e.g., Rushmer 1995, Rutter and Neumann 1995, Holyoke and Rushmer 2002). Such conditions may not be applicable to anatexis deep in the continental crust, although they may apply to partial melting in some shallow contact-aureoles.

In summary, the following key microstructures develop as partial melt forms in experiments: (1) convex, triangular or tetrahedral, cuspate patches of melt develop at the corners where the grains of the assemblage of reactant minerals touch, (2) narrow, cusp-shaped films of melt penetrate along the grain boundaries into the host away from pockets or pools of melt, (3) reactant minerals have corroded and embayed shapes, (4) pools or pockets of melt in the matrix have a rounded, cuspate outline, (5) the solid products of reaction (also called peritectic) have crystal faces where they grew into melt, and (6) if cracks or shears form in the partially molten material, then they are occupied by melt. The following key microstructures form during the crystallization of the melt: (1) the minerals that crystallize from the melt tend to be euhedral, (2) the mineral products of the partial-melting reaction (i.e., the peritectic phases) that grew partially, or wholly, in the melt commonly develop faces, (3) the pools or patches of melt have a blocky outline, (4) an overgrowth develops on the reactant minerals, and (5) an alignment of tabular or prismatic crystals is developed in the places where flow of the melt occurred.

Microstructures in the residual rocks, and evidence for partial melting

Generally, the melt fraction that forms in a migmatite is segregated into sites that become patches of leucosome. However, some melt (a few percent) remains in the residuum, where it occupies grain corners, or forms films along grain faces. Moreover, the residual parts of migmatites contain the evidence with which to identify the melt-producing reactions and to deduce the conditions of metamorphism; thus, they are the best place to start a study of migmatites.

There is a progressive change in the microstructure of migmatites from very shallow contact-metamorphism through deeper contact-aureoles to regional metamorphic terranes that reflects the systematic increase in the duration of the partial-melting event, and a decrease in the cooling rate. Fast cooling near the surface results in supercooling of the melt and a quench to glass, but as the degree of undercooling decreases with depth, there is a change from fine-grained to coarse-grained granophyre in progressively deeper contact-aureoles to coarse-grained, polycrystalline aggregates in the migmatites formed in regional metamorphic terranes.

The microstructures found in rapidly cooled partial melts formed in subsurface contact-metamorphism (e.g., Knopf 1938, Wyllie 1961, Frankel 1950, Kaczor et al. 1988, Knesel and Davidson 1999, Holness et al. 2005) are essentially identical to those found in quenched partial-melting experiments (see Figs. F5 and F6). Planar and cuspate, intragranular films of glass, more extensive patches of glass, and even veins of glass are commonly reported. Typically, varieties of glass of several different colors and compositions are found within individual samples. The glass contains corroded relics of reactant quartz and feldspar (e.g., Fig. F5), together with newly grown prismatic tridymite, ß-quartz, orthopyroxene (e.g., Fig. F6), cordierite, and magnetite. Glass also is found in some pelitic and felsic plutonic rocks that were partially melted at mid-crust depths (5–7 kbar), but then brought rapidly to the surface as xenoliths in erupting lavas (Le Maitre 1974, Bacon 1992, Cesare et al. 1997, Cesare and Maineri 1999). Some of the microstructures in these xenoliths (see Figs. F7–F10) are like those produced in many partial-melting experiments. As an example, Fig. F10 shows corroded grains of biotite in which euhedral spinel (a reaction product) is located in the embayments and is separated from the reactant biotite by a film of brown glass (Grapes 1986, Cesare 2000).

Platten (1982) found very fine-grained granophyre in partially melted quartzite from a shallow (0.5 kbar) contact-aureole. The granophyre forms small cuspate patches between rounded reactant phases, and some patches contain corroded quartz and feldspar. Thus, the microstructures in slightly deeper aureoles (Holness and Watt 2002, Holness and Isherwood 2003), and in some xenoliths (Bouloton and Gasquet 1995), where cooling was a little slower, are also similar to those found in melting experiments, except that the melt has crystallized to granophyre (see Figs. F11 and F12). Cracks that were filled with melt are a common feature in the partially melted rocks from shallow contact-aureoles (e.g., Figs. F13–F15), and commonly develop in relict K-feldspar and quartz that were adjacent to grains of reactant muscovite.

Neither glass nor granophyre occurs in deeper and more slowly cooled contact-aureoles. Partially melted rocks in the contact aureole of the Duluth Igneous Complex formed at a pressure between 1.5 and 2 kbar, and contain a wide range of microstructures (e.g., Figs. F17–F28). Granophyre occurs in some very quartz-rich rocks of the Biwabik Formation (e.g., Fig. F21), but not in the metapelitic rocks. Metapelitic rocks from the aureole contain small (0.1–20 mm), isolated, ellipsoidal leucocratic patches of neosome (e.g., Figs. F17–F20) in addition to small veinlike accumulations of leucosome (<5 mm wide). The leucocratic patches have a cuspate outline because the adjacent matrix quartz, feldspar, and biotite grains are corroded. Moreover, the patches comprise a few large anhedral grains of quartz and alkali feldspar or plagioclase, which contain rounded inclusions of quartz, biotite, and plagioclase and tiny euhedral inclusions of plagioclase, cordierite, and orthopyroxene. The microstructure suggests that the patches of neosome were sites where biotite melted incongruently; then, during slow cooling, the melt crystallized to give small prismatic crystals first, and then, as the temperature neared the solidus, coarse-grained quartz and feldspar (Sawyer 2001). Similar microstructures are found in the metapelitic rocks of the 3 kbar aureole around the Ballachulish Igneous Complex (Pattison and Harte 1988, Harte et al. 1991), and in the aureole of the Onawa pluton (Marchildon and Brown 2001, 2002). Thus, the characteristic shape of the pools of melt (Fig. F29), and the corroded form of the reactant minerals (Fig. F30), still remain in many deeper contact-aureoles, but the former melt is now represented by early-crystallized euhedral grains set in coarse-grained quartz, plagioclase, and K-feldspar that crystallized later. Unfortunately, the microstructural evidence for the former presence of melt is easily overlooked if the amount of melt present was small, or where the protolith is a quartzite. Holness and Clemens (1999) were able to establish that small amounts of melt existed in the Appin quartzite up to 500 m from the Ballachulish Igneous Complex by using several microstructures (see Figs. F31 and F32), such as (1) feldspar grains with cuspate extensions along quartz–quartz contacts, (2) narrow trains of feldspar blebs along quartz–quartz contacts, and (3) fine-grained quartz, albite, and K-feldspar along the grain boundaries.

The general belief (e.g., Hibbard 1987, Ashworth 1985) that the combined effects of deformation, recrystallization (grain growth), and retrograde metamorphism would have eliminated the microstructures generated during partial melting in slowly cooled, regional metamorphic terranes is proving to be incorrect. For example, microstructures (Figs. F33–F36) that resemble those produced in partial-melting experiments have survived in melt-depleted rocks in the Ashuanipi Subprovince, a regional granulite-facies (6 kbar, 875°C) terrane in northern Quebec (Sawyer 1999, 2001; Guernina and Sawyer 2003). Metagreywacke rocks there contain small domains (0.5 mm across) in which large irregularly shaped grains of K-feldspar (mostly), quartz, and plagioclase have cuspate extensions that penetrate along the grain boundaries between the adjacent smaller, equant grains in the matrix. These large interstitial crystals contain small, rounded, or embayed inclusions of quartz, plagioclase, and biotite, in addition to large grains of orthopyroxene, which have crystal faces (Figs. F33, F34). These domains are interpreted to be sites where biotite, plagioclase, and quartz reacted to produce orthopyroxene and melt; thus, some of the microstructure that formed at the peak metamorphic temperature is actually preserved in these rocks. Most (>80%) of the melt produced drained away, leaving a small amount (<3 vol.%) that became the coarse-grained K-feldspar, quartz, and plagioclase.

Kenah and Hollister (1983) showed similar microstructures preserved around small, remnant pockets of melt left after most of the melt that was generated had been expelled from partially melted Tertiary plutonic rocks in British Columbia. Very similar microstructures (Figs. F37, F38) have also been reported in migmatites developed from plutonic protoliths in Archean (Sawyer 2001), Mesoproterozoic (Timmermann et al. 2002, Slagstad et al. 2005), and Phanerozoic (Brown 2002) regional anatectic terranes. In some cases, the former presence of melt can still be inferred from the distribution of small grains of quartz, plagioclase, or K-feldspar in the matrix of larger residual grains, even after some later re-equilibration of the grain boundaries has occurred (e.g., Hartel and Pattison 1996, Sawyer 2001). Key microstructures are (1) small strings of connected grains of quartz, plagioclase, and in some cases, K-feldspar, along the grain boundaries of larger grains in the matrix (i.e., the "string of beads" microstructure described by Holness and Clemens 1999), (2) the presence of monomineralic films of these minerals on the matrix grains, and (3) small, cuspate grains of these minerals between large grains in the matrix (e.g., Fig. F32).

Figures F39–F46 show examples of the more typical microstructure found in many residual rocks. Most have undergone a considerable degree of recrystallization that has greatly modified the original size and shape of the grains. Thus, the evidence for the former presence of melt, such as cuspate-shaped pore space that was filled with melt (pools of melt), is largely erased.

Microstructures in the melt-rich parts of migmatites; evidence for crystallization of the melt

The melt-rich parts of migmatites are the most likely to contain microstructural evidence for the crystallization of the anatectic melt; Figs. F47–F58 show typical examples from the leucosome domains, and Figs. F59–F65 illustrate the microstructures from diatexite migmatites. Plagioclase and K-feldspar are the most voluminous products of crystallization of felsic anatectic melts, and in most cases, plagioclase starts to crystallize before quartz and K-feldspar. Consequently, most microstructures used for the identification of crystallized melts in migmatites involve either the feldspars, or the shape of crystals.

Platten (1983) studied partially melted semipelitic schists close to the Barnamuc intrusive complex in Scotland. He noted two features that are essentially igneous in cross-cutting pink leucosome: (1) the plagioclase and K-feldspar have crystal faces where in contact with interstitial quartz, and (2) small grains of quartz, plagioclase, biotite, and cordierite located in patches of granophyre, and in the coarse-grained parts of the leucosome, are commonly euhedral. These features indicate that the leucosome crystallized from an anatectic melt, or from a magma (see Figs. F47 and F48). Similar criteria for identifying minerals that crystallized from a melt were proposed by Cuney and Barbey (1982) and later by Pattison and Harte (1988) and Vernon and Collins (1988).

Some patches of leucosome, many of the narrow veins of leucosome, and domains the diatexite migmatites in the contact aureole around the Duluth Igneous Complex consist of large (1–2 mm), anhedral grains of unstrained quartz, plagioclase, and K-feldspar that contain inclusions of much smaller euhedral plagioclase, cordierite, and orthopyroxene crystals (see Figs. F23–F27). Moreover, as shown in Figs. F28 and F52, the prismatic inclusions are commonly strongly aligned (Sawyer 2001). Together, these microstructures indicate that the small prismatic grains crystallized in, or from, a melt, and were aligned by flow of that magma (melt + entrained crystals), before the melt finally crystallized.

The microstructures in leucosome and the melt-rich migmatites from regional metamorphic terranes are different from those in contact aureoles. The grain size is considerably larger and more uniform, and far fewer of the minerals have crystal faces. Nevertheless, the presence of euhedral crystals of feldspar in leucosome (Figs. F56, F57) and in diatexite migmatites (Figs. F59, F64) that have crystal faces against later-crystallizing minerals, such as quartz or K-feldspar, is the most widely used microstructure for inferring the former presence of a melt, or of magma, in regional migmatites. Euhedral, or idioblastic, unzoned ferromagnesian minerals that have few inclusions have also been used to infer crystallization in the presence of a melt (Stüwe and Powell 1989, Florence et al. 1995, Braun et al. 1996, Barbey et al. 1999, Berger and Kalt 1999, Sawyer et al. 1999, Alcock and Muller 2000, Berger and Roselle 2001, Johnson et al. 2001*a*). However, other criteria are needed to determine whether the mineral is the product of an incongruent melting reaction (i.e., a peritectic phase) and developed crystal faces because it grew in a melt, or whether the mineral crystallized from the melt (i.e., liquidus phase). The presence of euhedral accessory phases, such as zircon, in leucosome and diatexites also has been used as evidence for crystallization from a melt (e.g., Barbey et al. 1995).

A zonal distribution of minerals within a leucosome that is similar to that expected from *in situ* fractional crystallization of felsic melt has also been used to infer crystallization from a melt or magma (Sawyer 1987, Kohn et al. 1997). In particular, Kohn et al. (1997) noted that myrmekite occurs only at the edges of the leucosome they studied, and muscovite, only in the center; they interpreted this as the progressive inward crystallization of a melt in which the last volume of melt in the center finally became sufficiently hydrous to crystallize muscovite.

Magmatic and submagmatic foliations

Platy or tabular crystals suspended in a magma become aligned during flow of the magma, and form a foliation. It is important to identify such magmatic foliations in migmatites, especially in the field, but the problem is to distinguish them from tectonic foliations formed in the solid state. Criteria for identifying magmatic foliations have been developed (Blumenfeld and Bouchez 1988, Paterson et al. 1989, Arbaret et al. 1996). In essence, these methods compare the degree of intracrystalline deformation in the foliation-defining minerals with that in the large domains of interstitial quartz that represent the places where the last melt crystallized. There are three regimes, based on fraction of melt present, that are of interest; the magmatic ($M_f > 0.45$), the submagmatic ($0 < M_f < 0.45$), and the subsolidus ($M_f = 0$); the submagmatic regime has been subdivided still further (Vigneresse et al. 1996, Rosenberg 2001).

In the magmatic state, platy or tabular minerals are sufficiently widely spaced to rotate passively as the magma is deformed and to become aligned; local interaction between rotating crystals can occur, but is not significant (Paterson et al. 1989). There are microstructural criteria with which to identify magmatic foliations: the minerals that crystallized from, or were in the magma, must have a preferred orientation, and the quartz crystals and domains in the matrix around them should be equant, and there should be no intracrystalline deformation or recrystallization (e.g., undulose extinction, deformation bands, polygonal subgrains, kinks, bent cleavage-planes, or evidence of grain-boundary migration, such as lobate boundaries) in either the foliation-defining minerals, or in the interstitial quartz. Examples of the microstructure developed during the flow of magma in diatexite migmatites are shown in Figs. F66–F69.

As the fraction of melt decreases, interactions between rotating crystals becomes progressively more common, and this forms clusters of imbricated or "tiled" grains (Blumenfeld and Bouchez 1988; Ildefonse et al. 1992, 1997; Arbaret et al. 1996, 1997) and the submagmatic regime begins. A framework of crystals develops, and once formed, the framework may become deformed (Bouchez et al. 1992). A submagmatic foliation is one in which a framework of aligned crystals formed and may have become deformed before the interstitial magma crystallized. Hence, the microstructural criteria for identifying a submagmatic foliation are the presence of intracrystalline deformation-induced features in the foliation-defining and framework-forming minerals, but not in the interstitial domains of quartz, which remain equant in shape.

Intracrystalline deformation in both the foliation-defining minerals and the quartz domains, which may have become elongate, indicates deformation in the solid state. Deformation in the solid state is generally characterized by a decrease in grain size, and at high strains, by the development of mylonite. Unfortunately, microstructures formed in either the magmatic or submagmatic states are easily destroyed if deformation continues into the subsolidus regime.

Melt inclusions

Melt can be trapped as inclusions in the minerals that grew in, or crystallized from, a melt; examples are shown in Figs. F70–F73. Therefore, the identification of melt inclusions in minerals in the neosome is clear evidence for the presence of melt. Rapid cooling, as occurs in rapidly exhumed, partially melted crustal xenoliths (Cesare et al. 2003), will quench melt inclusions to glass. Melt inclusions that have quenched to glass may contain structures due to shrinkage, which resemble bubbles (Lowenstern 1995). Glass inclusions are important, as they may enable the composition of natural anatectic melts to be determined. However, with slower cooling, the melt crystallizes, and the former melt inclusions may be mistaken for solid mineral inclusions and be overlooked. The homogenization temperature at which the crystals in former melt inclusions disappear into a single phase (i.e., melt) has been used to infer the temperatures at which the magmas were trapped as inclusions in their hosts. Most notably, this approach has been applied to the study of volcanic and shallow plutonic rocks.

Cordierite–, garnet–, and orthopyroxene–quartz intergrowth microstructures

The intergrowth in large poikiloblasts of garnet and, to a lesser extent, of cordierite and orthopyroxene, with quartz and surrounded by feldspar, is a very common microstructure in neosome (Powell and Downes 1990; Tracy and Robinson 1983; Brown 1983; Waters and Whales 1984; Waters 1988, 2001; Barbey et al. 1990; Braun et al. 1996; Fitzsimons et al. 1996; Sawyer et al. 1999; Oliver et al. 1999; Alcock and Muller 2000; White et al. 2004). Field observations (see examples in Figs. B20–B25) indicate that this microstructure (Figs. F74–F77) does not occur in the neosome developed from H_2O-saturated melting, or muscovite-breakdown reactions (Tracy and Robinson 1983), but that it is characteristic of the biotite-dehydration-melting reactions that occur in the higher-grade parts of migmatite terranes. The details of this microstructure are remarkably similar for all reported occurrences. (1) The domains of neosome have a rather diffuse margin without a melanosome. (2) The neosome tends to be leucocratic rather than mesocratic in appearance, probably because the melanocratic minerals are concentrated into poikiloblasts, and do not include biotite. (3) Domains of neosome have a patch, or more commonly a net-like morphology composed of irregularly shaped (Tracy and Robinson 1983, White et al. 2004), or planar and regularly spaced segments (Braun et al. 1996). (4) Host and neosome have different mineral parageneses; hosts contain plagioclase, biotite, and quartz, with or without sillimanite, whereas the neosome contains plagioclase, K-feldspar, quartz and one or more of cordierite, garnet or orthopyroxene, and ilmenite or magnetite. Biotite is generally absent, but it may occur as a replacement of garnet, cordierite, or orthopyroxene. The absence of biotite in the neosome suggests an incongruent biotite-dehydration-melting reaction in which cordierite, garnet, and orthopyroxene are the solid products of reaction. (5) The bulk composition of the neosome is commonly the same as the host, a result suggesting *in situ* formation of the neosome in a closed system (e.g., Braun et al. 1996), which is the reason it is not truly leucocratic. However, in some cases the neosome has an excess of ferromagnesian components that is readily explained by the loss of some, or virtually all, of the melt (Fitzsimons 1996, Sawyer et al. 1999, Waters 2001, White et al. 2004); such neosome is melanocratic. (6) The ferromagnesian poikiloblasts or porphyroblasts are intergrown with blebs, patches (see Figs. F74 and F75), and lamellae of quartz (e.g., Waters 2001, White et al. 2004) and, more rarely, with feldspar (Barbey et al. 1999) also. In rare cases, the poikiloblasts have a distinct core that contains sillimanite inclusions, but no quartz blebs. Vernon and Collins (1988) interpreted such sillimanite-bearing cores to be pre-neosome relics. Powell and Downes (1990) and White et al. (2004), in contrast, suggested that the cores too could have grown during the melting reaction. (7) In many cases, the leucocratic part of the neosome is zoned. The ferromagnesian core is surrounded by an inner leucocratic annulus rich in K-feldspar, which is interpreted to be the solid product of the incongruent melting reaction. The outer zone consists of quartz, K-feldspar, and quartz; it is interpreted to have been derived from the anatectic melt. (8) Generally, the quartz in the poikiloblasts connects to a thin mantle (called a moat by some authors) of quartz that surrounds the poikiloblasts and separates it from feldspar in the neosome (see Figs. F76 and F77).

Models proposed for the formation of these intergrowths are based on kinetics, principally the difficulty of nucleating and growing the solid assemblage produced, or on the slow diffusion of Si and Al in melts. Stüwe and Powell (1989), Powell and Downes (1990), and White et al. (2004) regarded the location of the neosome as controlled by the location of the nuclei of ferromagnesian mineral produced (garnet in their examples), which explains the lack of a rim of melanosome. Barbey et al. (1999) proposed that the dissolution of biotite into the melt to make cordierite resulted in a local deficit in Al, which was compensated for by the incongruent dissolution into the melt of nearby feldspar, the resulting quartz being intergrown with the cordierite. Waters (2001) suggested that the microstructures result from the kinetic control imposed by Si and Al, the most slowly diffusing species; the growth of garnet results in a local excess of Si at its edge and in the incorporation of quartz. Brown (2004) suggested that local sites of lower stress may control the initial nucleation of the product minerals and the accumulation of melt that is produced. Importantly, all the models and petrological observations indicate that this type of microstructure can be used to infer biotite dehydration melting.

Hornblende dehydration melting in mafic protoliths may produce neosome with similar types of microstructures: dark cores consisting of pyroxene and garnet surrounded by a leucocratic mantle, or moat, of plagioclase and quartz. However, more work is required on migmatites derived from mafic protoliths.

Symplectitic intergrowths of quartz and plagioclase with mica

On cooling, there is the potential that minerals in a migmatite will react with the remaining melt. There are two types of reaction: (1) final melt reacting with the residual solid and (2) aqueous fluid reacting with residual solid. Both involve the formation of new hydrous minerals. Reactions between the final increments of melt left in the rock and the residual solid necessarily occur at temperatures at or above the solidus, and so are of interest here. One objective of this section is to identify the microstructures associated with this type of reaction. The fluid for solid – aqueous fluid reactions can be derived from the crystallization of the anatectic magma in which case the reactions are a near-solidus phenomenon. As an alternative, the fluid could come from an external source, and the process could take place at subsolidus temperatures, which could extend down through the amphibolite to the greenschist facies; these low-temperature reactions will not be considered further.

Thompson (1983) and Powell (1983) argued that the preservation of low-$a(H_2O)$ mineral assemblages in high-grade rocks is due to the formation and subsequent escape of anatectic melt. The removal of H_2O with the melt limits the extent to which rehydration reactions can take place (Spear et al. 1999, Brown 2002) as the migmatite terrane cools and the melt crystallizes. Although large volumes of melt can be generated in anatectic terranes, not all that melt is extracted from the source to make plutons at higher levels in the crust; small amounts of melt remain in pockets and along the boundaries between the minerals in the residual rocks (Guernina and Sawyer 2003), and a small amount of melt also remains in the leucosome (Caggianelli et al. 1991). This minor amount of residual melt reacts with the residual minerals during the final stages of crystallization and produces new microstructures in both the residual and melt-rich parts of the migmatite.

Although of minor abundance generally, the microstructures (Figs. F45–F48) are a characteristic symplectite-like intergrowth of skeletal or thinly bladed biotite with quartz or plagioclase (Ashworth 1976, Jones and Brown 1990, Brown and Raith 1996, Raith and Harley 1998, Kriegsman and Hensen 1998, Sawyer 1999, Guernina and Sawyer 2003), quartz and K-feldspar (Nyman et al. 1995), or alkali feldspar and plagioclase (Waters 2001); the intergrowth locally replaces small amounts of garnet, cordierite, or orthopyroxene. Because these are replacement microstructures, and as the product assemblage is essentially the same as the reactant assemblage in some prograde biotite-dehydration reactions, many authors have interpreted them as resulting from the reaction of anhydrous solid minerals with the remaining anatectic melt. Commonly cited reactions are Grt ± Crd + Kfs + melt = Bt + Pl + Qtz ± Sil and Opx + melt = Bt + Pl + Qtz. However, Waters (2001) expressed a different opinion, and proposed that this particular microstructure should be considered together with other, less obvious microstructures (e.g., Fig. F82) in the rocks that involve biotite and sillimanite (see examples in Büttner et al. 2005). These cases suggest that a more complex overall reaction occurred as the melt was crystallizing in the rock, although the melt was not directly involved in creating any particular microstructure.

Many low-grade migmatites in which melting occurred through the breakdown of muscovite contain analogous late-stage microstructures in which K-feldspar is replaced by an intergrowth of skeletal muscovite and quartz, generally associated with myrmekite (Ashworth 1972, Brown 1979, Ashworth and McLellan 1985, Weber et al. 1985). The symplectite-like microstructure is generally interpreted to be the result of the slow diffusion of Si and Al, and hence the Al:Si ratio of the reactant is inherited by the products and controls the nucleation of the product assemblage. Ashworth (1972) proposed a reaction between melt and K-feldspar, such as Kfs + melt + Sil + Pl^1 = Ms + Qtz + Pl^2 to account for the microstructures [the superscripts indicate reactant (^{1}) and product (^{2}) plagioclase, respectively].

Symplectite-like microstructures between the micas and quartz seem to indicate the reaction of residual minerals with melt; the extent to which these microstructures have developed depends upon how much melt was present. Therefore, mapping the distribution of microstructures produced at late stages of crystallization and those microstructures formed under high-temperature subsolidus conditions may enable the loci of net loss of melt to be distinguished from those where melt has accumulated (or was injected); the former should exhibit a predominance of microstructures indicative of reaction between the residual melt and solid, whereas the latter should contain more extensive evidence of rehydration involving reaction between the residual solids and an aqueous fluid (e.g., Spear et al. 1999, Brown 2002, White et al. 2005).

Composition and zoning of plagioclase

Bowen (1913) showed that during the melting of plagioclase, the albite component fractionates into the melt, and the remaining plagioclase is more calcic. Consequently, the composition of plagioclase was proposed as a test for the origin of the leucosome in migmatites (e.g., Yardley 1978). The leucosome generated by partial melting should have plagioclase more sodic than that in the associated melanosome and protolith. Some investigators found the plagioclase in the leucosome to be systematically more sodic than those in the melanosome (e.g., Weber et al. 1985, Weber and Barbey 1986, Alcock and Muller 2000). More commonly, the compositions of plagioclase in the leucosome and in the melanosome were found to be very similar (e.g., Härme 1962, Dougan 1979, Henkes and Johannes 1981, Gupta and Johannes 1982, Kenah and Hollister 1983), even in leucosome where other criteria point to an anatectic origin. The euhedral crystals of plagioclase in diatexite migmatites that represent the accumulation of plagioclase crystallized from the "initial" anatectic melt in the Ashuanipi and Opatica subprovinces of Quebec are more calcic (around An_{25}) than the euhedral crystals of plagioclase (around An_{17}) that occur in K-feldspar-rich diatexite migmatites that formed from evolved diatexite magmas. Similarly, large euhedral crystals of plagioclase that are part of the framework (e.g., Figs. F59–F64) of early-crystallized minerals in diatexite migmatites (and in some examples of leucosome, e.g., Fig. F56) are more calcic than the interstitial plagioclase that crystallized later.

Johannes (1978, 1980, 1983*b*) and Acosta-Vigil et al. (2006) have shown that the kinetics of dissolution of intermediate plagioclase at the temperature (<800°C) at which granitic melts start to form are very slow indeed. This leads to metastable melting, and the plagioclase in the leucosome and in the melanosome need not have different compositions. This effect may be important if melt separates rapidly from its residuum. Thus, the composition of plagioclase is not now considered a reliable criterion to evaluate the origin of leucosome in migmatites. However, plagioclase is still a crucial mineral in the interpretation of migmatites, but emphasis now rests on its morphology, the microstructures of which it is a part, and on the nature of its compositional zoning.

Plagioclase from most migmatites shows rather weak and irregular (nonconcentric) compositional zoning; both normal and reverse trends have been described. The strongest compositional zoning is generally found in plagioclase from the melanocratic residual rocks, particularly if the protolith was a mafic rock. The pattern of compositional zoning in residual plagioclase is strongly dependent on whether other Ca-bearing phases were present in the residuum, and the stage at which they grew (Skjerlie and Johnston 1996, Patiño Douce and Beard 1996).

Strongly developed concentric zoning that parallels the outline of a plagioclase crystal is generally considered as evidence of magmatic growth. However, the presence of oscillatory zoning, truncated growth-zoning, and complex internal boxlike, skeletal, or honeycomb structures in plagioclase is regarded as evidence of disequilibrium during growth (Wiebe 1968, Hibbard 1981), and as characteristic of regimes where magma mixing occurred. Plagioclase crystals containing disequilibrium structures are well known from granites (e.g., Barbarin 1990, Grogan and Reavy 2002). Petrographic studies of the Variscan peraluminous anatectic complexes in Spain have revealed the presence of plagioclase crystals containing disequilibrium microstructures in some of the migmatite bodies in these complexes. Plagioclase crystals with similar complex patterns of internal zoning are also found in leucosome from the migmatites of the Glen Elg region of southeastern Australia (see Fig. F83). Experimental work by Castro (2001) has shown that complex patterns of internal zoning can result from the interaction of a mantle-derived mafic magma and a crust-derived felsic magma. Thus, detailed study of the zoning and internal morphology of plagioclase may be a very useful way of distinguishing migmatites that are wholly crustal material from those with an admixed component of mantle-derived magma (e.g., Miller and Wooden 1994).

Biotite composition and microstructures

Biotite-dehydration reactions occur over a temperature interval that can exceed 100°C because of various solid solutions in the biotite (Montel and Vielzeuf 1997). Experimental studies show that besides the well-known increase in TiO_2 (Robert 1976), *Mg number* and fluorine content also increase across the reaction interval (Stevens et al. 1997; Patiño Douce and Harris 1998; Patiño Douce and Beard 1995, 1996; Patiño Douce and Johnston 1991; Pickering and Johnston 1998) and extend the stability field of biotite to more than 1000°C. The Al content of biotite tends to decrease relative to the starting composition of the biotite in bulk compositions lacking an aluminosilicate mineral, and to increase in those where one is present. Thus, there are several compositional trends in biotite that can be used to determine metamorphic conditions (e.g., Patiño Douce et al. 1994).

On the basis of morphological and microstructural differences, several different populations, or generations, of biotite have been identified in many migmatites. In some cases, there is no compositional difference between the morphological types (e.g., Solar and Brown 2001). However, in others, there are compositional differences that can be used to explore the metamorphic history of the migmatite. For example, in migmatites that underwent biotite dehydration melting, the residual biotite, with the highest *Mg number*, TiO_2, and F contents, commonly occurs as small, rounded, reddish inclusions in other phases, whereas the larger crystals of biotite in the matrix of these rocks have lower TiO_2 contents and lower *Mg numbers* (e.g., Braun et al. 1996, Fitzsimons 1996, Barbey et al. 1990, Percival 1991), indicating that they formed, or re-equilibrated, at lower temperatures. In contrast, the largest, bladed crystals of biotite in migmatites, where biotite dehydration melting was incipient, have the highest *Mg number* and are residual (Brown 1979, Milord et al. 2001), and the smaller grains, with much lower *Mg numbers* (around 20) and TiO_2 contents, grew from the anatectic melt (Sawyer 1998, Milord et al. 2001, Milord and Sawyer 2003). Large poikilitic crystals of biotite in leucosome from the lower-crust migmatites of southern Italy have the same morphology and *Mg number* as biotite in the melanosome, and were interpreted by Fornelli et al. (2002) as residual. In contrast, small grains of biotite in the leucosome are less magnesian and interpreted as possibly crystallized from the anatectic melt. Kenah and Hollister (1983) found that biotite (and hornblende) from leucosome have lower *Mg numbers* than those from adjacent melanosome, probably reflecting a greater proportion of magma-grown biotite in the leucosome they examined.

Contact between leucosome and melanosome in metatexite migmatites

Observations made in the field on the nature of the contacts between a leucosome and its melanosome are extremely important in determining whether the leucosome is *in situ*, in source, or simply a leucocratic vein. Critical details are whether the contacts are sharp or diffuse, concordant or discordant, and whether there is a rim of biotite between the two. In spite of the great significance attached to the contacts for the interpretation of leucosome, very little has been reported on their microstructure.

The contacts in partially melted quartzites from rapidly quenched contact-aureoles show a petrological continuity of either glass (Platten 1982) or granophyre (Fig. F84) from the leucosome into the residuum. Thin films of glass, or granophyre, extend along the grain boundaries of the quartz grains from the leucosome into the residuum for a distance of several grain-diameters (Fig. F84), which supports an *in situ* interpretation for the leucosome. The contacts between leucosome and melanosome in metapelitic rocks from rapidly cooled contact-aureoles are more complex than in the essentially monomineralic metaquartzites, but are also irregular at the scale of several grain-diameters (Figs. F85–F87). The outline of the large crystals in the *in situ* leucosome suggests that melt did extend into the residuum along grain boundaries and that lobes of melt may have almost completely surrounded pieces of the residuum (e.g., Fig. F85). At a larger scale, continuity of microstructure in the leucosome from a segment with one orientation to a segment with another (e.g., Fig. F88) can be used to infer that all segments of a leucosome contained melt at the same time. Alteration of ferromagnesian minerals (e.g., cordierite in Fig. F88) in the residuum close to the leucosome may indicate that H_2O was exsolved from the melt as it crystallized.

Highly irregular contacts are commonly preserved between leucosome and melanosome components of *in situ* patches of neosome formed at the start of partial melting in regional granulite-facies metamafic (Figs. F89–F92) and metapelitic (Figs. F93, F94) rocks. However, recrystallization in slowly cooled rocks generally eliminates the evidence for former films of melt. Grain size typically increases through a melanosome toward its contact with leucosome, but the concentration of strain (which reduces grain size), or the abundance of sillimanite (which inhibits grain growth) near the interface, can result in a smaller grain-size and straighter contacts (Figs. F95, F96).

Microstructure of schlieren in diatexite migmatites

Schlieren are planar aggregates of platy, tabular, or acicular minerals and are common in many diatexite migmatites. The presence or absence of schlieren is part of the nomenclature for naming the subtypes of diatexite migmatite, but very little is known about their microstructure or how they developed. All the examples shown in this book (Figs. F97–F100) are of the common biotite schlieren, but schlieren composed mainly of plagioclase, hornblende, orthopyroxene or sillimanite also occur in migmatites. The crystals of biotite that comprise the schlieren (see Figs. F97–F100) are typically strongly imbricated, or tiled, in the same sense as the imbrication of the tabular minerals (i.e., feldspar) that define the magmatic foliation in the host diatexite. Thus, schlieren were formed during the flow of the diatexite. Schlieren range in width from less than the length (<2 mm) of a component crystal of biotite (e.g., Figs. F97, F98), to aggregates (>15 mm) tens of crystals wide (e.g., Fig. F100, and examples in Milord and Sawyer 2003). In most cases, the crystals of biotite in the schlieren are larger, have a greater aspect-ratio and a stronger preferred orientation than biotite in the host, a feature that Milord and Sawyer (2003) attributed to the ease of recrystallization in the essentially monomineralic schlieren. Many schlieren are significantly enriched in accessory minerals such as apatite and zircon, but where these minerals occur in the schlieren and how they came to be concentrated there remain unknown.

Microstructure of biotite-rich selvedges in migmatites

Biotite-rich selvedges that develop around leucocratic veins and granitic dikes are another common feature in migmatites for which the conditions and mechanisms of formation are poorly understood. Typically, they occur as narrow (<5 mm) zones in the host rock that form at the contact with a leucocratic vein, or granitic dike. In most cases, biotite selvedges show a rapid gradation in microstructure and mineral proportions from the country rock; the proportion of quartz and plagioclase decreases, and there is a concomitant increase in the modal proportion and grain size of biotite (Figs. F101–F104) toward the contact. The increase in grain size of biotite is probably facilitated by easier diffusion in the essentially monomineralic selvedge. Generally, the crystals of biotite in the selvedge are oriented parallel to the contact with the vein or dike. However, the orientation of biotite in many selvedges around leucocratic veins has been locally modified by the bulbous growth of crystals of quartz and plagioclase outward from the leucocratic vein (e.g., Fig. F102), which may be interpreted as evidence for grain growth after the leucocratic vein has solidified.

6.

WHOLE-ROCK GEOCHEMISTRY IN MIGMATITE STUDIES

Most migmatites can be recognized simply by their distinctive morphologies. These morphologies are, to some degree, the result of deformation that occurred during anatexis. However, the underlying processes that change a protolith into a migmatite are petrological, not structural. Therefore, the basic requirement in understanding migmatite formation is the identification of these petrological processes, and one way that this can be done is by examining whole-rock compositions. This book is not the place for an exhaustive discussion of the geochemistry of migmatites. Nevertheless, some consideration of how whole-rock geochemistry can be used to understand these processes is useful. From a practical point of view, this summary also serves to outline which rocks should be sampled for follow-up geochemical study while mapping migmatites.

Whole-rock composition is commonly used to classify rocks, and as a means of discriminating between the tectonic settings in which rocks such as greywacke, basalt and granite formed. Neither use has much application to migmatites. Migmatites are classified by morphology, and they inherit their compositional traits from their protoliths. The protoliths may have become migmatites in a completely different tectonic setting to that in which they formed. Whole-rock geochemistry is used principally to investigate the processes that occurred during the formation of migmatites, to define the compositions of the protolith and the various parts of the neosome, and to identify the regions of material loss or material gain.

6.1 A possible sequence of processes and some relevant questions

A migmatite represents the sum of many processes. How far each process has been able to progress and the sequence in which they occur are major factors in determining the characteristics of a particular migmatite. The sequence of processes that make migmatites and granites in three different scenarios of partial melting is shown on Fig. 4.

In the first (Fig. 4*a*), partial melting occurs under the special conditions of lithostatic stress ("static melting"; see section 3.4) only, and melt forms and later crystallizes without segregation from its complementary solid fraction. In this scenario, anatexis occurs in a closed system, and the degree of partial melting (F) is, therefore, equal to the fraction of melt (M_f) in the rock. These conditions are most likely to occur in contact aureoles. If only a small amount of partial melting (dashed line in Fig. 4*a*) occurs, either because the protolith was not very fertile, or the metamorphic temperatures were not very high, then a patch migmatite is formed. However, if a larger degree of partial melting occurs (black line in Fig 4*a*), then a nebulitic diatexite migmatite forms; since anatexis takes place in a closed system, this is a primary diatexite.

In the general case of anatexis in the crust (Fig. 4*b*), partial melting occurs under conditions of differential stress ("dynamic melting"). Any melt fraction generated in excess of that required to establish permeability (generally <5%) in the melting rock is segregated and collects elsewhere, i.e., the partially molten rock is drained, and anatexis is an open-system process, at least on the scale of a hand sample and an outcrop. Overall, the migmatites formed under these conditions develop by a four-step process. The first is the onset of partial melting. Once the partially melted rocks become permeable, this step is followed by a second one, the extraction of the excess melt and its accumulation elsewhere. This process generates two rock types. A residual metatexite migmatite develops at the site of partial melting; it contains a low fraction of melt, even if the total degree of partial melting was substantial (typically 30–60%), and has a melt-depleted bulk composition. A diatexite magma forms at the site where the segregated melt accumulates; because the fraction of melt there is high (much greater than the degree of partial melting) and is not *in situ*, this can be called a secondary diatexite magma. Four lines are shown in Fig. 4*b* connecting the residual metatexite to the secondary diatexite migmatite; the first corresponds to loss of melt at the onset of permeability (lowest fraction of melt), and the others show segregation after some accumulation of melt in the metatexite. The third step affects the diatexite magma during anatexis and involves flow of

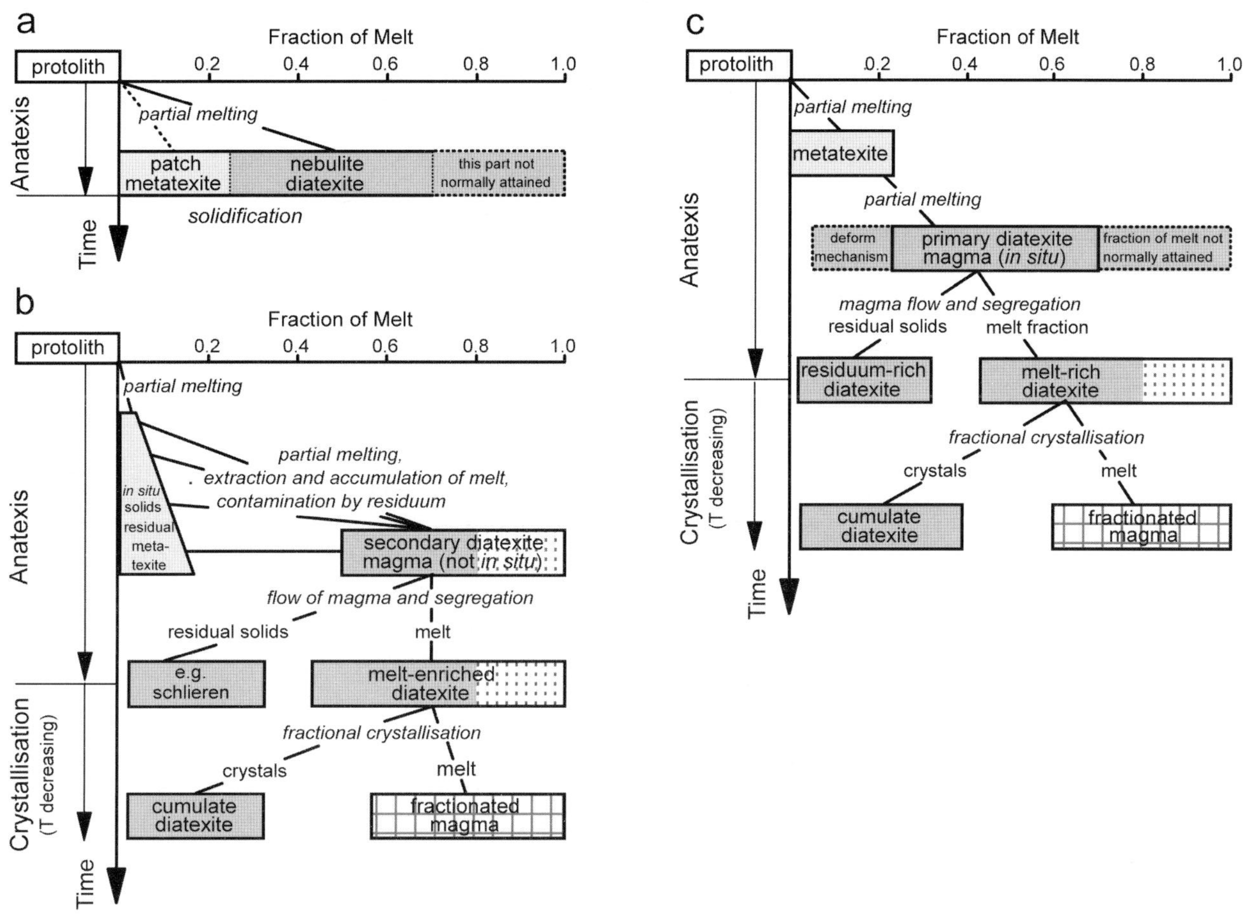

Fig. 4. Diagrams to illustrate the relationship between physical processes and the generation of migmatites (metatexite and diatexite) and granite during anatexis in three different scenarios. (*a*) The special case of partial melting under conditions of lithostatic stress (so-called "static melting"). These conditions are most likely to occur in contact aureoles. (*b*) The general case of partial melting under differential stress (so-called "dynamic melting"), which applies to regional anatexis in orogenic settings, such as melt-depleted granulite terranes and their associated granites. (*c*) A hybrid scenario in which melt accumulates in the source before the processes that separate the melt from the solid fraction become active, i.e., anatexis begins under conditions of hydrostatic stress, but ends in a differential stress regime. This scenario is most likely to apply to regional anatectic settings where large volumes of melt were able to form rapidly. Legend: pale grey, metatexite migmatite; dark grey, diatexite magma / diatexite migmatite; coarse dotted pattern, diatexite magma / migmatite with a very high fraction of melt fraction that may resemble granite in terms of composition and microstructure; grey, cross hachured pattern, fractionated granitic magma that, generally, crystallizes an igneous microstructure.

the diatexite magma. This generates a magmatic foliation and creates schlieren as well as some separation of the melt from the solid. The final step occurs after anatexis and is dominated by fractional crystallization of the secondary diatexite magma, which generates cumulate diatexite and melt-enriched diatexite, each with its own distinctive composition and microstructure. Granitic magma is a product of these processes. This scenario applies to regional anatexis in orogenic settings, notably the generation of melt-depleted granulite terranes and their associated granites, e.g., the Ashuanipi Subprovince in Canada.

Figure 4c depicts a hybrid scenario in which the melt fraction is able to accumulate at the site of partial melting before the differential stress regime starts; this scenario is equivalent to "static melting" followed by "dynamic melting." Consequently, as the degree of partial melting progressively increases, the anatectic rocks pass through a metatexite migmatite stage before becoming a diatexite magma. The diatexite magma in this case formed *in situ*, in a closed system; therefore, it is called a primary diatexite magma. The fraction of melt at which the diatexite magma is formed depends on the mechanism by which flow occurred. Assuming magmatic flow (i.e., a suspension of crystals in an anatectic melt), then a melt fraction in excess of 0.16 (NUP model) or 0.26 (URS model) is required; the upper limit of the diatexite field is the degree of partial melting attainable in regional metamorphism; this is not normally more than 70%. Under certain circumstances, the diatexite field might extend beyond the usual 0.26–0.70 range in terms of fraction of melt. Other mechanisms of deformation that produce bulk flow (see section 3.5) could generate diatexite migmatites at fractions of melt as low as 0.07; this potential extension of the diatexite field is labeled "deform mechanism" on Fig. 4c. In some circumstances, such as ultra-high-temperature metamorphism, or H_2O-fluxed melting, the degree of partial melting may approach 100% and extend the diatexite field to high fractions of melt. Flow of the diatexite magma in response to shear stresses leads to the development of schlieren and magmatic foliations, as well as some separation of the melt from the material entrained in it, to form residual diatexite and melt-rich diatexite magma. The third, and final step, is fractional crystallization of the melt-rich diatexite; this generates a fractionated magma of granitic composition that commonly develops a granitic microstructure on crystallization and cumulate rocks. This third scenario is most likely to apply to regional anatectic settings where large volumes of melt were able to form rapidly.

Some petrological questions, but by no means all, relevant to the processes that form migmatites are outlined next. The first process is partial melting of the protolith, and some relevant questions are as follows. (1) What was the protolith composition? (2) Are the migmatites the result of an open system or of closed-system (commonly called *in situ*) processes? (3) What is the composition of the neosome, and of its constituent parts? (4) Which reaction generated the melt, what was the degree of partial melting (F), and how does it compare to the fraction of melt (M_f) that the migmatite contained, which may possibly be derived from an estimate of the fraction of leucosome in the outcrop?

The next process is porous flow of the melt out of the source, i.e., melt segregation; here, the questions may include the following. (1) Did the melt segregate from the residuum, and if so, what proportion of the melt remained in the residuum? (2) How much residuum was entrained in the melt? (3) How variable is the composition of the residuum and what caused this variation (different protoliths, different melting reactions, different F, different degrees of segregation of the melt)? (4) Over what distance was the porous flow active, over centimeters or tens of centimeters?

The third step is transport and flow of the melt, or magma, in channels; questions relevant to this stage include the following. (1) Are there compositional variations that can be attributed to processes occurring during flow, such as the separation of suspended crystals? (2) Do the leucosome, veins, or diatexite migmatites that represent the bodies of former melt, or magma, that flowed in channels have compositions that are different compared to those of bodies of melt that remained *in situ*? (3) Did the wallrock contaminate the magma during flow?

The fourth step is crystallization, and it can, of course, begin at any time after partial melting has started. Cooling may be so rapid that separation of melt from its residuum before complete crystallization may not be possible in some contact aureoles. Some principal questions are as follows. (1) Did fractional crystallization occur, in other words, does the leucosome, or diatexite, have the composition of an anatectic melt? (2) Did the entrained residuum separate before or during crystallization? (3) Where did the separation of the entrained residuum take place? Finally, post-anatectic processes, such as retrograde metamorphism, rehydration, and veining can change the bulk composition of a migmatite, or its parts. Therefore, it is important that any altered samples be identified.

The following sections pertain to two factors that seriously affect the way in which whole-rock compositions can be used to investigate migmatites. First, in order to interpret the trends in whole-rock composition that arise from a petrogenetic process, some reference points are essential; the most commonly used are the protolith and neosome compositions (ideally both the melt and the residuum, but most commonly just the melt). How well these parameters can be constrained determines what kinds of petrological questions can be addressed, and what types of quantitative modeling can be applied. Secondly, it is generally very difficult to find samples affected by just one process. The investigation of any individual process requires finding ways to isolate its geochemical signature from the rest. Typically, this entails selecting specific combinations or ratios of elements.

6.2 Reference-point compositions

Morphological and compositional heterogeneity is typical of migmatites; hence, all its parts should be sampled, and the spatial and temporal relationships among samples should be recorded. Compositional gradients are another common problem with migmatites, especially in the rocks derived from melt-rich portions of migmatites and in the residual rocks. Care should be made to sample the full width of bodies of leucosome and melanosome so as to include the entire compositional gradient; this is especially important if detailed mass-balance calculations are contemplated. It is common practice to separate the melanosome from the leucosome for analysis, but it must not be forgotten that

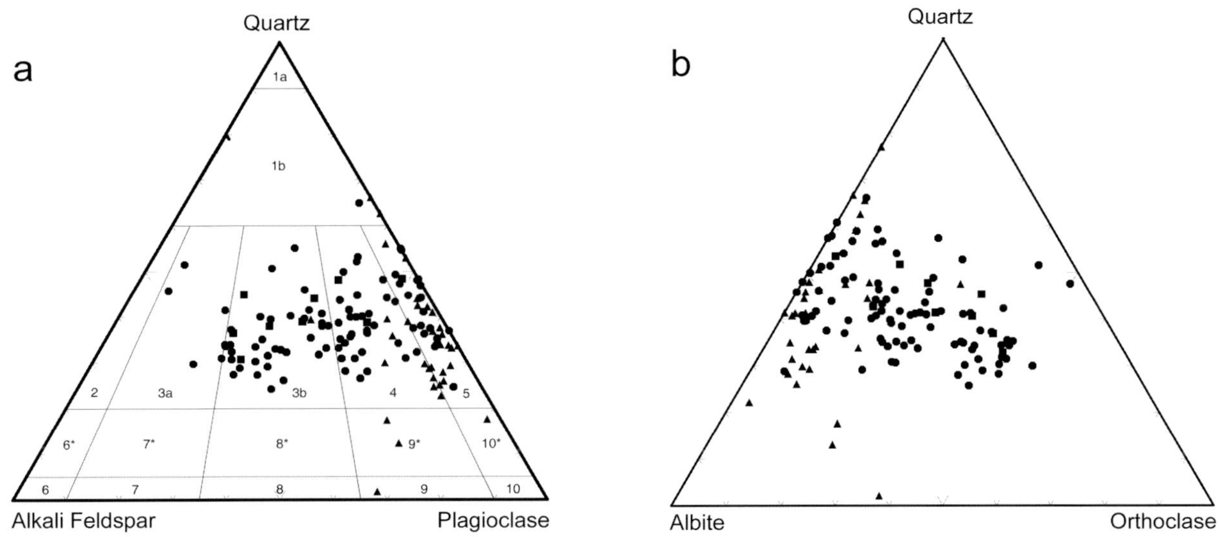

Fig. 5. Composition of 133 samples of leucosome from metatexite migmatites. (*a*) Quartz – alkali feldspar – plagioclase mesonorm plot with fields after Streckeisen (1976). (*b*) Quartz – albite – orthoclase mesonorm plot. Symbols: dots, leucosomes from metapelitic and metapsammitic (plagioclase + biotite + quartz) protoliths; squares, eight samples of leucosome from a single outcrop, metapelitic protolith; triangles, leucosome from metamafic protoliths. The leucosome from metapelitic and metapsammitic protoliths ranges in composition from syenogranite to tonalite, although most samples are monzogranitic or granodioritic in composition. In contrast, leucosome derived from metamafic rocks range in composition from monzodiorite to tonalite, but most are tonalitic. The leucosome collected from a single outcrop of metapelite show essentially the same range of composition as the entire set of leucosome samples from metapelitic and metapsammitic protoliths. The variability in composition of the leucosome thus is not simply a function of the bulk composition of the protolith and the *P–T* conditions at which anatexis occurred; processes that occur during crystallization of the melt also may be important.

such a methodology results in the loss of information on the composition and compositional gradients at, and close to, the contact. Some problems specific to protolith, melt, and residuum are discussed next.

Determining protolith compositions (the starting material)

At the incipient stage of partial melting, the rock between scattered patches of neosome in a fertile layer could be considered as being very close to the protolith composition. It may also be possible to trace a particular partially melted layer, or sequence of layers, laterally to lower metamorphic grades and to sample them just below the "melt-in" isograd that marks the onset of partial melting.

Determining what the protolith was is a challenge in the highest-grade parts of anatectic terranes where virtually all the rocks are partially melted and neosome abounds. It is an error to use the remaining paleosome (nonmelted and typically mesocratic) relics to represent the protolith composition. These mesocratic layers of paleosome are preserved specifically because they have infertile bulk compositions, and so escaped partial melting. There are three potential options in obtaining an estimate of the protolith composition. (1) Collect a very large bulk sample of the whole neosome; this works well if the migmatite formed in a closed system, and where the segregation distance was small, as in some contact aureoles. (2) Find, and sample, similar rocks in an adjacent part of the terrane not affected by partial melting. The problem then becomes which rocks to sample. Hydrous rocks progressively devolatilize during prograde metamorphism, and could simultaneously lose some major and trace elements. Thus, sampling the highest-grade nonmelted rocks could provide a better approximation of the protolith composition in a regional metamorphic terrane. Contact aureoles, and some regional metamorphic terranes in which low-grade country rocks were rapidly heated and melted, may be an exception if the fluids and any trace elements released by prograde devolatilization remain in the rocks and were incorporated wholesale into the migmatites. (3) The least satisfactory option, best considered as a last resort, is to adopt a compiled average composition (e.g., average upper continental crust) for the protolith; these averages invariably include both fertile and infertile rock types, and so preclude any meaningful mass-balance modeling.

Determining the "melt" composition

Melt composition here refers to that generated at partial melting. Because this melt composition serves as the starting point in interpreting leucosome and diatexite petrogenesis, it is in some cases called the "initial" anatectic melt composition (note that the term "primary melt" is generally reserved for partial melts of the mantle). The most widely used method to identify anatectic leucosome, or diatexite migmatites, that may be "initial" melts is to compare its position to that of the cotectics for haplogranitic melts on Ab–Or–Qtz or Ab–An–Or plots. Typically, CIPW norms are used, but mesonorms or weight-percent plots also have been used (e.g., Wyllie 1961, Brown 1979, Johannes 1983*b*, Gupta and Johannes 1986, Barbey et al. 1990, Symmes and Ferry 1995, Slagstad et al. 2005). A composition that plots on, or close to, the cotectic in the systems Ab–Or–Qtz–H_2O or Ab–An–Or–Qtz–H_2O, corresponding to the P–T–$a(H_2O)$ at which the natural melts formed, is consistent with the sample being an "initial" anatectic melt.

This approach is satisfactory if the melt was haplogranitic and crystallized in a closed system of smaller size than the sample. For rocks that do not satisfy these conditions, the Ab–Or–Qtz and Ab–An–Or plots should be used with caution. Additional components change the melt composition, and hence the position of the cotectic lines (e.g., Manning and Pichavant 1983). Furthermore, a position far from the cotectic does not necessarily exclude a magmatic origin for leucosome (or diatexite migmatites for that matter), although it probably excludes an "initial" melt composition. During **fractional crystallization**, the phases that have crystallized are separated from the remaining "**fractionated**" or "**evolved**" melt. Samples that represent accumulations of the early-formed crystals, for example, plagioclase, plot nearer to the plagioclase apex, or to the plagioclase–quartz join (Cuney and Barbey 1982, Sawyer 1987), whereas the samples of "fractionated" melt plot in the opposite direction, in this example, toward the quartz apex, the quartz–orthoclase join, or the orthoclase apex. Leucosome that contains accumulations of K-feldspar or quartz crystals also are possible. Furthermore, contamination of the initial anatectic melt by residual phases, notably quartz and plagioclase for these particular plots, can result in samples that plot away from a cotectic. The compositions of some domains of leucosome are shown in Fig. 5, on quartz – alkali feldspar – plagioclase and Ab–Qtz–Or mesonorm plots. The composition of the leucosome derived from metapelitic and metapsammitic protoliths extends from granite to tonalite, and from tonalite to monzodiorite for leucosome derived from metamafic protoliths. However, the composition of a leucosome is not related simply to the bulk composition of the protolith or the *P–T* conditions at which melting occurred. A range of leucosome compositions is commonly found within single outcrops; for example, the squares in Figs. 5*a* and 5*b* represent leucosome compositions from a single outcrop of migmatite derived from a metapelitic protolith. They show a range from granite to tonalite. This wide variation in the composition of leucosome is attributed to fractional crystallization of the anatectic melt from which the leucosome was derived.

An estimate of the major-element content of "initial" anatectic melts from more complex systems can be obtained from partial-melting experiments that use starting materials with compositions similar to the inferred protolith, and performed at *P–T–*$a(H_2O)$ conditions close to those at which the natural melting occurred. The experimentally produced melts show rather limited variations in TiO_2, FeO, and MgO contents, but the variation in CaO, Na_2O, and K_2O contents may be large and are related to the protolith composition, pressure, and temperature of partial melting, and whether melting occurred with, or without, an H_2O fluid present. The ability to measure volatile components, particularly Na_2O, correctly in experimental glasses is another factor that seriously affects any estimate of the composition of initial anatectic melts (Grant 2004). Notwithstanding the possible analytical problems in determining their compositions, I present in Fig. 6 a compilation of the composition of glasses (i.e., quenched melts) obtained from partial-melting experiments conducted on a variety of starting compositions. Figures 6*a–e* pertain to experiments without H_2O added (so-called "dehydration melting" experiments), and Figs. 6*f* and 6*g* pertain to experiments carried out with H_2O added to the experimental charge, although not necessarily with enough to saturate the resulting melt (so-called "H_2O-fluxed" melting). Samples of leucosome or diatexite plotting in the same compositional field as appropriate experimental glasses may represent compositions of "initial" melt.

Once potential samples of "initial" anatectic melt have been identified, their trace-element contents should be investigated for confirmation. The possibilities of contamination with the residuum and of fractional crystallization should be investigated. Fractional crystallization results in an enrichment in the cumulate of the trace elements compatible with the early-crystallized and accumulated minerals, and an enrichment in the fractionated melts of the trace elements incompatible with these minerals. Thus, the signature of a "initial" melt composition is that it should lie between the cumulate field and the fractionated melt field on bivariant plots of an incompatible constituent versus a compatible one (e.g., K_2O versus Na_2O or Rb versus Sr), or on rare-earth element (REE) diagrams (e.g., Sawyer 1987), where Eu is the compatible element and the other REEs are incompatible (see Fig. 7). This central position arises because the "initial" anatectic melt is **parental** to the crystal-rich cumulate and to the fractionated melt. More importantly, the compositions of the cumulate and fractionated melt show the same relationship to the "initial" melt irrespective of whether the melt formed as an equilibrium melt, a trace-element-undersaturated equilibrium melt, or a disequilibrium melt. If disequilibrium melting produces

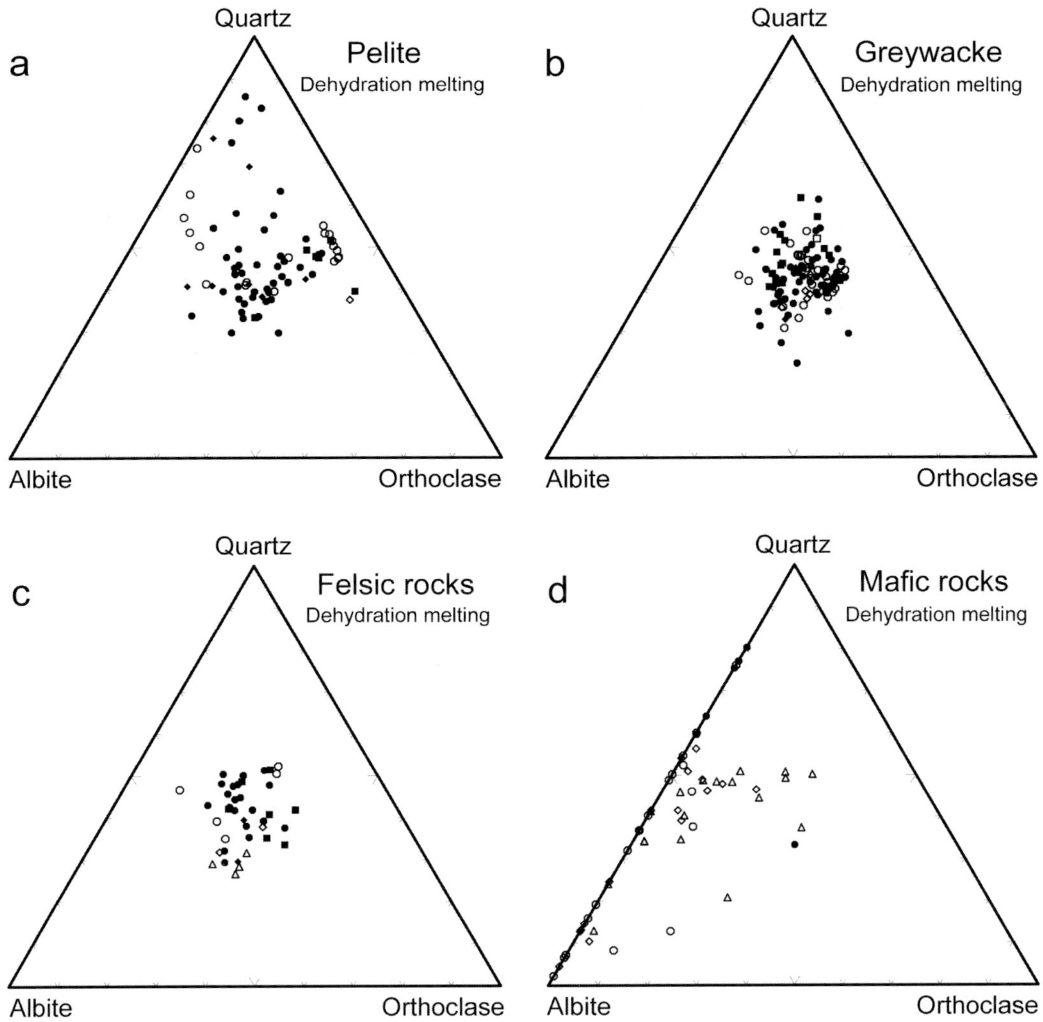

Fig. 6. Composition of glass (quenched partial melts) obtained from partial-melting experiments shown as a function of composition of starting material (protolith) and the type of melting. (*a*) Partial melting of pelitic protoliths without H_2O added (i.e., dehydration melting). Note the wide range of compositions. (*b*) Partial melting of greywacke (psammitic) protoliths without H_2O added; note the tight cluster of compositions corresponding to monzogranite. (*c*) Partial melting of felsic protoliths (granite, tonalite, etc.) without added H_2O; the compositional range is similar to greywackes. (*d*) Partial melting of mafic protoliths without H_2O added; most compositions of glass fall on the quartz–albite join, but some are granodioritic. (*e*) Glass compositions for H_2O-added partial-melting experiments of pelitic, greywacke, felsic, and mafic protoliths, shown on a quartz–albite–orthoclase plot. (*f*) All the glass compositions from *a* to *d* shown on a quartz – alkali feldspar – plagioclase mesonorm plot, shown with the fields from Streckeisen (1976). The glass compositions range from syenogranite to tonalite, but most are monzogranite. (*g*) Glass compositions for H_2O-added (i.e., H_2O-fluxed) partial-melting experiments on pelitic, greywacke, felsic, and mafic protoliths, shown on a quartz – alkali feldspar – plagioclase mesonorm plot, along with the fields of Streckeisen (1976). The glass compositions range from syenogranite to tonalite, but most lie in the granodiorite field. Legend: squares, P <3.5 kbar; circles, P 4–10 kbar; diamonds, P 10–16 kbar; triangles, P >20 kbar; filled symbols, T <925°C; open symbols, T >925°C. Data sources: metapelites (Vielzeuf and Holloway 1988, Le Breton and Thompson 1988, Patiño Douce and Johnston 1991, Patiño Douce and Harris 1998, Pickering and Johnston 1998, Koester et al. 2002); metagreywackes (Conrad et al. 1988; Vielzeuf and Montel 1994; Gardien et al. 1995, 2000; Patiño Douce and Beard 1995, 1996; Stevens et al. 1997; Montel and Vielzeuf 1997; Castro et al. 1999; Grant 2004); felsic rocks (Castro et al. 2000; Conrad et al. 1988; Johnston and Wyllie 1988; Litvinovsky et al. 2000; Rutter and Wyllie 1988; Skjerlie and Johnston 1992, 1996; Singh and Johannes 1996*a*, *b*); mafic rocks (Rushmer 1991; Rapp and Watson 1995; Wolf and Wyllie 1991, 1994; Springer and Seck 1997; Skjerlie and Patiño Douce 2002; Selbekk and Skjerlie 2002).

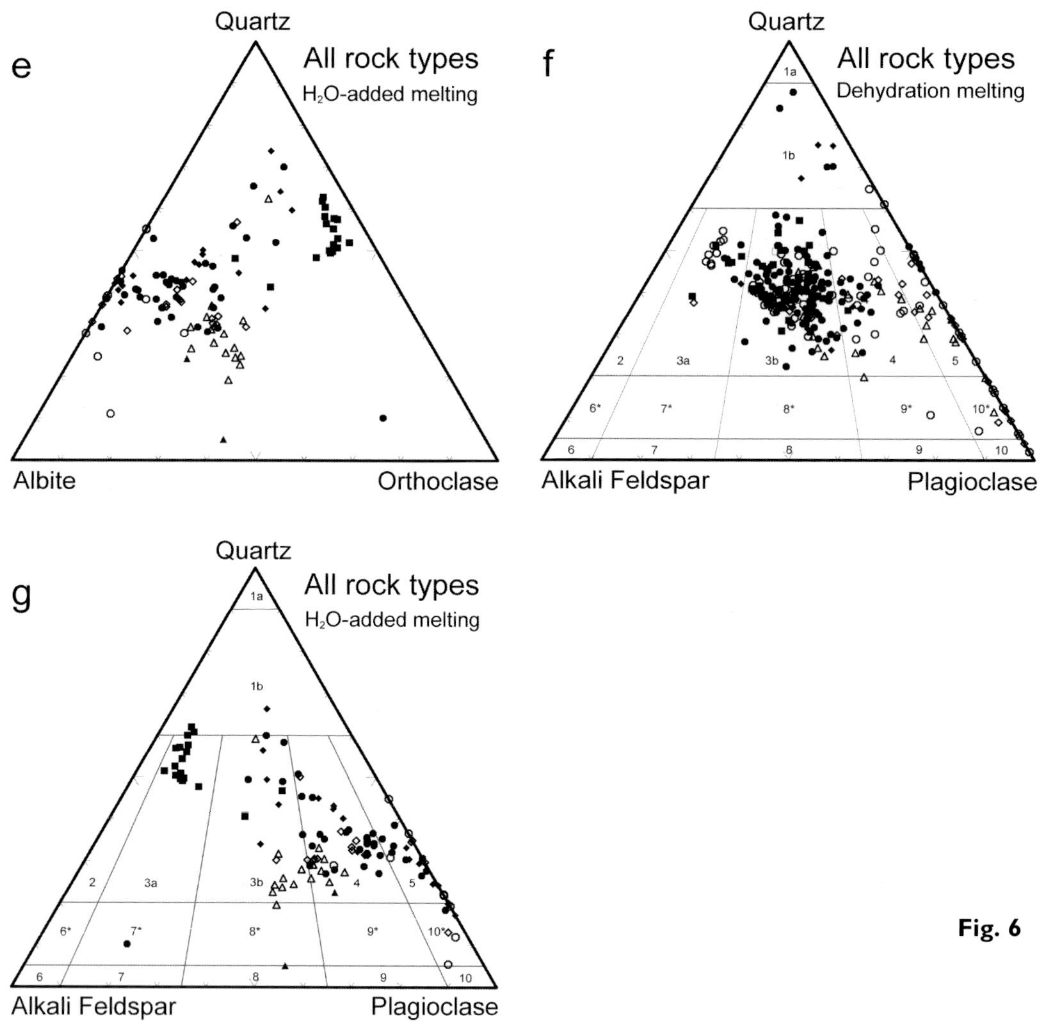

Fig. 6

an "initial" melt with very low concentrations of incompatible elements, then fractional crystallization of that melt will yield a cumulate with still lower levels.

If no "initial" anatectic melts can be identified from the migmatite samples collected, then a glass composition from an appropriate experimental study [same or similar starting composition and *P–T–a*(H_2O) conditions as the migmatites] could be used as a reference-point composition. However, using the compositions of glasses from partial-melting experiments has the disadvantage that trace-element contents have generally not been reported for them, although recent advances in analytical techniques will enable the trace-element contents of experimental glasses to be determined routinely.

A unique composition of "initial" anatectic melt may only exist at the very start of anatexis, when the first rock partially melts, or if the protolith was homogeneous. Successive increments of anatectic melt derived from a single source have different compositions because the composition of the source has changed, and the temperature of partial melting has increased. Anatectic melts of diverse compositions start to appear as soon as partial melting begins in other rocks or layers of different bulk composition. Thus, it is not a simple task to determine the composition of the "initial" anatectic melt from which a particular batch of leucosome, or diatexite migmatite, was derived.

Residual rocks

Rocks that contain the residuum left after partial melting can occur in many forms in migmatites and in associated autochthonous granites (see, for example, Barbey 1991). In the lowest-grade parts of migmatite terranes, the residual rocks are commonly, but not exclusively, found in the melanocratic zones around bodies of leucosome. Care should be taken to distinguish melanosome (i.e., melt-depleted residuum) from mafic selvedges formed by reaction between a leucosome, or vein, and its wallrocks. The

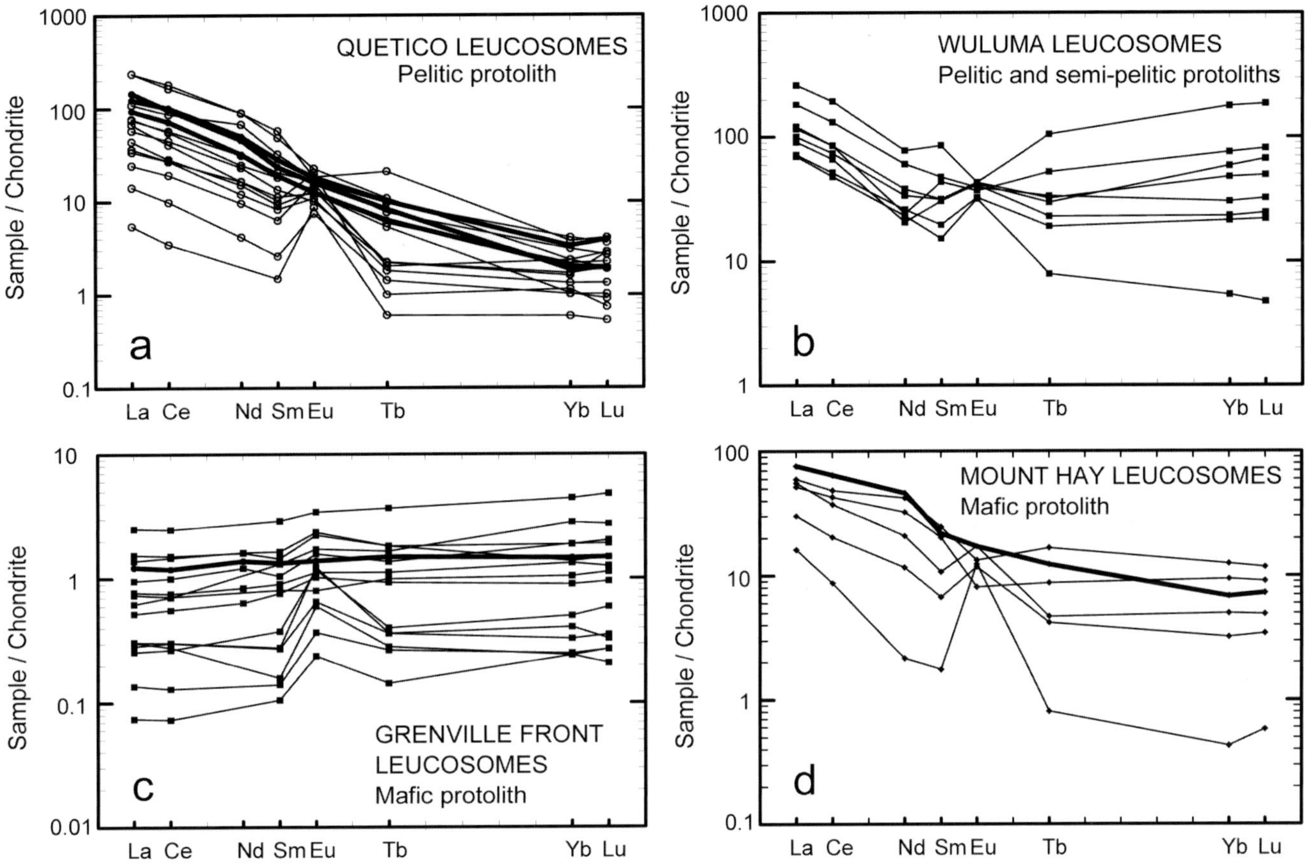

Fig. 7. Chondrite-normalized rare-earth element (REE) patterns for leucosome in four migmatite terranes. (*a*) Upper-amphibolite-facies anatexis, *T* 700–800°C, P 3–4 kbar. (*b*) Granulite-facies anatexis, *T* 825–875°C, *P* 6–7 kbar. (*c*) Upper-amphibolite- to lower-granulite-facies anatexis, *T* 800–850°C, *P* 8–10 kbar. (*d*) Granulite-facies anatexis *T* 825–875°C, *P* ca. 5 kbar. All the leucosome samples are from metatexite migmatites. Although they formed from different protoliths that underwent anatexis at different pressures and temperatures, the overall REE characteristics of the leucosome suites are very similar. There are three types of REE patterns within the leucosome data. (1) Most samples of leucosome have a REE pattern with a strong, positive Eu anomaly, but lower abundances of the total REE compared to the other patterns. (2) A few samples of leucosome have REE patterns with a negative Eu anomaly, and in general, the highest abundances of the REE. (3) A few samples of leucosome (shown as bold lines) have intermediate-level REE abundances and a smooth pattern without Eu anomaly. The presence of the three types of REE patterns within each of the leucosome suites is interpreted as strong evidence that fractional crystallization played a major role in the evolution of the anatectic melt from which the leucosome in a migmatite terrane was derived. The leucosome samples with a smooth pattern are interpreted to have crystallized from melts that had compositions close to that of the initial anatectic melt; leucosome samples with a positive Eu anomaly are interpreted to be feldspar accumulations derived from primary melts, and the leucosome samples with a negative Eu anomaly are interpreted to have crystallized from evolved anatectic melts that had already crystallized substantial amounts of feldspar, principally plagioclase.

latter are generally narrow and composed largely of biotite or hornblende. In higher-grade metatexite migmatites, such as those found in melt-depleted, granulite-facies terranes (Guernina and Sawyer 2003), conspicuously melanocratic rocks are rare, but virtually all the rocks are melt-depleted, and hence residual in bulk composition (Barbey and Cuney 1982, Maccarrone et al. 1983, Caggianelli et al. 1991). In places where diatexite migmatites are abundant, residual rocks may be rare, and a detailed geochemical study may reveal a deficit of residuum, i.e., a surplus of anatectic melt (e.g., Obata et al. 1994).

Mineral compositions

Many of the processes of interest in migmatites can be modeled as the addition of one or more minerals to a particular starting composition, or their subtraction from that starting composition. Therefore, the compositions of minerals present in migmatites are a very useful addition to whole-rock geochemical diagrams because they may indicate the vectors of compositional change associated with the processes that occur in migmatites (e.g., contamination of anatectic melt by residual biotite). With the increasing availability of ion probe and laser-ablation techniques, contents of major and trace elements are becoming routinely available for the main minerals in migmatites. These data may prove useful in identifying the provenance of residual minerals in leucosome and diatexite migmatites, as well as in providing further constraints on models of petrogenetic processes in migmatites, such as partial melting, melt extraction, and fractional crystallization.

6.3 Diagrammatic representation

Most geological processes yield suites of samples that define trends, or fields, on compositional diagrams, because the process has advanced farther in some samples than in others. To interpret such trends, the composition of the starting (i.e., pre-process) material, and either the end-product composition, or the changes in mineral assemblage and modal proportions that are involved in the process, are required. Useful quantitative modeling can only be done with this information. Changes in composition are best shown in diagrams that display the greatest dispersion of the product rock types about the starting composition. Bivariant diagrams in which an element that characterizes one of the products of the process considered (e.g., residuum) is plotted against an element that characterizes the other (e.g., anatectic melt) are generally sufficient. Both the major elements and trace elements should be treated.

Residual rocks are most commonly enriched in Fe–Mg silicate minerals relative to the protolith, but they may also be enriched in accessory minerals, plagioclase, K-feldspar, and even quartz, depending on the particular composition of the protolith. The Fe–Mg silicate fraction is commonly represented by the sum (FeO_T + MgO + TiO_2), where FeO_T represents all the iron expressed as FeO, but in some circumstances that choice of components is inappropriate. If TiO_2 is principally in minerals other than an Fe–Mg silicate (e.g., rutile, rather than biotite), it may be better to consider it separately from (FeO_T + MgO), because minerals that form very small grains could be segregated with the melt, much like inherited zircon, for example. The presence of Fe oxide and Fe sulfide minerals in the protolith means that some, and possibly most, of the Fe is not in the silicate minerals. In that circumstance, MgO alone may be the best way of depicting the ferromagnesian silicate fraction in the residuum. Alternatively, the whole-rock FeO_T content may be corrected by subtracting the amount of Fe in the sulfide and Fe oxide minerals, using a combination of modal abundance and mineral compositions determined with an electron-microprobe analyzer. Trace elements that can be used to represent the Fe–Mg silicate minerals fraction include Sc and Co; Cr, Ni, and V may be used where sulfide and Fe oxide minerals are absent. The biotite component of the residuum could, for example, be represented by Rb. Residua rich in sillimanite, cordierite, or spinel could be represented by Al_2O_3. Similarly, CaO may be used for residua rich in hornblende, clinopyroxene, and plagioclase; Sr may also be used for investigations of plagioclase-rich residua.

Many investigators have shown that a large part of the total budget of some trace elements, including the REE, Y, Th, U, Hf, Zr, Nb, and Ta, is located in accessory minerals, and the behavior of these minerals during partial melting largely controls the way trace elements are distributed between the anatectic melt and the residuum (Sawka 1988; Bea 1991, 1996*a*, *b*; Bea et al. 1994; Watt and Harley 1993). In some circumstances, the accessory minerals dissolve completely into the melt, resulting in a residuum depleted in the trace elements associated with those particular accessory phases, but in others, the accessory minerals were either in excess, or did not dissolve (e.g., not enough time), or could not dissolve (e.g., armored by other phases) into the melt, and an enrichment of the associated trace elements in the residuum results. However, recent results from ultra-high-temperature (UHT) metamorphic terranes suggest that accessory phases may control less of the trace-element budget in very-high-temperature anatexis. Hokada and Harley (2004) and Condie et al. (2004) found that the major phases in rocks formed at ultra-high temperatures can contain significant amounts of high-field-strength elements, such as Zr and the REE, in sufficient amounts that the major phases, because of their high modal abundance, can control the budget for many of these trace elements in the whole rock.

The bulk compositions of some protoliths result in residua that are very rich in K-feldspar, or in quartz; K_2O and SiO_2, respectively, may be an appropriate way to differentiate rocks derived from such residua from those derived from the melt fraction.

Anatectic melts generally range from granite to tonalite–trondhjemite in composition (see Figs. 5 and 6), although potassic anatectic melts of syenogranite composition may form from protoliths such as plagioclase-poor pelites (Pickering and Johnston 1998, Grant 2004). Generally, the anatectic melts are enriched in K_2O, Na_2O, and SiO_2 relative to their protoliths. The trace elements Cs, Ba, and Li are partitioned into the melt under most circumstances of partial melting in the continental crust (Icenhower and London 1995, 1996). Hence, these major oxides and trace elements are commonly used to represent the anatectic melt, or melt-derived rocks, in bivariant plots.

Plagioclase is generally one of the first minerals to start to crystallize from melts of granitic composition, and in the early part of the crystallization history, it may be the most voluminous product. Thus, a plagioclase cumulate formed by fractional crystallization may be represented by Na_2O, CaO, or Sr. The remaining evolved melt is enriched in the components incompatible with plagioclase; the rocks crystallized from such a melt could be represented by SiO_2, K_2O, Cs, Ba, Li, Pb, Sn, and even U. The crystallization of accessory phases, e.g., zircon, monazite, xenotime, and apatite, from anatectic melts commonly controls the high-field-strength elements (HFSE) and REE contents of the rocks (e.g., leucosome and some diatexite migmatites) derived from the anatectic melt (Miller and Mittlefehldt 1982, Sawka 1988, Wark and Miller 1993, Bea et al. 1994) and should be investigated. The papers by Bea (1991, 1996*a*, *b*) and Bea et al. (1994) are important resources for the trace-element contents of both accessory and major phases, and are useful for those wishing to model partial-melting and crystallization processes.

Combinations of elements may generally be found to produce a wide dispersion of the sample points in the compositional space of pelitic, greywacke, and mafic bulk compositions. However, in cases where the composition of the protolith and the anatectic melt are similar (e.g., partial melting of granite, arkoses, and some psammites), compositional changes can be very small and, consequently, are very difficult to illustrate diagrammatically.

Matched triplet sets of samples

Sample sets in which a particular protolith and the "initial" anatectic melt (leucosome) and residuum (melanosome) derived from it have been systematically sampled are few. Because of the known relationship among the samples, these triplet sets of samples offer unique opportunities to study the details of the processes that occurred in migmatites. They can be used for mass-balance techniques, such as the isocon method of Grant (1986; but see also Baumgartner and Olsen 1995), to investigate whether melting occurred in an open or a closed system, and to calculate the volume and composition of the material that was added, or lost (e.g., Olsen et al. 2004). The degree of partial melting (*F*) can also be constrained, and the various conceptual models for partial melting (e.g., equilibrium batch-melting, fractional melting, disequilibrium melting, etc.) can be tested by trying to replicate the measured compositions of the leucosome considered to represent the "initial" melt, although selecting which distribution coefficients to use is quite another problem (see comments by Icenhower and London 1995), and one outside the scope of this book.

Matched samples collected from different parts of individual bodies of leucosome could be used to investigate fractional crystallization, or the unmixing of the melt fraction from the residuum. The problem can also be inverted to investigate the extent to which contamination by the wallrocks has occurred during the formation of the leucosome, for example, by the movement of melt through the network of channels that drained melt from its source (see Fig. 4*b*, for example). Multi-element variation diagrams (the so-called "spidergrams") in which the products are normalized to the starting composition are a good way of showing compositional variations within matched sets of samples.

General sets of samples

In many migmatites, particularly if the proportion of neosome is high, or if the melt fraction has moved long (meter-scale) distances, it is impossible to be certain that the leucosome sampled is genetically related to its adjacent, or even the nearest, melanosome; moreover, determining what was the protolith may be impossible, which precludes sampling it. Although these general sets of samples are not suited for studies of the kinds of problem outlined in the previous section, they are, if a large number of samples are collected, ideal to examine the complete range of petrological processes that occur in making migmatites; recent examples of this approach were provided by Sawyer (1998, 1999), Sawyer et al. (1999), Milord et al. 2001, Solar and Brown (2001), Villaseca et al. (2001), Fornelli et al. (2002), and Guernina and Sawyer (2003).

Two types of diagram are particularly useful (Fig. 8). The first, MgO or (FeO_T + MgO) versus K_2O or SiO_2, can be used to examine partial melting and the unmixing of melt from the residuum. The second, CaO or (CaO + Na_2O) versus K_2O or SiO_2, is useful where plagioclase is a major residual phase; with it, one can investigate the processes that occurred during the crystallization of the anatectic melt, such as fractional crystallization in which an early-crystallized mineral, typically plagioclase, but in some cases K-feldspar, accumulates in one part of the system, and the complementary fractionated melt crystallizes elsewhere. Cumulate plagioclase could be represented by CaO or (CaO + Na_2O), K_2O could represent either the K-rich fractionated melt or cumulate K-feldspar, and SiO_2 could be used to represent the fractionated melt. In order to interpret the scatter and trends that a large set of general samples from a migmatite terrane generally show in composition space, some reference points and trends need to be established. Ideally, these are the compositions of the sampled protolith and the anatectic melt (or melts). If the composition of the melt (from the leucosome) is unavailable, then a suitable composition of an experimental melt (i.e., glass) is the next best option. The compositional range of felsic anatectic melts obtained from a particular type of protolith (e.g., greywacke, pelite, or felsic) in partial-melting experiments below 900°C and at pressures below 10 kbar is relatively small (Figs. 6*a–c*); major diversity in the composition of anatectic melts starts after the hydrous phases have disappeared. From a knowledge of the melting reactions and $P–T–a(H_2O)$ conditions under which the migmatites formed, and the type of protolith involved (e.g., pelite, greywacke, mafic, granite, or tonalite), glass compositions can be obtained from the literature with which to define a "field of anatectic melt." The compositions of both natural melt (from leucosome) and experimental melt are shown in Fig. 8 for comparison. Any samples of leucosome, or diatexite migmatite, that lie in (or close to) the experimental melt field may be "initial" melts, and should be further investigated using their trace-element (i.e., REE) contents; these samples could be useful in defining the trace-element characteristics of the melt. The leucosome compositions from metatexite migmatites derived from mafic protoliths (Figs. 8*a*, *b*) plot closer to experimental glasses (W) that were generated by H_2O-added partial melting, than to the glasses (D) obtained from experiments with no H_2O added. This may mean that melting in these migmatites was in fact H_2O-fluxed. Leucosome compositions from the Quetico (Ontario) migmatite terrane (Figs. 8*c*, *d*) partially overlap the compositional field for experimental glasses generated from dehydration melting (i.e., without H_2O added). For samples of diatexite migmatites (Figs. 8*e–j*), the appropriate experimental glass compositions from partial melting without H_2O added invariably coincide with those samples of diatexite with the lowest MgO, (FeO_T + MgO) and CaO contents, but with high K_2O and SiO_2 contents, which from their petrographic characteristics were essentially anatectic melt.

With the trend between melt and protolith compositions established, the melt-depleted or residual rocks (e.g., melanosome and some diatexite migmatites) are identified by their position along this trend on the opposite side of the protolith to the melt (Fig. 8). However, Fig. 8 shows that the compositional field of the protolith is much larger than the compositional field of the "initial" anatectic melts (i.e., excluding leucosome that is the product of fractional crystallization) in most migmatite terranes; consequently, the compositional field of the residual rocks is larger still. Nevertheless, assuming a representative sampling of melanosome, it could be argued that a tie line from the densest cluster of melanosome samples to the melt composition should pass through the most abundant composition of

Fig. 8. Examples of plots of compatible versus incompatible elements used in the interpretation of general sets of samples from migmatite terranes. (*a*) (FeO_T + MgO) versus SiO_2 and (*b*) CaO versus SiO_2 for migmatites from the Grenville Front near Chibougamau, Quebec; anatexis at *T* 800–850°C, *P* 8–10 kbar, involving the breakdown of hornblende and the formation of a tonalitic melt with clinopyroxene and garnet. (*c*) (FeO_T + MgO) versus SiO_2 and (*d*) (CaO + Na_2O) versus SiO_2 for migmatites from the Quetico Subprovince, Ontario; anatexis at *T* 700–800°C, *P* 3–4 kbar, involving the breakdown of biotite and sillimanite and the formation of melt plus cordierite. Note that some samples of diatexite plot in the protolith field, whereas others lie between the protolith and anatectic melt fields, and that the paleosome (resister) lithologies have compositions that fall outside the protolith field. (*e*) MgO versus K_2O and (*f*) (CaO + Na_2O) versus K_2O for migmatites from the Opatica Subprovince, Ontario: anatexis at *T* 750–800°C, *P* 5–7 kbar, principally via the H_2O-present melting of quartz, plagioclase, and K-feldspar. (*g*) MgO versus K_2O and (*h*) CaO versus K_2O for migmatites from the Ashuanipi Subprovince, Quebec: anatexis at *T* 825–875°C, *P* 6–7 kbar, through biotite + quartz + plagioclase = melt + orthopyroxene + ilmenite or magnetite. The plagioclase that initially crystallized from the melt had compositions of about An_{30}, whereas the last plagioclase crystallized from more evolved melt and is approximately An_{20}. Many samples plot along the trend of melt + residuum (composed of plagioclase + orthopyroxene). On this plot, however, that trend also coincides with the accumulations of plagioclase crystallized from the anatectic melt. (*i*) (FeO_T + MgO) versus K_2O and (*j*) CaO versus SiO_2 for migmatites from Saint-Malo, France; anatexis at a *T* of 750°C, *P* 4–7 kbar, involving the breakdown of muscovite, and with biotite generally stable. Whole-rock data are shown as points (the symbols are explained on the figures) and are combined with mineral compositions (determined by electron-microprobe analysis), which are located at the tips of the arrow (see text for further explanation). Solid arrows pertain to minerals belonging to the residuum and are drawn radiating out from a representative composition of the protolith. The dotted arrows indicate the composition of the minerals that crystallized from the anatectic melt, and they are drawn from a representative initial composition of the anatectic melt. The positions of anatectic melt (also marked AM where space is restricted), protolith, residuum, plagioclase accumulations, and fractionated melts are labeled on the diagrams. Labeled arrows: CAM, trend for melts contaminated with residuum; AP, trend for plagioclase accumulation; FM, trend for fractionated melts. D represents glass compositions from dehydration-melting experiments, and W represents glass compositions from H_2O-fluxed experiments.

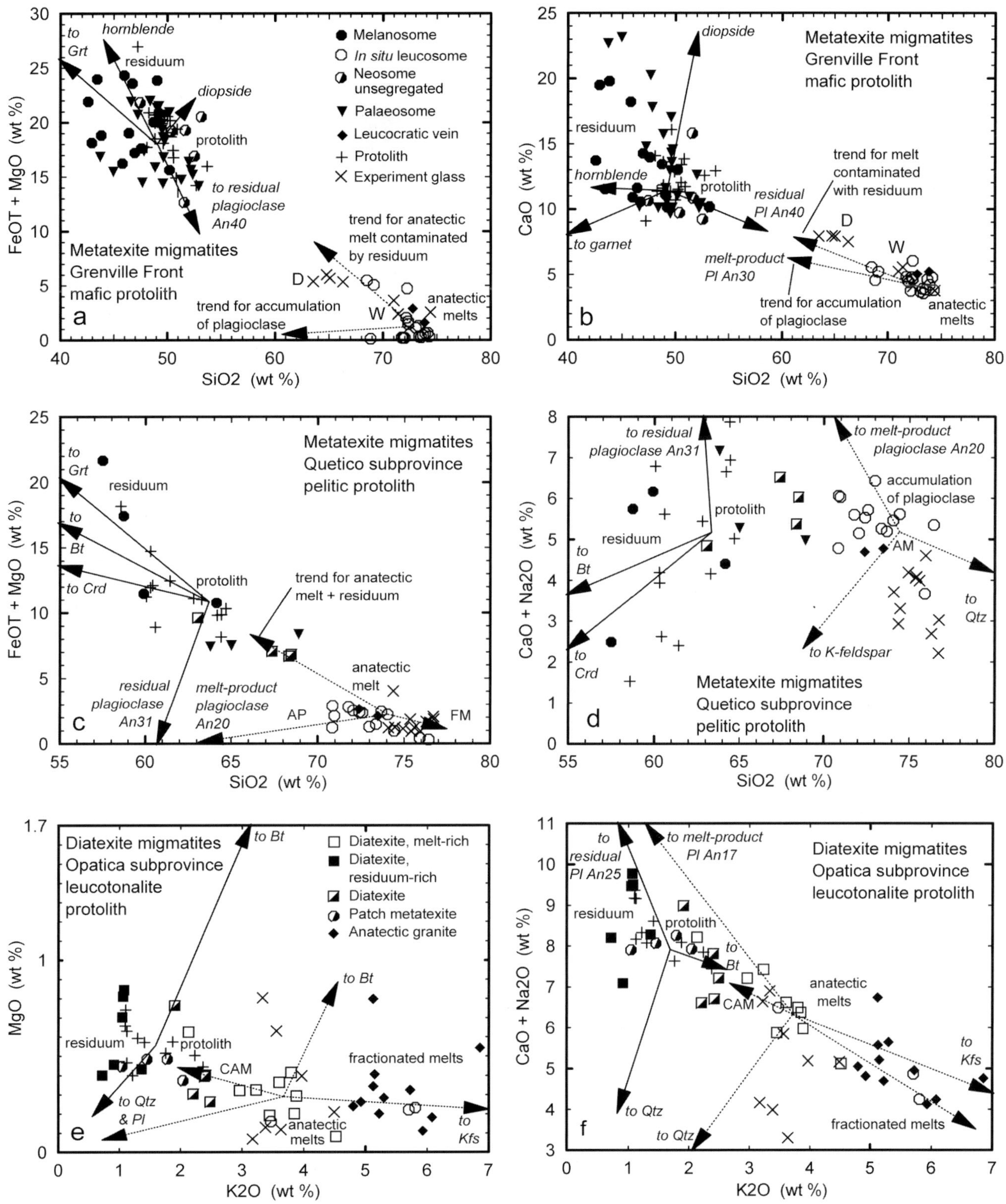
a
Metatexite migmatites
Grenville Front
mafic protolith
FeOT + MgO (wt %)
SiO2 (wt %)
to Grt
hornblende
residuum
diopside
protolith
to residual plagioclase An40
Melanosome
In situ leucosome
Neosome unsegregated
Palaeosome
Leucocratic vein
Protolith
Experiment glass
trend for anatectic melt contaminated by residuum
D
W
anatectic melts
trend for accumulation of plagioclase
b
Metatexite migmatites
Grenville Front
mafic protolith
CaO (wt %)
diopside
residuum
hornblende
protolith
residual Pl An40
to garnet
trend for melt contaminated with residuum
melt-product Pl An30
trend for accumulation of plagioclase
anatectic melts
c
Metatexite migmatites
Quetico subprovince
pelitic protolith
to Grt
residuum
to Bt
to Crd
protolith
trend for anatectic melt + residuum
anatectic melt
residual plagioclase An31
melt-product plagioclase An20
AP
FM
d
to residual plagioclase An31
to melt-product plagioclase An20
accumulation of plagioclase
protolith
residuum
AM
to Bt
to Crd
to Qtz
to K-feldspar
Metatexite migmatites
Quetico subprovince
pelitic protolith
CaO + Na2O (wt %)
e
Diatexite migmatites
Opatica subprovince
leucotonalite protolith
MgO (wt %)
K2O (wt %)
to Bt
Diatexite, melt-rich
Diatexite, residuum-rich
Diatexite
Patch metatexite
Anatectic granite
residuum
protolith
CAM
to Qtz & Pl
fractionated melts
anatectic melts
to Kfs
f
to residual Pl An25
to melt-product Pl An17
Diatexite migmatites
Opatica subprovince
leucotonalite protolith
residuum
protolith
to Bt
CAM
anatectic melts
to Qtz
to Kfs
fractionated melts

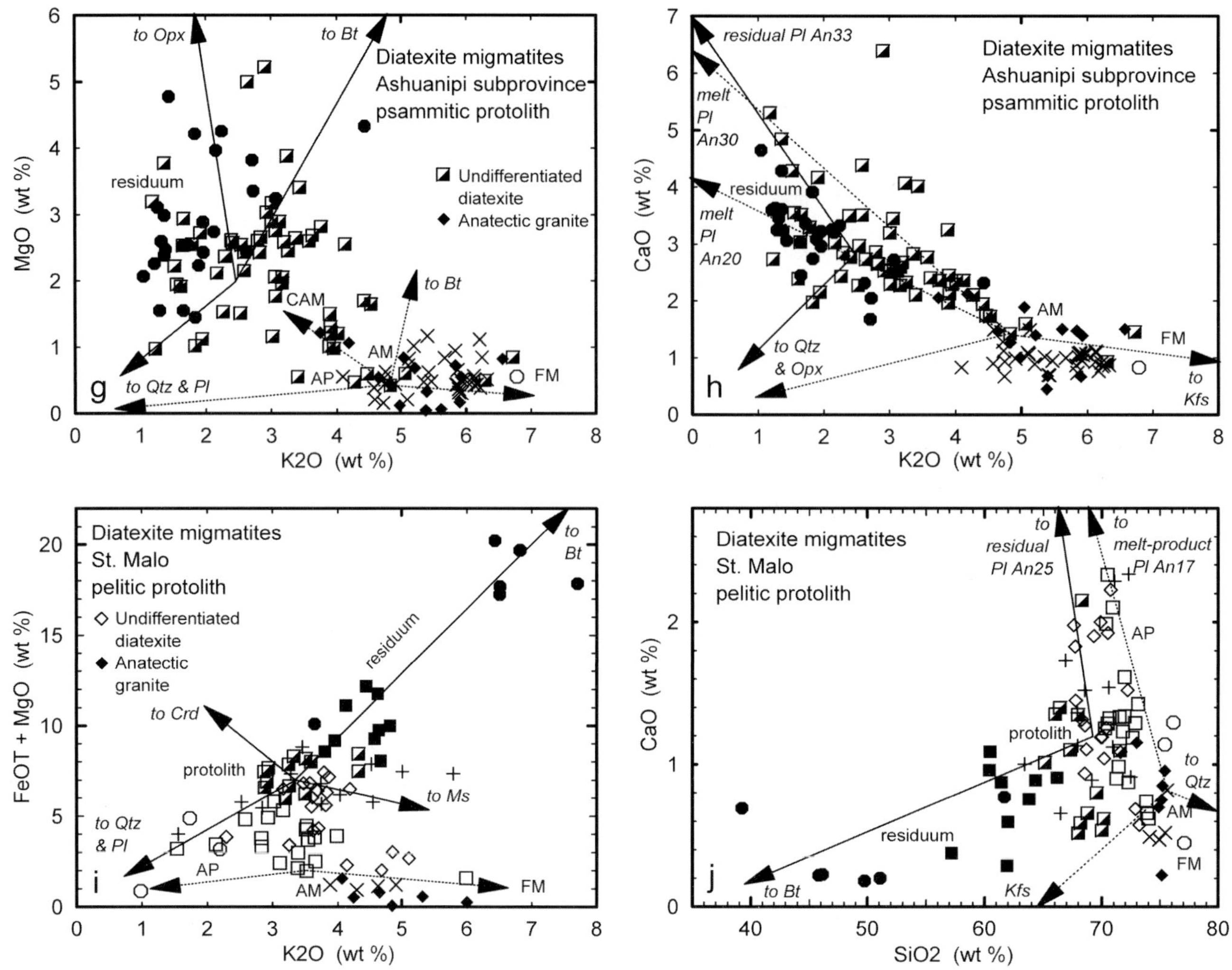

Fig. 8

protolith for the migmatite terrane. The orientation of the melt-residuum trend is controlled by which minerals are abundant in the residuum, and this is a function of both the composition of the protolith and the pressure–temperature conditions at which partial melting took place. A set of vectors (solid arrows) can be drawn for each protolith composition to each residual mineral; these vectors should contain the residuum composition derived from that protolith. For the sake of clarity, Fig. 8 only shows the vectors to the residual minerals from one typical protolith composition. The tip of the arrow rests on the mineral composition; if the mineral composition is not at the arrow (e.g., it lies off the plot), however, then the arrow is labeled "to." The melanocratic parts of the migmatite derived from a metamafic protolith shown in Figs. 8*a* and *b* are depleted in plagioclase, but are enriched in garnet and diopside relative to the protolith, which is consistent with a residual origin. The protolith field for the Saint-Malo migmatites is elongate in a direction approximately along the muscovite vector, but the samples of residuum from Saint-Malo follow the biotite trend (Figs. 8*i*, *j*), because anatexis resulted from the breakdown of muscovite, and biotite remained a stable mineral. However, the residuum trend for biotite dehydration melting must lie in the direction of either the orthopyroxene, cordierite, or garnet vectors, plus plagioclase, quartz, and K-feldspar, if they are present. For example, the residuum compositions (plagioclase + orthopyroxene ± biotite ± quartz) for the granulite-facies migmatites in the Ashuanipi Subprovince lie between the orthopyroxene and plagioclase vectors on Figs. 8*g* and *h*. Orthopyroxene-rich residua can commonly be distinguished from cordierite- and garnet-rich ones.

If the protolith field is well defined, in addition to the position of the residuum and melt relative to the protolith, then open- or closed-system behavior can be identified, and this is particularly useful in interpreting the processes that occurred in migmatites. Samples of neosome (e.g., the unsegregated neosome on Figs. 8*a* and *b*, and the patch metatexite on Figs. 8e and *f*) that plot within the field of their respective protoliths may have formed in a closed system, and are most probably generated *in situ*, which should be determined by field observation at the time of sampling. In contrast, the diatexite migmatites in Figs. 8e and *f* define a field that is far larger than that of the protolith, a feature that may indicate open-system behavior. In particular, some of the diatexite samples plot outside the protolith and in, or close, to the field of anatectic melt, whereas others plot beyond the protolith on the side opposite that of the anatectic melt; these two groups of diatexite samples are interpreted to have gained melt (i.e., they are melt-rich diatexites) and to have lost melt (i.e., they are residuum-rich diatexites), respectively. In contrast, the patch metatexites and some of the diatexite migmatites from the same migmatite terrane (Figs. 8e, *f*) plot within the protolith field and can, therefore, be interpreted to have neither lost, nor gained, anatectic melt, i.e., to have formed in a closed system.

The position of migmatite samples relative to the residuum, protolith, and melt fields has also been used to identify samples infiltrated (Sawyer et al. 1999) or veined (Guernina and Sawyer 2003) by leucocratic melt. A position to the K_2O-enriched side of the melt–protolith tie-line could indicate that the injected melt was a fractionated one.

The various processes such as contamination and fractional crystallization that affect the anatectic melt can also be investigated on these diagrams. Some trends or vectors useful in identifying processes are shown as dotted arrows emanating from a single "typical" initial composition of anatectic melt in Fig. 8; it should not be forgotten that similar vectors apply for all the compositions of anatectic melt. Using the melt–protolith–residuum tie-line as a reference, melt-derived rocks, such as diatexite migmatites and leucosome, that have become contaminated by the residuum, or perhaps by particular residual phases, can be distinguished by a trend (CAM) toward the residuum, e.g., in Figs. 8*a*, *b*, and *g*.

Plagioclase is one of the first minerals to crystallize abundantly from a granitic magma, and Fig. 8 contains a vector from the composition of a representative melt to the plagioclase that crystallized from it. Samples that lie along this trend (AP) may represent accumulations of plagioclase. Note that the plagioclase in the residuum and the cumulate plagioclase in the melt-derived rocks do not, generally, have the same composition. Many of the domains of leucosome (Fig. 8*d*, but see also Figs. 5 and 6) and diatexite migmatites (Figs. 8*g*–*j*) in migmatite terranes are accumulations of plagioclase that crystallized from the anatectic melt. If the plagioclase that crystallized first is separated, then the remaining melt has a more evolved, or fractionated, bulk composition relative to that of the initial melt. Rocks that crystallized from evolved melts are typically enriched in K-feldspar and quartz, and lie along a fractionated melt trend (FM) that lies between these two minerals. In some cases, leucosome (e.g., Figs. 8*d*–*f*) and diatexite migmatites (Figs. 8*g*–*j*) have compositions consistent with crystallization from a fractionated melt. Figure 8 shows that most of the small leucocratic veins or dikes and small bodies of anatectic granite that are a common, although not necessarily abundant, component of migmatite terranes, have compositions indicating that they crystallized from an evolved anatectic melt.

7.

MIGMATITE-LIKE ROCKS

Migmatites are produced by partial melting, and this is why they have complex structures and striking appearances. The anatectic melt has a lower viscosity than nonmelted rock, and this enables the melt to segregate from its residuum and to collect in low-pressure sites during deformation. This separation creates the light-colored (leucosome) and dark-colored (residuum) parts of migmatites. The differences in the fraction of melt from place to place in a migmatite results in a wide range in rock strength (or competency). Consequently, migmatites deform heterogeneously and develop complex morphologies, as shown in the illustrations of this book. The presence of the segregated leucocratic domains in the darker, more mafic-looking host accounts for the etymological root of migmatite, meaning "mixed rock."

However, there are processes other than partial melting that can produce rocks that have light-colored patches, layers, and veins in a darker host and complex morphologies. One is subsolidus segregation in which the mobile (generally felsic) components are separated from the less mobile (generally mafic) ones in a rock, and another is the injection of veins of felsic magma into a host of another composition, as occurs in syntectonic plutons and vein complexes. Unfortunately, some investigators have focused only on the "mixed rock" aspect of light-colored veins in a darker host when calling their rocks migmatites rather than consider the other requirements, that migmatites are produced by partial melting and are, therefore, found (only) in medium- and high-grade metamorphic areas. Consequently, some rocks have been called migmatite that are not. Therefore, as an aid to the identification of migmatites, some of the rocks that are commonly mistaken for migmatites and some criteria by which they can be distinguished from migmatites, are described in this section. Examples of rocks that can be mistaken for migmatites are illustrated in section G of the book.

7.1 Rocks formed by subsolidus segregation

Metamorphic differentiation is the general term, not specific to any grade of metamorphism, used to describe the movement of material in a rock during metamorphism; the process, or processes, whereby the material is mobilized is not specified. For Holmquist (1920) and Eskola (1932), metamorphic differentiation was accomplished by the movement of anatectic melt in high-grade rocks, whereas Turner (1941) argued that in low-grade chlorite schists, it was achieved by the movement of material through an intergranular pore-fluid. **Subsolidus segregation**, or **subsolidus differentiation**, refers specifically to the movement of material in metamorphic rocks by processes not involving a partial melt.

Many investigators (e.g., Eskola 1939, Turner 1941, Ramberg 1952, Williams 1972, Robin 1979) have noted the close association in low-grade metamorphic rocks of light-colored, quartz-rich segregations with dark-colored, mica-rich bands. Virtually all investigators show that planar subsolidus segregations grew during periods of penetrative deformation (e.g., Barr 1985; McLellan 1983, 1984, 1989; Sawyer and Robin 1986), which suggests that they result from a stress-driven phenomenon.

7.2 Models for the process of subsolidus segregation

There have been many investigations (e.g., Strömgård 1973; Rutter 1976; Robin 1978, 1979; Fletcher 1982; van der Molen 1985; Wheeler 1987) of stress-induced chemical transfer that address the mechanism of subsolidus segregation in rocks. Some investigators consider grain aggregates of one mineral species, whereas others consider layers of different compositions.

In an aggregate of grains subject to a differential stress, these differences in stress exist along grain interfaces because of the variation in the orientation of the grain boundaries relative to the principal direction of compressive stress. Such variation establishes chemical potential gradients that drive the "dissolution" of "soluble" material from the points of highest stress, and its movement to the low-stress interfaces, where it "precipitates" and leads to the growth of a new mineral or minerals. In polymineralic rocks, quartz, the feldspars, and carbonate are generally more susceptible to stress-induced chemical transfer, and hence are more "mobile" or "soluble," than the micas, amphiboles, and accessory phases, which are regarded as "insoluble."

Robin (1979) developed a model for subsolidus segregation between layers of different ratio of mica to quartz, and hence of different competency or viscosity. The model is equivalent to a sequence containing beds of different modal mineralogy. During deformation, differences in mean stresses between layers set up gradients in chemical potential, which drive the movement of the more "mobile" material (quartz and feldspar) to the most competent layers (richest in strong minerals, e.g., quartz and feldspar), in a process that accentuates the initial contrasts in competency inherited from sedimentary deposition (Sawyer and Robin 1986).

The grain and the layer models show that a driving force for dissolution and movement of material exists if the grains are subject to differential stress, but there are different mechanisms of transport by which the "mobile" material can move from sites of high to low stress (Walther and Wood 1983). From slowest to fastest, these are (1) diffusion through mineral grains, (2) diffusion along dry grain boundaries, (3) diffusion through a fluid film along a grain boundary, (4) solution in, and diffusion through, a static pore-fluid, and (5) solution into an intergranular pore-fluid that is advecting.

Models have also been developed for the formation of subsolidus segregations under conditions of hydrostatic stress (e.g., Fisher 1973, 1978; Joesten 1977) to describe the diffusion-controlled growth of reaction-rim microstructures and the growth of concretions. The driving forces in these isobaric and isothermal models are the chemical potential gradients resulting from small-scale variations in the mineral assemblages.

7.3 *P–T* conditions at which subsolidus segregation occurs

Subsolidus segregations form at low metamorphic grades and are commonly reported from metasedimentary rocks in greenschist-facies metamorphic terranes (e.g., Turner 1941, Vidale 1974, Sawyer and Robin 1986). The conditions of maximum grade of metamorphism at which they form are well constrained in several regional metamorphic terranes. The subsolidus segregations that give rise to stromatic, migmatite-like schists in central Massachusetts formed at temperatures between 550 and 625°C (Tracy 1985). In northeastern Scotland, the highest-grade subsolidus segregations occur between the kyanite and sillimanite isograds, which corresponds to temperatures between 550 and 600°C and a pressure of 6 kbar (McLellan 1989); Barr (1985) estimated that the temperatures were a little higher, 580–640°C, but still subsolidus. In the Quetico Subprovince of Canada, the highest-grade subsolidus segregations formed in rocks that reached temperatures close to

650°C and pressures of 3–4 kbar (Sawyer and Robin 1986), very close to where the first scattered, anatectic neosome starts to appear in metapelitic rocks. Blom (1988) studied amphibolite- and granulite-facies rocks in Finland, and found that the processes of subsolidus segregation operated only in the amphibolite-facies part of the region's metamorphic history, between 500 and 640°C (at about 5 kbar); above the amphibolite facies, they were simply heated up to about 800°C without any further growth.

The maximum temperatures at which subsolidus segregations formed by diffusion-controlled processes in a hydrostatic stress regime are not as well constrained. Trumbull (1988) reported that small, equant, subsolidus segregations in biotite and biotite + hornblende gneisses formed at a temperature between 600 and 700°C. Lindh et al. (1984) reported temperatures of 550–600°C for almost identical, equant, diffusion-controlled segregations.

7.4 The relationship between subsolidus segregation and migmatites

All the data presently available indicate that stress-driven subsolidus segregations grow in the low- and medium-grade parts of metamorphic terranes, and not in the high-grade parts. Therefore, rocks formed by stress-driven subsolidus segregation do not fall within Ashworth's (1985) definition of migmatite, and because they are not formed by partial melting, they do not fall within the definition of migmatite proposed in this book; consequently, they are not considered to be migmatites.

Subsolidus segregations consisting of quartz-rich veins and pods surrounded by melanocratic selvedges (see next section for a fuller description) may be found in many migmatite terranes. What then is the relationship between the process of subsolidus segregation and migmatites? Blom (1988) suggested that subsolidus segregations are not expected in migmatites and should, therefore, be regarded as part of the pre-anatectic history of the rock, i.e., they are part of the paleosome. Why should subsolidus segregations seem to stop growing when stress gradients still exist? An analogy with deformation-mechanism maps may provide an insight. Space on the maps (e.g., Ashby 1972, Poirier 1985) is divided on the basis of the fastest, that is, dominant, but not exclusive, mechanism of deformation, because it controls the rate at which bulk deformation can proceed. For processes occurring in high-grade metamorphic rocks, the important parameter may be rate of transport; it controls the rate of petrological change. Once partial melting begins, the melt offers by far the fastest rate of transport by which petrological changes can be accomplished, because diffusion in a melt is faster than in a solid. Moreover, melt is easily advected. The process of subsolidus segregation may not cease, but it is simply too slow to contribute to the petrological changes that occur once melting starts. Possibly, subsolidus segregation could occur in those layers not undergoing partial melting, but if these layers become fluid-absent because of melting nearby, then transport in them will be greatly inhibited, and only short-range diffusion-induced segregation of the type described by Fisher (1973, 1978) and Trumbull (1988) may operate, producing small, scattered leucocratic patches called **flecks**.

Because the process of subsolidus segregation separates the "mobile" quartz and feldspar from the "immobile" micas, it can produce rocks that resemble migmatites. Furthermore, since subsolidus segregations that formed at lower metamorphic grades are commonly preserved in the paleosome parts of migmatites, it is clearly desirable that subsolidus segregations be distinguished from neosome in migmatite. In the next section, I outline the principal features of subsolidus segregations that enable this to be done.

7.5 Small-scale features of subsolidus segregations

The mineralogical, microstructural, and morphological characteristics of subsolidus segregations are best known for those developed from metapelitic and metagreywacke protoliths. Consequently, these will serve as the type example.

The constituent parts

Two basic types of internal structure have been described for subsolidus segregations developed in metapelite and metagreywacke protoliths. The most common type is formed by stress-induced chemical transfer; they are thin, laterally extensive, high-aspect-ratio segregations. They have an inner leucocratic part that is bordered by an outer melanocratic rim, beyond which is the country rock (McLellan 1983, 1989; Sawyer and Robin 1986; Blom 1988). In most cases, the central, leucocratic part is comparatively homogeneous and dominated by quartz, but others are compositionally zoned (typically a quartz-rich core and a more feldspathic outer part) owing to the accretion of successive increments of different composition during prograde metamorphism. The dark-colored borders, called melanocratic selvedges, are rarely uniform; typically, there is a progressive decrease in the proportion of quartz and plagioclase that produces a progressive darkening toward the leucocratic core. The second type is much rarer and is formed without differential stresses. These diffusion-controlled subsolidus segregations are typically small (<20 mm) and pod-shaped; they have a melanocratic core surrounded by a leucocratic mantle, with the enveloping country-rock beyond that (Lindh et al. 1984, Trumbull 1988).

The ratio of leucocratic to melanocratic parts in the stress-induced segregations is a function of volume fraction of "mobile" minerals. A protolith with more quartz and feldspar will have a greater ratio of leucocratic to melanocratic parts in the segregation than a protolith with a smaller volume fraction of "mobile" minerals. For diffusion-controlled segregations formed under hydrostatic conditions, the ratio is fixed by the overall reaction (e.g., Lindh et al. 1984, Trumbull 1988).

Mineralogy of subsolidus segregation

Veins formed by subsolidus segregation in the lowest-grade clastic metasedimentary rocks consist almost entirely of quartz, but the plagioclase to quartz ratio in the veins increases systematically with metamorphic grade from the lower greenschist to the middle amphibolite facies (Vidale 1974, Sawyer and Robin 1986). Furthermore, Sawyer and Robin (1986) noted an increase in the plagioclase to quartz ratio in successive increments of material added to individual segregations as they grew during conditions of rising metamorphic temperature. The changing plagioclase to quartz ratio might be due to the depletion of quartz near the segregation, which increases the transport distance of quartz and so reduces its effective "mobility," or to an increase in the solubility of plagioclase faster than that of quartz as metamorphic temperature rises.

Most investigators have found the same mineral parageneses in each part of the segregation as in the protolith; the modal proportions differ greatly, however. In general, there is very little variation in the composition of minerals from the various parts of the segregation and the adjacent protolith. This property contrasts with partial melting, where melanosome and leucosome in migmatites commonly have different assemblages of minerals. In the majority of subsolidus segregations, the leucocratic parts contain mostly quartz + plagioclase, some consist of quartz + K-feldspar; consequently, they tend to plot far from the cotectic lines in the system Ab–An–Or–Qtz–H_2O on CIPW norm diagrams. The melanocratic borders are characterized by an enrichment of "insoluble" minerals, typically biotite and the accessory minerals, relative to the protolith. Garnet, staurolite, hornblende, magnetite, andalusite, kyanite, and sillimanite also may occur in the mafic selvedges. Orthopyroxene has not been reported from subsolidus segregations.

Different assemblages of minerals can develop in the leucocratic and melanocratic parts of the subsolidus segregations if the melanocratic part becomes depleted in one of the "mobile" minerals, generally quartz. Different assemblages can also develop if a metamorphic dehydration-type reaction occurs as the segregation grows and the "immobile" phases produced remain in the melanocratic part, but the "mobile" phase produced appears in the part that is leucocratic. Curiously, all the reported cases (Lindh et al. 1984, Sawyer and Robin 1986, Trumbull 1988) seem to involve the same basic reaction: Fe-rich biotite = magnetite + K-feldspar + fluid, although in some cases, hornblende also is a product phase.

Results from mass-balance studies, whether based on modal analyses or on whole-rock compositions, show that subsolidus segregations generally form *in situ* by an essentially isochemical process (Sawyer and Robin 1986, Sawyer and Barnes 1988, McLellan 1989, Blom 1988, Trumbull 1988). Relative to the protolith, the melanocratic selvedge represents the *in situ* residue of insoluble or immobile material from which the soluble or mobile components that comprise the leucocratic part were extracted.

Microstructure

The leucocratic parts are coarser grained than the melanocratic domains, and they are coarser than the protolith, unless reduced by later mylonitization or cataclasis. Although the leucocratic parts are coarse-grained, their microstructure is metamorphic and dominated by equant grains; they do not contain pegmatite, or other igneous textures and microstructures (McLellan 1983). Minerals in the melanocratic selvedges generally have a strong preferred orientation; quartz and plagioclase have tabular shapes controlled by the preferred orientation of the more abundant biotite. The overall microstructure is, therefore, lepidoblastic or nematoblastic, and is progressively more strongly developed toward the melanocratic part of the segregation, i.e., with progressive loss of mobile minerals (mostly quartz).

7.6 Outcrop-scale morphology

The morphology of subsolidus segregations is controlled by their bulk rheology, which remains that of solid rock. This represents the major difference between rocks formed by subsolidus segregation and migmatites. Structures due to magmatic flow thus do not develop, and older structures are widely preserved. Furthermore, because of their quartz-rich mineralogy, the leucocratic parts of subsolidus segregations are the most competent part of the rock; consequently, they commonly become boudinaged during growth, in contrast to the leucosome in migmatites.

Stromatic, or layered, subsolidus segregations are characterized by thin (mm–cm wide) but laterally extensive segregations (meters or more) located in the plane of mechanical anisotropy in the host, either the foliation planes or the compositional layering (see Figs. G1 and G2). The segregations comprise of two parts: a quartz-rich, veinlike core and an outer melanocratic border. The segregations are concordant to host-rock foliation or bedding, and not discordant (Yardley 1978; Barr 1985; McLellan 1983, 1984; Sawyer and Barnes 1988). Single-layer segregations are common, but the most striking layered structure develops when numerous thin, parallel segregations develop in protoliths that either were originally thinly bedded or else contain a closely spaced cleavage or foliation. Layered subsolidus segregations generally do not form in thick sequences of homogeneous rocks. The overall appearance of the outcrops, therefore, depends on the distribution of layers of a suitable protolith.

Dilatant structures, such as interboudin partitions, can be favorable places for subsolidus segregations to develop. Typically, the "mobile" quartzofeldspathic constituents migrate to the interboudin partition, leaving the "immobile" minerals to form a melanocratic border in the adjacent wallrock. Hence, the leucocratic parts can be discordant to the foliation in the host, but are quite short. The formation of this type of segregation is analogous to pressure shadows (Strömgård 1973). The appearance of outcrops containing subsolidus segregations and of metatexite migmatites in which leucosome is located in dilatant structures can be similar; they can be distinguished by differences in the microstructure and mineral assemblages.

Fleck structures are small, round subsolidus segregations (Figs. G3, G4) that commonly have a melanocratic core surrounded by a leucocratic rim, or mantle (Lindh et al. 1984, Trumbull 1988). Trumbull (1988) suggested that flecks only develop in protoliths of certain bulk compositions (notably K-enriched), and are randomly distributed in their host. Their location is not directly controlled by planes of mechanical anisotropy, such as foliation. The equant shape of the flecks and the lack of a penetrative foliation in some hosts suggested to Lindh et al. (1984) that the fleck structure is an example of diffusion-controlled subsolidus segregation formed in the absence of differential stress; hence, they resemble patch-type migmatites. Past use of the term *fleck* has been reserved for segregations of subsolidus origin; this practice should be continued.

7.7 Rocks formed in syntectonic plutons and plutonic complexes

The injection of one magma into another of different composition can produce a wide variety of magma-mingling and magma-mixing structures that have been reported from a large number of plutonic complexes (e.g., Mason 1985, Barbarin 1988). Some of these rocks can be mistaken for diatexite migmatites if the lobate and cuspate boundaries to the enclaves are not found; commonly, they may be lost as a result of high strain. The injection of one magma into another of different composition that has already crystallized sufficiently (but not completely) that it will fracture during intrusion of the second can produce net-vein arrays (e.g., Morogan and Sørensen 1994) that may resemble some net-structured metatexite migmatites.

The widest range of migmatite-like igneous rocks are found where magmas of felsic or intermediate composition were either injected into, or crystallized in, places of active penetrative deformation. These correspond to regimes of syntectonic emplacement and syntectonic crystallization of magma. The similarity in appearance between igneous rocks produced under these circumstances and migmatites is not surprising. A crystallizing pluton is a two-phase system (melt + solid) like a migmatite, and the melt fraction can separate from the solid if there is an applied stress. Once the magma has crystallized sufficiently ($M_f < 0.4$) that a rigid skeleton of crystals is formed, subsequent deformation of the magma produces structures (see Figs. G5–G8) that resemble those that develop where partially melted rocks are deformed during anatexis (e.g., Fig. 2). The point at which a rigid framework of crystals is formed as magma crystallizes is referred to as the rigid percolation threshold (RPT) by Vigneresse et al. (1996). In migmatites, the particular geometry and morphology of the leucosome are controlled by the planar anisotropy of the protolith and the paleosome, which may be of primary (e.g., bedding) or tectonic origin, and by strain. The location of felsic veins in syntectonic plutons is also controlled by the anisotropy of the host and strain, but in this case, the anisotropy is generally a foliation acquired by shearing in the submagmatic state.

Syntectonic injection of magma

Subparallel bands of lighter- and darker-colored rocks that resemble stromatic migmatites can form where magmas of different compositions are injected into zones of high shear strain in plutons. Neves and Vauchez (1995) reported that a stromatic-like alternation of rocks of different compositions and microstructures developed only where a synmagmatic shear zone traverses the plutonic complex they studied in Brazil. They attributed the stromatic layering to the attenuation and disruption of successive, compositionally different batches of magma as they were intruded into the shear zone.

Rocks that resemble stromatic migmatites can also form by the repeated diking of a partially crystallized, darker magmatic host-rock by a more felsic magma during noncoaxial deformation (see Figs. G9 and G10). Shearing in the magmatic and submagmatic states orients the platy and tabular minerals in the darker magma, and this fabric can then control the orientation of subsequent dikes of the more felsic magmas (e.g., John and Stunitz 1997) to produce compositionally layered rocks that resemble stromatic migmatites. In some examples, the felsic magma is genetically different from its host. However, in others it is the fractionated, residual melt expelled from its darker host by shear-induced compaction of its crystal framework in the submagmatic state (Cuney et al. 1990, Pons et al. 1995).

If the shear zone remained active after complete crystallization, then the magmatic and submagmatic fabrics in the darker host and the felsic dikes are overprinted by solid-state mylonitic fabrics, and all the felsic dikes show some degree of rotation into parallelism with the trend of the shear zone. However, if the shear zone ceased to be active before complete crystallization occurred, then the darker host and early light-colored felsic dikes preserve submagmatic fabrics, and the late dikes cross-cut the earlier ones and have more igneous microstructures (e.g., the Skagit Gneiss and rocks in the nearby Chelan area of Washington). Babcock and Misch (1989) showed a CIPW norm plot in which the late, cross-cutting dikes plot on the cotectic line, whereas the lighter-colored, stromatic layers plot nearby, but are displaced slightly toward the albite apex. This relationship could be used to infer that the lighter-colored layers in the Skagit gneiss are in fact magmatic, and result from the syntectonic crystallization of a magma that resembled the late dikes from which a residual, evolved melt was expelled after the RPT was reached, leaving behind light-colored, plagioclase-dominant felsic dikes.

Syntectonic crystallization of felsic plutonic rocks

A variety of nonstromatic migmatite-like structures can develop in compositionally uniform granitic plutons undergoing noncoaxial deformation during the final stages of crystallization. For magmas of granitic composition, most crystallization occurs over a relatively narrow interval of temperature, just above the solidus. Consequently, the fraction of melt decreases sharply near the solidus, and a crystal framework containing the remaining melt in its interstices forms (Bryon et al. 1994). At fractions of melt below 0.25, the crystal framework deforms, partly by crystal-plastic mechanisms, and partly by fracture in response to applied stresses, so that small (cm-scale) dilatant structures (shear bands, extension gashes, boudins) form in specific orientations relative to the foliation and principal stresses (Pons et al. 1995, Sawyer 2000).

Fractionated, residual melt that is in the interstices of the crystal framework is forced out as porosity is destroyed and the crystal framework compacts, and migrates into the dilatant low-pressure sites that are growing nearby. Hence, the segregation of residual melt can be an essentially *in situ* process, just as in many migmatites. The volume of melt that is segregated and forced to migrate out of the framework of crystals by deformation must decrease as crystallization progresses; therefore, the morphology of the segregations of residual melt changes also. In the early stages, when the local volume of residual melt that has been squeezed out of the framework is large, the domains of expelled melt closely resemble nebulites and diatexite migmatites (e.g., Figs. G11–G14). As the volume of residual melt declines, the orientation and morphology of the individual structures that are filled with residual melt change and are analogous to some types of leucosome that form in metatexite migmatites at the beginning of partial melting (e.g., Figs. G5–G8). If the melt-filled dilatant sites become joined, then the resulting array of light-colored felsic veins can resemble the net and dilatant types of metatexite migmatite. Locally, pooling of expelled residual melt may result in the development of foliation-parallel veins (Sawyer 2000).

7.8 Vein complexes

In some geological settings, metasedimentary or plutonic rocks have become veined with felsic melts and magmas derived from lower structural levels. These rocks can look very much like some migmatites (Figs. G15–G20), particularly if the lighter-colored, felsic veins have narrow mafic selvedges, formed owing to reaction of the wallrock with the magma, or with an aqueous fluid evolved from it, and are mistaken for melanosome. The felsic melts may be derived by partial melting of much deeper rocks or by the syntectonic expulsion of residual melt from deeper crystallizing plutons. However, because the felsic melts and magmas are not genetically related to their hosts, these rocks do not meet the definition of migmatite. Using the same type of reasoning as Brown (1973), these injected rocks should be called vein complexes. If the host rock was a migmatite, then the addition of felsic veins or dikes would make the term "veined migmatite" applicable.

7.9 Rocks formed in syntectonic plutonic and vein complexes compared with migmatites

Similarities

Multiple injections of magma in syntectonic plutonic complexes can produce rocks with a number of migmatite-like characteristics. (1) Light-colored veins in a darker host can resemble either leucosome in a paleosome, or leucosome and melanosome. (2) If the light-colored veins have mafic selvedges formed by reaction with their wallrocks, then the rocks could be mistaken for a migmatite if these are interpreted as leucosome and melanosome, respectively, and the host rock is treated as paleosome. (3) The alternation of parallel and laterally persistent light and dark layers may superficially resemble stromatic metatexite migmatites. (4) The attenuation of structures due to magma mingling and the development of flow banding in zones of high synmagmatic strain can resemble enclaves of paleosome, or melanosome, in schlieric and schollen diatexite migmatites.

The segregation of felsic residual melt by syntectonic deformation at a late stage of pluton crystallization can be a local, *in situ*, process that generates isolated, lighter-colored felsic veins bordered by melt-depletion halos, or more extensive arrays of linked light-colored felsic veins within a broader, melt-depleted zone. Both are morphologically and geometrically similar to leucosome located in dilatant and net-structured metatexite migmatites.

Differences

The most obvious difference between the migmatite-like rocks from syntectonic plutons and true migmatites rests with the genetic relationship between their constituent parts. The definition of migmatite requires that there be a genetic connection through partial melting. In metatexite migmatites, the melanocratic rocks contain petrographic evidence for a partial-melting reaction, and geochemical and petrographic evidence to indicate that the leucosome (or diatexite migmatite) is, or was, derived at least in part from that anatectic melt. In contrast, the darker-colored rocks from migmatite-like rocks in syntectonic plutonic complexes do not contain petrographic evidence for partial-melting reactions, and although the light-colored, felsic rocks might be genetically related to the darker rocks, it is not through partial melting. In most cases, the felsic rocks have compositions that lie at the far end of the fractional crystallization trend of the associated darker-colored rocks, compatible with an origin as expelled residual melt, not as a partial melt. In other cases of syntectonic diking, the light and dark parts are related by magma mixing; in some, they are not genetically related at all.

Complexes having multiple syntectonic dikes have some additional characteristics that are unlike stromatic migmatites. (1) Most of the felsic veins are not surrounded by zones of melanocratic melt-depleted rocks, or they have melt-depletion halos that are far too small to be compatible with *in situ* segregation of melt. (2) Many felsic veins are, in detail, cross-cutting. (3) The typical width of stromatic felsic veins is much greater than the width of the leucosome domains in stromatic migmatites. (4) A significant proportion of the felsic veins in most complexes have true pegmatitic textures, whereas a true pegmatitic texture is rare in migmatite leucosome. (5) Overall, many migmatite-like complexes of syntectonic dikes have a far higher ratio of felsic veins to darker host than most migmatites. Together, these criteria indicate an origin by diking rather than *in situ* melting and segregation processes.

8.

WORKING WITH MIGMATITES

The final section of this book deals with the problems of working with and mapping migmatites, and these are addressed in two parts. The first deals with the question of what divisions to depict on a map, and the second is concerned with what should be noted at each outcrop during field work that is of use in later, follow-up work in the laboratory. This material is condensed and presented as a checklist in Appendix 9.1.

Migmatites have been studied for many reasons. In the recent past, many studies were in some way or another related to the petrological questions of the origin of granitic magmas and the processes by which the continental crust became compositionally differentiated; these questions remain of great interest. In geodynamics, there is presently much interest in the rheology of the continental crust, and the effect that weak layers in it have on the evolution of orogens and on the morphology of mountain chains and plateaus. As zones of partially molten rocks are candidates for weak layers in modern orogens, new studies on migmatites will no doubt focus on structural and rheological aspects as a means to study the distribution and geometry of former weak layers and zones of so-called "channel flow" in ancient, eroded orogens. No matter what the purpose is for studying migmatites, some part of the work will likely involve examining and mapping them in the field.

The task is to decide what is critical to observe and map in the field, what needs to be sampled, and which aspects of the rocks that crop out at the surface should be portrayed on a map. Not only does this entail showing the location and distribution of the various migmatites and the adjacent rocks, but also the internal features and relationships within the migmatites. The choice of what units to show is critical to the usefulness of the map, and although that depends to a large extent on the size of the map area and the scale of the final map, the principal ingredient to success is a clear understanding of what the purpose and objectives of the study are before field work begins.

8.1 First-level map units

For regional-scale maps, the map units are normally the petrologically distinguishable rock-types. For migmatites, these should be the first-order morphological divisions, i.e., metatexite and diatexite with, if necessary, a transitional migmatite unit between them. Other units that could be shown include the protolith (or whatever is on the low-grade side of the migmatites) and the nonmigmatite rocks on the high-grade side, most commonly, plutonic rocks of one sort or another. Paleosome, or resister, lithologies could be differentiated and shown within the migmatites, if scale permits. Resister lithologies can be an effective way of showing some types of pre-anatectic structures in migmatites, notably folds, and how their geometry may have changed from the non-anatectic rocks to the migmatites and from low to high grade within the migmatites. The first appearance of metatexite migmatites and the passage from metatexite to diatexite are easily identified and mapped using the definitions of both given earlier. Whether or not these are parallel to isograds in the lower-grade rocks may provide critical information in determining why melting began or whether the metatexite–diatexite transition is structurally controlled or not. However, the upper limit of diatexite migmatites can be difficult to define, especially if, as in many regional migmatite terranes, the diatexite migmatites tend to grade into anatectic granites that are largely *in situ* (autochthonous granite). Placing the boundary in such cases is subjective, and depends on the individual mapper's tolerance to uniformity in diatexite, or to heterogeneity in granites. The upper limit of diatexite migmatites in contact aureoles is commonly much easier to map, even where against granite, because migmatites in contact aureoles generally have a much finer grain-size than the rocks in felsic plutons, or migmatites from regional metamorphic terranes.

In terranes where the protolith is compositionally diverse, further divisions of both the protolith and migmatites may be possible. If a particular protolith generates migmatites of sufficiently different appearance or mineral assemblage that they can be distinguished and mapped separately, then that should also be shown on the map. The additional information concerning the protolith type can be added as a prefix to the migmatite type to give map units such as pelitic metatexite, pelitic diatexite; psammitic metatexite, psammitic diatexite; granitic metatexite and granitic diatexite. Some workers have distinguished the parts of migmatites derived from cordierite-bearing melt from those derived from garnet- and orthopyroxene-bearing melts (Johnson et al. 2001*b*), or separated biotite diatexites from orthopyroxene-bearing diatexites (Mäkitie 2001).

Choosing map units on the basis of the percentage of leucosome and leucocratic veins present is not recommended for first-level maps, because of the possibility that some of the "leucosome" is in fact later vein material, superimposed on some other migmatite morphology, as discussed in the section on vein-structured migmatites. The distinction between metatexite and diatexite, a division that already incorporates the fraction of melt present in the migmatite, is preferred.

8.2 Second-level map units

This level of map units is intended to add a further level of detail to the metatexite and diatexite divisions already made on the regional scale, but they could be the primary divisions on a map legend under some circumstances. For example, in the detailed mapping of smaller areas, the scientific objectives and the opportunities are different compared to those in regional mapping. Small areas in migmatite terranes are typically mapped in detail to examine some specific petrological or structural problem, such as how the local variations in morphology of migmatites are spatially related to the structures (e.g., Oliver and Barr 1997, Brown and Solar 1998*a*, *b*), because deformation appears to control the migration and accumulation of melt. The mapping may be required as the basis for a petrogenetic or geochemical study in which case it should be concerned with obtaining as much information as possible on how and where each individual sample collected fits into the overall migmatite, in other words, to provide context.

For mapping of areas where the variation in migmatite morphology is related to anisotropy in the protolith or the paleosome, or to deformation style and strain intensity, as well as to fraction of melt present, then the second-order morphological terms in Fig. 1*b* (plus vein-structured migmatite) are very useful. If used together with the systematic mapping of regional structures, such as folds, shear zones, and large boudins, the distribution of these migmatite morphologies is a powerful aid in understanding the movement of melt within anatectic terranes and orogenic belts; they lead to the identification of the domains where net loss of melt or net gain of anatectic melt occurred. Zones of relatively high syn-anatectic strain may be distinguished from regions of lower-strain by the morphology of the migmatites present; for example, Brown and Solar (1998*a*, *b*) found that stromatic migmatites are characteristic of regions of higher strain. At these scales, variation in the proportion of leucosome from outcrop to outcrop may be a very useful parameter to show, although it would be equally useful to know whether this was the "initial" anatectic melt, an evolved or fractionated melt, or an accumulation of early-crystallizing phases from the anatectic melt.

For the most detailed level of mapping done at the scale of individual outcrops, simple units that are directly related to petrology, e.g., leucosome for melt product, melanosome for residuum, and paleosome for nonpartially melted rocks in a migmatite, can be particularly effective in revealing the spatial details of the processes that combine to form migmatites (e.g., Oliver and Barr 1997; Sawyer et al. 1999; Sawyer 2001; Guernina and Sawyer 2003; Marchildon and Brown 2002, 2003). In order to best understand how the migmatites formed, it is very important that the spatial relationship between these petrological features and the tectonic structures in the migmatites are shown.

8.3 Other considerations in mapping migmatites

Determining the relative chronology of events in migmatites is an important first step in understanding how they formed. In most cases, the timing of events in migmatites is determined by the cross-cutting relationships between domains of leucosome or more commonly by determining the age (i.e., the generation) of the structural element in which the leucosome is located. These structures might be foliation planes, axial surfaces, shear bands, or fold hinges. It is not surprising, therefore, that the sequential nomenclature used in structural geology (e.g., D_1, D_2 . . . S_1, S_2 . . . F_1 and F_2, etc.) has influenced the way the formation of leucosome, and by extension, regional partial melting, have been viewed. The erection of a chronology of sequential "generations" of leucosome over a large area, because it occupies a similar structural site within a mapped area, must be done with considerable caution; the formation of leucosome might be diachronous from place to place or even of completely different age between one structural level, or domain, and another. Some investigators have gone as far as to ascribe each "generation" of leucosome in a regional anatectic terrane to a separate melting event. The common observation of one domain of leucosome cutting another, and each located in a different "generation" of structure in contact aureoles where there was only one anatectic event, strongly suggests that this approach may not the best way to interpret the temporal information obtained from outcrops in regional migmatite terranes.

Conduction is the principal mechanism of heat transfer in the continental crust. Because heat is conducted very slowly in rocks, the Earth's crust is slow to heat up and slow to cool. Virtually all the numerical models of crustal heating since that of England and Thompson (1984) show slow cooling in the deep crust, except where it has been exhumed rapidly, e.g., in metamorphic core complexes. The slow response-time of the continental crust means that it is very unlikely that the complex timing deduced from the relationship between domains of leucosome in migmatites is the result of the metamorphic temperature repeatedly cycling back and forth across the solidus to generate multiple partial-melting events. Recent geochronological studies from granulite-facies metamorphic terranes (e.g., Zaleski et al. 1999, Rubatto et al. 2001, Willigers et al. 2001) show that metamorphic temperatures remained above the granite solidus between 30 and 50 My. This suggests that once formed, anatectic melt can exist in the middle and lower crust for a very long period of time. Moreover, 30–50 My is much longer than most deformation events in orogens; thus, it is to be expected that during the period when the middle and lower crust is hot and contains melt, the stress orientations and hence deformation patterns will change.

The sequence in which each "generation" of leucosome forms could reflect each melting reaction that occurred on the prograde *P–T* path; for example, the first could have formed by H_2O-present melting, the next by muscovite dehydration melting, and a third by a biotite-dehydration-melting reaction. Melting could be further spread out in time (and temperature) because each protolith layer has a different composition. However, the problem with this explanation is that all "generations" of leucosome coexisted as anatectic melt during the prograde part of the *P–T* trajectory, and yet remained in their separate sites; neither does this explanation allow for commonly inferred differences in competency between the different "generations" or domains of leucosome. An alternative, and perhaps more likely explanation, is that each successive "generation" of leucosome simply records the next of many episodes in which the melt fraction segregated from the solid fraction, and these are brought about by changes in the local pressure gradients during progressive deformation.

The segregation of melt during the prograde part of the metamorphic history ($T >$ solidus) involves separating the melt fraction from the residuum and the subsequent migration of the anatectic melt down local pressure-gradients with little or no crystallization; much of the network of channels through which the melt moved at this stage may be lost (Sawyer 2001). However, as the melt starts to crystallize, it becomes a magma consisting of fractionated melt + crystals; typically this occurs once temperature starts to decline. If at any time after cooling and crystallization have started, a new episode of deformation occurs (perhaps F_2 folds start to form), both the distribution and magnitude of the pressure gradients will change, and new low-pressure sites are created in the migmatite. The melt fraction in the magma will then be driven to migrate again to the new, lowest-pressure sites and will leave the minerals that have already crystallized behind; they become the first "generation" of leucosome and mark the older, low-pressure sites. Notice that at this stage, melt segregation is not about separating anatectic melt from its residuum; it is now the separation of the evolved melt fraction from the solid fraction that has crystallized from the parental anatectic melt. This segregation process does two things. First, it increases the viscosity of the existing leucosome from that of a magma (crystals suspended in a melt) to that of a framework of crystals, simply by removing the evolved melt from it; the effect of this is to create a mechanically competent leucosome without the temperature having to go below the solidus. Secondly, the migration of the expelled melt fraction to the new low-pressure site creates a new leucosome there that is largely melt and, therefore, of low viscosity; these new domains of leucosome may have a different orientation determined by the stresses that led to the new episode of deformation, an F_2 fold in our example. This segregation process during cooling and crystallization may be repeated several times; each subsequent deformation event creates a new low-pressure site to which the melt migrates to form the next "generation" of leucosome, and it makes the older "generations" of leucosome stiffer because of the loss of melt from them. In this scheme, the oldest batches of leucosome should be plagioclase-rich (i.e., they are accumulations of plagioclase), and the youngest should have the most evolved, or fractionated, compositions, and this is generally what is observed in migmatite terranes. In general, the network of leucosome in regional migmatites marks where melt was crystallizing as temperature declined. However, the array of leucosome in rapidly cooled contact-metamorphic aureoles may record the distribution of melt that existed at temperatures much closer to the peak.

The oldest "generation" of leucosome in many migmatite terranes is located parallel to the compositional layering (commonly attenuated bedding), which is itself parallel to the principal foliation in the rock; together, these loci are commonly described as a composite S_0/S_1 banding. Characteristically, this "leucosome" is thin, contains few mafic minerals, and importantly, has a very fine grain-size that is much less than in the other domains of leucosome in the migmatite. The interpretation of this generation of "leucosome" depends on its petrographic features. If it is plagioclase-rich and does not have a melanosome, then it could have formed from a melt generated by H_2O-present partial melting, and injected into country rocks that had not yet melted. It pre-dates partial melting in the hosts but belongs to the same anatectic event, having formed at a very different time within it; the host melted later when the isotherms moved higher into the crust. In this interpretation, the felsic material forms leucocratic veins but not leucosome; it was competent and fine-grained because it crystallized in cool rocks before the migmatite formed. Examples of fine-grained, layer-parallel, and discordant leucocratic veins formed from a melt generated by fluid-present partial melting and injected into the rocks just below, and just above, the "melt-in" isograd occur at Saint-Malo, France (Brown 1979, Weber et al. 1985, Milord et al. 2001), and in the Quetico Subprovince of Ontario. If the "leucosome" has a melanocratic selvedge and is quartz-rich, then it may be a subsolidus segregation, not anatectic in origin. Alternatively, such early "leucosome" may be leucocratic veins that do not belong to the anatectic event that created the migmatite; in this case, they should be significantly older than the leucosome in the migmatite.

9.

APPENDICES

9.1 Checklist of observations for each outcrop of migmatites

The purpose of the observations outlined below is to properly identify and name the migmatite, and to gather field information critical to interpreting the petrological and geochemical data that will come from later studies in the laboratory.

Observations on the neosome and paleosome

(1) If both neosome and paleosome can be identified, note their relative proportions. Decide whether it is a metatexite or diatexite migmatite; it is a metatexite if paleosome dominates and contiguous pre-anatectic structures are present, but a diatexite if neosome dominates and pre-partial-melting structures are lost.

(2) For metatexites, (*a*) if the neosome has segregated, note the form, size, location, and proportions of the leucosome and the melanosome in the outcrop, and note any cross-cutting relationships between the domains of leucosome. Is there petrological continuity between domains of leucosome? Decide what type of metatexite on the basis of the form of the domains of leucosome. The relationship

between the host and former segregated melt is next; is the leucosome *in situ* or in-source, or is it a leucocratic vein or dike? This information can be critical to the interpretation of geochemical data. (*b*) If the neosome has not segregated, note how its mineralogy, microstructure, and grain size compare with its host. Decide whether or not this is a patch metatexite migmatite. (*c*) For the paleosome, identify the rock types present, their mineralogy and microstructures, and identify the structures that were present before partial melting.

(3) For diatexite migmatites, (*a*) if the neosome is unsegregated and nonfoliated, it is a nebulitic diatexite. (*b*) If the neosome has flow structures or a magmatic foliation, examine the form of the paleosome or melanocratic material; decide if the diatexite migmatite could be a schollen or a schlieric type. (*c*) If possible, note whether the diatexite is leucocratic or melanocratic relative to the "typical" diatexite; describe any leucocratic veins in it. Decide whether it is a melt-rich, residuum-rich, or cumulate-rich diatexite migmatite.

Petrological observations in the study of migmatites

The mineralogy, microstructure, and grain size of each part of the neosome and paleosome are extremely useful and should be recorded; note any variations between and within the different domains of leucosome and melanosome. Field observations should aim to allow one to deduce *P–T* conditions and melting reactions, to identify which parts of the outcrop were melt, solid, and partially melted solid, and to identify the regions of melt loss (residual rocks) and of melt gain (vein migmatites, some diatexites). They will provide a possible aid in the later quantification of M_f to compare with *F*. It is also important to identify the parts of the migmatite where no loss, or gain, of melt occurred, i.e., where the formation of migmatite was essentially a closed-system process.

Structural observations in the study of migmatites

(1) Record, describe, and measure all the structural elements (foliations, lineations, folds, shear zones, displacement senses, etc.), noting their location and their geometrical and timing relationships within each part of the migmatite. (2) Identify the parts with magmatic, or submagmatic, foliations and structures. (3) Identify which parts of the migmatite terrane represent low, relative syn-anatectic strain and which are high, and those where there is significant transposition; the parts of high relative strain or transposition tend to be dominated by layered or stromatic structures in both metatexite and diatexite migmatites. (4) Determine differences in relative competence between the different "generations" of leucosome.

The purpose of these structural observations is to distinguish pre-partial-melting from syn-partial-melting structures and to compare their distribution. Identify the structures that formed after partial melting. Distinguish melt-rich rocks from melt-poor ones. Finally, interpret the structural and tectonic history of the migmatites.

Way-up criteria in migmatites

Determining the way-up direction, or the structural facing direction, in rocks is vital for understanding the structural and tectonic history of deformed terranes. Unfortunately, the structures used to determine these geometric properties are progressively lost during partial melting. Consequently, tectonic reconstruction of migmatite terranes is very difficult, especially in diatexite migmatites. Fortunately, structures that form during the movement and crystallization of anatectic melts and that can be used to determine the paleovertical, and hence way-up in migmatites, have been described by Burg and Vanderhaeghe (1993) and Vanderhaeghe (1999, 2001). However, caution may be required in using some of the way-up criteria, as the local pressure-gradients down which melt moved could have been lateral, or even downward-directed. Branching arrays of leucosome indicate flow direction, but the direction of flow may not have been vertical.

Sampling of migmatites

The objectives of the study need to be clear before sampling begins in order to ensure that appropriate material is collected. Residual material is taken for the determination of the *P–T* conditions, reaction history, reaction progress (i.e., *F*), and for microstructural evidence of partial melting. On the other hand, samples from the melt-derived rocks (leucosome and diatexite) are best for the study of processes such as flow of melt, fractional crystallization, and crystal accumulation. Oriented samples are required for some types of quantitative microstructural analysis.

Geochemical samples should be fresh and from parts of the migmatite that have been fully identified in the field; ideally, they should include some from the protolith. One purpose of the samples is to define the key reference-points (protolith, melt, and residuum) that are necessary for a geochemical interpretation of the dataset. These are also required for most quantitative modeling, such as estimating *F*, making mass-balance calculations, and investigating the various petrogenetic processes. Well-constrained samples serve to guide the interpretation of samples from the more difficult outcrops.

9.2 Glossary

Anatexis: a general term used to describe partial melting of the continental crust (in this case) without specific reference to the degree of partial melting; hence, anatexis applies to all stages, from incipient partial melting right up to complete fusion. Crustal anatexis is generally accompanied by deformation, which enables many other processes to occur, such as segregation of the melt from the solid, migration of the melt (by porous flow and by channeled flow), fractional crystallization, and magma flow.

Anatexite: a rock formed by anatexis and the associated processes; an under-used synonym for migmatite.

Bedded migmatite: a term for the overall morphology of a migmatite in which partial melting and the formation of neosome are confined to certain beds, or layers. The term should be abandoned. *See* Layer-confined migmatite.

Degree of partial melting (F): the weight fraction of melt produced in a rock by melt-producing reactions.

Diatexis: defunct term meaning complete fusion, now replaced in common usage by the more useful term "anatexis."

Diatexite migmatite: a migmatite in which neosome is dominant and melt was pervasively distributed throughout. Pre-partial-melting structures are absent from the neosome, and are commonly replaced by syn-anatectic flow structures (e.g., magmatic or submagmatic foliations, schlieren), or by isotropic neosome. The neosome is diverse in appearance, reflecting a large range in the fraction of melt. It can range from predominantly leucocratic to predominantly mesocratic (e.g., unsegregated melt and residuum) to predominantly melanocratic. Paleosome occurs as rafts, schollen, and enclaves, but may be absent. In certain circumstances, it may be useful to divide diatexites into two types. (1) *Primary diatexite*: a diatexite migmatite formed in a closed system, where the melt fraction present is equal to the degree of partial melting. (2) *Secondary diatexite*: a diatexite migmatite in which the fraction of melt required to pass the solid-to-melt transition was reached by the ingress of melt (i.e., in an open system where the fraction of melt present exceeds the degree of partial melting).

Dilation-structured migmatite: a type of metatexite migmatite in which the location of the domains of leucosome, or neosome, is controlled by the distribution of dilational structures that develop in the competent layers as the migmatite is deformed.

Dynamic melting: anatexis under differential stress conditions. This is the general condition during anatexis of anisotropic rocks undergoing tectonic deformation (for the opposite, *see* Static melting).

Evolved melt: the fraction of melt left during, or after, the process of fractional crystallization. The evolved melt is depleted in the elements compatible with, and enriched in the elements incompatible with, the crystallized and separated minerals. Synonym: fractionated melt.

Fold-structured migmatite: migmatite that was folded while it contained melt.

Fraction of melt (M_f): the fraction of melt that was present in a migmatite, or part of a migmatite.

Fractional crystallization: crystallization of a magma in which the melt becomes separated from the minerals that have crystallized, resulting in a change in composition of the remaining melt (it becomes evolved). The separation of the melt fraction from the crystals may be brought about by either growth of a certain mineral at a specific site (e.g., growth of a mineral on the walls of a dike) or the relative movement of the crystals of a particular mineral (or minerals) relative to the remaining melt (e.g., the accumulation of plagioclase).

Granite (**or granodiorite, tonalite**, etc.) **dike** (**or sill**): the product of crystallization of a felsic melt that has migrated out of its source region completely, and is injected into host rocks of lower metamorphic grade, or into unmetamorphosed rocks.

Initial melt: an anatectic melt formed by a partial-melting reaction, without modification of its composition by other processes such as fractional crystallization, contamination, or alteration.

Layer-confined or bed-confined migmatite: term for a migmatite in which partial melting and the formation of neosome are confined to certain beds, or layers, in which the bulk composition was appropriate (fertile) for partial melting. The term is scale-dependent, as it considers only individual fertile layers in a migmatite; thus, its use cannot be recommended. The neosome viewed at the scale of an individual layer may display a wide range of morphologies (some layers retain their pre-anatectic structures, whereas in others they are completely overprinted by syn-anatectic structures and microstructures). However, viewed at the appropriate scale, i.e., one that is representative of the whole migmatite, layer-confined or bed-confined migmatites are clearly metatexite migmatites because pre-anatectic structures, such as layering or bedding, are preserved.

Leucosome: general term for the lighter-colored part of the neosome in a migmatite, consisting dominantly of feldspar and quartz. The leucosome is that part of the migmatite derived from segregated partial melt; it may contain microstructures that indicate crystallization from a melt, or a magma. Leucosome may not necessarily have the composition of an anatectic melt; fractional crystallization and separation of the fractionated melt may have occurred. Because the anatectic melt can be mobile in migmatites, leucosome may or may not be found *in situ*; additional terms are used to indicate how far the melt from which the leucosome was derived moved within the migmatite.

***In situ* leucosome**: the product of crystallization of an anatectic melt, or part of an anatectic melt, that has segregated from its residuum, but has remained at the site where it formed.

In-source leucosome: the product of crystallization of an anatectic melt, or part of an anatectic melt, that has migrated away from the place where it formed, but is still within the confines of its source layer.

Leucocratic vein or dike: the product of crystallization of an anatectic melt, or part of an anatectic melt, that has migrated out its source layer and been injected into another rock, which may be nearby, or farther away, but is still in the region affected by the anatectic event.

Magma: a silicate liquid that contains crystals, which may have formed from the melt (liquidus phases), be the solid products of the melting reaction (so-called peritectic products), or be minerals in excess.

Magmatic foliation: preferred orientation of minerals in a rock derived from a magma, acquired during flow when the crystals suspended in the magma were free to rotate with no, or very little, interaction with nearby crystals, i.e., when the magma contained less than 55% crystals. *See* Submagmatic foliation.

Mafic selvedge: a particular variety of selvedge in which mafic minerals, most commonly biotite or hornblende, predominate and form a thin rim (a few millimeters wide) around leucosome, leucocratic vein, or granite dike.

Melanosome: a residuum that is composed predominantly of dark-colored minerals. Definition of melanosome: the darker-colored part of the neosome in a migmatite that is rich in dark minerals such as biotite, garnet, cordierite, orthopyroxene, hornblende, clinopyroxene, and even olivine. The melanosome is the solid, residual fraction (i.e., it is residuum) left after some, or all, of the melt fraction has been extracted. Microstructures indicating partial melting may be present.

Melt: a silicate liquid without crystals.

Mesosome: a part of a migmatite that is intermediate in color between leucosome and melanosome. Because mesosome can occur in the neosome or in the paleosome, it should be specified from which part of the migmatite the mesosome originates. Mesosome should be used only in a descriptive sense.

Metatexis: defunct term for partial melting in the continental crust; the term "anatexis" has precedence and should be used.

Metatexite migmatite: a migmatite that is heterogeneous at the outcrop scale, and in which coherent pre-partial-melting structures are widely preserved in the paleosome (where the microstructure appears unchanged), and possibly in the melanosome (residuum) part of the neosome, where the fraction of melt was low. The neosome part is generally segregated into leucosome and melanosome, but neosome in which melt and residuum did not segregate may also occur.

Microstructure: term used to describe the orientation, distribution (i.e., spatial arrangement), relative size, and the internal features of the grains that comprise a rock.

Migmatite: a rock formed by partial melting; a synonym is anatexite. Definition of migmatite: a rock found in medium- and high-grade metamorphic areas that can be heterogeneous at the microscopic to macroscopic scale, which consists of two or more petrographically different parts. One of these parts must have formed by partial melting and contains rocks that are petrogenetically related to each other (called the neosome) and to their protolith through partial melting and the segregation of the melt from the solid fraction. The partially melted part typically contains pale-colored rocks that are quartzofeldspathic, or feldspathic, in composition, and dark-colored rocks that are enriched in ferromagnesian minerals. However, the partially melted part may simply have changed mineralogy, microstructure, and grain size without developing separate light or dark parts. Migmatites are classified and described on the basis of their morphology using a two-tier system. The primary-level division leads us to metatexite migmatites and diatexite migmatites.

Nebulite migmatite: a type of migmatite in which the neosome is diffuse and difficult to differentiate from the paleosome.

Neosome: the parts of a migmatite newly formed, or reconstituted, by partial melting. The neosome may or may not have undergone segregation in which the melt and solid fractions are separated.

Net-structured migmatite: a type of metatexite migmatite in which the neosome, or more commonly the leucosome, or leucocratic veins, form a net-like pattern enclosing paleosome or residuum. In some cases, this pattern arises because the domains of leucosome or neosome have very irregular forms, but in most cases, it arises because the domains of neosome or leucosome follow two or more systematic orientations.

Paleosome: term meaning "old rock" in a migmatite; unfortunately, previous usage is not consistent on which rocks these are. Here, paleosome is defined as the non-neosome part of a migmatite that was not affected by partial melting, and in which structures (such as foliations, folds, layering) older than the partial melting are preserved. The microstructure (size, form, and orientation of grains) is either unchanged, or only slightly coarsened, compared to that in similar rocks just outside the region affected by anatexis.

Parent melt: the anatectic melt from which the other melts in a migmatite were derived, most commonly by fractional crystallization; generally, the parent melt is an initial melt, although there are exceptions.

Parent rock: synonym for protolith.

Patch migmatite: a type of metatexite migmatite in which the neosome occurs *in situ* as small, discrete patches. Although typical of the onset of anatexis, it also occurs in many granulite-facies migmatites.

Protolith: the rock from which the neosome in a migmatite was derived. Protolith cannot be present in a migmatite because it has been converted to neosome. Thus, protolith can only exist in the lower-grade rocks that did not undergo anatexis. Synonym: parent rock.

Raft migmatite: synonym for schollen migmatite.

Residuum: the part of the neosome that is predominantly the solid fraction left after partial melting and the extraction of some, or all, of the melt fraction. Microstructures indicating partial melting may be present. Some residua are melanocratic; those rich in quartz or feldspar can be leucocratic or mesocratic. *See also* Melanosome.

Resisters: rocks, typically competent ones, in the paleosome that are especially resistant to microstructural change. Quartzites, calc-silicates, and metamafic rocks are common resisters.

Schliere: a thin layer comprising aligned platy, tabular, or prismatic minerals in a diatexite migmatite. Commonly, a schliere consists mainly of biotite, but examples rich in sillimanite, orthopyroxene, hornblende, or plagioclase also may occur. Schlieren is the plural.

Schlieric migmatite: a diatexite migmatite characterized by the presence of schlieren, but few schollen, or rafts, of paleosome material.

Schollen migmatite: a diatexite migmatite that contains enclaves (called schollen, or rafts) of paleosome material or, more rarely, residuum. In many terranes, the schollen show a progression from blocky to rounded or elongate shapes, and a progressively better alignment in the direction of a higher fraction of melt and syn-anatectic strain. Raft migmatite is a synonym.

Secondary migmatites: a term used to describe migmatites formed around a pluton as a result of partial melting triggered by the influx of H_2O from the crystallizing pluton into the country rocks.

Segregation: term used to describe the overall process in which anatectic melt is separated from the residuum in a neosome. However, not all examples of neosome have undergone segregation of the melt.

Selvedge: general term for a rim, or border zone, that is not residuum and is compositionally, mineralogically, or microstructurally different from the host and that occurs around a component of a migmatite. Selvedges can be leucocratic, mesocratic, or melanocratic; the most common are mafic and are composed of biotite; see Mafic selvedge.

Static melting: anatexis that occurs under conditions of lithostatic stress without deviatoric stress; conditions that approximate this state may occur in some contact aureoles, or perhaps in competent layers (see Dynamic melting).

Stromatic migmatite: a type of metatexite migmatite in which the neosome (leucosome and melanosome), or just the leucosome, occurs as laterally continuous, parallel layers oriented along the compositional layering (e.g., bedding) or foliation. Stromatic migmatites are common in high-strain, strongly transposed environments, such as crust-scale shear zones, but not restricted to them.

Submagmatic foliation: preferred orientation of minerals in a rock acquired when the magma it formed from contained sufficient crystals that interactions occurred widely between them as they rotated in the flow, i.e., a foliation formed when the fraction of melt was greater than zero, but less than 0.45. Interactions between crystals initially form a "tiling" microstructure, then a framework of crystals that may subsequently become deformed before the magma solidified. See Magmatic foliation.

Vein-structured migmatite: a metatexite or diatexite migmatite in which leucocratic veins are conspicuous and abundant.

Ultrametamorphism: term used to describe the higher-than-normal degree of metamorphism required for partial melting in the continental crust.

10

REFERENCES

ACOSTA-VIGIL, A., LONDON, D. & MORGAN, G.B., VI (2006): Experiments on the kinetics of partial melting of a leucogranite at 200 MPa and 690–800°C: compositional variability of melts during the onset of H_2O-saturated crustal anatexis. ***Contributions to Mineralogy and Petrology* 151**, 539–557.

ALCOCK, J. & MULLER, P. (2000): Anatexis at 700 to 1000 MPa in the aureole of the Marcy anorthosite, Adirondack Highlands, New York. ***Contributions to Mineralogy and Petrology* 139**, 643–654.

ALLIBONE, A.H. & NORRIS, R.J. (1992): Segregation of leucogranite microplutons during synanatectic deformation: an example from the Taylor Valley, Antarctica. ***Journal of Metamorphic Geology* 10**, 589–600.

ARBARET, L., DIOT, H. & BOUCHEZ, J.-L. (1996): Shape fabrics of particles in low concentration suspensions: 2D analog experiments and application to tiling in magma. ***Journal of Structural Geology* 18**, 941–950.

ARBARET, L., DIOT, H., BOUCHEZ, J.-L., LESPINASSE, P. & DE SAINT-BLANQUAT, M. (1997): Analogue 3D simple shear experiments of magmatic biotite subfabrics. ***In*** Granite: from Segregation of Melt to Emplacement Fabrics (J.-L. Bouchez, D.H.W. Hutton & W.E. Stephens, eds.). Kluwer Academic Publisher, Dordrecht, Netherlands (129–143).

ARZI, A.A. (1978): Critical phenomena in the rheology of partially melted rocks. ***Tectonophysics* 44**, 173–184.

ASHBY, M.F. (1972): A first report on deformation-mechanism maps. ***Acta Metallurgica* 20**, 887–897.

ASHWORTH, J.R. (1972): Myrmekites of exsolution and replacement origin. ***Geological Magazine* 109**, 45–62.

ASHWORTH, J.R. (1976): Petrogenesis of migmatites in the Huntly–Portsoy area, north-east Scotland. ***Mineralogical Magazine* 40**, 661–682.

ASHWORTH, J.R. (1985): Introduction. ***In*** Migmatites (J.R. Ashworth, ed.). Blackie, Glasgow, U.K. (1–35).

ASHWORTH, J.R. & MCLELLAN, E.L. (1985): Textures. ***In*** Migmatites (J.R. Ashworth, ed.). Blackie, Glasgow, U.K. (180–203).

BABCOCK, R.S. & MISCH, P. (1989): Origin of the Skagit migmatites, North Cascades Range, Washington State. ***Contributions to Mineralogy and Petrology* 101**, 485–495.

BACKLUND, H.G. (1946): The granitization problem. ***Geological Magazine* 83**, 105–117.

BACON, C.R. (1992): Partially melted granodiorite and related rocks ejected from Crater Lake caldera, Oregon. ***Transactions of the Royal Society of Edinburgh* 83**, 27–47.

BARBARIN, B. (1988): Field evidence for successive mixing and mingling between the Piolard diorite and the Saint-Julien-la-Vêntre monzogranite (Nord-Forez, Massif Central, France). ***Canadian Journal of Earth Sciences* 25**, 49–59.

BARBARIN, B. (1990): Plagioclase xenocrysts and mafic magmatic enclaves in some granitoids of the Sierra Nevada Batholith, California. ***Journal of Geophysical Research* 95**, 17747–17756.

BARBEY, P. (1991): Restites in migmatites and autochthonous granites: their main features and their genesis. ***In*** Enclaves and Granite Petrology (J. Didier & B. Barbarin, B., eds.). Elsevier, Amsterdam, Netherlands (479–492).

BARBEY, P., ALLÉ, P., BROUAND, M. & ALBARÈDE, F. (1995): Rare-earth patterns in zircons from the Manaslu granite and Tibetan Slab migmatites (Himalaya): insights in the origin and evolution of a crustally-derived granite magma. ***Chemical Geology*, 125**, 1–17.

BARBEY, P., BROUAND, M., LE FORT, P. & PÊCHER, A. (1996): Granite-migmatite genetic link: example of the Manaslu Granite and Tibetan slab migmatites in central Nepal. ***Lithos* 38**, 63–79.

BARBEY, P. & CUNEY, M. (1982): K, Rb, Sr, Ba, U and Th geochemistry of the Lapland granulites (Fennoscandia). LILE fractionation controlling factors. ***Contributions to Mineralogy and Petrology* 81**, 304–316.

BARBEY, P., MACAUDIÈRE, J. & NZENTI, J.P. (1990): High-pressure dehydration melting of metapelites: evidence from migmatites of Yaoundé (Cameroon). ***Journal of Petrology* 31**, 401–427.

BARBEY, P., MARIGNAC, C., MONTEL, J.-M., MACAUDIÈRE, J., GASQUET, D. & JABBORI, J. (1999): Cordierite growth textures and the conditions of genesis and emplacement of crustal granitic magmas: the Velay Granite Complex (Massif Central, France). ***Journal of Petrology* 40**, 1425–1441.

Barnes, C.G., Yoshinobu, A.S., Prestvik, T., Nordgulen, Ø., Karlsson, H.R. & Sundvoll, B. (2002): Mafic magma intraplating: anatexis and hybridization in arc crust, Bindal Batholith, Norway. ***Journal of Petrology* 43**, 2171–2190.

Barr, D. (1985): Migmatites in the Moines. ***In*** Migmatites (J.R. Ashworth, ed.). Blackie, Glasgow, U.K. (225–264).

Barraud, J., Gardien, V., Allemand, P. & Grandjean, P. (2004): Analogue models of melt-flow networks in folding migmatites. ***Journal of Structural Geology* 26**, 307–324.

Barrois, C. (1884): Le granite de Rostrenen, ses apophyses et ses contacts. ***Annales Soc. géol. Nord* 12**, 1–119.

Baumgartner, L.P. & Olsen, S.N. (1995): A least-squares approach to mass transport calculations using the isocon method. ***Economic Geology* 90**, 1261–1270.

Bea, F. (1991): Geochemical modeling of low melt-fraction anatexis in a peraluminous system: the Peña Negra Complex (central Spain). ***Geochimica et Cosmochimica Acta* 55**, 1859–1874.

Bea, F. (1996*a*): Residence of REE, Y, Th and U in granites and crustal protoliths: implications for the chemistry of crustal melts. ***Journal of Petrology* 37**, 521–552.

Bea, F. (1996*b*): Controls on the trace element composition of crustal melts. ***Transactions of the Royal Society of Edinburgh* 87**, 33–42.

Bea, F., Pereira, M.D. & Stroh, A. (1994): Mineral/leucosome trace-element partitioning in a peraluminous migmatite (a laser ablation – ICP–MS study). ***Chemical Geology* 117**, 291–312.

Beaumont, C., Jamieson, R.A., Nguyen, M.H. & Medvedev, S. (2004): Crustal channel flows. I. Numerical models with applications to the tectonics of the Himalayan–Tibetan orogen. ***Journal of Geophysical Research* 109**, B06406, 29 p.

Berger, A. & Kalt, A. (1999): Structures and melt fractions as indicators of rheology in cordierite-bearing migmatites of the Bayerische Wald (Variscan Belt, Germany). ***Journal of Petrology* 40**, 1699–1719.

Berger, A. & Roselle, G. (2001): Crystallization processes in migmatites. ***American Mineralogist* 86**, 215–224.

Blom, K.A. (1988): Subsolidus migmatization in high-grade meta-tuffs (Kurkijärvi, southwest Finland). ***Lithos* 21**, 263–278.

Blumenfeld, P. & Bouchez, J.-L. (1988): Shear criteria in granite and migmatite deformed in the magmatic and solid states. ***Journal of Structural Geology* 10**, 361–372.

Bouchez, J.-L., Delas, C., Gleizes, G., Nédélec, A. & Cuney, M. (1992): Submagmatic microfractures in granites. ***Geology* 20**, 35–38.

Bouloton, J. & Gasquet, D. (1995): Melting and under-cooled crystallisation of felsic xenoliths from minor intrusions (Jebilet massif, Morocco). ***Lithos* 35**, 201-219.

Bowen, N.L. (1913): The melting phenomena of the plagioclase feldspars. ***American Journal of Science* 185**, 577–599.

Bowen, N.L. (1915): Crystallization-differentiation in silicate liquids. ***American Journal of Science* 189**, 175–191.

Bowen, N.L. (1928): ***The Evolution of the Igneous Rocks***. Princeton University Press, Princeton, New Jersey.

Braun, I., Raith, M. & Ravindra Kumar, G.R. (1996): Dehydration-melting phenomena in leptynitic gneisses and the generation of leucogranites: a case study from the Kerala Khondalite Belt, southern India. ***Journal of Petrology* 37**, 1285–1305.

Breaks, F.W., Bond, W.D. & Stone, D. (1978): Preliminary geological synthesis of the English River Subprovince, northwestern Ontario and its bearing on mineral exploration. ***Ontario Geological Survey, Miscellaneous Paper* 72**.

Brown, M. (1973): The definition of metatexis, diatexis and migmatite. ***Proceedings of the Geologistsí Association* 84**, 371–382.

Brown, M. (1979): The petrogenesis of the St-Malo migmatite belt, Armorican Massif, France, with particular reference to the diatexites. ***Neues Jahrbuch fur Mineralogie, Abhandlungen* 135**, 48–74.

Brown, M. (1983): The petrogenesis of some migmatites from the Presquíile de Rhuys, southern Brittany, France. ***In*** Migmatites, Melting and Metamorphism (M.P. Atherton & C.D. Gribble, eds.). Shiva, Nantwich, U.K. (174–200).

Brown, M. (1994): The generation, segregation, ascent and emplacement of granitic magma: the migmatite-to-crustally-derived granite connection in thickened orogens. ***Earth Science Reviews* 36**, 83–130.

BROWN, M. (1995): Late-Precambrian geodynamic evolution of the Armorican segment of the Cadomian Belt (France): distortion of the active continental margin during south-west directed convergence and subduction of a bathymetric high. ***Géologie de France* 3**, 3–22.

BROWN, M. (2002): Retrograde processes in migmatites and granulites revisited. ***Journal of Metamorphic Geology* 20**, 25–40.

BROWN, M. (2004): The mechanism of melt extraction from lower continental crust of orogens. ***Transactions of the Royal Society of Edinburgh: Earth Sciences* 95**, 35–48.

BROWN, M., AVERKIN, Y.A., MCLELLAN, E. & SAWYER, E.W. (1995): Melt segregation in migmatites. ***Journal of Geophysical Research* 100**, 15, 655–679.

BROWN, M.A., BROWN, M., CARLSON, W.D. & DENISON, C. (1999): Topology of syntectonic meltflow networks in the deep crust: inferences from three-dimensional images of leucosome geometry in migmatites. ***American Mineralogist* 84**, 1793–1818.

BROWN, M. & DALLMEYER, R.D. (1996): Rapid Variscan exhumation and role of magma in core complex formation: Southern Brittany metamorphic belt, France. ***Journal of Metamorphic Geology* 14**, 361–379.

BROWN, M. & RAITH, M. (1996): First evidence of ultrahigh-temperature decompression from the granulite province of southern India. ***Journal of the Geological Society of London* 153**, 819–822.

BROWN, M. & RUSHMER, T. (1997): The role of deformation in the movement of granitic melt; views from the laboratory and the field. ***In*** Deformation-Enhanced Fluid Transport in the Earthís Crust and Mantle (M.B. Holness, ed.). The Mineralogical Society, Series **8**. Chapman and Hall, London, U.K. (111–144).

BROWN, M. & SOLAR, G.S. (1998***a***): Shear-zone systems and melts: feedback relations and self-organisation in orogenic belts. ***Journal of Structural Geology* 20**, 211–227.

BROWN, M. & SOLAR, G.S. (1998***b***): Granite ascent and emplacement during contractional deformation in convergent orogens. ***Journal of Structural Geology* 20**, 1365–1393.

BRYON, D.N., ATHERTON, M.P. & HUNTER, R.H. (1994): The description of the primary textures of "Cordilleran" granitic rocks. ***Contributions to Mineralogy and Petrology* 117**, 66–75.

BUICK, I.S., STEVENS, G. & GIBSON, R.L. (2004): The role of water retention in the anatexis of metapelites in the Bushveld Complex aureole, South Africa: an experimental study. ***Journal of Petrology* 49**, 1777–1797.

BURG, J.-P. & VANDERHAEGHE, O. (1993): Structures and way-up criteria in migmatites, with applications to the Velay Dome (French Massif Central). ***Journal of Structural Geology* 15**, 1293–1301.

BÜSCH, W., SCHNEIDER, G. & MEHNERT, K.R. (1974): Initial melting at grain boundaries. II. Melting in rocks of granodioritic, quartzdioritic and tonalitic composition. ***Neues Jahrbuch für Mineralogie, Monatshefte* 1974**, 345–370.

BÜTTNER, S.H., GLODNY, J., LUCASSEN, F., WEMMER, K., ERDMANN, S., HANDLER, R. & FRANZ. G. (2005): Ordovician metamorphism and plutonism in the Sierra de Quilmes metamorphic complex: implications for the tectonic setting of the northern Sierras Pampeanas (NW Argentina). ***Lithos* 83**, 143–181.

CAGGIANELLI, A., DEL MORO, A., PAGLIONICO, A., PICCARRETA, G., PINARELLI, L. & ROTTURA, A. (1991): Lower crustal granite genesis connected with chemical fractionation in the continental crust of Calabria (southern Italy). ***European Journal of Mineralogy* 3**, 159–180.

CARRINGTON, D.P. & WATT, G.R. (1995): A geochemical and experimental study of the role of K-feldspar during water-undersaturated melting of metapelites. ***Chemical Geology* 122**, 59–76.

CASTRO, A. (2001): Plagioclase morphologies in assimilation experiments. Implications for disequilibrium melting in the generation of granodiorite rocks. ***Mineralogy and Petrology* 71**, 31–49.

CASTRO, A. (2004): The source of granites: inferences from the Lewisian complex. ***Scottish Journal of Geology* 40**, 49–65.

Castro, A., Corretgé, L.G., El-Biad, M., El-Hmidi, H., Fernández, C. & Patiño Douce, A.E. (2000): Experimental constraints on Hercynian anatexis in the Iberian Massif, Spain. ***Journal of Petrology* 41**, 1471–1488.

Castro, A., Patiño Douce, A.E., Corretgé, L.G., de la Rosa, J.D., El-Biad, M. & El-Hmidi, H. (1999): Origin of peraluminous granites and granodiorites, Iberian Massif, Spain: an experimental test of granite petrogenesis. ***Contributions to Mineralogy and Petrology* 135**, 255–276.

Cesare, B. (2000): Incongruent melting of biotite to spinel in a quartz-free restite at El Joyazo (SE Spain): textures and reaction characterization. ***Contributions to Mineralogy and Petrology* 139**, 273–284.

Cesare, B. & Maineri, C. (1999): Fluid-present anatexis of metapelites at El Joyazo (SE Spain): constraints from Raman spectroscopy of graphite. ***Contributions to Mineralogy and Petrology* 135**, 41–52.

Cesare, B., Marchesi, C., Hermann, J. & Gómez-Pugnaire, M.T. (2003): Primary melt inclusions in andalusite from anatectic graphitic metapelites: implications for the position of the Al_2SiO_5 triple point. ***Geology* 31**, 573–576.

Cesare, B., Salvioli Mariani, E. & Venturelli, G. (1997): Crustal anatexis and melt extraction during deformation in the restitic xenoliths at El Joyazo (SE Spain). ***Mineralogical Magazine* 61**, 15–27.

Collins, W.J. & Sawyer, E.W. (1996): Pervasive granitoid magma transfer through the lower-middle crust during non-coaxial compressional deformation. ***Journal of Metamorphic Geology* 14**, 565–579.

Condie, K.C., Cox, J., O'Reilly, S.J., Griffin, W.L. & Kerrich, R. (2004): Distribution of high field strength and rare earth elements in mantle and lower crust xenoliths from the southwestern United States: the role of grain boundary phases. ***Geochimica et Cosmochimica Acta* 68**, 3919–3942.

Conrad, W.K., Nicholls, I.A. & Wall, V.J. (1988): Water-saturated and -undersaturated melting of metaluminous and peraluminous crustal compositions at 10 kb: evidence for the origin of silicic magmas in the Taupo Volcanic Zone, New Zealand, and other occurrences. ***Journal of Petrology* 29**, 765–803.

Cuney, M. & Barbey, P. (1982): Mise en évidence de phénomènes de cristallisation fractionée dans les migmatites. ***C. R. Acad. Sci. Paris* 295**, 37–42.

Cuney, M., Friedrich, M., Blumenfeld, P., Bourguignon,, A., Boiron, M.C., Vigneresse, J.-L. & Poty, B. (1990): Metallogenesis in the French part of the Variscan orogen. I. U-preconcentrations in the pre-Variscan formations – a comparison with Sn, W and Au. ***Tectonophysics* 177**, 39–57.

Dell'Angelo, L.N. & Tullis, J. (1988): Experimental deformation of partially melted granitic aggregates. ***Journal of Metamorphic Geology* 6**, 495–516.

Dietrich, R.V. & Mehnert, K.R. (1960): Proposal for the nomenclature of migmatites and associated rocks. ***In*** Symposium on Migmatite Nomenclature (H. Sørensen, ed.). ***Int. Geol. Congress* (*Norden*), *XXI Sess., Part XXVI, Suppl. Vol. Sect.*** 1–21, 56–67.

Dougan, T.W. (1979): Compositional and modal relationships and melting reactions in some migmatitic metapelites from New Hampshire and Maine. ***American Journal of Science* 279**, 897–935.

Dougan, T.W. (1983): Textural relations in melanosomes of selected specimens of migmatitic pelitic schists: implications for leucosome-generating processes. ***Contributions to Mineralogy and Petrology* 83**, 82–98.

Edelman, N. (1949): Structural history of the eastern part of the Gullkrona Basin, SW Finland. ***Bulletin de la Commision géologique de Finlande* 148**.

Edelman, N. (1973): Tension cracks parallel with the axial plane. ***Bulletin of the Geological Society of Finland* 45**, 61–65.

England, P.C. & Thompson, A.B. (1984): Pressure – temperature – time paths of regional metamorphism. I. Heat transfer during the evolution of regions of thickened continental crust. ***Journal of Petrology* 25**, 894–928.

Eskola, P. (1932): On the principles of metamorphic differentiation. ***Compt. Rend. Soc. Géol. Finlande* 5**, 68–77.

Eskola, P. (1939): Die metamorphen Gesteine. ***In*** Die Enstehung der Gesteine (T.F.W. Barth, C.W. Correns & P. Eskola, eds.). Springer, Berlin, Germany (263–407).

Finger, F. & Clemens, J.D. (1995): Migmatization and "secondary" granitic magmas: effects of emplacement and crystallization of "primary" granitoids in southern Bohemia, Austria. ***Contributions to Mineralogy and Petrology* 120**, 311–326.

Fisher, G.W. (1973): Nonequilibrium thermodynamics as a model for diffusion-controlled metamorphic processes. ***American Journal of Science* 273**, 897–924.

Fisher, G.W. (1978): Rate laws in metamorphism. ***Geochimica et Cosmochimica Acta* 42**, 1035–1041.

Fitzsimons, I.C.W. (1996): Metapelitic migmatites from Brattstrand Bluffs, East Antarctica – metamorphism, melting and exhumation of the mid crust. ***Journal of Petrology* 37**, 395–414.

Fletcher, R.C. (1982): Coupling of diffusional mass transport and deformation in a tight rock. ***Tectonophysics* 83**, 275–291.

Flood, R.H. & Vernon, R.H. (1978): The Cooma Granodiorite, Australia: an example of ***in situ*** crustal anatexis? ***Geology* 6**, 81–84.

Florence, F.P., Darling, R.S. & Orrell, S.E. (1995): Moderate pressure metamorphism and anatexis due to anorthosite intrusion, western Adirondack Highlands, New York. ***Contributions to Mineralogy and Petrology* 121**, 424–436.

Fornelli, A., Piccarreta, G., Del Moro, A. & Acquafredda, P. (2002): Multi-stage melting in the lower crust of the Serre (southern Italy). ***Journal of Petrology* 43**, 2191–2217.

Frankel, J.J. (1950): A note on the vitrification of Karroo sediment by dolerite intrusions. ***Transactions of the Royal Society of South Africa* 32**, 287–293.

Gardien, V., Thompson, A.B., Grujic, D. & Ulmer, P. (1995): Experimental melting of biotite + plagioclase + quartz → muscovite assemblages and implications for crustal melting. ***Journal of Geophysical Research* 100**, 15581–15591.

Gardien, V., Thompson, A.B. & Ulmer, P. (2000): Melting of biotite + plagioclase + quartz gneisses: the role of H_2O in the stability of amphibole. ***Journal of Petrology* 41**, 651–666.

Grant, J.A. (1985): Phase equilibria in partial melting of pelitic rocks. ***In*** Migmatites (J.R. Ashworth, ed.). Blackie, Glasgow, U.K. (86–144).

Grant, J.A. (1986): The isocon diagram – a simple solution to Gresens' equation for metasomatic alteration. ***Economic Geology* 81**, 1976–1982.

Grant, J.A. (2004): Liquid compositions from low-pressure experimental melting of pelitic rock from the Morton Pass, Wyoming, USA. ***Journal of Metamorphic Geology* 22**, 65–78.

Grant, J.A. & Frost, B.R. (1990): Contact metamorphism and partial melting of pelitic rocks in the aureole of the Laramie Anorthositic Complex, Morton Pass, Wyoming. ***American Journal of Science* 290**, 425–472.

Grapes, R.H. (1986): Melting and thermal reconstitution of pelitic xenoliths, Wehr Volcano, east Eifel, West Germany. ***Journal of Petrology* 27**, 343–396.

Greenfield, J.E., Clarke, G.L., Bland, M. & Clarke, D.L. (1996): ***In situ*** migmatite and hybrid diatexite at Mt. Stafford, central Australia. ***Journal of Metamorphic Geology* 14**, 413–426.

Greenly, E. (1903): The diffusion of granite into crystalline schists. ***Geological Magazine* 10**, 207–212.

Grogan, S.E. & Reavy, R.J. (2002): Disequilibrium textures in the Leinster Granite Complex, SE Ireland: evidence for acid–acid magma mixing. ***Mineralogical Magazine* 66**, 929–939.

Guernina, S. & Sawyer, E.W. (2003): Large-scale melt-depletion in granulite terranes: an example from the Archaean Ashuanipi subprovince of Quebec. ***Journal of Metamorphic Geology* 21**, 181–201.

Gupta, L.N. & Johannes, W. (1982): Petrogenesis of a stromatic migmatite (Neulag, southern Norway). ***Journal of Petrology* 23**, 548–567.

Gupta, L.N. & Johannes, W. (1986): Genetic model for the stromatic migmatites of the Rantasalmi–Sulkava area, Finland. ***Journal of Petrology* 27**, 521–539.

Gürich, G. (1905): Granite und Gneis. ***Himmel Erde* 17**, 241–251.

Haack, U., Heinrichs, H., Boneß, M. & Schneider, A. (1984): Loss of metals from pelites during regional metamorphism. ***Contributions to Mineralogy and Petrology* 85**, 116–132.

Hand, M. & Dirks, P.H.G.M. (1992): The influence of deformation on the formation of axial-planar leucosomes and the segregation of small melt bodies within the migmatitic Napperby Gneiss, central Australia. ***Journal of Structural Geology* 14**, 591–604.

Hansen, J., Skjerlie, K.P., Pederson, R.B. & De La Rosa, J. (2002): Crustal melting in the lower parts of island arcs: an example from the Bremanger Granitoid Complex, west Norwegian Caledonides. ***Contributions to Mineralogy and Petrology* 143**, 316–335.

Härme, M. (1962): An example of anatexis. ***Bulletin de la Commision géologique de Finlande* 204**, 113–125.

Harris, N., McMillan, A., Holness, M., Uken, R., Watkeys, M., Rogers, N. & Fallick, A. (2003): Melt generation and fluid flow in the thermal aureole of the Bushveld complex. ***Journal of Petrology* 44**, 1031–1054.

Harte, B., Pattison, D.R.M. & Linklater, C.M. (1991): Field relations and petrography of partially melted pelite and semi-pelitic rocks. ***In*** Equilibrium in Contact Metamorphism: the Ballachulish Igneous Complex and its Aureole (G. Voll, J., Topel, D.R.M. Pattison & F. Seifert, eds.). Springer-Verlag, Heidelberg, Germany (181–209).

Hartel, T.H.D. & Pattison, D.R.M. (1996): Genesis of the Kapuskasing (Ontario) migmatitic mafic granulites by dehydration melting of amphibolite: the importance of quartz to reaction progress. ***Journal of Metamorphic Geology* 14**, 591–611.

Henkes, L. & Johannes, W. (1981): The petrology of a migmatite (Arvika, Värmland, western Sweden). ***Neues Jahrbuch fur Mineralogie, Abhandlungen* 141**, 113–133.

Hibbard, M.J. (1981): The magma mixing origin of mantled feldspars. ***Contributions to Mineralogy and Petrology* 76**, 158–170.

Hibbard, M.J. (1987): Deformation of incompletely crystallized magma systems: granitic gneisses and their implications. ***Journal of Geology* 95**, 543–561.

Higgins, M.D. (1998): Origin of anorthosite by textural coarsening: quantitative measurements in a natural sequence of textural development. ***Journal of Petrology* 39**, 1307–1323.

Hirth, G. & Kohlstedt, D.L. (1995): Experimental constraints on the dynamics of the partially molten upper mantle. 2. Deformation in the dislocation creep regime. ***Journal of Geophysical Research* 100**, 15441–15449.

Hobson, A., Bussy, F. & Hernandez, J. (1998): Shallow-level migmatization of gabbros in a metamorphic contact aureole, Fuerteventura basal complex, Canary Islands. ***Journal of Petrology* 39**, 1025–1037.

Hokada, T. & Harley, S. (2004): Zircon growth in UHT leucosome: constraints from zircon–garnet rare earth element (REE) relations in Napier Complex, East Antarctica. ***Journal of Mineralogical and Petrological Sciences* 99**, 180–190.

Holland, T.B.J. & Powell, R. (1998): An internally consistent thermodynamic dataset for phases of petrological interest. ***Journal of Metamorphic Geology* 16**, 309–343.

Holland, T.B.J. & Powell, R. (2001): Calculation of phase relations involving haplogranitic melts using an internally consistent thermodynamic dataset. ***Journal of Petrology* 42**, 673–683.

Holmquist, P.J. (1916): Swedish Archaean structures and their meaning. ***Bulletin of the Geological Institute Upsala* 15**, 125–148.

Holmquist, P.J. (1920): Om pegmatitpalingenes och ptygmatisk vecking. ***Geologiska Föreningens i Stockholm Förhandlingar* 42**, 191–213.

Holmquist, P.J. (1921): Typen und nomenklatur der adergesteine. ***Geologiska Föreningens i Stockholm Förhandlingar* 43**, 612–631.

Holness, M.B. (1999): Contact metamorphism and anatexis of Torridonian arkose by minor intrusions of the Rum Igneous Complex, Inner Hebrides, Scotland. ***Geological Magazine* 136**, 527–542.

Holness, M. & Clemens, J.D. (1999): Partial melting of the Appin quartzite driven by fracture-controlled H_2O infiltration in the aureole of the Ballachulish Igneous Complex, Scottish Highlands. ***Contributions to Mineralogy and Petrology* 136**, 154–168.

Holness, M.B., Dane, K., Sides, R., Richardson, C. & Caddick, M. (2005): Melting and melt segregation in the aureole of the Glenmore Plug, Ardnamurchan. ***Journal of Metamorphic Geology* 25**, 29–44.

Holness, M.B. & Isherwood, C.E. (2003): The aureole of the Rum Tertiary Igneous Complex, Inner Hebrides, Scotland. ***Journal of the Geological Society of London* 160**, 15–27.

Holness, M.B. & Watt, G.R. (2002): The aureole of the Traigh Bhàn na Sgùrra sill, Isle of Mull: reaction-driven microcracking during pyrometamorphism. ***Journal of Petrology* 43**, 511–534.

Holtz, F. & Johannes, W. (1991): Genesis of peraluminous granites. I. Experimental investigation of melt composition at 3 and 5 kbar and various H_2O activities. ***Journal of Petrology* 32**, 935–958.

Holyoke, C.W., III & Rushmer, T. (2002): An experimental study of grain scale melt segregation mechanisms in two common crustal rock types. ***Journal of Metamorphic Geology* 20**, 493–512.

HUTTON, J. (1795): ***Theory of the Earth with Proofs and Illustrations*** (in two volumes). Edinburgh, U.K.

ICENHOWER, J. & LONDON, D. (1995): An experimental study of element partitioning between biotite, muscovite and coexisting peraluminous silicic melt at 200 MPa (H_2O). ***American Mineralogist* 80**, 1229–1251.

ICENHOWER, J. & LONDON, D. (1996): Experimental partitioning of Rb, Cs, Sr and Ba between alkali feldspar and peraluminous melt. ***American Mineralogist* 81**, 719–734.

ILDEFONSE, B., ARBARET, L. & DIOT, H. (1997): Rigid particles in simple shear flow: is their preferred orientation periodic or steady state? ***In*** Granite: from Segregation of Melt to Emplacement Fabrics (J.-L. Bouchez, D.H.W. Hutton & W.E. Stephens, eds.). Klüwer, Dordrecht, Netherlands (177–185).

ILDEFONSE, B., LAUNEAU, P., BOUCHEZ, J.-L. & FERNANDEZ, A. (1992): Effect of mechanical interactions on the development of shape preferred orientations: a two dimensional experimental approach. ***Journal of Structural Geology* 14**, 73–83.

JAMIESON, R.A. (1984): Low pressure cordierite-bearing migmatites from Kelly's Mountain, Nova Scotia. ***Contributions to Mineralogy and Petrology* 86**, 309–320.

JAMIESON, R.A., BEAUMONT, C., MEDVEDEV, S. & NGUYEN, M.H. (2004): Crustal channel flows. 2. Numerical models with applications for metamorphism in the Himalayan–Tibetan orogen. ***Journal of Geophysical Research* 109**, B06407, 24 p.

JOESTEN, R. (1977): Evolution of mineral assemblage zoning in diffusion metasomatism. ***Geochimica et Cosmochimica Acta* 41**, 649–670.

JOHANNES, W. (1978): Melting of plagioclase in the system Ab–An–H_2O and Qz–Ab–An–H_2O at P_{H2O} = 5 kbar, an equilibrium problem. ***Contributions to Mineralogy and Petrology* 66**, 295–303.

JOHANNES, W. (1980): Metastable melting in the granite system Qz–Ab–Or–An–H_2O. ***Contributions to Mineralogy and Petrology* 72**, 73–80.

JOHANNES, W. (1983***a***): On the origin of layered migmatites. ***In*** Migmatites, Melting and Metamorphism (M.P. Atherton & C.D. Gribble, eds.). Shiva, Nantwich, U.K. (234–248).

JOHANNES, W. (1983***b***): Metastable melting in granite and related systems. ***In*** Migmatites, Melting and Metamorphism (M.P. Atherton & C.D. Gribble, eds.). Shiva, Nantwich, U.K. (27–36).

JOHANNES, W. (1985): The significance of experimental studies for the formation of migmatites. ***In*** Migmatites (J.R. Ashworth, ed.). Blackie, Glasgow, U.K. (36–85).

JOHANNES, W. & GUPTA, L.N. (1982): Origin and evolution of a migmatite. ***Contributions to mineralogy and Petrology* 79**, 114–123.

JOHANNES, W., HOLTZ, F. & MÖLLER, P. (1995): REE distribution in some layered migmatites: constraints on their petrogenesis. ***Lithos* 35**, 139–152.

JOHN, B.E. & STUNITZ, H. (1997): Magmatic fracturing and small-scale melt segregation during pluton emplacement: evidence from the Adamello Massif (Italy). ***In*** Granite: from Segregation of Melt to Emplacement Fabrics (J.-L. Bouchez, D.H.W. Hutton & W.E. Stephens, eds.). Kluwer, Dordrecht, Netherlands (55–74).

JOHNSON, T.E. & BROWN, M. (2004): Quantitative constraints on metamorphism in the Variscides of Southern Brittany – a complementary pseudosection approach. ***Journal of Petrology* 45**, 1237–1259.

JOHNSON, T.E., GIBSON, R.L., BROWN, M., BUICK, I.S. & CARTWRIGHT, I. (2003): Partial melting of metapelitic rocks beneath the Bushveld Complex, South Africa. ***Journal of Petrology* 44**, 789–813.

JOHNSON, T.E., HUDSON, N.F.C. & DROOP, G.T.R. (2001***a***): Melt segregation structures within the Inzie Head gneisses of the northeastern Dalradian. ***Scottish Journal of Geology* 37**, 59–72.

JOHNSON, T.E., HUDSON, N.F.C. & DROOP, G.T.R. (2001***b***): Partial melting in the Inzie Head gneisses: the role of water and a petrogenetic grid in KFMASH applicable to anatectic pelitic migmatites. ***Journal of Metamorphic Geology* 19**, 99–118.

JOHNSTON, A.D. & WYLLIE, P.J. (1988): Constraints on the origin of Archean trondhjemites based on phase relationships of Nûk gneiss with H_2O at 15 kbar. ***Contributions to Mineralogy and Petrology* 100**, 35–46.

JONES, K.A. & BROWN, M. (1990): High-temperature "clockwise" ***P–T*** paths and melting in the development of regional migmatites: an example from southern Brittany, France. ***Journal of Metamorphic Geology* 8**, 551–578.

JUNG, S., HOERNES, S., MASBERG, P. & HOFFER, E. (1999): The petrogenesis of some migmatites and granites (central Damara Orogen, Namibia): evidence for disequilibrium melting, wall-rock contamination and crystal fractionation. ***Journal of Petrology* 40**, 1241–1269.

JUREWICZ, S.R. & WATSON, E.B. (1984): Distribution of partial melt in a felsic system: the importance of surface energy. ***Contributions to Mineralogy and Petrology* 85**, 25–29.

JUREWICZ, S.R. & WATSON, E.B. (1985): The distribution of partial melt in a granitic system: the application of liquid phase sintering theory. ***Geochimica et Cosmochimica Acta* 49**, 1109–1121.

KACZOR, S.M., HANSON, G.N. & PETERMAN, Z.E. (1988): Disequilibrium melting of granite at the contact of a basic plug: a geochemical and petrographic study. ***Journal of Geology* 96**, 61–78.

KAWAKAMI, T. (2001): Boron depletion controlled by the breakdown of tourmaline in the migmatite zone of the Aoyama area, Ryoke metamorphic belt, southwestern Japan. ***Can. Mineral.* 39**, 1529–1546.

KAWAKAMI, T. & IKEDA, T. (2003): Boron in metapelites controlled by the breakdown of tourmaline and retrograde formation of borosilicates in the Yanai area, Ryoke metamorphic belt, SW Japan. ***Contributions to Mineralogy and Petrology* 145**, 131–150.

KENAH, C. & HOLLISTER, L.S. (1983): Anatexis in the Central Gneiss Complex, British Columbia. ***In*** Migmatites, Melting and Metamorphism (M.P. Atherton & C.D. Gribble, eds.). Shiva, Nantwich, U.K. (142–162).

KNESEL, K.M. & DAVIDSON, J.P. (1999): Sr isotope systematics during melt generation by intrusion of basalt into continental crust. ***Contributions to Mineralogy and Petrology* 136**, 285–295.

KNOPF, A. (1938): Partial fusion of granodiorite by intrusive basalt, Owens Valley, California. ***American Journal of Science* 236**, 373–376.

KOESTER, E., PAWLEY, A.R., FERNANDES, L.A.D., PORCHER, C.C. & SOLIANI, E., JR. (2002): Experimental melting of cordierite gneiss and the petrogenesis of syntranscurrent peraluminous granites in southern Brazil. ***Journal of Petrology* 43**, 1595–1616.

KOHN, M.J., SPEAR, F.S. & VALLEY, J.W. (1997): Dehydration-melting and fluid recycling during metamorphism: Rangeley Formation, New Hampshire, USA. ***Journal of Petrology* 38**, 1255–1277.

KORZHINSKI, D.S. (1971): ***Theory of Metasomatic Zoning***. Oxford University Press, Oxford, U.K.

KRANCK, E.H. (1953): Interpretation of gneiss structures with special reference to Baffin Island. ***Proceedings of the Geological Association of Canada* 6**, 59–68.

KRIEGSMAN, L.M. & HENSEN, B.J. (1998): Back reaction between restite and melt: implications for geothermobarometry and pressure – temperature – paths. ***Geology* 26**, 1111–1114.

LACROIX, M.A. (1898): Le granite des Pyrénées et ses phénomènes de contact. ***Bull. Serv. Carte géol. France* 9**(64), 1–68.

LACROIX, M.A. (1900): Le granite des Pyrénées et ses phénomènes de contact. ***Bull. Serv. Carte géol. France* 11**(71).

LAPADU-HARGUES, P. (1945): Sur l'existence et la nature de l'apport chimique dans certains séries cristallophylliennes. ***Bull. Soc. Géol. France***, 5e sér., **15**, 255–310.

LAPORTE, D., RAPAILLE, C. & PROVOST, A. (1997): Wetting angles, equilibrium melt geometry, and the permeability threshold of partially molten crustal protoliths. ***In*** Granite: from Segregation of Melt to Emplacement Fabrics (J.-L. Bouchez, D.H.W. Hutton & W.E. Stephens, eds.). Kluwer Academic Publisher, Dordrecht, Netherlands (31–54).

LE BRETON, N. & THOMPSON, A.B. (1988): Fluid-absent (dehydration) melting of biotite in metapelites in the early stages of crustal anatexis. ***Contributions to Mineralogy and Petrology* 99**, 226–237.

LEJEUNE, A.-M. & RICHET, P. (1995): Rheology of crystal-bearing silicate melts: an experimental study at high viscosities. ***Journal of Geophysical Research* 100**, 4215–4230.

LE MAITRE, R.W. (1974): Partially fused granite blocks from Mt. Elephant, Victoria, Australia. ***Journal of Petrology* 15**, 403–412.

LINDH, A., HOEFS, J. & MALMSTRÖM, L. (1984): Oxygen isotope ratios in magnetite core-and-mantle structures. ***Journal of Metamorphic Geology* 2**, 343–348.

LITVINOVSKY, B.A. & PODLADCHIKOV, YU.YU. (1993): Crustal anatexis during the influx of mantle volatiles. ***Lithos* 30**, 93–107.

LITVINOVSKY, B.A., STEELE, I.M. & WICKHAM, S.M. (2000): Silicic magma formation in overthickened crust: melting of charnockite and leucogranite at 15, 20 and 25 kbar. ***Journal of Petrology* 41**, 717–737.

LOWENSTERN, J.B. (1995): Applications of silicate melt inclusions to the study of magmatic volatiles. ***In*** Magmas, Fluid and Ore Deposits (J.F.H. Thompson, ed.). Mineralogical Association of Canada, Short Course **23**, 71–99.

LUPULESCU, A. & WATSON, E.B. (1999): Low melt fraction connectivity of granitic and tonalitic melts in a mafic crustal rock at 800°C and 1 GPa. ***Contributions to Mineralogy and Petrology* 134**, 202–216.

LYELL, C. (1855): ***Manual of Elementary Geology*** (5th ed.). John Murray, London, U.K.

MAALØE, S. (1982): Geochemical aspects of permeability controlled partial melting and fractional crystallization. ***Geochimica et Cosmochimica Acta* 46**, 43–57.

MACCARRONE, E., PAGLIONICO, A., PICCARRETA, G. & ROTTURA, A. (1983): Granulite–amphibolite facies metasediments from Serre (Calabria, southern Italy): their protoliths and the processes controlling their chemistry. ***Lithos* 16**, 95–111.

MACGREGOR, M. & WILSON, G. (1939): On granitization and associated processes. ***Geological Magazine* 76**, 193–215.

MÄKITIE, H. (2001): Eastern margin of the Vaasa Migmatite Complex, Kauhava, western Finland: preliminary petrography and geochemistry of the diatexites. ***Bulletin of the Geological Society of Finland* 73**, 35–46.

MANCKTELOW, N.S. (2002): Finite-element modelling of shear zone development in viscoelastic materials and its implication for the localisation of partial melting. ***Journal of Structural Geology* 24**, 1045–1053.

MANNING, D.A.C. & PICHAVANT, M. (1983): The role of fluorine and boron in the generation of granitic melts. ***In*** Migmatites, Melting and Metamorphism (M.P. Atherton & C.D. Gribble, eds.). Shiva, Nantwich, U.K. (95–109).

MARCHILDON, N. & BROWN, M. (2001): Melt segregation in late syn-tectonic anatectic migmatites; an example from the Onawa contact aureole, Maine, U.S.A. ***Physics and Chemistry of the Earth* A 26**, 225–229.

MARCHILDON, N. & BROWN, M. (2002): Grain-scale melt distribution in two contact aureole rocks: implications for controls on melt localization and deformation. ***Journal of Metamorphic Geology* 20**, 381–396.

MARCHILDON, N. & BROWN, M. (2003): Spatial distribution of melt-bearing structures in anatectic rocks from southern Brittany, France: implications for melt transfer at grain- to orogen-scale. ***Tectonophysics* 364**, 215–235.

MASON, G.H. (1985): The mineralogy and textures of the Coastal Batholith, Peru. ***In*** Magmatism at a Plate Edge: the Peruvian Andes (W.S. Pitcher, A.M. Atherton, E.J. Cobbing & R.D. Beckinsale, eds.). Blackie, Glasgow, U.K. (156–166).

MCKENZIE, D. (1984): The generation and compaction of partially molten rock. ***Journal of Petrology* 25**, 713–765.

MCLELLAN, E.L. (1983): Contrasting textures in metamorphic and anatectic migmatites; an example from the Scottish Caledonides. ***Journal of Metamorphic Geology* 1**, 241–262.

MCLELLAN, E.L. (1984): Deformational behaviour of migmatites and problems of structural analysis in migmatite terrains. ***Geological Magazine* 121**, 339–345.

MCLELLAN, E.L. (1988): Migmatite structures in the Central Gneiss Complex, Boca De Quadra, Alaska. ***Journal of Metamorphic Geology* 6**, 517–542.

MCLELLAN, E.L. (1989): Sequential formation of subsolidus and anatectic migmatites in response to thermal evolution, eastern Scotland. ***Journal of Geology* 97**, 165–182.

MEHNERT, K.R. (1968): ***Migmatites and the Origin of Granitic Rocks.*** Elsevier, Amsterdam, Netherlands.

MEHNERT, K.R., BÜSCH, W. & SCHNEIDER, G. (1973): Initial melting at grain boundaries of quartz and feldspar in gneisses and granulites. ***Neues Jahrbuch für Mineralogie, Monatshefte***, 165–183.

MICHEL-LÉVY, A. (1893): Contribution à l'étude du granite de Flamanville et des granites français en général. ***Bull. Serv. Carte géol. France* 5**(36), 1–41.

MILLER, C.F. & MITTLEFEHLDT, D.W. (1982): Depletion of light rare-earth elements in felsic magmas. ***Geology* 10**, 129–133.

MILLER, C.F. & WOODEN, J.L. (1994): Anatexis, hybridization and the modification of ancient crust: Mesozoic plutonism in the Old Woman Mountains area, California. ***Lithos* 32**, 111–133.

MILORD, I. & SAWYER, E.W. (2003): Schlieren formation in diatexite migmatite: examples from the St. Malo migmatite terrane, France, ***Journal of Metamorphic Geology* 21**, 347–362.

MILORD, I., SAWYER, E.W. & BROWN, M. (2001): Formation of diatexite migmatite and granite magma during anatexis of semi-pelitic metasedimentary rocks: an example from St. Malo, France. ***Journal of Petrology* 42**, 487–505.

MISCH, P. (1949): Metasomatic granitization of batholithic dimensions. ***American Journal of Science* 247**, 209–245.

MONTEL, J.-M. & VIELZEUF, D. (1997): Partial melting of metagreywackes. II. Compositions of minerals and melts. ***Contributions to Mineralogy and Petrology* 128**, 176–196.

Morogan, V. & Sørensen, H. (1994): Net veined complexes in the Oslo Rift, southeast Norway. ***Lithos* 32**, 21–45.

Nair, R. & Chacko, T. (2002): Fluid-absent melting of high-grade semi-pelites: P–T constraints on orthopyroxene formation and implications for granulite genesis. ***Journal of Petrology* 43**, 2121–2142.

Nédélec, A., Minyem, D. & Barbey, P. (1993): High-P – high-T anatexis of Archaean tonalitic grey gneiss: the Eseka migmatites, Cameroon. ***Precambrian Research* 62**, 191–205.

Neves, S.P. & Vauchez, A. (1995): Successive mixing and mingling of magmas in a plutonic complex of northeast Brazil. ***Lithos* 34**, 275–299.

Nicolas, A. & Poirier, J.-P. (1976): ***Crystalline Plasticity and Solid State Flow in Metamorphic Rocks***. Wiley Interscience, London, U.K.

Nyman, M.W., Pattison, D.R.M. & Ghent, E.D. (1995): Melt extraction during formation of K-feldspar + sillimanite migmatites, west of Revelstoke, British Columbia. ***Journal of Petrology* 36**, 351–372.

Obata, M., Yoshimura, Y., Nagakawa, K., Odawara, S. & Osanai, Y. (1994): Crustal anatexis and melt migrations in the Higo metamorphic terrane, west-central Kyushu, Kumamoto, Japan. ***Lithos* 32**, 135–147.

Oliver, N.H.S. & Barr, T.D. (1997): The geometry and evolution of magma pathways through migmatites of the Halls Creek Orogen, Western Australia. ***Mineralogical Magazine* 61**, 3–14.

Oliver, N.H.S., Bodorkos, S, Nemchin, A.A., Kinny, P.D. & Watt, G.R. (1999): Relationships between zircon U–Pb SHRIMP ages and leucosome type in migmatites of the Halls Creek Orogen, Western Australia. ***Journal of Petrology* 40**, 1553–1575.

Olsen, S.N. (1982): Open- and closed-system migmatites in the Front Range, Colorado. ***American Journal of Science* 282**, 1596–1622.

Olsen, S.N. (1984): Mass-balance and mass-transfer in migmatites from the Colorado Front Range. ***Contributions to Mineralogy and Petrology* 85**, 30–44.

Olsen, S.N. (1985): Mass balance in migmatites. ***In*** Migmatites (J.R. Ashworth, ed.). Blackie, Glasgow, U.K. (145–179).

Olsen, S.N. (1987): The composition and role of the fluids in migmatites: a fluid inclusion study of the Front Range rocks. ***Contributions to Mineralogy and Petrology* 96**, 104–120.

Olsen, S.N. & Grant, J.A. (1991): Isocon analysis of migmatization in the Front Range, Colorado, USA. ***Journal of Metamorphic Geology* 9**, 151–164.

Olsen, S.N., Marsh, B.D. & Baumgartner, L.P. (2004): Modelling mid-crustal migmatite terrains as feeder zones for granite plutons: the competing dynamics of melt transfer by bulk versus porous flow. ***Transaction of the Royal Society of Edinburgh* 95**, 49–58.

Otamendi, J.E. & Patiño Douce, A.E. (2001): Partial melting of aluminous metagreywackes in the northern Sierra de Comechingones, central Argentina. ***Journal of Petrology* 42**, 1751–1772.

Park, A.K. (1983): Lit-par-lit migmatite fabrics in a metagabbro–anorthosite complex, Sygnefjell, Jotunheim, south Norway. ***In*** Migmatites, Melting and Metamorphism (M.P. Atherton & C.D. Gribble, eds.). Shiva, Nantwich, U.K. (296).

Park, Y. & Means, W.D. (1996): Direct observation of deformation processes in crystal mushes. ***Journal of Structural Geology* 18**, 847–858.

Park, Y. & Means, W.D. (1997): Crystal rotation and growth during grain flow in a deforming crystal mush. ***In*** Evolution of Geological Structures in Micro- and Macro-Scales (A. Sengupta, ed.). Chapman and Hall, London, U.K. (245–258).

Paterson, S.R., Vernon, R.H. & Tobisch, O.T. (1989): A review of criteria for the identification of magmatic and tectonic foliations in granitoids. ***Journal of Structural Geology* 11**, 349–363.

Patiño Douce, A.E. (2005): Vapor-absent melting of tonalite at 15–32 kbar. ***Journal of Petrology* 46**, 275–290.

Patiño Douce, A.E. & Beard, J.S. (1995): Dehydration-melting of biotite gneiss and quartz amphibolite from 3 to 15 kbar. ***Journal of Petrology* 36**, 707–738.

Patiño Douce, A.E. & Beard, J.S. (1996): Effects of P, $f(O_2)$ and Mg/Fe ratio on dehydration melting of model metagreywackes. ***Journal of Petrology* 37**, 999–1024.

Patiño Douce, A.E. & Harris, N. (1998): Experimental constraints on Himalayan anatexis. ***Journal of Petrology* 39**, 689–710.

PATIÑO DOUCE, A.E. & JOHNSTON, A.D. (1991): Phase equilibria and melt productivity in the pelitic system: implications for the origin of peraluminous granitoids and aluminous granulites. ***Contributions to Mineralogy and Petrology* 107**, 202–218.

PATIÑO DOUCE, A.E., JOHNSTON, A.D. & RICE, J.M. (1994): Octahedral excess mixing properties in biotite: a working model with applications to geobarometry and geothermometry. ***American Mineralogist* 78**, 113–131.

PATTISON, D.R.M. (1991): Infiltration-driven dehydration and anatexis in granulite facies metagabbro, Grenville Province, Ontario, Canada. ***Journal of Metamorphic Geology* 9**, 315–332.

PATTISON, D.R.M. & HARTE, B. (1988): Evolution of structurally contrasting migmatites in the 3-kbar Ballachulish aureole, Scotland. ***Journal of Metamorphic Geology* 6**, 475–494.

PERCIVAL, J.A. (1991): Granulite-facies metamorphism and crustal magmatism in the Ashuanipi complex, Quebec–Labrador, Canada. ***Journal of Petrology* 32**, 1261–1297.

PERRIN, R. & ROUBAULT, M. (1937): Les réactions à l'état solide et la géologie. ***Bull. Serv. Carte géol. Algerie, 5e sér., Pétrographie* 1**.

PERRIN, R. & ROUBAULT, M. (1939): Le granite et les réactions à l'état solide. ***Bull. Serv. Carte géol. Algerie, 5e sér., Pétrographie* 4**.

PETFORD, N. (1995): Segregation of tonalitic–trondhjemitic melts in the continental crust: the mantle connection. ***Journal of Geophysical Research* 100**, 15735–15744.

PICKERING, J.M. & JOHNSTON, A.D. (1998): Fluid-absent melting behavior of a two-mica metapelite: experimental constraints on the origin of Black Hills granite. ***Journal of Petrology* 39**, 1787–1804.

PITCHER, W.S. (1952): The migmatitic older granodiorite of Thorr District, Co. Donegal. ***Quarterly Journal of the Geological Society of London* 108**, 413–446.

PLATT, J.P . & VISSERS, R.L.M. (1980): Extensional structures in anisotropic rocks. ***Journal of Structural Geology* 2**, 397–410.

PLATTEN, I.M. (1982): Partial melting of feldspathic quartzite around late Caledonian minor intrusions in Appin, Scotland. ***Geological Magazine* 119**, 413–419.

PLATTEN, I.M. (1983); Partial melting of semipelite and the development of marginal breccias around a late Caledonian minor intrusion in the Grampian Highlands of Scotland. ***Geological Magazine* 120**, 37–49.

POIRIER, J.-P. (1985): ***Creep of Crystals***. Cambridge University Press, Cambridge, U.K.

PONS, J., BARBEY, P., DUPUIS, D. & LÉGER, J.M. (1995): Mechanisms of pluton emplacement and structural evolution of 2.1 Ga juvenile continental crust: the Birimian of southwestern Niger. ***Precambrian Research* 70**, 281–301.

POWELL, R. (1983): Processes in granulite-facies metamorphism. ***In*** Migmatites, Melting and Metamorphism (M.P. Atherton & C.D. Gribble, eds.). Shiva, Nantwich, U.K. (127–139).

POWELL, R. & DOWNES, J. (1990): Garnet porphyroblast-bearing leucosomes in metapelites: mechanisms, phase diagrams and an example from Broken Hill, Australia. ***In*** High Temperature Metamorphism and Crustal Anatexis (J.R. Ashworth & M. Brown, eds.). Mineralogical Society, Ser. **2**, Unwin Hyman, London, U.K. (105–123).

POWELL, R. & HOLLAND, T.B.J. (1988): An internally consistent thermodynamic dataset with uncertainties and correlations. 3. Applications to geobarometry, worked examples and a computer program. ***Journal of Metamorphic Geology* 6**, 173–204.

RAITH, J.G. & HARLEY, S.L. (1998); Low-P/high-T metamorphism in the Okiep copper district western Namaqualand, South Africa. ***Journal of Metamorphic Geology* 16**, 281–305.

RAMBERG, H. (1952): ***The Origin of Metamorphic and Metasomatic Rocks***. University of Chicago Press, Chicago, Illinois.

RAPP, R.P. & WATSON, E.B. (1995): Dehydration melting of metabasalt at 8–32 kbar: implications for continental growth and crust–mantle recycling. ***Journal of Petrology* 36**, 891–931.

READ, H.H. (1957): ***The Granite Controversy***. Thomas Murby & Co., London, U.K.

RENNER, J., EVANS, B. & HIRTH, G. (2000): On the rheologically critical melt fraction. ***Earth and Planetary Science Letters* 181**, 585–594.

ROBERT, J.-L. (1976): Titanium solubility in synthetic phlogopite solid solutions. ***Chemical Geology* 17**, 213–227.

ROBIN, P.-Y.F. (1978): Pressure solution at grain-to-grain contacts. ***Geochimica et Cosmochimica Acta* 42**, 1383–1389.

ROBIN, P.-Y.F. (1979): Theory of metamorphic segregation and related processes. ***Geochimica et Cosmochimica Acta* 43**, 1587–1600.

Rosenberg, C.L. (2001): Deformation of partially molten granite: a review and comparison of experimental and natural case studies. ***International Journal of Earth Sciences* 90**, 60–76.

Rosenberg, C.L. & Handy, M.R. (2000): Syntectonic melt pathways during simple shearing of a partially molten rock analogue (norcamphor–benzamide). ***Journal of Geophysical Research* 105**, 3135–3149.

Rosenberg, C.L. & Handy, M.R. (2005): Experimental deformation of partially melted granite revisited: implications for the continental crust. ***Journal of Metamorphic Geology* 23**, 19–28.

Rubatto, D., Williams, I.S. & Buick, I.S. (2001): Zircon and monazite response to prograde metamorphism in the Reynolds Range, Central Australia. ***Contributions to Mineralogy and Petrology* 140**, 458–468.

Rushmer, T. (1991): Partial melting of two amphibolites: contrasting experimental results under fluid-absent conditions. ***Contributions to Mineralogy and Petrology* 107**, 41–59.

Rushmer, T. (1995): An experimental deformation study of partially molten amphibolite: application to low-melt fraction segregation. ***Journal of Geophysical Research* 100**, 15681–15695.

Rutter, E.H. (1976): The kinetics of rock deformation by pressure solution. ***Philosophical Transactions of the Royal Society of London* A283**, 203–219.

Rutter, E.H. (1997): The influence of deformation on the extraction of crustal melts: ***In*** Deformation-Enhanced Melt Segregation and Metamorphic Fluid Transport (M. Holness, ed.). ***Mineralogical Society, Series* 8**. Chapman and Hall, London, U.K. (82–110).

Rutter, E.H. & Neumann, D.H.K. (1995): Experimental deformation of partially molten Westerly granite under fluid-absent conditions, with implications for the extraction of granitic magmas. ***Journal of Geophysical Research* 100**, 15697–15715.

Rutter, M.J. & Wyllie, P.J. (1988): Melting of vapour-absent tonalite at 10 kbar to simulate dehydration melting in the deep crust. ***Nature* 331**, 159–160.

Sawka, W. (1988): REE and trace element partitioning in accessory minerals and hornblende from the strongly zoned McMurry Meadows Pluton, California. ***Transactions of the Royal Society of Edinburgh* 79**, 157–168.

Sawyer, E.W. (1986): The influence of rock source type, chemical weathering and sorting on the geochemistry of clastic sediments from Quetico Metasedimentary Belt, Superior Province, Canada. ***Chemical Geology* 55**, 77–95.

Sawyer, E.W. (1987): The role of partial melting and fractional crystallization in determining discordant migmatite leucosome compositions. ***Journal of Petrology* 28**, 445–473.

Sawyer, E.W. (1991): Disequilibrium melting and the rate of melt–residuum separation during migmatization of mafic rocks from the Grenville Front, Quebec. ***Journal of Petrology* 32**, 701–738.

Sawyer, E.W. (1994): Melt segregation in the continental crust. ***Geology* 22**, 1019–1022.

Sawyer, E.W. (1996) Melt-segregation and magma flow in migmatites: implications for the generation of granite magmas. ***Transactions of the Royal Society of Edinburgh: Earth Sciences* 87**, 85–94.

Sawyer, E.W. (1998): Formation and evolution of granite magmas during crustal reworking: the significance of diatexites. ***Journal of Petrology* 39**, 1147–1167.

Sawyer, E.W. (1999): Criteria for the recognition of partial melting. ***Physics and Chemistry of the Earth* A 24**, 269–279.

Sawyer, E.W. (2000): Grain-scale and outcrop-scale distribution and movement of melt in a crystallizing granite. ***Transactions of the Royal Society of Edinburgh: Earth Sciences* 91**, 73–85.

Sawyer, E.W. (2001): Melt segregation in the continental crust: distribution and movement of melt in anatectic rocks. ***Journal of Metamorphic Geology* 19**, 291–309.

Sawyer, E.W. & Barnes, S.-J. (1988): Temporal and compositional differences between solidus and anatectic migmatite leucosomes from the Quetico metasedimentary belt, Canada. ***Journal of Metamorphic Geology* 6**, 437–450.

Sawyer, E.W., Dombrowski, C. & Collins, W.J. (1999): Movement of melt during synchronous regional deformation and granulite-facies anatexis, an example from the Wuluma Hills, central Australia. ***In*** Understanding Granites; Integrating New and Classical Techniques (A. Castro, C. Fernández, & J.-L. Vigneresse, eds.). ***Geological Society, Special Publication* 168**, 221–237.

Sawyer, E.W. & Robin, P.-Y.F. (1986): The subsolidus segregation of layer-parallel quartz–feldspar veins in greenschist to upper amphibolite facies metasediments. ***Journal of Metamorphic Geology* 4**, 237–260.

Scaillet, B., Holtz, F. & Pichavant, M. (1998): Phase equilibrium constraints on the viscosity of silicic magmas. I. Volcanic–plutonic comparison. ***Journal of Geophysical Research* 103**, 27257–27266.

Scheumann, K.H. (1936): Zur Nomenklatur migmatitischer und verwandter Gesteine. ***Tschermaks Mineral Petrogr. Mitt.* 48**, 297–302.

Sederholm, J.J. (1907): Om granit och gneis, deras uppkomst, uppträdande och utbredning inom urberget i Fennoskandia. ***Bulletin de la Commission géologique de Finlande* 23**.

Sederholm, J.J. (1923): On migmatites and associated Precambrian rocks of southwestern Finland. I. The Pellinge region.***Bulletin de la Commission géologique de Finlande* 58**.

Sederholm, J.J. (1926): On migmatites and associated Precambrian rocks of southwestern Finland. II. The region around the Barösundsfjärd west of Helsingfors and neighbouring areas. ***Bulletin de la Commission géologique de Finlande* 77**.

Selbekk, R.S. & Skjerlie, K.P. (2002): Petrogenesis of the anorthosite dyke swarm of Tromsø, north Norway: experimental evidence for hydrous anatexis of an alkaline mafic complex. ***Journal of Petrology* 43**, 943–962.

Shaw, D.M. (1954): Trace elements in pelitic rocks. ***Geological Society of America, Bulletin* 65**, 1151–1182.

Shaw, D.M. (1956): Geochemistry of pelitic rocks. III. Major elements and general geochemistry. ***Geological Society of America, Bulletin* 67**, 919–934.

Simakin, A. & Talbot, C. (2001*a*): Tectonic pumping of pervasive granitic melts. ***Tectonophysics* 332**, 387–402.

Simakin, A. & Talbot, C. (2001*b*): Transfer of melt between microscopic pores and macroscopic veins in migmatites. ***Physics and Chemistry of the Earth* 26**, 363–367.

Singh, J. & Johannes, W. (1996*a*): Dehydration melting in tonalites. I. Beginning of melting. ***Contributions to Mineralogy and Petrology* 125**, 16–25.

Singh, J. & Johannes, W. (1996*b*): Dehydration melting in tonalites. II. Composition of melts and solids. ***Contributions to Mineralogy and Petrology* 125**, 26–44.

Skelton, A.D.L. (1996): The timing and direction of metamorphic fluid flow in Vermont. ***Contributions to Mineralogy and Petrology* 125**, 75–84.

Skjerlie, K.P. & Johnston, A.D. (1992): Vapour-absent melting at 10 kbar of a biotite- and amphibole-bearing tonalitic gneiss: implications for the generation of A-type granites. ***Geology* 20**, 263–266.

Skjerlie, K.P. & Johnston, A.D. (1996): Vapour-absent melting from 10 to 20 kbar of crustal rocks that contain multiple hydrous phases: implications for anatexis in the deep to very deep continental crust and active continental margins. ***Journal of Petrology* 37**, 661–691.

Skjerlie, K.P. & Patiño Douce, A.E. (2002): The fluid-absent partial melting of a zoisite-bearing quartz eclogite from 1.0 to 3.2 GPa; implications for melting in thickened continental crust and for subduction-zone processes. ***Journal of Petrology* 43**, 291–314.

Slagstad, T., Jamieson, R.A. & Culshaw, N.G. (2005): Formation, crystallisation and migration of melt in the mid-orogenic crust: Muskoka domain migmatites, Grenville Province, Ontario. ***Journal of Petrology* 46**, 893–919.

Solar, G.S. & Brown, M. (2001): Petrogenesis of migmatites in Maine, USA: possible source of peraluminous leucogranite in plutons? ***Journal of Petrology* 42**, 789–823.

Spear, F.S., Kohn, M.J. & Cheney, J.T. (1999): P–T paths from anatectic pelites. ***Contributions to Mineralogy and Petrology* 134**, 17–32.

Spicer, E.M., Stevens, G. & Buick, I.S. (2004): The low-pressure partial melting behaviour of natural boron-bearing metapelites from the Mt. Stafford area, central Australia. ***Contributions to Mineralogy and Petrology* 148**, 160–179.

Springer, W. & Seck, H.A. (1997): Partial fusion of basic granulites at 5 to 15 kbar: implications for the origin of TTG magmas. ***Contributions to Mineralogy and Petrology* 127**, 30–45.

Stevens, G., Clemens, J.D. & Droop, G.T.R. (1997): Melt production during granulite-facies anatexis: experimental data from "primitive" metasedimentary protoliths. ***Contributions to Mineralogy and Petrology* 128**, 352–370.

Stevenson, D.J. (1989): Spontaneous small-scale melt segregation in partial melts undergoing deformation. ***Geophysical Research Letters* 16**, 1067–1070.

Streckeisen, A. (1976): To each plutonic rocks its proper name. ***Earth Science Reviews* 12**, 1–33.

Strömgård, K.-E. (1973): Stress distribution during formation of boudinage and pressure shadows. ***Tectonophysics* 16**, 215–248.

STÜWE, K. & POWELL, R. (1989): Metamorphic segregations associated with garnet and orthopyroxene porphyroblast growth: two examples from the Larsemann Hills, East Antartica. ***Contributions to Mineralogy and Petrology* 103**, 523–530.

SYMMES, G.H. & FERRY, J.M. (1995): Metamorphism, fluid-flow and partial melting in pelitic rocks from the Onawa contact aureole, Central Maine, USA. ***Journal of Petrology* 36**, 587–612.

TAKEDA, Y.-T. & OBATA, M. (2003): Some comments on the rheologically critical melt percentage. ***Journal of Structural Geology* 25**, 813–818.

THOMPSON, A.B. (1983): Fluid-absent metamorphism. ***Journal of the Geological Society, London* 140**, 533–547.

TIMMERMANN, H., JAMIESON, R.A., PARRISH, R.R. & CULSHAW, N.G. (2002): Coeval migmatites and granulites, Muskoka domain, southwestern Grenville Province, Ontario. ***Canadian Journal of Earth Sciences* 39**, 239–258.

TINKHAM, D.K., ZULUGA, C.A. & STOWELL, H.H. (2001): Metapelite phase equilibria modeling in MnNCKFMASH: the effect of variable Al_2O_3 and MgO / (MgO + FeO) on mineral stability. ***Geological Materials Research* 3**, 1–42.

TRACY, R.J. (1985): Migmatite occurrences in New England. ***In*** Migmatites (J.R. Ashworth, ed.). Blackie, Glasgow, U.K. (204–224).

TRACY, R.J. & ROBINSON, P. (1983): Acadian migmatite types in pelitic rocks of central Massachusetts. ***In*** Migmatites, Melting and Metamorphism (M.P. Atherton & C.D. Gribble, eds.). Shiva, Nantwich, U.K. (163–173).

TRUMBULL, R.B. (1988): Petrology of flecked gneisses in the northern Wet Mountains, Fremont County, Colorado. ***Geological Society of America, Bulletin* 100**, 247–256.

TURNER, F.J. (1941): The development of pseudo-stratification by metamorphic differentiation in the Otago schists. ***American Journal of Science* 239**, 1–16.

VANDERHAEGHE, O. (1999): Pervasive melt migration from migmatites to leucogranite in the Shuswap metamorphic core complex, Canada: control of regional deformation. ***Tectonophysics* 312**, 35–55.

VANDERHAEGHE, O. (2001): Melt segregation, pervasive melt migration and magma mobility in the continental crust: the structural record from pores to orogens. ***Physics and Chemistry of the Earth* 26**, 213–223.

VAN DER MOLEN, I. (1985): Interlayer material transport during layer-normal shortening. I. The model. ***Tectonophysics* 115**, 275–295.

VAN DER MOLEN, I. & PATERSON, M.S. (1979): Experimental deformation of partially-melted granite. ***Contributions to Mineralogy and Petrology* 70**, 299–318.

VERNON, R.H. (1981): Optical microstructure of partly recrystallized calcite in some naturally deformed marbles. ***Tectonophysics* 78**, 601–612.

VERNON, R.H. & COLLINS, W.J. (1988): Igneous microstructures in migmatites. ***Geology* 16**, 1126–1129.

VERNON, R.H. & PATERSON, S.R. (2001): Axial-surface leucosomes in anatectic migmatites. ***Tectonophysics* 335**, 183–192.

VIDALE, R.J. (1974): Vein assemblages and metamorphism in Dutchess County, New York. ***Geological Society of America, Bulletin* 85**, 303–306.

VIELZEUF, D. & HOLLOWAY J.R. (1988): Experimental determination of the fluid-absent melting relations in the pelitic system: consequences for crustal differentiation. ***Contributions to Mineralogy and Petrology* 98**, 257–276.

VIELZEUF, D. & MONTEL, J.-M. (1994): Partial melting of metagreywackes. I. Fluid-absent experiments and phase relationships. ***Contributions to Mineralogy and Petrology* 117**, 375–393.

VIGNERESSE, J.-L., BARBEY, P. & CUNEY, M. (1996): Rheological transitions during partial melting and crystallization with application to felsic magma segregation and transfer. ***Journal of Petrology* 37**, 1579–1600.

VIGNERESSE, J.-L. & TIKOFF, B. (2000): Strain partitioning during partial melting and crystallizing felsic magmas. ***Tectonophysics* 312**, 117–132.

VILLASECA, C., ROMERA, C.M. & BAREBERO, L. (2001): Melts and residua geochemistry in a low-to-mid crustal section (central Spain). ***Physics and Chemistry of the Earth* A 26**, 273–280.

WAFF, H.S. & BULAU, J.R. (1979): Equilibrium fluid distribution in an ultramafic partial melt under hydrostatic stress conditions. ***Journal of Geophysical Research* 84**, 6109–6114.

WALTHER, J.V. & WOOD, B.J. (1983): Rate and mechanism in prograde metamorphism. ***Contributions to Mineralogy and Petrology* 88**, 246–259.

WARK, D.A. & MILLER, C.F. (1993): Accessory mineral behavior during differentiation of a granitic suite: monazite, xenotime and zircon in the Sweetwater Wash pluton, southeastern California, U. S. A. ***Chemical Geology* 110**, 49–67.

WATERS, D.J. (1988): Partial melting and the formation of granulite facies assemblages in Namaqualand, South Africa. ***Journal of Metamorphic Geology* 6**, 387–404.

WATERS, D.J. (2001): The significance of prograde and retrograde quartz-bearing intergrowth microstructures in partially melted granulite-facies rocks. ***Lithos* 56**, 97–110.

WATERS, D.J. & WHALES, C.J. (1984): Dehydration melting and the granulite transition in metapelites from southern Namaqualand, S. Africa. ***Contributions to Mineralogy and Petrology* 88**, 269–275.

WATT, G.R. & HARLEY, S.L. (1993): Accessory phase controls on the geochemistry of crustal melts and restites produced during water-undersaturated partial melting. ***Contributions to Mineralogy and Petrology* 114**, 550–566.

WEBER, C. & BARBEY, P. (1986): The role of water, mixing processes and metamorphic fabric in the genesis of the Baume migmatites (Ardèche, France). ***Contributions to Mineralogy and Petrology* 92**, 481–491.

WEBER, C., BARBEY, P., CUNEY, M. & MARTIN, H. (1985): Trace element behaviour during migmatization. Evidence for a complex melt – residuum – fluid interaction in the St. Malo migmatitic dome (France). ***Contributions to Mineralogy and Petrology* 90**, 52–62.

WHEELER, J. (1987): The significance of grain-scale stresses in the kinetics of metamorphism. ***Contributions to Mineralogy and Petrology* 97**, 397–404.

WHITE, R.W. & POWELL, R. (2002): Melt loss and the preservation of granulite facies mineral assemblages. ***Journal of Metamorphic Geology* 20**, 621–632.

WHITE, R.W., POWELL, R. & CLARKE, G.L. (2003): Prograde metamorphic assemblage evolution during partial melting of metasedimentary rocks at low pressures: migmatites from Mt. Stafford, Central Australia. ***Journal of Petrology* 44**, 1937–1960.

WHITE, R.W., POWELL, R. & HALPIN, J.A. (2004): Spatially-focussed melt formation in aluminous metapelites from Broken Hill, Australia. ***Journal of Metamorphic Geology* 22**, 825–845.

WHITE, R.W., POWELL, R. & HOLLAND, T.J.B. (2001): Calculation of partial melt equilibria in the system Na_2O–CaO–K_2O–FeO–MgO–Al_2O_3–SiO_2–H_2O (NCKFMASH). ***Journal of Metamorphic Geology* 19**, 139–153.

WHITE, R.W., POMROY, N.E. & POWELL, R. (2005): An ***in situ*** metatexite–diatexite transition in upper amphibolite facies rocks from Broken Hill, Australia. ***Journal of Metamorphic Geology* 23**, 579–602.

WHITE, S. (1977): Geological significance of recovery and recrystallization processes in quartz. ***Tectonophysics* 39**, 143–170.

WHITNEY, D.L. & IRVING, A.J. (1994): Origin of K-poor leucosomes in a metasedimentary migmatite complex by ultrametamorphism, syn-metamorphic magmatism and subsolidus processes. ***Lithos* 32**, 173–192.

WICKHAM, S.M. (1987***a***): Crustal anatexis and granite petrogenesis during low-pressure regional metamorphism: the Trois Seigneurs Massif, Pyrenees, France. ***Journal of Petrology* 28**, 127–169.

WICKHAM, S.M. (1987***b***): The segregation and emplacement of granitic magmas. ***Journal of the Geological Society, London* 144**, 281–297.

WIEBE, R.A. (1968): Plagioclase stratigraphy: a record of magmatic conditions and events in a granitic stock. ***American Journal of Science* 266**, 690–703.

WILLIAMS, M.L., HANMER, S., KOPF, C. & DARRACH, M. (1995): Syntectonic generation and segregation of tonalitic melts from amphibolite dikes in the lower crust, Striding–Athabasca mylonite zone, northern Saskatchewan. ***Journal of Geophysical Research* 100**, 15717–15734.

WILLIAMS, P.F. (1972): Development of metamorphic layering and cleavage in low-grade metamorphic rocks at Bermagui, Australia. ***American Journal of Science* 272**, 1–47.

WILLIGERS, B.J.A., KROGSTAD, E.J. & WIJBRANS, J.R. (2001): Comparison of thermochronometers in a slowly cooled granulite terrain, Nagssugtoqidian Orogen, west Greenland. ***Journal of Petrology* 42**, 1729–1749.

Wolf, M.B. & Wyllie, P.J. (1991): Dehydration-melting of solid amphibolite at 10 kbar: textural development, liquid interconnectivity and applications to the segregation of magmas. ***Mineralogy and Petrology* 44**, 151–179.

Wolf, M.B. & Wyllie, P.J. (1994): Dehydration-melting of amphibolite at 10 kbar: the effects of temperature and time. ***Contributions to Mineralogy and Petrology* 115**, 369–383.

Wyllie, P.J. (1961): Fusion of Torridonian sandstone by a picrite sill in Soay (Hebrides). ***Journal of Petrology* 2**, 1–37.

Wynne-Edwards, H.R. (1963): Flow folding. ***American Journal of Science* 261**, 793–814.

Yardley, B.W.D. (1978): Genesis of the Skagit Gneiss migmatites, Washington, and the distinction between possible mechanisms of migmatization. ***Geological Society of America, Bulletin* 89**, 941–951.

Yardley, B.W.D. & Barber, J.P. (1991): Melting reactions in the Connemara Schists: the role of water infiltration in the formation of amphibolite facies migmatites. ***American Mineralogist* 76**, 848–856.

Zaleski, E., van Breemen, O. & Peterson, V.L. (1999): Geological evolution of the Manitouwadge greenstone belt and Wawa–Quetico subprovince boundary, Superior Province, Ontario, constrained by U–Pb zircon dates of supracrustal and plutonic rocks. ***Canadian Journal of Earth Sciences* 36**, 945–966.

The Photographs

Migmatites are the product of the process of partial melting, but their morphology is to a considerable extent determined by other factors. The most important among these are (1) the nature of the protolith, i.e., what rock types it contained and how they were distributed, (2) whether or not the rocks were being deformed when they contained partial melt, as this deformation drives the separation of the melt from the solid fraction, and also changes the shape of the rocks through folding, shearing, and transposition, and (3) time, as the length of the interval between melting and solidification is critical, e.g., it determines how far melt can separate from its residuum and how far the microstructure progresses to a fully equilibrated state. Thus, no two migmatites are the same; the study of migmatites is not like the study of fossils, and there is no equivalent to a "type specimen". Compare, for example, petrological specimens in museums; samples of diorite look much the same from one collection to another, but the samples labeled migmatite are all utterly different. The older literature on migmatites does contain line drawings that represent idealized versions of the relationships between melanosome, leucosome, and paleosome found in migmatites. However, one can scour outcrop after outcrop of migmatite and find nothing that resembles these so-called "classic relationships". What is found in outcrops, or in thin section, is variety. The line drawings are a construct, a simplified interpretation of the connection between morphology and process obtained by the study of naturally complex migmatites. The photographs in this book illustrate what is actually in migmatites.

The photographs in the following sections of this book are intended to give the reader an appreciation of the range of appearance and microstructure that can be found in migmatites. All the sections have a brief introduction and all, except the first, have further subdivisions to highlight specific key topics, for example, neosome or diatexite migmatite or melt inclusions. The photographs have been selected with two objectives in mind; the first photograph in a section shows a representative example of the topic at hand, and the subsequent ones illustrate the range of variations that can occur.

A.

SOME EXAMPLES OF MIGMATITES [FIGS. A1–A3]

The photographs in this short first section are intended to introduce the reader to the range of morphology that migmatites can have. The three migmatites shown were derived from protoliths that were compositionally layered, but it is whether or not the migmatites were deformed during anatexis that largely determines the appearance in the three cases.

Figure A1

Fig. A1. The absence of differential stresses during anatexis means that there is little driving force for the separation of the less viscous and less dense melt from the residual solids. Consequently, the migmatites that result, such as the one shown in this figure, have a very different appearance to the migmatites that form when there are differential stresses, such as those shown in Figs. A2 and A3. In this migmatite, partial melting occurred in the absence of differential stresses and, moreover, affected only the metapelitic beds, because the metamorphic temperatures were too low for melting to start in the interlayered beds of psammite. As a result, the pelite beds have changed color, become much coarser grained, and lost any pre-anatectic fine-scale structures that they may have had. Distinct leucosome and melanosome did not form in the pelitic beds because there was no driving force for the effective separation of the melt from the solid (i.e., no movement of the melt occurred). Thus, overall this migmatite consists of neosome, which was derived from the metapelitic beds, and paleosome, which are the dark psammitic beds that did not partially melt. However, the neosome in the center does show slight separation of the melt from residuum at a scale of 1–2 cm.

Location: Mount Stafford, Australia. *Rock type*: metatexite migmatite consisting of alternating layers of neosome and paleosome; pelitic protolith interbedded with psammite, partial melting at T ca. 775–785°C, P 3.3–4.0 kbar, on the boundary between local metamorphic zones 3 and 4. *Image and caption*: Richard White.

Further reading: Greenfield, J.E., Clarke, G.L., Bland, M. & Clarke, D.L. (1996): *In situ* migmatite and hybrid diatexite at Mt. Stafford, central Australia. *Journal of Metamorphic Geology* **14**, 413–426.

Greenfield, J.E., Clarke, G.L. & White, R.W. (1998): A sequence of partial melting reactions at Mt. Stafford, central Australia. *Journal of Metamorphic Geology* **16**, 363–378.

White, R.W., Powell, R. & Clarke, G.L. (2003): Prograde metamorphic assemblage evolution during partial melting of metasedimentary rocks at low pressures: migmatites from Mt. Stafford, central Australia. *Journal of Petrology* **44**, 1937–1960.

Figure A2

Fig. A2. Exhumation of upper amphibolite facies from a mid-crust depth in the thickened crust in British Columbia began with crustal extension in the early Tertiary. Partial melting and the upward migration of melt occurred during exhumation and led to a major weakening of the crust, which contributed to the rise of metamorphic core complexes and domes, and to the late collapse of the orogen. The weak, partially melted rocks in the core complexes, the migmatites, were highly strained as the core complexes developed and were exhumed by detachment faulting. The migmatites in the photograph are from the Thor–Odin dome and have a morphology that is characteristic of rocks that were highly strained while they contained melt. The most obvious feature is that the majority of the leucocratic domains are parallel to the strong tectonic foliation and the attenuated compositional banding in the mesocratic, grey paleosome host. The preponderant morphology of this migmatite is stromatic. Closer inspection indicates that there are two types of leucocratic stromata, a set of narrow (<3 cm) greyish ones (A) that do not appear to have either a conspicuous melanosome or a mafic selvedge associated with them, and a later group of wider, slightly discordant white domains (B) that commonly do have a mafic selvedge. The older grey set may have formed *in situ* and, therefore, be leucosome, but the later set was injected and, therefore, best termed leucocratic veins. Shorter and thicker leucocratic bodies (C) that are highly discordant to the foliation occur in extensional structures and small oblique shear zones throughout the migmatite. Such leucocratic domains have a semiregular spacing between 50 cm and 1 m, and commonly have a conspicuous melanosome or a mafic selvedge around them; they may also be leucosome. Melt may have flowed from the stromatic leucosome to the discordant leucosome as the migmatite was deformed. In order to test this possibility, it would be necessary to determine whether there is petrological continuity from the stromatic to the discordant leucosome in order to establish whether they contained melt at the same time.

Location: Thor–Odin dome, British Columbia, Canada. *Rock type*: stromatic metatexite migmatite, partially melted at upper-amphibolite-facies conditions. The white area in the upper left is snow. *Scale*: hammer near the center of the photograph for scale. *Image*: Olivier Vanderhaeghe.

Further reading: Vanderhaeghe, O., Teyssier, C. & Wysoczanski, R. (1999): Structural and geochronological constraints on the role of partial melting during the formation of the Shuswap metamorphic core complex at the latitude of the Thor–Odin dome, British Columbia. *Canadian Journal of Earth Sciences* **36**, 917–943.

Vanderhaeghe, O. & Teyssier, C. (2001): Partial melting and flow in orogens. *Tectonophysics* **342**, 451–472.

Figure A3

Fig. A3. Rocks at Sand River in the Limpopo Mobile Belt of southern Africa have been deformed and partially melted in the deep parts of an orogen at least once and possibly three times. Many of the high-grade rocks there are highly strained. This photograph shows a migmatite which is mostly paleosome that has strongly attenuated banding consisting of layers of trondhjemitic, granodioritic, and mafic composition. The more competent layers in the paleosome are boudinaged. Laterally persistent, centimeter- and millimeter-wide leucocratic bands (A) occur in the incompetent trondhjemitic and granodioritic layers that did not boudinage, but they are so strongly attenuated that it is not possible to determine whether they are felsic dikes or leucosome material. However, there are networks of conspicuous, coarser-grained leucosome associated with each of the boudinaged layers. The appearance of the leucosome in these layers depends on structural position. Leucosome located in pressure shadows and interboudin partitions are thickest, have the coarsest grain-size and are not foliated. Tracing along the leucosome domains reveals that they become finer grained, thinner, and foliated where located on the long sides of boudins, or are in shear zones, i.e., they are strongly attenuated where not in protected structural sites. Some anatectic melt thus has moved from foliation planes and collected between the boudins. There are a few bodies of leucosome with a coarse grain-size and no foliation in dilatant fractures that cross several layers. Their presence indicates that anatectic melt was still present as the deformation ended. This migmatite and that shown in Fig. A2 were both highly strained while melt was present, and they have similar overall morphologies in which stromatic leucosome dominates. However, the appearance of the two migmatites is very different, and there are two reasons for this. Firstly, the Thor–Odin migmatite contained much more melt, and so has more and wider bodies of leucosome. Secondly, the paleosome in the Sand River migmatites is leucocratic, and consequently, the leucosome is far less conspicuous.

Location: Sand River Gneisses, Causeway locality, South Africa. *Rock type*: metatexite migmatite; trondhjemite, granodiorite, and mafic protoliths. Anatexis in the granulite facies, *T* 800–850°C, *P* 7–10 kbar. *Scale*: the pocket knife is 11 cm long. *Image*: E.W. Sawyer.

B.

THE PARTS OF A MIGMATITE

Neosome and paleosome [Figs. B1–B4]

The parts of a migmatite created by and during anatexis are called neosome. The part not affected by partial melting, which commonly retains some characteristics of the pre-partial-melting state, is called the paleosome. The photographs in this section illustrate some of the characteristics used to identify and describe neosome and paleosome.

Neosome with leucosome and melanosome [Figs. B5–B14]

The best-known constituents of neosome are leucosome and melanosome, and they form when the melt fraction generated by partial melting becomes separated, or segregated, from the solid fraction. The leucosome is derived from the melt fraction, whereas the melanosome represents the solid products of the melting reaction, typically as a result of incongruent melting, and it occurs with or without excess reactant phases. Some of the common relationships between leucosome and melanosome that are found in neosome are shown on the following photographs. The examples also show some of the morphological variety displayed by leucosome and melanosome; such variety is in large measure due to protolith anisotropy and the way in which it responds to deformation.

Neosome without distinct leucosome or melanosome [Figs. B15–B26]

Not all examples of neosome develop conspicuous and separate domains of leucosome and melanosome, and this can make the recognition of anatexis and therefore migmatites difficult. However, in most cases, there is little difficulty in recognizing the neosome. One particular morphology of neosome (Figs. B20–B25) appears to be diagnostic of reactions involving dehydration melting of biotite, and perhaps of hornblende as well.

Neosome in open-system migmatites [Figs. B27–B34]

Migmatites that form without the addition, or loss, of material are said to have formed in a "closed system". If material was added to, or removed from, the migmatite during anatexis, then the migmatite formed in an "open system". Once a low-viscosity phase (melt) has formed during anatexis, open-system behavior very commonly ensues, because the melt fraction has a low viscosity and is able to move out of the more viscous solid residuum. Scale is an important factor in deciding whether the system was "open" or "closed". For practical purposes, if all the melt and residual parts can be identified at the scale of a hand specimen, or the outcrop, then one can regard the processes of forming the migmatite as a "closed system". However, if the scale of the separation between the melt and residuum is greater than the scale of observation, one regards the migmatite-forming processes as having occurred in an "open system". Migmatites that formed in closed systems are easier to understand because the relative proportions of the melt and residuum are preserved in the neosome. In open-system migmatites, the relative proportions of the melt and residuum have been changed. Migmatites, or regions within migmatites, that have lost melt are recognized by their higher proportion of residual material relative to melt, or melt products (e.g., leucosome). Conversely, migmatites that have had melt added to them can be recognized by a paucity of residuum. This group of photographs show some examples of "open-system" migmatites.

Variations within neosome [Figs. B35–B42]

In many migmatites, there is a considerable range in the morphology, microstructure and composition within the leucosome and melanosome they contain. However, it is the leucosome that shows the most obvious variability within individual domains, between generations of leucosome, or between the structural sites they occupy in the migmatite. This section illustrates some of this variability.

From leucosome to leucocratic dikes in migmatites [Figs. B43–B48]

The process of melt segregation in migmatites starts with the movement of melt from the grain boundaries where it formed toward channels. The first channels into which melt flows are located in the source area where the melt formed; they are surrounded by melt-depleted rocks that are, typically, melanocratic. The location of channels is controlled by such factors as the nature of the rock's anisotropy, and the way in which the rock is deformed. The channels serve as both the sites where melt may collect and as conduits through which melt flows. Where the melt in the channels crystallizes, it is preserved as the leucosome in a migmatite. For anatectic melt to leave its source area and reach higher levels in the crust, the flow of melt must become focused into progressively fewer, but larger, channels. The larger channels do not have a melanosome around them because they are no longer fed from the adjacent rocks. In addition, the channels are commonly discordant; consequently, what they contain no longer is called leucosome; rather, they are leucocratic veins, or where larger, leucocratic dikes. Some examples of the transition from leucosome to leucocratic dikes in migmatites are shown in this series of photographs.

Selvedges in migmatites [Figs. B49–B54]

In many migmatites, there are narrow, compositionally and microstructurally distinct rims around the leucosome and leucocratic veins; these are called *selvedges*. The compositionally distinct rims that are the residual material left after the extraction of the anatectic melt that makes up some, most, or all of the associated leucosome could be called *melt-depleted selvedges*. However, the residual material is most commonly melanocratic, and the term *melanosome* is reserved exclusively for these. Therefore, the term *selvedge* applies to the narrow, compositionally distinct rim that occurs around leucosome and leucocratic veins and that is not formed by the extraction of anatectic melt. There is no requirement that selvedges be melanocratic. However, the most common type of selvedge is melanocratic, and consists of a very thin biotite- or hornblende-rich rim around leucocratic veins; such a rim is called a *mafic selvedge*, not a melanosome. Mafic selvedges in migmatite terranes are most commonly developed around late, discordant leucocratic veins or dikes; they can also develop around granitic dikes intruded into host rocks that are not migmatites (see Figs. G13, G15, G16, and G19).

Figure B1

Fig. B1. The migmatite shown consists of scattered patches of neosome in the paleosome. The paleosome is a mesocratic plagioclase–quartz–K-feldspar orthogneiss with <5% biotite + sillimanite, and is recognized by the presence of a foliation that predates the partial melting. The small patches of neosome consist of a melanocratic (garnet + hercynite) part, generally in the center, and an outer leucocratic rim with the assemblage K-feldspar + quartz + plagioclase. Thus, the neosome appears to have undergone segregation of the melt from the residual solids. Note that some patches of neosome do not have a melanocratic core, probably because the plane of the rock surface does not pass near the center of those patches. Unlike their host, the patches of neosome do not have a foliation, and they are generally equidimensional and somewhat irregularly distributed. However, a band of neosome to the left of the lens cap is subparallel to the foliation in the paleosome, and its origin may be due to the infiltration of a fluid along the direction of the foliation plane in the host.

Location: Baume migmatite, Ardèche, France. *Rock type*: patch metatexite migmatite; quartzofeldspathic gneiss protolith partially melted under upper-amphibolite-facies conditions, T 700 ± 50°C, P 3 kbar. *Scale*: the lens cap is 6 cm across. *Image*: Pierre Barbey.

Further reading: Weber, C. & Barbey, P. (1986): The role of water, mixing processes and metamorphic fabric in the genesis of the Baume migmatites (Ardèche, France). *Contributions to Mineralogy and Petrology* **92**, 481–491.

Figure B2

Fig. B2. This migmatite has a compositionally uniform and weakly foliated paleosome in which scattered patches of light-colored neosome have a somewhat irregular shape that might be related to fractures that developed in the paleosome at the time of partial melting. The paleosome is a medium-grained garnet–hornblende–plagioclase metagabbro; the neosome has the mineral assemblage plagioclase + quartz + clinopyroxene + garnet and is much coarser grained. Hence, a possible reaction to describe melting is hornblende + plagioclase = clinopyroxene + melt. The minerals are rather uniformly distributed in the neosome in the upper left. However, the larger patch of neosome in the center is more complex, as the residual ferromagnesian minerals (mostly clinopyroxene) are at the margin in some places, and in the center elsewhere. Because the paleosome contains hornblende, it is darker colored than the residual part of the neosome, illustrating that color is not invariably reliable as a basis for distinguishing paleosome from residuum, and hence for genetic interpretations in migmatites.

Location: Central Metasedimentary Belt, Grenville Province near Wakefield, Quebec, Canada. *Rock type*: patch metatexite migmatite; metagabbro protolith, anatexis at conditions of the upper amphibolite to lower granulite facies, T ca. 800°C, P 6–9 kbar. *Scale*: the ruler is 15 cm long. *Image*: E.W. Sawyer.

Figure B3

Fig. B3. Only the fine-grained, grey, lens-shaped body, which can be called either a raft, scholle, or enclave (labeled P), to the right of the center is paleosome in this migmatite; all the rest is neosome of one sort or another. The paleosome scholle is fine-grained and mesocratic; it represents the remains of a plagioclase + quartz + biotite resister lithology (a psammite or metagreywacke) that did not melt, but it did undergo boudinage. The rest of the metasedimentary rocks, of pelitic or semipelitic composition, underwent partial melting and has been modified. The neosome is morphologically complex, as some parts (top of the photograph) have a layered appearance, whereas others are more irregular in form. Because the melt fraction has been segregated from its residuum to various extents, the neosome is compositionally complex also. The melt-rich parts of the neosome are evident as leucosome (L). The parts rich in residual minerals are melanocratic (M), but the parts of the melanosome enriched in biotite are far darker colored than those parts where garnet is abundant. Furthermore, there are coarse-grained, mesocratic parts to the neosome; in these parts, the melt did not completely segregate from the residuum. This migmatite illustrates that no specific genetic connotation should be attached to the parts of a migmatite that are mesocratic.

Location: Nemiscau Subprovince, Quebec, Canada. *Rock type*: schollen diatexite migmatite; siliciclastic metaturbidite protolith partially melted at T 750–820°C, P 4–5 kbar. *Scale*: the ruler is 15 cm long. *Image*: E.W. Sawyer.

Figure B4

Fig. B4. This migmatite consists almost entirely of neosome; it is a diatexite migmatite. Paleosome occurs only as scattered, dark, rounded rafts or schollen in the neosome (e.g., between the hammer and the ruler). Although leucocratic relative to the paleosome, the neosome cannot be described as leucosome because it contains too high a proportion of ferromagnesian minerals; biotite and orthopyroxene in this case. The neosome is characterized by a coarse grain-size and by the loss of structures, such as bedding, that predate the partial melting. The very uniform appearance of this migmatite could arise because (1) the protolith had a very homogeneous bulk composition, so that the degree of partial melting was everywhere similar, or (2) there was very little separation of the melt fraction from the solid fraction, and consequently, the neosome did not develop conspicuous leucosome and melanosome. However, some places are a little lighter colored, or a little darker colored, than others. Several dark bands (S), which are concentrations of residual biotite and orthopyroxene (the melting reaction in these rocks was biotite + plagioclase + quartz = orthopyroxene + ilmenite + melt) are present at the lower right. These thin bands, ribbons, and foliae of melanocratic material are called schlieren.

Location: Ashuanipi Subprovince, Quebec, Canada. *Rock type*: diatexite migmatite; siliciclastic metaturbidite protolith, granulite-facies anatexis; T 825–875°C and P 6–7 kbar. *Scale*: the hammer is 40 cm long. *Image*: E.W. Sawyer.

Figure B5

Fig. B5. The protolith to this migmatite is a foliated metamafic rock, and the neosome shown has a concentrically zoned structure; there is a rim of melanosome, and an inner domain of leucosome. The leucosome is comparatively uniform; it consists of coarse-grained leucotonalite. The melanosome, however, is not uniform. It consists of a clearly visible hornblende-rich inner part that is essentially devoid of plagioclase and quartz, and a wider and less obvious outer part in which the abundant plagioclase and quartz progressively increase outward to the level found in the host, which is the paleosome where it did not partially melt. It may not be easy to locate precisely where the outer edge of the melanosome is in the field; consequently, petrographic examination of thin sections may be required to do this if mass-balance calculations are contemplated. Geochemical mass-balance of this and other areas of neosome from the outcrop indicates that the neosome formed in a closed system, i.e., the melt fraction that created the leucosome came from the adjacent melanosome, which therefore represents a melt-depleted halo around the leucosome. Anatexis in a closed system in which the melt and residuum have separated, but remained in place, is referred to as *in situ*; hence, this is an example of an *in situ* leucosome.

Location: Abitibi Subprovince south of Chibougamau, Quebec, Canada. *Rock type*: metatexite migmatite; metamafic protolith, anatexis at T 800–850°C and P 8–10 kbar. *Scale*: the ruler is 15 cm long. *Image*: E.W. Sawyer.

Further reading: Sawyer, E.W. (1991): Disequilibrium melting and the rate of melt–residuum separation during migmatization of mafic rocks from the Grenville Front, Quebec. *Journal of Petrology* **32**, 701–738.

Figure **B6**

Fig. B6. This migmatite consists mostly of a strongly foliated and compositionally banded, grey, titanite–K-feldspar–hornblende–biotite–quartz–plagioclase metadiorite paleosome. There are two types of neosome present. A large domain of low-aspect-ratio *in situ* neosome is located in a dilatant structure that has developed oblique to the foliation and the compositional banding in the paleosome. This neosome consists of an inner leucosome and an outer melanosome, which is developed around most, but not all of the leucosome. Both parts of the neosome are coarser grained than the paleosome. They do not have the foliation or the banding present in the adjacent paleosome and, therefore, postdate these structures. The leucosome portion is tonalitic in composition (plagioclase + quartz + minor K-feldspar), and some of the plagioclase crystals that occur within quartz are euhedral, a microstructure interpreted to indicate crystallization from a melt. The melanosome belonging to the leucosome has the mineral assemblage plagioclase + hornblende and, in places, hornblende + plagioclase (i.e., quartz and K-feldspar are absent). Furthermore, it is much wider, and darker, in some places than in others; consequently, this melanosome is more irregular and less symmetrical about its leucosome than that in Fig. B5. The second type of neosome consists of thin bands (<10 mm) of plagioclase + quartz leucosome with a very narrow hornblende-rich melanosome. Because it occurs parallel to the foliation and gives the migmatite a layered appearance, it can be described as stromatic. The stromatic leucosome has a smaller grain-size than the leucosome in the dilatant structure, but where the two are in contact, the transition from one to the other is transitional. In neither case can the melanosome borders of the stromatic leucosome be traced into the low-aspect-ratio leucosome. These relationships could mean that both were partially molten at the same time.

Location: Abitibi Subprovince south of Chibougamau, Quebec, Canada. *Rock type*: metatexite migmatite; metadiorite protolith, anatexis at T 800–850°C and P 8–10 kbar. *Scale*: the ruler is 15 cm long. *Image*: E.W. Sawyer.

Figure B7

Fig. B7. The paleosome part of this migmatite is a biotite–quartz–hornblende–plagioclase gneiss in which folded grey leucocratic plagioclase + quartz layers 1–5 mm wide without melanocratic borders is evident. The neosome part (around the pen) consists of quartz–plagioclase–K-feldspar leucosome enveloped by melanosome, which consist mainly of hornblende. The concentration of hornblende to form the melanosome suggests that hornblende was either not involved in the melting reaction and was in excess during partial melting, or that hornblende was a solid product of reaction. The presence of inclusion-free, presumably melt-precipitated hornblende crystals in the leucosome suggests that hornblende was indeed involved in the melting reaction, which may have been biotite + plagioclase + quartz + hornblende^{1} = hornblende^{2} + melt. Poorly aligned flakes of poikiloblastic biotite in the melanosome probably formed by reversal of the melt-producing reaction during crystallization of the leucosome.

Location: Arunta Inlier, Australia. *Rock type*: metatexite migmatite; gneiss protolith of intermediate composition partially melted under conditions of the upper amphibolite to lower granulite facies. *Scale*: the pen is 14 cm long. *Image and caption*: Tony I.S. Kemp.

Figure **B8**

Fig. B8. A band of quartz + plagioclase + K-feldspar leucosome about 15 mm wide can be seen in the center of this photograph. Although it is not evident in this close-up, the leucosome is oriented parallel to the compositional layering (bedding) in the migmatite; therefore, the leucosome is described as stromatic. There are prominent, narrow (about 4 mm) borders of melanosome containing large crystals of garnet, orthopyroxene and biotite developed along both sides of the leucosome. The very low modal content of quartz and feldspar in the melanocratic rims is consistent with the interpretation that they contain the residuum left after the extraction of the melt that subsequently crystallized to become the leucosome. The grey mesocratic part of the migmatite beyond the melanosome is a quartz + plagioclase + biotite + garnet + K-feldspar ± orthopyroxene metapsammite, which on the basis of its mineral assemblage has undergone partial melting and is mostly residuum. Some porphyroblasts of garnet in the mesocratic part have a narrow leucocratic rim, and these are interpreted to be K-feldspar (a solid product of the melt-producing reaction) and crystallized melt. Therefore, the mesocratic part of this migmatite is neosome, not paleosome. All the biotite in the melanosome and some in the mesocratic part replace garnet and orthopyroxene. This replacement may be due to the loss of melt, a reaction between anhydrous residual phases and melt during cooling, or to reaction with an aqueous fluid.

Location: Wuluma Hills, Arunta Inlier, Australia. *Rock type*: metatexite migmatite; psammite protolith, partially melted at T 825–875°C, P ca. 5 kbar. *Scale*: the ruler is graduated in cm and mm. *Image*: E.W. Sawyer.

Figure B9

Fig. B9. The neosome in this migmatite occurs in a layer of medium-grained (2–4 mm) garnet–plagioclase–hornblende metamafic schist, on which the scale rests. The neosome is coarse grained (5–12 mm), and consists of buff-colored, irregularly shaped leucosome of tonalite composition that contains minor amounts of garnet and clinopyroxene, and a brownish purple, plagioclase–hornblende–clinopyroxene–garnet melanosome. If it is assumed that the host layer has a composition similar to the protolith of the neosome, then the inferred reaction responsible for partial melting was hornblende + plagioclase = melt + clinopyroxene + garnet. The leucosome occupies dilatant structures that formed as the host layer (protolith) underwent layer-parallel extension. Contacts between the melanosome and leucosome are diffuse; melanosome is absent where the leucosome shows a sharp contact with its host metamafic schist. Some of the melt thus was able to migrate a short distance away from its *in situ* melanosome when brittle fracturing occurred. The leucosome contains large, euhedral crystals of clinopyroxene, whereas the clinopyroxene in the melanosome is generally xenoblastic or subidioblastic; the clinopyroxene in the leucosome thus crystallized in a melt. Many clinopyroxene crystals have a narrow rim of hornblende.

Location: Abitibi Subprovince south of Chibougamau, Quebec, Canada. *Rock type*: metatexite migmatite; metamafic protolith, anatexis at *T* 800–850°C and *P* 8–10 kbar. *Scale*: the ruler is 15 cm long. *Image*: E.W. Sawyer.

Figure **B10**

Fig. B10. This strongly foliated metatexite migmatite from southern Brittany has a complex neosome. The grey, fine-grained, schistose layers (P) contain the mineral assemblage quartz + oligoclase + biotite (some layers contain minor amounts of garnet) and probably underwent little, if any, partial melting; they are possibly quartzofeldspathic paleosome. The coin rests on one of several coarser-grained layers that contain the mineral assemblage plagioclase + quartz + biotite ± garnet ± cordierite ± sillimanite and may be residual; these were derived from more aluminous metapelitic beds. Domains of leucosome up to a few centimeters wide occur within the coarser-grained layers and are either trondhjemitic (quartz + oligoclase + biotite) or granitic (quartz + K-feldspar + oligoclase + biotite) in composition; both may contain small amounts of garnet and sillimanite. Many of the leucosome areas have conspicuous, coarse-grained melanosome from 1 to 10 mm wide around them that contains the mineral assemblage biotite + cordierite + sillimanite ± garnet ± quartz ± plagioclase. The assemblage is consistent with a residuum derived by the loss of anatectic melt from a pelitic protolith. An unusual feature of the melanosome in this migmatite is the development of prominent biotite-rich bands within it. These consist almost entirely of decussate crystals of biotite up to 2 cm long. The coarse-grained biotite-rich bands could be part of the melanosome, or they could be a mafic selvedge that formed later by reaction between the leucosome and its host. The microstructures associated with the bands provide the critical information. Some biotite in the bands is partially replaced by a rim of cordierite that separates it from sillimanite in the rest of the melanosome. This microstructure is interpreted to have formed during the incongruent melting of biotite. Consequently, the biotite bands were present before dehydration melting began. Therefore, the biotite bands are interpreted to have developed at the very beginning of anatexis, at which stage H_2O-present melting reactions occurred and did not involve biotite. The biotite bands are an integral part of the melanosome and are not mafic selvedges. Most of the leucosome domains in this migmatite are oriented parallel to the compositional layering and foliation. Thus, this is a stromatic metatexite migmatite. However, leucosome is present between the boudins. The migmatite thus was deformed while melt was still present.

Location: Port Navalo area, southern Brittany, France. *Rock type*: stromatic metatexite migmatite; quartzofeldspathic and aluminous metasedimentary protolith, anatexis at *T* ca. 800°C, *P* ca. 8 kbar, then decompression to 4 kbar and cooling. *Scale*: the coin is 24 mm across. *Image and caption*: Mike Brown.

Further reading: Jones, K.A. & Brown. M. (1990): High-temperature 'clockwise' *P–T* paths and melting in the development of regional migmatites: an example from southern Brittany, France. *Journal of Metamorphic Geology* **8**, 551–578.

Marchildon, N. & Brown, M. (2003): Spatial distribution of melt-bearing structures in anatectic rocks from southern Brittany, France: implications for melt transfer at grain- to orogen-scale. *Tectonophysics* **364**, 215–235.

Figure B11

Fig. B11. Migmatites in which the neosome forms parallel layers are called stromatic; single layers, as in this example, are called stroma. The prominent layering in the paleosome formed when the metamafic protolith underwent noncoaxial shearing before and during partial melting, which attenuated the pre-existing compositional layering. The neosome is parallel to the layering and consists of two parts: a white leucosome of leucotonalite composition (plagioclase + quartz), and a hornblende-rich, plagioclase-depleted melanosome. The leucosome contains subhedral crystals of plagioclase that have crystal faces against quartz, and large euhedral crystals of randomly oriented hornblende. Both microstructures indicate crystallization from a melt. This stromatic neosome is unusual in that melanosome is developed only on one side of the leucosome and not on both, as is more common. Mass-balance calculations using the whole-rock major- and trace-element compositions indicate that the neosome probably formed in a closed system and is, therefore, an example of *in situ* melting and subsequent segregation of the melt from the residual solids. This is an example of an *in situ* neosome; the leucosome part may also be called an *in situ* leucosome.

Location: Abitibi Subprovince, south of Chibougamau, Quebec, Canada. *Rock type*: metatexite migmatite, metamafic protolith partially melted at *T* 800–850°C and *P* 8–10 kbar. *Scale*: the ruler is 15 cm long. *Image*: E.W. Sawyer.

Further reading: Sawyer, E.W. (1991): Disequilibrium melting and the rate of melt–residuum separation during migmatization of mafic rocks from the Grenville Front, Quebec. *Journal of Petrology* **32**, 701–738.

Figure **B12**

Fig. B12. The neosome in this migmatite is much coarser grained than the hornblende + plagioclase + garnet paleosome upon which the scale rests. Most of the larger domains of neosome are located in and around shear zones and interboudin partitions. The melanosome is composed of hornblende + garnet + clinopyroxene with minor plagioclase, and the leucosome mostly consists of plagioclase + quartz, but locally contains large, euhedral crystals of light green clinopyroxene. The melting reaction may have been hornblende + plagioclase = clinopyroxene + melt. Although the leucosome and melanosome are in contact, the distribution of melanosome is irregular and asymmetrical relative to the leucosome. The darkest parts of the melanosome are rich in hornblende, which appears to replace clinopyroxene. The hornblende in the melanosome thus results from a late- or post-anatexis rehydration reaction. Small amounts of leucosome also are present parallel to the foliation planes in the paleosome. Some of this leucosome has a conspicuous melanocratic border (mafic selvedge) resulting from the late replacement of clinopyroxene by hornblende.

Location: Central Metasedimentary Belt, Grenville Province near Wakefield, Quebec, Canada. *Rock type*: metatexite migmatite; metagabbro protolith, anatexis under conditions of the lower granulite facies. *Scale*: the ruler is 15 cm long. *Image*: E.W. Sawyer.

Figure B13

Fig. B13. This photograph shows a thin (10 cm) layer of plagioclase–garnet–clinopyroxene metamafic rock with quartz–plagioclase *in situ* and in source tonalitic leucosome that occurs within a garnet-bearing plagioclase–hornblende metabasic rock, visible in the upper right. The photograph is taken normal to the plane of the layering, and shows that most leucosome is disposed in an array of linked extension-induced fractures. There are also small, round halos of leucosome, however, about 8 mm across, surrounding some of the garnet crystals. The borders of the leucosome are not planar; rather, they are highly irregular on the grain scale, which suggests that they have not been subsequently deformed. The protolith of this layer is interpreted to have had just the optimal bulk composition for partial melting to have consumed virtually all its hornblende to generate abundant garnet and clinopyroxene, both of which are very strong minerals. Consequently, as melting progressed, and the melt fraction was removed, this layer became more competent than its neighbors and developed many extensional fractures. The last increments of anatectic melt in the layer migrated into the extensional fractures as they developed and, when temperatures declined, crystallized there to form the leucosome. A possible melting reaction is hornblende + plagioclase = melt + garnet + clinopyroxene.

Location: Abitibi Subprovince south of Chibougamau, Quebec, Canada. *Rock type*: metatexite migmatite with dilatant structures; metamafic protolith, anatexis at *T* 800–850°C and *P* 8–10 kbar. *Scale*: the ruler is 15 cm long. *Image*: E.W. Sawyer.

Figure **B14**

Fig. B14. This migmatite was derived from metapelite and consists of only two parts, leucosome and melanosome; it is all neosome. However, the adjacent outcrops contain psammitic, calc-silicate, and mafic resister lithologies that can be traced laterally for some distance; thus, in the larger context, this is a metatexite migmatite. The migmatite is polyphase, as it has experienced anatexis and deformation in the Archean (twice) and again in the Paleoproterozoic. Hence, the bodies of leucosome display a complex history of deformation and have been boudinaged and folded at various times. The domains of leucosome in the photograph are located parallel to the foliation and compositional layering derived from the original bedding; they are thin, and have a high aspect-ratio. Hence, the migmatite can be described as stromatic. The leucosome consists of quartz, plagioclase, and K-feldspar in various proportions, and contains only minor amounts of garnet and biotite. The melanocratic, metapelitic rocks between the domains of leucosome are rich in biotite and garnet, consistent with a melt-depleted, or residual, bulk composition. However, the metapelitic layers show differences in modal mineralogy, grain size, and microstructure from layer to layer. These differences probably reflect the initial variation in protolith composition from bed to bed.

Location: Beit Bridge Complex, Verbaard Farm, South Africa. *Rock type*: stromatic metatexite migmatite; pelitic protolith, anatexis at T 800–850°C, P 7–10 kbar. *Scale*: the pocket knife is 11 cm long. *Image*: E.W. Sawyer.

Figure B15

Fig. B15. Migmatites derived from felsic igneous protoliths such as granites, leucotonalites and trondhjemites can be very difficult to recognize in the field because the neosome has a composition, mineralogy, and microstructure similar to those of the paleosome. The paleosome in this migmatite (P) is a grey, foliated leucotrondhjemite, part of a late Archean, calc-alkaline plutonic arc in the Superior Province of Canada. The neosome (N) forms lighter, pinkish and buff-colored, rather nebulitic patches that are slightly coarser grained and, most notably, devoid of a foliation, relative to the paleosome. Petrographic and geochemical evidence indicates that no segregation of melt from residual solids occurred as the neosome patches formed; hence, there is no leucosome and no melanosome, i.e., the products of anatexis remained *in situ*. Locally, there are thin, pink leucocratic veins (LV) parallel to the foliation in the paleosome; these represent injections of anatectic melt (i.e., they are not *in situ* or in-source leucosome). The whitish streaks oriented parallel to the scale are scratches on the surface made when clearing the outcrop.

Location: Opatica Subprovince, Quebec, Canada. *Rock type*: patch metatexite migmatite; leucotrondhjemite protolith partially melted under upper-amphibolite-facies conditions, T ca. 750°C, and P 5–7 kbar. *Scale*: the ruler is 15 cm long. *Image*: E.W. Sawyer.

Further reading: Sawyer, E.W. (1998): Formation and evolution of granite magmas during crustal reworking: the significance of diatexites. *Journal of Petrology* **39**, 1147–1167.

Figure **B16**

Fig. B16. The protolith for this migmatite was a strongly foliated, coarse-grained garnet- and orthopyroxene-bearing monzogranite to syenogranite that contained abundant megacrysts of K-feldspar. The granite occurs as a large dike in a thick sequence of mafic granulites. The small degree of partial melting at this locality has generated scattered lenticular patches of neosome (N) located in crenulation bands that are oriented at a high angle to the pre-existing foliation. The neosome in this migmatite is very difficult to see largely because it is practically the same color as the host. One reason for this uniformity is the very felsic composition of the paleosome. The residuum left after the extraction of the small amount of melt generated in this migmatite is dominated by K-feldspar, plagioclase, and quartz and, consequently, not particularly enriched in ferromagnesian minerals relative to the melt-derived portion. Hence, there is no distinct leucosome, nor melanosome, to assist the eye in detecting the neosome. The best criterion to identify neosome in such circumstances is its truncation of older structures in the paleosome, the foliation in this example.

Location: Mount Hay, Arunta Inlier, Australia. *Rock type*: metatexite migmatite; foliated charnockite protolith, partial melting at T 825–875°C, P 6–7 kbar. *Scale*: the ruler is 15 cm long. *Image*: E.W. Sawyer.

Figure B17

Fig. B17. The center of this photograph shows a pelitic layer that has undergone significant partial melting. It is bordered at the top (beneath the pen knife) and the bottom by psammitic layers in which there was little or no partial melting. As the entire pelitic layer has been affected by partial melting, it can be called neosome, and three components can be identified within it. The melanocratic part consists of scattered dark grey andalusite porphyroblasts (A); these are surrounded by a dark rim, or corona, of cordierite + spinel symplectite that also contains minor amounts of K-feldspar and plagioclase. The light grey mesocratic component (R) of the neosome consists principally of cordierite and K-feldspar with minor amounts of quartz and biotite, and is interpreted to be the residuum left after partial melting of a pelitic protolith and the removal of most of the anatectic melt generated. Pink leucocratic material consisting mostly of K-feldspar and quartz is scattered unevenly throughout the neosome; it forms a rim around the andalusite porphyroblasts, as well as rather nebulous patches and small veins in the mesocratic part of the neosome. Because the K-feldspar is commonly prismatic and occurs in larger crystals of quartz (see Figs. F47 and F48), the leucocratic part of the neosome is interpreted to have crystallized from a melt, and can be termed leucosome. Although the melt and residual fractions did separate, the striking aspect of this migmatite is that the scale of segregation was millimetric only, and the leucosome and residuum remained intimately associated, a relationship suggestive of anatexis in the absence of differential stresses. A dark-colored layer separates the pelite from the psammite; this selvedge has formed as a result of interaction between the psammite and the anatectic melt in the pelite.

Location: Mount Stafford, Australia. *Rock type*: unsegregated neosome in a metatexite migmatite; the protolith was a pelitic layer in a thinly bedded sequence of psammite and pelite, partial melting at T ca. 700°C, P 3.2 kbar in local metamorphic zone 3. *Scale*: the pen knife is 11 cm long. *Image and caption*: Richard White.

Further reading: Greenfield, J.E., Clarke, G.L., Bland, M. & Clarke, D.L. (1996): *In situ* migmatite and hybrid diatexite at Mt. Stafford, central Australia. *Journal of Metamorphic Geology* **14**, 413–426.

Greenfield, J.E., Clarke, G.L. & White, R.W. (1998): A sequence of partial melting reactions at Mt. Stafford, central Australia. *Journal of Metamorphic Geology* **16**, 363–378.

White, R.W., Powell, R. & Clarke, G.L. (2003): Prograde metamorphic assemblage evolution during partial melting of metasedimentary rocks at low pressures: migmatites from Mt. Stafford, central Australia. *Journal of Petrology* **44**, 1937–1960.

Figure B18

Fig. B18. The migmatite in this photograph developed in a rock that contained layers of different composition and texture; consequently, the morphology of the neosome differs subtly from layer to layer. The pen rests on neosome that developed from a layer of aluminous metapelite that contained large porphyroblasts of andalusite; a similar layer occurs at the bottom of the photograph. The aluminous metapelite layers underwent a significant degree of partial melting, and the neosome that developed is complex, as it contains two types of leucocratic and melanocratic domains. The leucocratic domains are rich in K-feldspar and quartz. On the basis of the microstructures that they contain, they are interpreted to have contained anatectic melt; therefore, they can be termed leucosome. Fine-grained leucosome (FL) occurs as narrow layers a few millimeters wide that alternate with thin layers of fine-grained melanosome that is rich in cordierite and interpreted to be the residuum; they are melanosome. Coarser-grained leucosome (CL) forms somewhat irregular layers that locally overgrow, and even truncate, the alternation of fine-grained leucocratic and melanocratic layers. The coarse leucosome contains melanocratic patches (A) that are up to 2 cm across. These are porphyroblasts of aluminosilicate (aggregates of sillimanite crystals that partially, or completely, replace andalusite) that have been partially, or wholly, replaced by a symplectitic intergrowth of cordierite and spinel where there was no quartz. The protolith to this neosome was very similar to that of the migmatite shown in Fig. B17, but perhaps because the degree of partial melting was a little higher in this migmatite, the melt fraction in the neosome began to segregate from the residuum and formed the irregular and slightly discordant layers of coarse-grained leucosome. However, the process did not advance sufficiently to separate the leucosome from the melanosome by more that a few millimeters. Consequently, at the outcrop scale, the leucosome and residuum remain interspersed in this migmatite. The finer-grained and more prominently layered neosome that occupies the center of the photograph (below the tip of the pen) was derived from a subaluminous, cordierite-rich metapelite that contained a far lower fraction of melt, and did not contain porphyroblasts of aluminosilicate. The millimeter-scale layering in this neosome is due to the alternation of leucocratic layers rich in K-feldspar and quartz with melanocratic layers that are rich in cordierite; the layering in the neosome is interpreted to mimic the fine-scale sedimentary layering of the protolith. The neosome in each of the layers in this migmatite is quite different in terms of its grain size and the proportion of leucocratic material that it contains; such characteristics are controlled by the composition and the metamorphic textures that developed in the protolith.

Location: Mount Stafford, Australia. *Rock type*: partially segregated neosome in a metatexite migmatite; thinly bedded pelitic protolith, partial melting at T ca. 750°C, P 3.2 kbar, on the boundary between local metamorphic zones 3 and 4. *Scale*: the pen is 16 cm long. *Image and caption*: Richard White.

Further reading: see Fig. B17.

Figure B19

Fig. B19. Close-up of a partially melted metapelite in which the melt did not separate significantly from the residuum. The result is a migmatite in which the neosome is without conspicuous leucosome or melanosome. However, it has a much coarser grain-size and a different microstructure than a metapelite. The dark-colored wisps are the remains of psammitic beds that have become folded. Note the absence of either a foliation or evidence for folding within the pelite-derived neosome. The melting reaction in this quartz-bearing metapelite was biotite + sillimanite + quartz = cordierite + K-feldspar + melt.

Location: Mount Stafford, Australia. *Rock type*: nebulitic diatexite migmatite; pelitic protolith, partial melting at *T* ca. 700°C, *P* 3.2 kbar, on the boundary between local metamorphic zones 2 and 3. *Scale*: the ruler is 15 cm long. *Image*: E.W. Sawyer.

Further reading: see Fig. B17.

Figure **B20**

Fig. B20. This photograph shows an outcrop of metatexite migmatite that contains patches of coarser-grained, nonfoliated neosome in a matrix of finer-grained, foliated, leucogranite gneiss. The neosome consists of irregular patches and veins, some of which have locally joined together to form a vague, net-like pattern. Neosome is readily distinguished from the host gneiss by its coarser grain-size, and by its truncation of the biotite-defined layering in the gneiss. The neosome contains conspicuous crystals of garnet in a light-colored, quartzofeldspathic matrix; the garnet crystals tend to occur toward the center of the patches of neosome and are commonly intergrown with quartz. There are no rims of melanosome around the neosome. There is no garnet in the host gneiss, and there is no biotite in the neosome. This is an important observation; from it, one can infer that the melting reaction involved the incongruent breakdown of biotite in a reaction similar to biotite + plagioclase + quartz = garnet + K-feldspar + melt. In that case then, the foliated leucogneiss represents the assemblage of reactant minerals and could, consequently, be viewed as both the protolith and the paleosome that would have melted, if the metamorphic temperatures had been higher. Because the leucogranite gneiss predominates, anatexis was not particularly advanced in this migmatite. The neosome, on the other hand, represents both the solid and melt products of the melting reaction. It can be divided into leucosome and melanosome only with some difficulty because the two are intimately distributed. The parts of the neosome that consist entirely of quartz, plagioclase, and K-feldspar could be termed leucosome. Other parts consist of large irregular poikiloblasts, or aggregates, of garnet intergrown with quartz, domains that could be described as melanosome.

Location: Broken Hill area, New South Wales, Australia. *Rock type*: net-structured metatexite migmatite; quartzofeldspathic protolith, anatexis in the upper amphibolite facies, T ca. 800°C, P 4–5 kbar. *Scale*: the ruler is 15 cm long. *Image*: E.W. Sawyer.

Further reading: Braun, I., Raith, M. & Ravindra Kumar, G.R. (1996): Dehydration-melting phenomena in leptynitic gneisses and the generation of leucogranites: a case study from the Kerala Khondalite Belt, southern India. *Journal of Petrology* **37**, 1285–1305.

White, R.W., Powell, R. & Halpin, J.A. (2004): Spatially-focussed melt formation in aluminous metapelites from Broken Hill, Australia. *Journal of Metamorphic Geology* **22**, 825–845.

Figure B21

Fig. B21. The light-colored neosome in this migmatite has developed in a biotite gneiss of semipelitic bulk composition, and consists of streaks and patches of coarse-grained quartz and feldspar with large crystals of garnet. There is neither melanosome nor leucosome. The neosome has indistinct or slightly diffuse boundaries. The minerals in the neosome have no shape-preferred orientation, and there is no foliation. The streaks of neosome are parallel and follow, approximately, the foliation in the host paleosome, giving an overall stromatic, or layered, appearance to the migmatite. Some neosome is locally discordant.

Location: Namaqualand, South Africa. *Rock type*: stromatic metatexite migmatite; semipelitic protolith partially melted in the granulite facies. *Scale*: the hammer head is 10 cm across. *Image and caption*: Dave Waters.

Further reading: Waters, D.J. (1988): Partial melting and the formation of granulite facies assemblages in Namaqualand, South Africa. *Journal of Metamorphic Geology* **6**, 387–404.

Figure **B22**

Fig. B22. The neosome in this photograph has developed at the end of a boudin derived from a resister layer of slightly darker-colored intermediate gneiss (paleosome) hosted by a grey leucocratic gneiss. The lighter-colored gneiss has locally undergone partial melting and has generated neosome that dominantly consists of feldspar and quartz, but that also contains conspicuous, euhedral crystals of pyroxene and some of amphibole. Two characteristics in particular serve to distinguish the neosome in this migmatite from the paleosome; it is considerably coarser grained, and it does not have a foliation. Although the neosome is lighter in color than the paleosome and could, therefore, be described as leucocratic, it should not be called leucosome because it contains a high proportion of ferromagnesian minerals that are the product of the melting reaction. The melt fraction thus did not segregate from the solid fraction (residuum), and hence the term leucosome is inappropriate for this migmatite; neosome is correct. This photograph illustrates a small part of an extensive pavement on which pre-anatectic compositional layering and cross-cutting dike relations can be seen in the orthogneiss paleosome; hence, this is a metatexite migmatite.

Location: Sand River Gneisses, Causeway locality, South Africa. *Rock type*: metatexite migmatite; leucocratic, orthogneiss protolith, partially melted in the granulite facies. *Scale*: the pocket knife is 11 cm long. *Image*: E.W. Sawyer.

Figure B23

Fig. B23. This migmatite consists of two main parts, one dark-colored and the other light-colored, and is a good example of a case in which the leucocratic and melanocratic parts present do not correspond wholly to the usual genetic relationship found between leucosome (melt-derived part) and melanosome (residual part) in migmatites. The dark-colored part is a sillimanite-rich, aluminous metapelitic schist or gneiss that contains the mineral assemblage garnet + cordierite + sillimanite + biotite + quartz + plagioclase + ilmenite. The light-colored part contains large, prominent poikiloblasts of garnet in a medium-grained leucocratic matrix that is rich in K-feldspar and quartz. The large poikiloblasts of garnet occur only in the leucocratic material, whereas cordierite occurs only in the dark metapelitic gneiss. The leucocratic material is inferred to have formed *in situ* around the growing poikiloblasts, and is dominantly composed of peritectic K-feldspar, together with some K-feldspar, quartz and plagioclase that crystallized from the anatectic melt. Hence, the light-colored part of this migmatite contains both the products of a peritectic reaction (K-feldspar and garnet), which would be considered part of the residuum assemblage, and the products of the crystallization of the anatectic melt (K-feldspar, quartz, and plagioclase). Consequently, the leucocratic portion of the migmatite represents neosome that has not undergone much, if any, separation of melt from residuum. The aluminous metapelite that makes up the dark (melanocratic) part of the outcrop is not paleosome, and neither is it wholly residuum, as it does not contain the garnet porphyroblasts or the peritectic K-feldspar; it may most closely correspond to the protolith. However, there is no doubt that the light color of the neosome, due principally to the elimination of disseminated biotite, is the dominant characteristic of this migmatite in outcrop; consequently, these domains have been referred to as "leucosome".

Location: Round Hill, Broken Hill area, Australia. *Rock type*: metatexite migmatite; aluminous metapelite protolith partially melted at *T* 800°C and *P* 5–6 kbar. *Scale*: the hand lens is 3 cm long. *Image and caption*: Richard White. Image previously published as fig. 2a in White et al. (2004) and reproduced with the permission of Blackwell Publishing.

Further reading: White, R.W., Powell, R. & Halpin, J.A. (2004): Spatially-focussed melt formation in aluminous metapelites from Broken Hill, Australia. *Journal of Metamorphic Geology* **22**, 825–845.

Figure **B24**

Fig. B24. This migmatite is from the same area as that in Fig. B23. It too contains dark-colored ilmenite–plagioclase–quartz–biotite–sillimanite–cordierite–garnet aluminous metapelite, but there are some lighter-colored layers in it that represent thin, aluminum-poor beds. Domains of leucocratic neosome are abundant, and they contain conspicuous poikiloblasts of garnet. However, there are many more small grains of garnet scattered throughout the neosome in this migmatite than in that shown in Fig. B23. As the garnet poikiloblasts do not occur in the host aluminous metapelite, the aluminous metapelite in this case may correspond more closely to protolith, rather than residuum or paleosome. The neosome is inferred to have formed *in situ* around the growing poikiloblasts of garnet and is dominantly composed of peritectic K-feldspar with lesser amounts of K-feldspar, quartz, and plagioclase that crystallized from an anatectic melt. A tight F_2 antiform folds both the neosome and the compositional layering in the aluminous metapelite; an S_2 foliation defined by an alignment of biotite and sillimanite occurs in the aluminous metapelite and is oriented parallel to the axial plane of the F_2 fold. In contrast, the S_2 structure in the neosome is a fracture cleavage, and this is consistent with neosome already being in the solid state during the D_2 event.

Location: Round Hill, Broken Hill area, Australia. *Rock type*: folded metatexite migmatite; aluminous metapelite protolith, anatexis at *T* 800°C and *P* 5–6 kbar. *Scale*: the pen is 14 cm long. *Image and caption*: Richard White. Image previously published as fig. 2b in White et al. (2004) and reproduced with the permission of Blackwell Publishing.

Further reading: same as for Fig. B23.

Figure B25

Fig. B25. This photograph shows a migmatite developed in a strongly foliated garnet–K-feldspar–plagioclase–sillimanite–quartz–biotite aluminous metapelite gneiss. The neosome consists of several diffuse, almost skeletal, coarse-grained patches that have a core of pink garnet intergrown with quartz and K-feldspar. The garnet-rich cores are surrounded by a rim consisting of roughly equal proportions of K-feldspar, quartz, and plagioclase. Locally, some coarse-grained biotite replaces garnet and K-feldspar in the neosome. The neosome is interpreted to be the *in situ* product of the incongruent melting of biotite by a reaction such as biotite + sillimanite + quartz + plagioclase = garnet + K-feldspar + granitic melt. Some of the patches of neosome have a high proportion of garnet and K-feldspar (the solid products of the melting reaction, considered "peritectic" phases by some authors), which suggests that these regions of neosome have lost their anatectic melt. This interpretation is supported by the presence of coarse-grained, granitic leucosome without garnet (e.g., center of the top edge) located in dilatant structures within the gneiss, which are interpreted to have crystallized from the segregated anatectic melt. Thus, the migmatite in this photograph shares the diffuse and patch morphology of the neosome shown in Fig. B15, but also shares the characteristic of neosome formed by the incongruent melting of biotite shown in Figs. B20–B24, or of amphibole.

Location: Saint-Fulgence, Grenville Province, Quebec, Canada. *Rock type*: patch metatexite migmatite; aluminous pelite protolith, granulite-facies anatexis T 800–850°C, P 5–8 kbar. *Scale*: the width of the photograph corresponds to 16 cm. *Image*: E.W. Sawyer.

Figure **B26**

Fig. B26. The leucosome in this migmatite is quite evident. Some domains are parallel to the bedding (stromatic morphology), whereas others are discordant (dilatant morphology), but most of them have somewhat diffuse borders. In contrast, there are no conspicuously melanocratic parts in the migmatite, such as melanosome around a patch of leucosome. However, the mineral assemblage (plagioclase + biotite + quartz + garnet), modal proportions, and the whole-rock composition of the rock around the leucosome indicate that it is the residuum left after the extraction of melt from a protolith of semipelitic bulk composition. Therefore, the migmatite shown in this photograph consists entirely of neosome. The melt fraction, or part of it, is represented by the leucosome, but the rest of the rock in the photograph is not paleosome, despite its medium (mesocratic) color. It is the residuum left after the melt was segregated.

Location: Nemiscau Subprovince, northern Quebec, Canada. *Rock type*: metatexite migmatite; semipelite protolith partially melted under lower-granulite-facies conditions, *T* ca. 800°C, *P* ca. 5 kbar. *Scale*: the ruler is 15 cm long. *Image*: E.W. Sawyer.

Figure **B27**

Fig. B27. This photograph shows the neosome developed in the neck region of a boudinaged layer in a metamafic paleosome (P). The macroscopic aspects of the three stages of melt movement that occur in migmatites are well illustrated. Therefore, this example serves as an introduction to "open-system" migmatites. The domains of leucosome trace out a complex pattern of linked segments that are parallel to the layering, orthogonal to the layering, and in a conjugate set oriented at about 45° to the layering. The orientations of the dilatational sites are consistent with layer-parallel extension occurring during anatexis. However, the location of melanosome (M) is more restricted. It generally forms layers oriented parallel, or subparallel, to the foliation. This arrangement suggests that certain sublayers have lost much more melt than others; indeed, some may not have lost melt. The expelled melt moved through the linked system of dilatant structures and may ultimately have drained out from the layer at the boudin neck. Thus, in general, leucosome represents melt frozen in the channel system, and not necessarily the sites where melting occurred. Some isolated, irregularly shaped domains of leucosome with melanosome around them (just less than half way down on the right-hand edge) are *in situ* domains of leucosome, and are loci where melt has collected after a centimeter or so of porous flow. The domains of leucosome that form the linked array are examples of in-source leucosome. Note that some parts of these domains do not have a prominent melanosome. The large triangular domain of leucosome in the center of the boudin neck might have been part of the system of larger channels that belonged to the tertiary stage of melt movement through which melt moved out of its source layer and fed leucocratic dikes elsewhere in the migmatite. The top end of the scale rests on a pegmatite vein, not leucosome.

Location: Abitibi Subprovince, south of Chibougamau, Quebec, Canada. *Rock type*: metatexite migmatite; metamafic protolith partially melted at *T* 800–850°C and *P* 8–10 kbar. *Scale*: the ruler is 15 cm long. *Image*: E.W. Sawyer.

Further reading: Sawyer, E.W. (1991): Disequilibrium melting and the rate of melt–residuum separation during migmatization of mafic rocks from the Grenville Front, Quebec. *Journal of Petrology* **32**, 701–738.

Sawyer, E.W. (2001): Melt segregation in the continental crust: distribution and movement of melt in anatectic rocks. *Journal of Metamorphic Geology* **19**, 291–309.

Figure **B28**

Fig. B28. At first glance, this outcrop appears to show a metamafic rock containing many discordant felsic veins in fractures; therefore, one may infer that it is not a migmatite. However, in a few places, the presence of *in situ* neosome (N), consisting of coarse-grained melanosome that contains minor, thin (<2 mm) leucosome, indicates that partial melting and separation of the melt fraction from the solid have occurred and that the rock is indeed a migmatite. Most of the rock can be considered to be paleosome. In general (e.g., just above and to the left of the scale), the passage from paleosome to melanosome is gradational, without the presence of leucosome, and this suggests the loss of melt on a local scale. The striking aspect of this outcrop is the abundance of medium-grained, leucocratic veins that cross the paleosome, melanosome and *in situ* neosome. These veins lack adjacent melanosome, which suggests that they result from fracturing and injection of melt. Thus, the history of this migmatite involved partial melting, then loss of melt, and finally, addition of melt.

Location: Abitibi Subprovince, south of Chibougamau, Quebec, Canada. *Rock type*: veined metatexite migmatite; metamafic protolith partially melted at *T* 800–850°C and *P* 8–10 kbar. *Scale*: the ruler is 15 cm long. *Image*: E.W. Sawyer.

Further reading: see Fig. B27.

Figure **B29**

Fig. B29. This migmatite was derived from the same layered metamafic rock as in Fig. B27, but at a locality that underwent a higher degree of partial melting and a much greater loss of melt. The migmatite is stromatic in that wide, dark bands with the mineral assemblage hornblende + plagioclase + garnet + clinopyroxene alternate with thin, light-colored plagioclase-rich layers. The bulk composition of the migmatite is melt-depleted relative to the likely protolith composition, confirming its residual appearance. The outcrop contains a linked array of very narrow, discordant stringers of leucosome (indicated by the arrows) that contain discontinuous pockets and scattered individual crystals of subhedral plagioclase, and very little quartz. These stringers of in-source leucosome consist of accumulations of plagioclase, and are interpreted as the collapsed (deflated) remains of a channel system through which the melt generated during anatexis moved out of its residuum. This is a melt-depleted migmatite. The preservation of the compositional layering that predates the partial melting indicates this to be a metatexite migmatite. In choosing another descriptive term for this migmatite, one has to decide which trait to emphasize. If morphology is important, then "stromatic metatexite migmatite" might be appropriate. However, if the bulk composition of the migmatite is important, then "melt-depleted" or "residual" would be appropriate.

Location: Abitibi Subprovince, south of Chibougamau, Quebec, Canada. *Rock type*: melt-depleted metatexite migmatite; metamafic protolith partially melted at *T* 800–850°C and *P* 8–10 kbar. *Scale*: the ruler is 15 cm long. *Image*: E.W. Sawyer.

Further reading: see Fig. B27.

Figure **B30**

Fig. B30. This close-up photograph shows a patch of neosome developed in a grey garnet–biotite–plagioclase–quartz aluminous metapsammite. The neosome has two parts. An inner portion consists of a single, large brown poikiloblast of garnet that is intergrown with quartz (see Fig. F74 for an example of the microstructure of the garnet–quartz intergrowths), and an outer part that consists predominantly of medium-grained K-feldspar, with a small proportion of quartz and plagioclase. The neosome is interpreted to be the solid products (i.e., garnet + K-feldspar) of a melt-producing, biotite-breakdown reaction. The absence from this neosome of a leucocratic part that has a granitic composition leads to the interpretation that the melt fraction generated by the incongruent breakdown of biotite has migrated away. Consequently, the remaining neosome consists essentially of residuum.

Location: Wuluma Hills, Arunta Inlier, Australia. *Rock type*: melt-depleted patch metatexite migmatite; aluminous psammite protolith, partially melted at *T* 825–875°C, *P* ca. 5 kbar. *Scale*: the ruler is graduated in cm and mm. *Image*: E.W. Sawyer.

Figure B31

Fig. B31. The large-scale morphology of this outcrop shows a medium-grained, grey schistose rock that contains a few thin veins of granite. The granite veins occur in two sets that are discordant to the foliation, and a locally preserved compositional layering in the host. Together, both sets of veins combine to form a very open, net-like pattern. Furthermore, the granite veins have no melanosome, or mafic selvedge, associated with them. Consequently, the veins may be thought of as injected along two sets of fractures in the host. Hence, at the scale of this photograph, there is no evidence for partial melting, and it is not at all evident that this is an outcrop of migmatite. The evidence for partial melting comes from the close-up, detailed examination of the contacts between the veins and the host, as shown in the next figure (Fig. B32).

Location: Broken Hill area, New South Wales, Australia. *Rock type*: metatexite migmatite; quartzofeldspathic protolith (Hores gneiss), anatexis in the upper amphibolite facies, *T* ca. 800°C, *P* 4–5 kbar. *Scale*: the ruler is 15 cm long. *Image*: E.W. Sawyer.

Figure B32

Fig. B32. This close-up shows the area close to the left edge and just above half way up in Fig. B31. The country rock to the "granitic vein" is garnet–biotite–quartz–K-feldspar–plagioclase schist, or perhaps gneiss, in which garnet occurs as scattered porphyroblasts that attain 6 mm across. The garnet porphyroblasts do not have a quartzofeldspathic rim, which suggests that they grew as a result of a subsolidus prograde reaction, and not as a result of a melt-producing reaction. Thus, the grey host rock does not appear to have undergone partial melting. Turning to the leucocratic "veins," there are four key observations. (1) The contacts with the grey host are diffuse, not sharp, and therefore a relationship inconsistent with a "vein" of granitic melt or magma injected into subsolidus rocks, but consistent with *in situ* formation. (2) The size and distribution of garnet porphyroblasts in the "veins" are similar to those in the grey host; the grains of garnet in the "veins" are thus inherited from the host, and as the "veins" formed, they overgrew or engulfed the grains of garnet. (3) The central parts of the "veins" are quartzofeldspathic, more leucocratic, and have a grain size greater than at the margins or in the host; this suggests a greater fraction of melt in the center of the "veins" than at the margins. (4) In many places, the garnet porphyroblasts located in the "veins" are completely, or extensively, replaced by biotite, although there are places where this has not occurred; overall, more H_2O seems to have been available to alter anhydrous minerals where the "veins" formed, compared to the matrix. These observations, together with those made at the larger, outcrop scale (Fig. B31), lead to the interpretation that the "veins" are in fact *in situ* stringers of leucosome formed by H_2O-fluxed partial melting. When the host rocks were at a high temperature, just below their solidus, an aqueous fluid entered them through a system of fractures and caused H_2O-fluxed partial melting of the quartz + feldspar assemblage in the host. Such H_2O-fluxed partial melting of quartz + plagioclase + K-feldspar in the protolith also explains the lack of melanosome associated with the leucosome. The narrowness, together with the wide spacing of the stringers of leucosome, suggests that comparatively little fluid entered the rocks, so that partial melting was only able to occur adjacent to the fracture. The degree of partial melting decreased to zero in the wallrocks a centimeter or two from the fracture. The garnet porphyroblasts that were present where partial melting occurred were replaced by biotite, either as the ingress of fluid occurred, or as the anatectic melt crystallized. Because of the small amount of aqueous fluid introduced and the corresponding small extent of partial melting that it caused, the distribution of the leucosome closely mirrors the fracture network through which the aqueous fluid was transported. The occurrence of partial melting in this outcrop is the evidence required to call this a migmatite. In that case, the grey host rock to the *in situ* leucosome is paleosome. However, at the time of partial melting, the grey host rock plus the aqueous fluid was the protolith.

Location: Broken Hill area, New South Wales, Australia. *Rock type*: metatexite migmatite; quartzofeldspathic metavolcanic protolith (Hores gneiss), anatexis in the upper amphibolite facies, *T* ca. 800°C, *P* 4–5 kbar. *Scale*: the ruler is 15 cm long. *Image*: E.W. Sawyer.

Figure B33

Fig. B33. The morphology of the rocks in the right half of the photograph under the maps indicate a metatexite migmatite with a layered or stromatic structure that is almost transitional to a diatexite migmatite. The layered structure in the migmatite is due to alternations of leucosome, melanosome, and mesocratic paleosome. Petrographic and geochemical analyses of these rocks indicate that the degree of *in situ* partial melting in the fertile layers was only about 20 vol.%, but some of the leucocratic layers may have formed from injected anatectic melt. The pen rests upon the most remarkable feature of this migmatite, a quartz–plagioclase–K-feldspar–kyanite–biotite–garnet granofels that, because it truncates the layering in the host metatexite migmatite (e.g., in center of the photograph), could be interpreted as a dike intruded after the formation of the layering in its host. A petrographic (see Fig. F46 for a photomicrograph of this rock) and geochemical analysis of the garnet granofels in the dike are consistent with the rock being the residuum left after more than 50% partial melting of a pelitic protolith and the almost complete separation of the melt fraction from the solid. The question is, how did the residuum (which is normally considered as remaining *in situ*) from a high degree of partial melting get to be in a host that experienced a far lower degree of partial melting? One scenario is to move the residuum. For this to happen, a magma (melt + residuum) would have to have formed elsewhere, where the combination of rock fertility and metamorphic temperature enabled >50% dehydration-induced melting to occur, and then be intruded into the layered metatexite migmatite. In this scenario, the melt fraction could have separated either after emplacement of the magma into this outcrop, leaving behind the entrained residual material and a little anatectic melt trapped between the garnet crystals, or the melt fraction may have progressively separated from the residuum as the magma moved through the transfer-dike system from its source region. It may even be possible that the garnet-rich residuum itself was capable of granular flow (garnet crystals are "rounded" in shape) if all grain boundaries were in contact with melt, and if it was able to intrude as a dike; this may occur with as little as 7% melt present (Rosenberg and Handy 2005). In a second scenario, the residuum is produced *in situ*. This could be accomplished if the layered metatexite had stopped dehydration melting and was cooling, but was still at a temperature above its solidus for H_2O-present melting. The introduction of H_2O along a fracture then may provoke further melting through H_2O-present reactions; the degree of melting that occurs depends on how much H_2O was introduced into the rock. Separation of the melt fraction results in residual rocks with a cross-cutting geometry, simply because they trace out the fracture through which the H_2O entered the rock. In either scenario, this migmatite has lost melt and is an example of an open-system migmatite.

Location: Abaukoma, Yaoundé, Cameroon. *Rock type*: dike-like residuum in a metatexite migmatite; metapelitic and metagreywacke protoliths, anatexis at T 800–850°C, P 10–12 kbar. *Scale*: the pen is 15 cm long. *Image*: Pierre Barbey. Previously published as fig. 3a in Barbey et al. (1990) and reproduced with the permission of Oxford University Press.

Further reading: Nzenti, J.P., Barbey, P., Macaudière, J. & Soba, J. (1988): Origin and evolution of the late Precambrian high-grade Yaoundé gneisses. *Precambrian Research* **38**, 91–109.

Barbey, P., Macaudière, J. & Nzenti, J.P. (1990): High-pressure dehydration melting of metapelites: evidence from migmatites of Yaoundé (Cameroon). *Journal of Petrology* **31**, 401–427.

Figure B34

Fig. B34. The paleosome in this migmatite is a light grey biotite–quartz–feldspar gneiss. The neosome comprises leucocratic layers, containing quartz, feldspar, and minor amounts of biotite, that are lined on each side by very narrow, melanocratic, biotite-rich and microcline-poor borders. Results of mass-balance calculations on these migmatites (Olsen 1982, 1984, 1985) indicate that the domains of leucosome are too large relative to their melanocratic borders to consist wholly of melt derived from the melt-depleted melanosome; hence, one may conclude that the neosome formed in an open system. Some domains of leucosome contain as little as 5% felsic material added from an external source to the locally derived melt, whereas other domains of leucosome contain as much as 90% added material. Thus, the neosome in this migmatite formed by a combination of mechanisms; some felsic material was introduced from an external source as veins, and some partial melt of local origin segregated from the host into the veins, thereby forming the melanocratic borders. It is the presence of the local melt that makes this a migmatite. A foliation defined by biotite in the grey gneiss is oriented parallel to the trace of the axial planes of the folds that deformed the leucosome.

Location: Colorado Front Range, U.S.A. *Rock type*: metatexite migmatite; biotite–quartz–feldspar gneiss protolith metamorphosed at T 650–700°C and P 4–6 kbar. *Scale*: the coin is 19 mm across. *Image and caption*: S.N. Olsen.

Further reading: Olsen, S.N. (1987): The composition and role of the fluid in migmatites: a fluid inclusion study of the Front Range rocks. *Contributions to Mineralogy and Petrology* **96**, 104–120.

Olsen, S.N. & Grant, J.A. (1991): Isocon analysis of migmatization in the Front Range Colorado, U.S.A. *Journal of Metamorphic Geology* **9**, 151–164.

Figure B35

Fig. B35. This polished slab of a migmatite has been chemically etched and stained to color the K-feldspar yellow. The staining process reveals that K-feldspar is not uniformly distributed throughout the leucosome; some parts lack K-feldspar altogether. The polished surface also reveals thin, melanocratic, biotite-rich layers located between the leucosome and the mesocratic, finer-grained paleosome. These melanocratic layers are far too thin to have supplied all the quartzofeldspathic material present in the leucosome, although they may have supplied some of it. This relationship suggests that some of the leucosome material has been intruded into the migmatite sample. An alternative possibility is that the melanocratic layer did not supply any material to the leucosome at all, but formed as a result of the interaction between the injected melt (or an aqueous fluid derived from the crystallization of the leucosome) and the adjacent paleosome. In this case, the biotite-rich layer could be called a mafic selvedge.

Location: Baume migmatites, Ardèche, France. *Rock type*: metatexite migmatite; quartzofeldspathic gneiss protolith partially melted under upper-amphibolite-facies conditions at T 700 ± 50°C and P 3 kbar. *Image*: Pierre Barbey.

Further reading: Weber, C. & Barbey, P. (1986): The role of water, mixing processes and metamorphic fabric in the genesis of the Baume migmatites (Ardèche, France). *Contributions to Mineralogy and Petrology* **92**, 481–491.

Figure B36

Fig. B36. This stromatic migmatite shows some differences in morphology and microstructure among the various constituent batches of leucosome. The isoclinally folded leucosome in the center is finer grained than the others, and has rather smooth margins against the paleosome. Although the leucosome at the top also has smooth margins, it is noticeably coarser grained than the folded one. Finally, the thickest leucosome at the bottom has the largest grain-size, and has margins that are conspicuously irregular in outline. Other important features within the outcrop are the following: (1) the isoclinally folded leucosome is an F_2 feature and rotates the S_1 foliation, (2) elsewhere, the leucosome at the top cuts across the main S_2 layering or foliation, and (3) the coarse leucosome at the bottom grades into a small pegmatitic pod that contains rotated enclaves that preserve F_3 structures. On the basis of these observations, the differences among batches of leucosome may indicate that they belong to different generations relative to the deformation events that have occurred in the migmatite. As an alternative, if they are all of the same generation, then the differences in grain size and microstructure may indicate that each had a different fraction of melt present at the time of deformation.

Location: Glenelg River Complex, Victoria, Australia. *Rock type*: metatexite migmatite; quartzofeldspathic schist protolith, partially melted at T 650–680°C, P 4–6 kbar. *Scale*: the pencil is 15 cm long. *Image and caption*: Tony I.S. Kemp.

Further reading: Kemp, A.I.S. & Gray, C.M. (1999): Geological context of crustal anatexis and granitic magmatism in the northeastern Glenelg River Complex, western Victoria. *Australian Journal of Earth Sciences* **46**, 407–420.

Kemp, A.I.S. & Hawkesworth, C.J. (2003): Granitic perspective on the generation and secular evolution of the continental crust. *In* The Crust. Treatise on Geochemistry 3 (R. Rudnick, ed.). Elsevier Science, Oxford, U.K. (349–410).

Figure **B37**

Fig. B37. This close-up shows two different types of neosome in a migmatite with a metamafic paleosome. The neosome at the center of the photograph contains large, euhedral crystals of clinopyroxene in a leucocratic matrix of finer-grained quartz and plagioclase. The neosome is interpreted to be the *in situ* product of a small degree of partial melting from the incongruent breakdown of amphibole in the metamafic host. Locally, orthopyroxene is present with the clinopyroxene. Larger areas of pink-colored leucosome flank the *in situ* clinopyroxene-bearing neosome. The pink leucocratic veins are granitic in composition and contain quartz, K-feldspar, and plagioclase with minor amounts of biotite and garnet. The leucocratic veins are surrounded by a halo, or mafic selvedge, in which the metamafic host has been modified; hornblende replaces clinopyroxene (e.g., hornblende rim on clinopyroxene in the *in situ* neosome), and some hornblende in the metamafic rock is replaced by biotite. The metamafic rock occurs as a large (10 m) boudin in pelitic metasediments. The metapelites have undergone a greater degree of partial melting, and have lost much of the melt they generated. They now have a residual assemblage of minerals, quartz + plagioclase + biotite + garnet + sillimanite + K-feldspar ± cordierite (e.g., Fig. B38). During regional deformation, the metamafic rock was considerably more competent than the surrounding metasedimentary rocks and developed a series of fractures, and some granitic melt from the metapelites migrated into these. An aqueous fluid was exsolved from the pink leucocratic veins as they crystallized, and as a result, the metamafic rock around them underwent rehydration and some K-metasomatism that produced the hornblende rims and the biotite.

Location: Saint-Fulgence, Grenville Province, Quebec, Canada. *Rock type*: metatexite migmatite; metamafic protolith, anatexis at lower-granulite-facies conditions, T 800–850°C and P 5–8 kbar. *Scale*: the ruler is marked in cm. *Image*: E.W. Sawyer.

Figure B38

Fig. B38. The grey part of this migmatite contains two rock types: a K-feldspar–garnet–sillimanite–biotite–plagioclase–quartz metapelite (Pe) that has prominent dark foliae rich in biotite and sillimanite, and more massive, lighter-colored garnet–biotite–plagioclase–quartz metasedimentary material (Ps) of initially psammopelitic composition. The modal mineralogy of the metapelite layers indicates that they are residual after the loss of anatectic melt, but it is uncertain whether the psammopelitic layers have lost melt. The morphology of the portions derived from the anatectic melt in the different layers of this migmatite is interesting. The metapelitic layers contain large, prominent porphyroblasts of garnet and cordierite (Crd), some of which are partially surrounded by a narrow rim, or moat, of leucosome. Some of the garnet crystals, but especially the cordierite, are poikiloblasts intergrown with quartz and feldspar; this microstructure suggests that the leucosome and garnet (or cordierite) are the melt and solid products, respectively, of the incongruent melting of biotite. A possible reaction for these rocks is biotite + sillimanite + quartz + plagioclase = garnet (or cordierite) + K-feldspar + melt. The lenticular shape of the garnet, or cordierite, plus leucosome pods, their sharp contacts with the residual metasedimentary material, and the low ratio of leucosome to garnet (or cordierite), suggest that these domains have lost melt during the shearing event experienced by the migmatite. Lens-shaped domains of leucosome (L) that have sharp edges, but lack either garnet or cordierite, also occur within the metapelitic layers; these may represent some of the escaped melt. The absence of a foliation in the leucosome, and the presence of crystal faces on cordierite and feldspar where against quartz, suggest that melting and shearing were synchronous. The neosome (N) in the psammopelitic layers consists of very small euhedral crystals of garnet in a quartz + plagioclase + K-feldspar groundmass. The melting reaction thus was essentially the same as that which occurred in the pelite layers. However, the morphology of the neosome is different; the ratio of melt to garnet was much higher, and the domains of neosome have a diffuse margin. Both these features are consistent with *in situ* formation of the neosome and no loss of melt. Thus, the melting reaction was essentially the same in both types of layer, but the final morphology of the migmatite that developed in each layer is controlled by the relative competency of each lithology.

Location: Saint-Fulgence, Grenville Province, Quebec, Canada. *Rock type*: metatexite migmatite; metapelitic protolith, anatexis at lower-granulite-facies conditions T 800–850°C and P 5–8 kbar. *Scale*: the ruler is 15 cm long. *Image*: E.W. Sawyer.

Figure B39

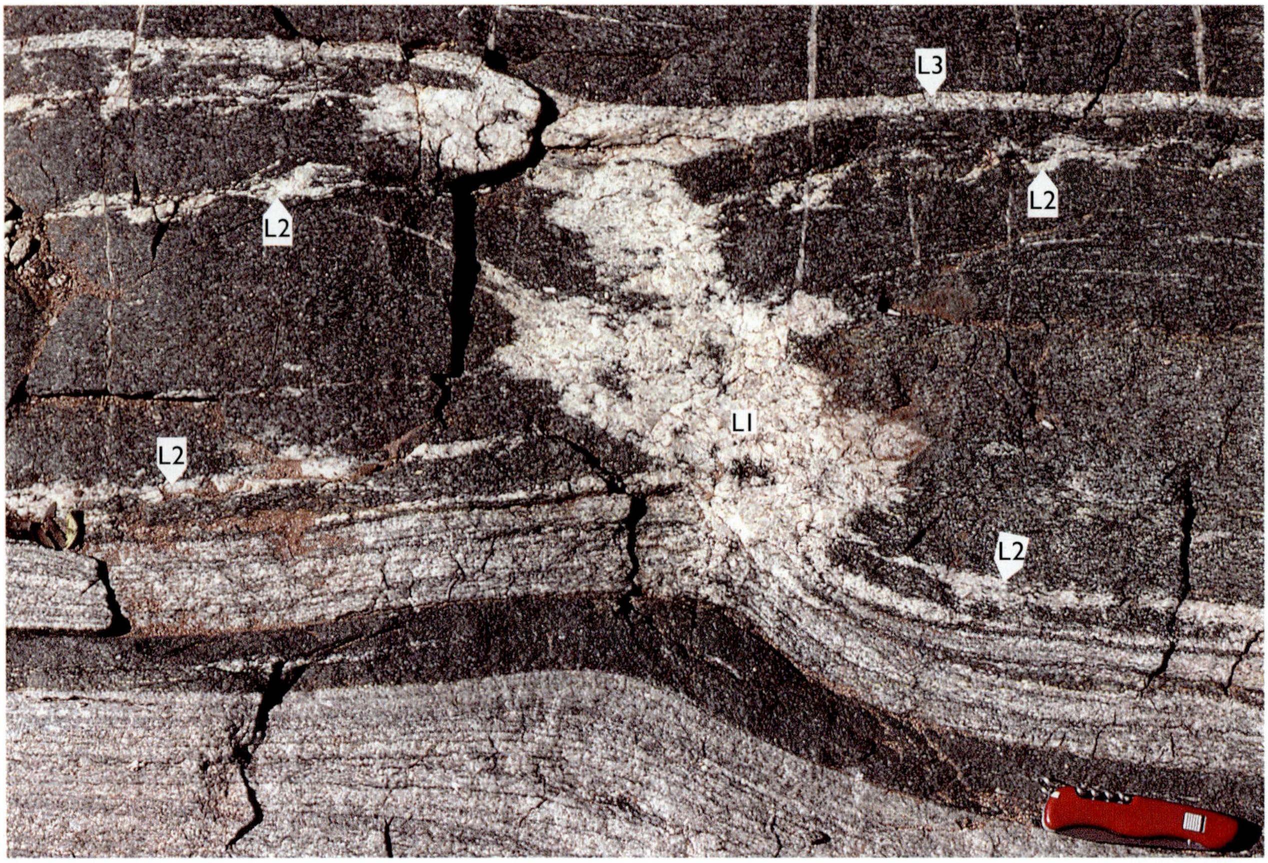

Fig. B39. The light grey rock in the lower part of the photograph is a biotite orthogneiss paleosome, whereas the dark-colored metamafic rock is in part paleosome and in part melanosome. The domains of leucosome hosted by the metamafic rock have the same mineralogy, essentially plagioclase + quartz, with minor amounts of hornblende and pyroxene. The point of interest in this migmatite is the difference in grain-size of the various bodies of leucosome. The coarsest grain-size is shown by the leucosome (L1) located in the interpartition space between the boudins. A finer grain-size exists in the four offshoots of leucosome (L2) that occur parallel to the layering within the two boudins. These offshoots have an irregular outline, and at least one of them is in contact with melanosome, evidence that suggests a local derivation for some of the melt that made the leucosome. The domains of leucosome (L3) with the finest grain-size occur on the outside edge of the boudinaged layer; they have the smoothest edges, and locally, the crystals in them have a shape-defined preferred orientation. Close examination of all these domains of leucosome shows that there are no cross-cutting relations among them; rather there is a smooth passage from one to the next over which the grain-size changes rapidly. Thus, one can conclude that all contained melt at the same time. Consequently, it is unrealistic to regard their solidification as occurring at different times, or that the differences in grain-size are the result of each having a different rate of cooling. However, the spaces that the domains of leucosome occupy may have started to form at different times, and may well have undergone different histories of deformation as the melt within was crystallizing. For example, the leucosome in the interpartition space has the coarsest grain-size because that site was dilatant during crystallization, whereas the domains of leucosome with the finest grain-size were located in sites where the deformation varied during crystallization and the growing crystals were periodically deformed and their grain-size reduced. Differences in grain-size and microstructure from one type of leucosome to the next might not indicate that the various types belong to different generations; their local structural position should always be considered.

Location: Beit Bridge Complex, Baklykraal Farm, South Africa. *Rock type*: metatexite migmatite; metamafic protolith, partially melted in the granulite facies. *Scale*: the pocket knife is 11 cm long. *Image*: E.W. Sawyer.

Further reading: Marchildon, N. & Brown, M. (2003): Spatial distribution of melt-bearing structures in anatectic rocks from southern Brittany, France: implications for melt transfer at grain- to orogen-scale. *Tectonophysics* **364**, 215–235.

Figure B40

Fig. B40. This migmatite is made up entirely of neosome, and shows a wide range of leucocratic through mesocratic to melanocratic parts. Coarse-grained melanocratic rafts in the upper center and left contain the mineral assemblage orthopyroxene + garnet + clinopyroxene, with a very small amount of plagioclase; these rafts (R) are interpreted to be residuum left after the extraction of anatectic melt. The lower right corner contains the most leucocratic part of the migmatite: a coarse-grained hornblende–pyroxene–quartz–plagioclase tonalitic leucosome (L) in which retrograde hornblende partially replaces clinopyroxene that is interpreted to be crystallized anatectic melt. Coarse-grained leucosome can be traced throughout much of the migmatite in a vague, net-like pattern that encloses various types of mesocratic rock. The thinnest segments of the leucosome network contain the highest proportions of clinopyroxene and orthopyroxene, and may have crystallized from melt that had a substantial residual component entrained in it. Many of the pyroxene crystals are euhedral, which suggests that they crystallized in the melt. The rounded mesocratic domain (M) below the center of the photograph contains the mineral assemblage plagioclase + clinopyroxene + hornblende ± garnet ± orthopyroxene. It has the smallest grain-size (lobe on the right) and the highest contents of hornblende and plagioclase. This mesocratic domain in the migmatite may be closest in composition to the metamafic protolith. However, parts of this domain are locally coarser grained and richer in pyroxene and, consequently, are interpreted as residuum. Therefore, this mesocratic domain may be a raft of paleosome, or less fertile protolith, that has been somewhat modified during partial melting, but did not become as melt-depleted as the melanocratic rafts above it. Much of the rest of the migmatite can also be described as mesocratic, but it has a substantially different morphology and microstructure; it does not occur as rafts (also called schollen). Rather, it is the matrix (M2) that contains the melanocratic and mesocratic rafts. The mesocratic matrix decreases in grain-size and changes its mineral assemblage from orthopyroxene + quartz + clinopyroxene + plagioclase near the leucosome to hornblende + clinopyroxene + plagioclase farther away. This gradation in the mesocratic matrix and the network of leucosome it contains are interpreted to represent the partially melted source and the network of channels by which anatectic melt was segregated from it. The extraction of melt from the mesocratic domain was incomplete, as it did not progress sufficiently to form melanocratic residuum, but enough melt was extracted so that coarse-grained leucosome could develop. The presence of schollen in the migmatite suggests that sufficient partial melt was present that layers of more competent material (e.g., residuum and less fertile units) were disrupted and rotated in the matrix; hence, this is a diatexite migmatite.

Location: Ashuanipi Subprovince, Quebec, Canada. *Rock type*: schollen diatexite; metamafic protolith, granulite-facies anatexis. *Scale*: the sticker with the number is 2.5 cm wide. *Image*: E.W. Sawyer.

Figure **B41**

Fig. B41. This migmatite contains coherent layers (30–50 cm thick) of different composition that are interpreted to have originally been bedding. The grey-colored layers have high modal proportions of cordierite and are interpreted to have lost most of the anatectic melt that formed in them, even though they are not melanocratic. The parts of this migmatite that were derived from anatectic melt exhibit a wide range of morphologies. Some anatectic melt collected parallel to the bedding and formed high-aspect-ratio domains of leucosome. Elsewhere, anatectic melt collected into smaller, lens-shaped bodies located in shear bands and formed leucosome that is approximately parallel to the axial planes of small-scale folds; these are oriented approximately parallel to the ruler. Both these types of leucosome have a relatively small proportion of ferromagnesian minerals in them, which indicates that they represent accumulations of anatectic melt that had segregated from the residuum. However, the point of interest here is the scattered, irregularly shaped patches of neosome that consist of a conspicuous melanocratic core surrounded by a leucocratic rim. The melanocratic cores consist of one, or perhaps a few large crystals of orthopyroxene that are intergrown with elongate blebs of quartz, whereas the leucocratic rims contain quartz, plagioclase, and K-feldspar. It is evident from the photograph that the ratio of melanocratic core to leucocratic rim differs from one neosome to the next. Re-integration of the melanocratic and leucocratic parts of the patches of neosome shows that some are the product of closed-system *in situ* partial melting. However, most have too large a melanocratic component relative to the stoichiometry of the melting reaction in these rocks; in these cases, the neosome formed *in situ*, but lost much of its anatectic melt, leaving it with an overall residual bulk composition. This type of melt-depleted neosome in patches is common in the higher-grade, granulite-facies parts of migmatite terranes. The melt lost from the neosome may have contributed to the two types of high-aspect-ratio leucosome present elsewhere in the outcrop. This metatexite migmatite is not dominated by any particular type of leucosome or neosome morphology. This fact makes it difficult to choose an additional descriptive term for it; neither patch, stromatic, nor net-structured is really representative of its overall appearance in outcrop.

Location: Wuluma Hills, Arunta Inlier, Australia. *Rock type*: metatexite migmatite; Al-poor psammite protolith, partially melted at T 825–875°C, P ca. 5 kbar. *Scale*: the ruler is 15 cm long. *Image*: E.W. Sawyer.

Figure B42

Fig. B42. The protolith to this migmatite was a leucogranodiorite that contained mafic layers (probably mafic dikes). During the subsequent anatexis, the leucogranodiorite partially melted, but the mafic layers did not; they were resisters. Where the fraction of melt in the leucogranodiorite was sufficient that it could flow, a diatexite migmatite was formed, and the resister layers were disrupted and became enclaves in the neosome, as in this image. Diffusional exchange between the anatectic melt and the fragments of the mafic resister occurred. Some hornblende at the edges of the resisters was converted to biotite, and the adjacent neosome was enriched in plagioclase, possibly by transformation of K-feldspar to plagioclase, which formed a white rim around the layers of resister. Elsewhere, the resister layers became disaggregated, and the adjacent neosome is contaminated with xenocrystic hornblende. This outcrop illustrates that there can be a considerable range in the composition of the melt-derived parts of migmatites arising from contamination, or diffusive exchange with resister lithologies; this possibility should be considered when sampling.

Location: Opatica Subprovince, Quebec, Canada. *Rock type*: diatexite migmatite; leucogranodiorite protolith, anatexis at *T* ca. 750°C, and *P* 5–7 kbar. *Scale*: the ruler is 15 cm long. *Image*: E.W. Sawyer.

Further reading: Sawyer, E.W. (1998): Formation and evolution of granite magmas during crustal reworking: the significance of diatexites. *Journal of Petrology* **39**, 1147–1167.

Figure B43

Fig. B43. This granulite-facies migmatite has prominent domains of melanosome that follow the foliation and compositional layering in the mafic gneiss precursor. The abundant areas of leucosome, however, display a more complex and hierarchical arrangement. Thin bodies of *in situ* leucosome (L1) are associated directly with the melanosome and are oriented either parallel, orthogonal, or oblique to the layering; these form an interconnected network. Some more continuous, layer-parallel domains of in-source leucosome (L2) with a thin melanosome occur outside the most melanocratic layers, and appear to have been fed by the network of thinner, orthogonal and oblique domains of leucosome. These layer-parallel domains then connect to much wider bodies (L3), which cross the layering in the migmatite. The absence of melanocratic borders to these wide, cross-cutting bodies suggests that they did not attract melt from their hosts, but are simply conduits for the migration of melt out of the source region. Hence, the wide felsic bodies should be called leucocratic veins and not leucosome.

Location: Kapuskasing Structural Zone, Ontario, Canada. *Rock type*: metatexite migmatite; mafic gneiss protolith partially melted at granulite-facies conditions, T ca. 850°C, P 11 kbar. *Scale*: the Brunton compass is 8 cm across. *Image*: D.R.M. Pattison.

Further reading: Hartel, T.H.D. & Pattison, D.R.M. (1996): Genesis of the Kapuskasing (Ontario) migmatitic mafic granulites by dehydration melting of amphibolite: the importance of quartz to reaction progress. *Journal of Metamorphic Geology* **14**, 591–611.

Sawyer, E.W. (2001): Melt segregation in the continental crust: distribution and movement of melt in anatectic rocks. *Journal of Metamorphic Geology* **19**, 291–309.

Figure B44

Fig. B44. The initial compositional heterogeneity, due to the intrusion of mafic and felsic igneous rocks that comprised the protolith of this migmatite, has been strongly attenuated by high strain synchronous with anatexis. Domains of leucosome are especially evident in the mafic parts of this migmatite, because the color contrast between paleosome, melanosome, and leucosome parts is greatest there. The leucosome exhibits a range of morphologies, which correlates with its structural position. Bodies of leucosome oriented parallel to the compositional layering (L1) and the foliation have the highest aspect-ratio and could be called stromatic leucosome. Most bodies of leucosome (L2) are located either between boudins (lowest aspect-ratio) or in sets of conjugate extensional shears and fractures (intermediate aspect-ratio). However, most striking in this outcrop is the combination of all domains of leucosome to form an ordered, branched array. The thinnest domains of leucosome form the *in situ* parts of the network, whereas the thickest domains of leucosome (L3) in the array clearly cross-cut the layering in the paleosome and melanosome at the lower right of the photograph. If the melt in this leucosome (L3) was derived locally, then this is an in-source leucosome; if it is not, then it is a leucocratic vein. Detailed examination of the contacts between the different segments of leucosome indicates the absence of abrupt changes in grain size or composition; there is continuity throughout the array. This finding is interpreted to mean that the entire array of leucosome contained melt at the same time and, hence, that melt migrated out of the migmatite through the array along progressively fewer and wider channels, finally migrating through the 10-cm-wide channel marked by the leucosome in the lower right. A similar, wider leucocratic vein on the left may represent the proximal end of another array of leucosome that drained melt from the part of the migmatite farther to the left.

Location: Sand River Gneisses, Causeway locality, South Africa. *Rock type*: metatexite migmatite; metamafic and metagranodioritic to tonalitic protolith, partially melted in the granulite facies. *Scale*: the pocket knife is 11 cm long. *Image*: E.W. Sawyer.

Figure B45

Fig. B45. Migmatites in southern Brittany are located south of the transcurrent South Armorican Shear Zone. Shearing during anatexis is probably the reason why the southern Brittany migmatites have a predominantly stromatic morphology (see also Fig. B10). Typically, thin bodies of feldspar-rich leucosome oriented parallel to the foliation alternate with biotite-rich melanocratic layers and rarer, fine-grained biotite–quartz–plagioclase schists. However, not all the bodies of leucosome in the migmatites are parallel to the foliation and compositional layering. In many places, the stromatic migmatites contain cross-cutting bodies of leucosome located in interboudin partitions and extensional shear structures, and are also cut by discordant veins of leucogranite that may be several meters to tens of meters long. Although the cross-cutting relationships indicate a clear progression in "structural" age from stromatic leucosome through discordant leucosome to granitic vein, there is no progression in "petrological" age. This photograph shows the contact relationships between a granitic vein 10–15 cm wide and the host stromatic metatexite migmatite. The granitic vein truncates the layering defined by the melanosome and the biotite–quartz–plagioclase schist, but it does not appear to truncate the stromatic leucosome. The passage from stromatic leucosome (most evident in the wider domains) to discordant granite vein is petrologically continuous; there is no mineralogical or microstructural discontinuity between the material in the leucosome and that in the vein. This relationship indicates that when the granitic vein formed, the stromatic leucosome still contained melt. The stromatic leucosome, the discordant leucosome, and the granitic veins together constituted a continuous melt-bearing network in the migmatite. Since the stromatic domains of leucosome are associated with melt-depleted melanosome, but the granitic veins are not, anatectic melt is interpreted to have moved subhorizontally a short distance, from where it formed along channels now represented by the stromatic leucosome, to the steeply dipping cross-cutting structures now represented by the leucogranitic dikes, through which it was then able to escape and feed higher-level granitic plutons.

Location: Port Navalo area, southern Brittany, France. *Rock type*: metatexite migmatite; quartzofeldspathic and aluminous gneiss protolith partially melted at T ca. 800°C, P ca. 8 kbar, then decompressed to 4 kbar, followed by cooling. *Scale*: the coin is 23 mm across. *Image and caption*: Mike Brown. Image previously published as fig. 10.11(b) in Brown (2006) and reproduced with the permission of Cambridge University Press.

Further reading: Jones, K.A. & Brown. M. (1990): High-temperature "clockwise" P–T paths and melting in the development of regional migmatites: an example from southern Brittany, France. *Journal of Metamorphic Geology* **8**, 551–578.

Marchildon, N. & Brown, M. (2003): Spatial distribution of melt-bearing structures in anatectic rocks from southern Brittany, France: implications for melt transfer at grain- to orogen-scale. *Tectonophysics* **364**, 215–235.

Brown, M. (2006): Melt segregation from lower continental crust of orogens: the field evidence. In Evolution and Differentiation of the Continental Crust (M. Brown & T. Rushmer eds.). Cambridge University Press, U.K. (331–383).

Figure **B46**

Fig. B46. This outcrop of the southern Brittany migmatites show features that may have formed during the waning stages of the system shown in Fig. B45. The migmatite is dominated by thin stromatic leucosome and its adjacent melanosome, but also contains discordant bodies of leucosome and dikes of leucogranite (under the ruler). The discordant leucosome and leucogranitic veins commonly display a change in the contact relationships with the host stromatic migmatites along their length. Most of the discordant bodies of leucosome appear to root in small shear zones, shear bands, or asymmetrical boudin structures, where they have diffuse contacts with the host (discordant leucosome on the right) stromatic migmatite. Farther along, the discordant bodies of leucosome widen and develop sharp margins against the layers of melanosome, but are petrologically continuous with the stromatic leucosome, as shown in Fig. B45. However, a few veins of leucogranite show a different relationship along their length, and cross-cut both the stromatic leucosome and the border of melanosome in the host migmatite (vein on the left), and may even contain fragments of the wallrock. These particular discordant veins thus are late and formed as the migmatites solidified; they are rooted in melt-bearing layers, but then propagated into parts of the migmatite that either contained far less melt, or were solid, so that the melt in the vein could not interact with melt in the wallrocks.

Location: Port Navalo area, southern Brittany, France. *Rock type*: metatexite migmatite; quartzofeldspathic and aluminous gneiss protolith partially melted at *T* ca. 800°C, *P* ca. 8 kbar, then decompressed to 4 kbar, followed by cooling. *Scale*: the ruler is 15 cm long. *Image*: E.W. Sawyer.

Further reading: see Fig. B45.

Figure B47

Fig. B47. This photograph is taken looking down into the core of a meter-scale S-shaped synform, and shows the root of two granitic dikes. The core of the fold (top right of the photograph) consists of metapelite (Pe) with abundant leucosome, which indicates that it contained a high fraction of melt, and the outer part (center left) consists of a thick layer of massive quartz-rich metapsammite (Ps). The metapsammite has undergone little or no partial melting, and can be regarded as a paleosome resister lithology in the migmatite. In contrast, the pelite has undergone extensive partial melting and has developed melanocratic garnet- and orthopyroxene-rich residual rocks (R). The massive metapsammite served as an impermeable layer above the pelite, and enabled anatectic melt to accumulate in the pelite layer under it, in both the synform and adjacent antiform hinges. Locally, the melt fraction became sufficiently voluminous in the pelite that the foliation and melanocratic residual layers in it became disrupted as the folds tightened. Thus, the morphology that developed in the pelite layer at the fold hinge is transitional from metatexite to diatexite migmatite. An extensive dark rim (S) that consists principally of late biotite occurs at the boundary between the melt-rich pelite layer and the massive psammite. This rim, together with the extensive replacement of garnet and orthopyroxene by biotite in the residual rocks, provide further evidence that anatectic melt accumulated in the fold hinges. Some of the melt in the hinges was locally derived (giving rise to *in situ* and in-source leucosome), but some may also have migrated up to the hinge from a lower position on the fold limbs (to form leucocratic veins or patches). The late biotite is interpreted to have formed when H_2O, released as the accumulated melt crystallized, hydrated the residual minerals. Some of the accumulated melt was able to escape from the pelite layer and cross the massive metapsammite layer through the 10-cm-wide dike (it could also be called a vein) that crosses the middle of the photograph (D1). The dike is rooted in the synform and is oriented subparallel to the axial plane of the fold. A second leucocratic dike (D2) that penetrated no more than 50 cm into the metapsammite occurs to the right, and confirms that the source of the melt was the pelite layer; hence, the term leucocratic dike or vein is appropriate. An increase in pressure on the magma in the pelite layer as the fold tightened may have enabled the formation and propagation of the fracture through which some of the magma was able to escape. The dike contains several large crystals of garnet, some of which have a rim of biotite; some aqueous fluid thus was released as the magma in the dike crystallized and reacted with the garnet to form the biotite rim. The scale rests on a shear zone that is located on the steep, right flank of the synform.

Location: Wuluma Hills, Arunta Inlier, Australia. *Rock type*: leucocratic dike in a metatexite migmatite; metapelitic protolith, partially melted at *T* 825–875°C, *P* ca. 5 kbar. *Scale*: the ruler is 15 cm long. *Image*: E.W. Sawyer.

Further reading: Sawyer, E.W., Dombrowski, C. & Collins, W.J. (1999): Movement of melt during synchronous regional deformation and granulite-facies anatexis, an example from the Wuluma Hills, central Australia. *In* Understanding Granites; integrating New and Classical Techniques (A. Castro, C. Fernández & J.-L. Vigneresse, eds.). *Geological Society, Special Publication* **168**, 221–237.

Figure **B48**

Fig. B48. Overall view of the migmatite outcrop shown in Fig. B47, taken from above and looking down the hinge of the folds. The two dikes (D1 and D2) that emerge from the pelitic layer in the synform can be seen in the left half of the photograph, just to the right of the scale. Part of the complementary antiform is just visible at the extreme right. Tight, minor folds can be seen in the metapelite; the layering that is folded involves a combination of biotite-rich pelite, leucosome, and biotite + garnet + orthopyroxene melanosome material.

Location: Wuluma Hills, Arunta Inlier, Australia. *Rock type*: leucocratic dike in a metatexite migmatite; metapelitic protolith, partially melted at *T* 825–875°C, *P* ca. 5 kbar. *Scale*: the ruler is 15 cm long. *Image*: E.W. Sawyer.

Figure **B49**

Fig. B49. A prominent mesocratic selvedge is developed in this migmatite between the discordant, feldspar-rich leucocratic vein and its host. The dominant pinkish grey portion of this outcrop is medium-grained neosome that was derived from a semipelitic protolith. The grey neosome has two parts. A darker-colored part with the mineral assemblage quartz + plagioclase + K-feldspar + biotite + garnet ± cordierite is interpreted to be residual; the lighter, K-feldspar + quartz + plagioclase granitic parts occur as small, diffuse, and elongate patches. These pinkish stringers are interpreted to be *in situ* leucosome. The neosome is intruded by a discordant, quartz-poor leucocratic vein that has the mineral assemblage K-feldspar + plagioclase + garnet + quartz + biotite; the biotite replaces garnet. Prominent mesocratic selvedges that are 1–2 cm wide occur between the host and the leucocratic veins. Contacts between the selvedges and the pinkish grey host are gradational, and their grain sizes and microstructures are similar, which suggests that the selvedge may have formed from, or overprinted, the host. The selvedges are not enriched in ferromagnesian minerals relative to the host, but do appear to be enriched in quartz. Consequently, they are not interpreted as residua. Thus, the selvedges are interpreted as a diffusion halo between the host and the leucocratic vein intruded into it. Possibly, a gradient in silica activity between the quartz-rich host and the intruded quartz-poor magma existed and drove the diffusion of silica from the host toward the leucocratic vein, producing the mesocratic selvedge.

Location: Wuluma Hills, Arunta Inlier, Australia. *Rock type*: selvedge around leucocratic vein in a metatexite migmatite; metapelitic protolith, partially melted at T 825–875°C, P ca. 5 kbar. *Scale*: the ruler is 15 cm long. *Image*: E.W. Sawyer.

Figure B50

Fig. B50. This photograph shows several bodies of garnet-bearing, quartz–K-feldspar leucosome developed in a medium-grained ilmenite–plagioclase–quartz–biotite–sillimanite–cordierite–garnet gneiss. All the domains of leucosome are separated from their host by a very narrow, melanocratic rim that is rich in biotite. The rim is probably a retrograde feature that formed as the melt in the leucosome crystallized; hence, it is a mafic selvedge and not a melanosome. The domains of leucosome in this photograph differ markedly from others present in migmatites of the same area (shown in Figs. B23 and B24) in that they do not contain large poikiloblasts of garnet. This difference is interpreted to indicate that the leucosome in this image may have formed in a different way to the others in the area. Here, the leucosome probably formed by the physical segregation of anatectic melt from the adjacent gneiss. Thus, it is essentially wholly derived from anatectic melt, whereas the leucosome shown in Figs. B23 and B24 formed by the *in situ* growth of the solid products garnet and K-feldspar and, consequently, was mostly solid as it grew, with a lesser proportion of anatectic melt. Reaction between the evolved melt generated, or an aqueous fluid exsolved as the leucosome crystallized, and residual minerals such as cordierite in the host formed the biotite-rich mafic selvedge.

Location: Round Hill, Broken Hill, Australia. *Rock type*: metatexite migmatite; semipelitic protolith, partial melting at *T* 800°C and *P* 5–6 kbar. *Scale*: the pen is 14 cm long. *Image and caption*: Richard White.

Further reading: White, R.W., Powell, R. & Halpin, J.A. (2004): Spatially-focussed melt formation in aluminous metapelites from Broken Hill, Australia. *Journal of Metamorphic Geology* **22**, 825–845.

Figure B51

Fig. B51. Leucocratic veins intruded into migmatite terranes during the waning stages of anatexis commonly develop thin mafic selvedges. This photograph shows such a thin (1–3 mm wide), biotite-rich mafic selvedge developed between a discordant leucocratic vein and its grey metasedimentary host. The leucocratic vein, a peraluminous leucogranite, contains the mineral assemblage quartz + plagioclase + K-feldspar + biotite + garnet, with minor amounts of muscovite, apatite, and tourmaline. The whole-rock composition of the leucocratic vein suggests that it crystallized from a fractionated anatectic melt derived from a metapelitic protolith. Partial melting has probably not occurred in the biotite–quartz–plagioclase psammitic schist that hosts the granitic vein. However, the interbedded pelitic layers do contain a residual mineral assemblage: sillimanite + K-feldspar + biotite + cordierite + quartz + plagioclase (e.g., Fig. F43). The mafic selvedge is only a few grains wide and consists predominantly of biotite that has a much coarser grain-size than the biotite in either the vein or the host schist. Garnet crystals occur locally within the selvedge, and closely resemble the garnet crystals in the vein. The mafic selvedge is slightly thicker in concavities at the edge of the vein. The biotite-rich rim is too narrow to be the complementary extraction zone of the melt from which the leucocratic vein crystallized; hence, it is not a melanosome. Moreover, the host probably did not partially melt. The mafic selvedge is interpreted to have formed as a result of reaction between the evolved felsic melt from which the vein crystallized and the host schist.

Location: Quetico Subprovince, Canada. *Rock type*: leucocratic vein with mafic selvedge in a psammitic layer; metamorphic conditions T 700–800°C and P 3–4 kbar. *Scale*: the ruler is in centimeters and millimeters. *Image*: E.W. Sawyer.

Figure B52

Fig. B52. The dark grey rock in this photograph is a garnet–diopside–plagioclase–hornblende metamafic rock. It contains a few thin (<10 mm) leucocratic veins that are subparallel to the compositional layering. However, the most striking feature of the outcrop is the array of subparallel lenses and elongate patches of coarse-grained leucotonalite that are oriented at about 70° to the compositional layering. Examination of nearby outcrops reveals that the lenses of leucotonalite are oriented parallel to the axial planes of a series of tight folds. A melanocratic rim is developed around each lens. Thus, at first sight, the outcrop appears to contain leucosome with melanosome and could, consequently, be considered a metatexite migmatite. However, the dark rims are narrow and consist predominantly of coarse-grained hornblende and biotite. There are two generations of hornblende; one is of similar morphology to the hornblende in the metamafic rock farther away from the leucocratic lenses, and is interpreted to be prograde, whereas the other replaces diopside and is retrograde. Biotite partially replaces hornblende close to the leucocratic lenses. These lenses are interpreted to have crystallized from a felsic melt intruded into the metamafic rock parallel to the axial planes of local folds, and the dark rim around them seems to be mafic selvedge formed by reaction between the metamafic host and the felsic melt, or an aqueous fluid exsolved from it during crystallization. Many of the larger lenses of leucotonalite exhibit a wide, lighter-colored selvedge or halo outside the mafic selvedge. This feature is attributed to the alteration of plagioclase caused by the aqueous fluid from the felsic melt. Melanocratic bands also occur subparallel to the layering in the metamafic schist and appear to form lateral bridges between chains of small lenses of leucotonalite. The dark color of these bands is due to the replacement of diopside by hornblende, and is interpreted as evidence that the exsolved aqueous fluid was locally able to infiltrate parallel to the foliation of the metamafic rock. The widespread development of retrograde hornblende has obscured any melanosome that may have existed in the metamafic host. Hence, it is not certain whether the metamafic part of the outcrop is a migmatite. However, the grey rocks in the top right corner of the photograph are a metatexite migmatite; the sillimanite–garnet–biotite–quartz–plagioclase pelitic schists contain patches of neosome that have cores of poikiloblastic garnet intergrown with quartz that are surrounded by a plagioclase + K-feldspar + quartz leucosome, consistent with the incongruent melting of biotite. Some of the garnet poikiloblasts contain scattered inclusions of sillimanite.

Location: Central Metasedimentary Belt, Grenville Province, near Wakefield, Quebec, Canada. *Rock type*: lenses of leucotonalite with mafic selvedges in a metatexite migmatite; metagabbro protolith, lower granulite-facies conditions *T* ca. 800°C, *P* 6–9 kbar. *Scale*: the ruler is 15 cm long. *Image*: E.W. Sawyer.

Figure B53

Fig. B53. This complex migmatite provides a good example of a mafic selvedge. The prominent layering in the migmatite is pre-anatectic; the light grey layers are biotite–plagioclase–quartz psammitic schists (Ps), the medium grey layers are garnet–cordierite–quartz–K-feldspar–plagioclase–sillimanite–biotite pelitic schists (Pe), and the dark grey layers are biotite–plagioclase–hornblende schists (Hs) that were possibly mafic volcanic rocks, or sills. The evidence for partial melting and, therefore, for considering this a migmatite, can be found in the pelitic layers. At the bottom left, stromatic, pinkish grey granitic leucosome (Ll) occurs in a grey schist that has the mineral assemblage plagioclase + sillimanite + biotite + cordierite + K-feldspar + quartz, with minor garnet. The leucosome is interpreted to be derived from the anatectic melt of an aluminous pelite via a reaction such as biotite + sillimanite + quartz + plagioclase = melt + cordierite + garnet + K-feldspar, and the host schist is interpreted to be the residuum, but it is not melanocratic. Another pelitic layer, located immediately above the dark grey layer that crosses the center of the photograph, contains additional evidence for partial melting. Diffuse patches of neosome (N) have a core consisting of quartz + garnet surrounded by quartz + plagioclase + K-feldspar in a leucocratic rim, which suggests that the neosome formed *in situ* by a similar melting reaction, but in a slightly more iron-rich pelite. Conspicuous, narrow mafic selvedges (MS) composed of biotite occur at the margins of a folded, pink leucocratic vein 5–10 cm wide that occurs in the upper half of the photograph. Although the vein is hosted by pelite, the mineral assemblage in the mafic selvedge is not similar to the residua associated with either the stromatic leucosome or the patches of neosome. Consequently, the mafic selvedge is interpreted to result from reaction between an injected vein of granitic magma and its pelitic host, and not to be a residuum generated by partial melting. This interpretation is supported by observations in adjacent outcrops, where a thin mafic selvedge occurs adjacent to discordant pink granitic veins. Because the stromatic leucosome is folded by the mesoscale tight folds evident in the photograph, the patches of neosome are equant and have diffuse edges, even though located in the core of the folds. Both types of leucosome have igneous microstructures, and the folding is considered to have occurred while anatectic melt was present in the rock. As another interesting feature in this migmatite, garnet developed at contacts between the mafic and pelitic layers, but not at the contacts between mafic and psammitic layers.

Location: Saint-Fulgence, Grenville Province, Quebec, Canada. *Rock type*: leucocratic vein with mafic selvedge in a metatexite migmatite; metasedimentary protolith, granulite-facies anatexis at T 800–850°C, P 5–8 kbar. *Scale*: the ruler is marked in cm. *Image*: E.W. Sawyer.

Figure B54

Fig. B54. This photograph is a close-up of the interior of a meter-wide granitic dike, one of several that are oriented parallel to local shear zones and the axial surfaces of mesoscale folds developed in metatexite migmatites. In general, these dikes are interpreted to be part of the transfer network of conduits through which anatectic melt moved from the migmatites to small granite plutons growing in larger dilatant sites nearby. Because these dikes are still within their broad source-volume, they belong to the category of leucocratic dikes. The photograph shows several thin (1–1.5 cm) leucocratic veins, oriented roughly parallel to the scale, and each has very narrow biotite-rich mafic selvedges. The selvedges are interpreted to have formed as a result of interaction between with veins and their host. A striking feature of the host is that it contains numerous biotite-rich foliae that are regularly spaced about 1.5 cm apart, the same spacing as the mafic selvedges at the edges of the leucocratic veins. Both the thin veins and their host dike rocks have granitic bulk compositions, similar grain-sizes, and similar microstructures. The host dike thus probably was itself constructed by the multiple injection of many thin leucocratic veins, each of which developed a mafic selvedge. The two more prominent leucocratic veins just happen to be the last increments of melt injected into the dike. If the thin veins represent the entire history of flow of the larger dike, then the narrowness of the individual veins may mean that this particular dike was not a successful conduit for the transfer of melt. Alternatively, the repeated diking may simply be a record of the waning pulses of melt in the wider dike.

Location: Wuluma Hills, Arunta Inlier, Australia. *Rock type*: leucocratic vein with mafic selvedge, metatexite migmatite; metapelitic protolith, partially melted at *T* 825–875°C, *P* ca. 5 kbar. *Scale*: the ruler is 15 cm long. *Image*: E.W. Sawyer.

C.

METATEXITE AND DIATEXITE, THE FIRST-ORDER DIVISION OF MIGMATITES

The photographs in this section show the striking change in the morphology of migmatites that takes places between the lowest- and the highest-grade parts of most migmatite terranes. The same changes occur whether the migmatites formed in a regional, or in a contact metamorphic setting.

Migmatites from the contact aureole of the Ballachulish Igneous Complex [Figs. C1–C4]

The Ballachulish Igneous Complex (BIC), in the Scottish Highlands, was emplaced into Dalradian metasedimentary rocks at about 425 Ma. The metasedimentary rocks were regionally deformed and metamorphosed to the greenschist facies before the intrusion of the BIC. Intrusion occurred at a pressure of 3 ± 0.5 kbar (corresponding to a depth of about 10 km) and a contact-metamorphic aureole, locally up to 2 km wide, developed around the BIC. Rocks in the inner part of the contact aureole were partially melted.

Further reading: Fraser, G.L., Pattison, D.R.M. & Heaman, L.M. (2004): Age of the Ballachulish and Glencoe Igneous Complexes (Scottish Highlands), and paragenesis of zircon, monazite and baddeleyite in the Ballachulish aureole. *Journal of the Geological Society of London* **161**, 447–462.

Harte, B., Pattison, D.R.M. & Linklater, C.M. (1991): Field relations and petrography of partially melted pelite and semi-pelitic rocks. *In* Equilibrium in Contact Metamorphism: the Ballachulish Igneous Complex and its Aureole (G. Voll, J. Topel, D.R.M. Pattison & F. Seifert, eds.). Springer-Verlag. Heidelberg, Germany (181–209).

Holness, M. & Clemens, J.D. (1999): Partial melting of the Appin quartzite driven by fracture-controlled H_2O infiltration in the aureole of the Ballachulish Igneous Complex, Scottish Highlands. *Contributions to Mineralogy and Petrology* **136**, 154–168.

Pattison, D.R.M. & Harte, B. (1997): Geology and evolution of the Ballachulish Igneous Complex and aureole. *Scottish Journal of Geology* **33**, 1–29.

Migmatites from the contact aureole of the Duluth Igneous Complex [Figs. C5, C6]

Mafic rocks of the Duluth Igneous Complex (DIC), in northern Minnesota, were intruded over a very narrow time period around 1098 Ma, and produced a considerable contact-aureole in the Archean and Proterozoic country rocks. The inner part of the aureole locally reached the pyroxene hornfels facies, and anatexis is particularly evident in the pelitic and semipelitic rocks of the Virginia Formation, on the western side of the DIC. The depth of intrusion was relatively shallow; estimates of the pressure of the contact metamorphism are in the range 1.5–2 kbar.

Upper amphibolite facies, regional migmatites from Saint-Malo, France [Figs. C7, C8]

The Saint-Malo Terrane consists largely of metasedimentary rocks of Brioverian age; greywacke and pelitic compositions predominate. These rocks were partially melted during an upper-amphibolite-facies metamorphic event at 536 ± 14 Ma. Melting occurred through the breakdown of muscovite, and biotite generally remained stable. Diatexite migmatites occur in the central part of the terrane and are partially surrounded by an envelope of metatexite migmatites. Nonmelted Brioverian metasediments are well exposed south of the migmatites in the estuary of the Rance.

Further reading: Brown, M. (1979): The petrogenesis of the St-Malo migmatite belt, Armorican Massif, France, with particular reference to the diatexites. *Neues Jahrbuch für Mineralogie, Abhandlungen* **135**, 48–74.

Milord, I., Sawyer, E.W. & Brown, M. (2001): Formation of diatexite migmatite and granite magma during anatexis of semi-pelitic metasedimentary rocks: an example from St. Malo, France. *Journal of Petrology* **42**, 487–505.

Weber, C., Barbey, P., Cuney, M. & Martin, H. (1985): Trace element behaviour during migmatization. Evidence for a complex melt–residuum–fluid interaction in the St. Malo migmatitic dome (France). *Contributions to Mineralogy and Petrology* **90**, 52–62.

Upper amphibolite facies, regional migmatites from the Opatica Subprovince, Quebec [Figs. C9, C10]

The Opatica Subprovince, in central Quebec, consists principally of tonalite–trondhjemite–granodiorite (TTG) plutonic rocks with ages between 2825 and 2702 Ma, into which plutons belonging to a monzodiorite–tonalite–granodiorite suite were intruded between 2697 and 2693 Ma. A subsequent high-grade metamorphic event led to widespread partial melting of rocks of the tonalite–trondhjemite–granodiorite suite in the center of the subprovince, and to the generation of migmatites and anatectic leucogranites between 2681 and 2678 Ma. Partial melting occurred under conditions of the upper amphibolite facies (T ca. 750°C, P 5–7 kbar), and probably involved the influx of aqueous fluid.

Further reading: Davis, W.J., Machado, N., Gariépy, C., Sawyer, E.W. & Benn, K. (1995): U–Pb geochronology of the Opatica tonalite–gneiss belt and its relationship to the Abitibi greenstone belt, Superior Province, Quebec. *Canadian Journal of Earth Sciences* **32**, 113–127.

Sawyer, E.W. (1998): Formation and evolution of granite magmas during crustal reworking: the significance of diatexites. *Journal of Petrology* **39**, 1147–1167.

Sawyer, E.W. & Benn, K. (1993): Structure of the high-grade Opatica Belt and adjacent low-grade Abitibi Subprovince, Canada: an Archaean mountain front. *Journal of Structural Geology* **15**, 1443–1458.

Granulite-facies, regional migmatites from the Ashuanipi Subprovince, Quebec [Figs. C11, C12]

The migmatites in the Ashuanipi Subprovince, Quebec–Labrador, were derived from Archean siliciclastic sedimentary material deposited by turbidity-current action. The beds in the sedimentary sequence range up to a meter thick, but the typical thickness is 10–20 cm. Some beds were graded during deposition, and this has resulted in a compositional gradation within the beds; the tops have a broadly pelitic composition, whereas the bottoms are greywacke (psammitic). Plagioclase-rich metagreywacke with a simple plagioclase + quartz + biotite + quartz mineral assemblage dominates (ca. 85%) the sedimentary pile and was volumetrically the principal protolith to the migmatites. Partial melting in these rocks occurred through the biotite-dehydration reaction: biotite + plagioclase + quartz = orthopyroxene + melt ± ilmenite. The average degree of partial melting was about 31 vol.%.

Further reading: Guernina, S. & Sawyer, E.W. (2003): Large-scale melt-depletion in granulite terranes: an example from the Archaean Ashuanipi subprovince of Quebec. *Journal of Metamorphic Geology* **21**, 181–201.

Percival, J.A. (1991): Granulite-facies metamorphism and crustal magmatism in the Ashuanipi complex, Quebec–Labrador, Canada. *Journal of Petrology* **32**, 1261–1297.

Sawyer, E.W. (1994): Melt segregation in the continental crust. *Geology* **22**, 1019–1022.

Sawyer, E.W. (2001): Melt segregation in the continental crust: distribution and movement of melt in anatectic rocks. *Journal of Metamorphic Geology* **19**, 291–309.

Figure C1

Fig. C1. One of the first indications of the onset of partial melting in the field is the presence of very narrow (<2 mm) feldspar-rich leucocratic veins in pelitic and semipelitic hornfelses. This photograph shows that the leucocratic veins are located parallel to the foliation (indicated by the pencil), and in orientations that are oblique to the layering in the hornfelses. Locally, the veins form a net-like pattern in their hosts. Very narrow biotite selvedges are present around some of the veins. In thin section, the veins show K-feldspar grains with crystal faces in contact with "xenomorphic-interstitial" quartz, a microstructure indicative of crystallization from a melt. The main macroscopic characteristic of this migmatite is the continuity of pre-anatectic structures (bedding and foliation) in the paleosome; hence, it is a metatexite migmatite.

Location: east side of the aureole. *Rock type*: metatexite migmatite; the protolith was interbedded pelite and semi-pelite of the Appin Phyllite, anatexis at T 650–730°C. *Scale*: the pencil is 15 cm long. *Image*: D.R.M. Pattison. Previously published as Photo 20 in Pattison and Harte (2001) and is reproduced with the permission of The Edinburgh Geological Society.

Further reading: Pattison, D.R.M. & Harte, B. (2001): The Ballachulish Igneous Complex and Aureole: a Field Guide. Edinburgh Geological Society Guidebook Series, Edinburgh, U.K. 148 p.

Figure C2

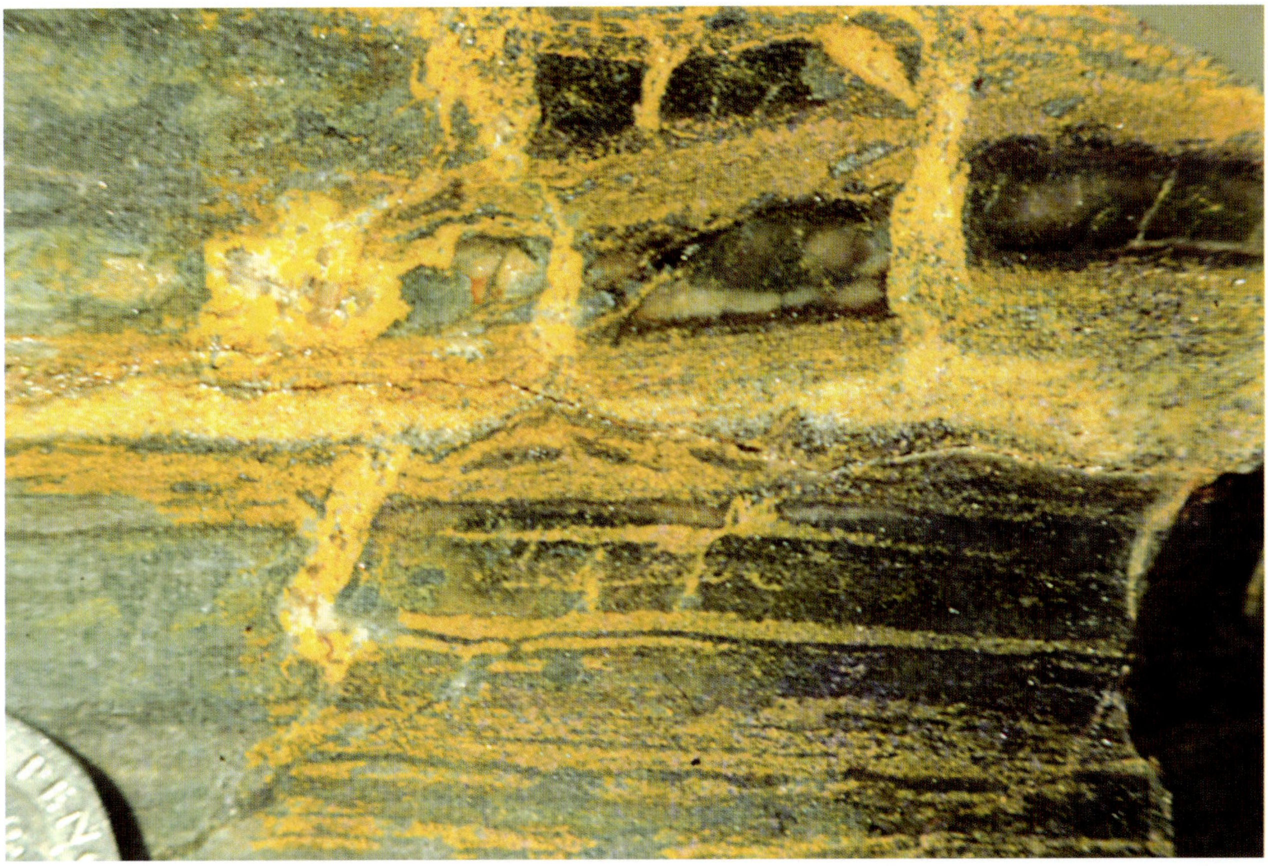

Fig. C2. Migmatites with a very distinctive appearance are produced at slightly higher grades from protoliths consisting of thin pelitic and semipelitic beds. The photograph shows a sawn slab of this type of migmatite, which has been etched and stained to reveal the K-feldspar. The competent layers, typically rich in cordierite, have been extended and become trains of rectangular boudins, whereas the less competent units (partially melted semipelitic layers) have deformed by ductile flow and remain more or less continuous, but are thinned. Coarser-grained leucocratic material extracted from the partially melted semipelitic layers occupies the dilatant spaces between the rectangular boudins, but also occurs as veins that cross the layering. Although the proportion of leucosome is locally higher, and this migmatite is more strained than that shown in Fig. C1, the pre-anatectic structures remain continuous. Thus, this too is a metatexite migmatite.

Location: east side of contact aureole, 100 m from contact. *Rock type*: metatexite migmatite; the protolith was thinly bedded pelite and semipelite of the Appin Phyllite, partially melted at T 650–730°C. *Scale*: the lens cap is 65 mm across. *Image*: D.R.M. Pattison. Previously published as fig. 3 in Pattison and Harte (1988) and is reproduced with the permission of Blackwell Publishing.

Further reading: Pattison, D.R.M. & Harte, B. (1988): Evolution of structurally contrasting migmatites in the 3-kbar Ballachulish aureole, Scotland. *Journal of Metamorphic Geology* **6**, 475–494.

Figure C3

Fig. C3. The migmatites located in the parts of the aureole that reached the highest metamorphic temperatures, and consequently generated the highest fractions of melt, have a very different appearance. This cut and stained slab shows rectangular boudins generated by the pulling apart of the more competent (more viscous, or stiffer) metasedimentary layers enclosed in an anatectic melt-rich matrix. In contrast to Fig. C2, the blocks of the competent rock have been moved apart and rotated such that the original orientation and continuity of the pre-partial-melting structures (bedding and foliation) are disrupted and destroyed; hence, this is a diatexite migmatite.

Location: screen within the intrusive body, east side of the complex. *Rock type*: diatexite migmatite; protolith was thinly bedded pelite and semipelite of the Appin Phyllite, partially melted at T 750–800°C. *Scale*: the coin is 19 mm in diameter. *Image*: G.L. Fraser.

Figure C4

Fig. C4. The migmatites closest to the intrusive body on its west side show the greatest disruption of the pre-partial-melting layering, and were mapped as the "chaotic zone" by Pattison and Harte (1988). The outcrop appearance of the "chaotic zone" migmatites is a jumbled mass of psammite, semipelite, and pelite blocks (i.e., schollen or rafts) within a coarser-grained, lighter-colored semipelitic matrix. Some schollen have rectilinear outlines, but others are rounded. The matrix is generally without a foliation or compositional layering, but where present, it appears to bend around the schollen in a manner suggestive of flow. The semipelitic matrix has a granular microstructure and contains patches of interstitial-xenomorphic quartz and K-feldspar that, together with the flow-like foliation, strongly suggest that the matrix contained a substantial fraction of anatectic melt. However, the semipelitic bulk composition of the matrix, and the scarcity of both leucosome and melanosome in the "chaotic zone," suggest that the melt fraction did not segregate from the solid residuum in the matrix. The fraction of melt in many parts of the "chaotic zone" migmatite was sufficiently high that bulk flow of the migmatite could occur. The stronger layers (with a lower fraction of melt) were disrupted during the flow and became the schollen in the portions with a higher fraction of melt. Even though the "chaotic zone" experienced the highest extent of melting in the aureole, the temperature of melting was relatively low (ca. 650–700°C). This situation is attributed to the influx of H_2O at peak metamorphic conditions released from the crystallization of the underlying intrusive body.

Location: "Chaotic Zone" on the western side of the intrusive body. *Rock type*: diatexite migmatite; the protolith was pelite and semipelite of the Leven schist; anatexis at *T* 650–700°C. *Scale*: the lens cap is 65 mm across. *Image*: D.R.M. Pattison. Previously published as fig. 13 in Pattison and Harte (1988) and is reproduced with the permission of Blackwell Publishing.

Further reading: Pattison, D.R.M. & Harte, B. (1988): Evolution of structurally contrasting migmatites in the 3-kbar Ballachulish aureole, Scotland. *Journal of Metamorphic Geology* **6**, 475–494.

Figure C5

Fig. C5. This migmatite is typical of those developed in the outer part of the anatectic zone. The original sedimentary layering in the hornfels derived from the Virginia Formation remains evident and can be traced laterally across outcrops; hence, this a metatexite migmatite. Two sets of thin discordant domains of leucosome cross the compositional layering in the hornfels and locally form a net-like pattern. A systematic sense of displacement across the oblique domains of leucosome suggests that they occupy a conjugate set of extensional shears related to shortening perpendicular to the layering. This is possibly due to loading of the footwall by the DIC as it was emplaced. Another set of somewhat thicker, stromatic domains of leucosome that are oriented parallel to the bedding also is evident. All the leucosome material is fine-grained (0.5 mm), but still much coarser grained than the paleosome (0.1 mm). Although not visible in outcrop, the domains of leucosome are bordered by a narrow melanocratic selvedge that results from the alteration of cordierite in the wallrock (see Fig. F88). After local segregation of melt from the wallrocks into nearby low-pressure sites (the extensional shears) to create the leucosome, crystallization of the melt appears to have released some H_2O, which then reacted with cordierite in the wallrock.

Location: Linwood Lake, area 35m from the western contact with the DIC. *Rock type*: net-structured metatexite migmatite; pelitic and semipelitic protolith; anatexis at *T* ca. 750°C. *Scale*: the ruler is 15 cm long. *Image*: E.W. Sawyer.

Figure C6

Fig. C6. The migmatites from the highest-grade parts of the Duluth aureole no longer contain recognizable and coherent structures that predate the partial melting and are, therefore, diatexite migmatites. Like the one in this photograph, they are typically very uniform, medium-grained (0.5–2 mm), mesocratic rocks that locally contain small elongate fragments (schollen) of hornfels that are, in places, aligned. In many cases, the migmatites have a foliation defined by the parallel orientation of tabular minerals, typically plagioclase, cordierite, and orthopyroxene; typically, there are no leucocratic segregations or mafic schlieren developed. Locally, deflections in the trend of the foliation define small shear zones that formed in the magmatic state. Furthermore, cordierite, plagioclase, and K-feldspar commonly have crystal faces against abundant pockets of coarser-grained, equant but irregularly shaped quartz, K-feldspar and plagioclase. The microstructures indicate that these migmatites contained a significant fraction of melt, and that they developed their foliation by the alignment of crystals in a flowing magma. The modal proportions of plagioclase, quartz, biotite, cordierite, K-feldspar, and orthopyroxene, together with the whole-rock composition of most of the diatexite migmatites in the Duluth aureole, indicate that partial melting of the pelitic and semipelitic protoliths, and subsequent flow of the resulting diatexite magma, occurred without any significant segregation of the melt fraction from the residuum fraction. In other words, most of these diatexite migmatites represent a closed anatectic system.

Location: Linwood Lake, area 5 m from the western contact with the DIC. *Rock type*: diatexite migmatite; pelitic and semipelitic protolith; anatexis at *T* 800–825°C. *Scale*: the ruler is 15 cm long. *Image*: E.W. Sawyer.

Figure C7

Fig. C7. Neosome in the migmatites located close to the melt-in isograd on the southeastern side of the Saint-Malo migmatite terrane, in Brittany, formed first in the muscovite-rich semipelitic layers. The bodies of neosome are stromatic and consist of thin, coarse-grained muscovite + K-feldspar + plagioclase + quartz leucosome bordered by biotite-rich melanosome. The start of partial melting in the migmatites at Saint-Malo involved the breakdown of muscovite; biotite remained stable and was concentrated in the melanosome. Beds of psammitic and calc-silicate composition were not as strongly affected by partial melting, and hence these are paleosome resister lithologies at this stage. Consequently, the continuity of structures that predate the partial melting, sedimentary layering in this case, is preserved in these low-grade migmatites, and hence this is a metatexite migmatite.

Location: La Landrais, Saint-Malo migmatite terrane, France. *Rock type*: metatexite migmatite; Brioverian semipelitic protolith from just on the high-grade side of the "melt-in" isograd, T ca. 650°C, P 4–7 kbar. *Scale*: the ruler is 15 cm long. *Image*: E.W. Sawyer.

Figure C8

Fig. C8. Regional metamorphic temperatures were higher farther into the Saint-Malo migmatite terrane; consequently, anatexis was far more extensive. A feature common of virtually all of these migmatites is the absence of any laterally coherent structures that predate the partial melting; hence, these are diatexite migmatites, and they consist largely of neosome. The relative proportions of leucocratic to melanocratic material in the neosome, and the ratio of neosome to paleosome, vary considerably from outcrop to outcrop. In the example shown here, the neosome is mesocratic overall, and consists of wider leucocratic layers, or bands, that alternate with rather discontinuous, curviplanar biotite-rich foliae, called schlieren. In addition, the remains of a large schollen of a resister lithology also can be seen.

Location: Saint-Malo migmatite terrane, France. *Rock type*: mesocratic diatexite migmatite; semipelitic and pelitic protolith, partially melted at T ca. 750°C, P 4–7 kbar. *Scale*: the side of the yellow box is 17 cm long. *Image*: E.W. Sawyer.

Figure C9

Fig. C9. A continuous foliation, or compositional layering, that predates the partial melting can be seen in all the migmatites developed in the lowest-grade parts of the anatectic zone in the Opatica Subprovince; thus they are metatexite migmatites. Two principal morphologies of metatexite migmatite have developed from the tonalite–trondhjemite–granodiorite plutonic protoliths. Patch migmatites (see Figs. B15 and D12) formed in the Opatica Subprovince where there was no penetrative deformation of the protolith during partial melting, i.e., they occur in low-strain environments. This photograph shows another morphology, which developed as the protolith was sheared during partial melting. The light-grey and darker grey layers are a compositional banding that existed before partial melting and are due to changes in the ratio of plagioclase to quartz + K-feldspar. These layers were deformed into north-vergent, asymmetrical folds during partial melting. Light-grey, coarser-grained, diffuse areas of neosome that do not have a foliation are located in shear planes on the limbs of the folds or subparallel to the axial surfaces of the folds. The neosome is slightly lighter colored than the rock around it, but not really leucocratic. The diffuse edges suggest that this is *in situ* neosome. The outcrop is also cut by thin, pink leucocratic veins that tend to have diffuse margins; some of these veins are folded. Commonly, the veins occur in shear zones located on the long limbs of the asymmetrical folds. The pink veins have a composition suggestive of a probable derivation from a similar protolith; they may represent injections of anatectic melt.

Location: Opatica Subprovince, Quebec, Canada. *Rock type*: metatexite migmatite; tonalite protolith, partially melted under conditions of the upper amphibolite facies, T ca. 750°C, and P 5–7 kbar. *Scale*: the ruler is 15 cm long. *Image*: E.W. Sawyer.

Figure C10

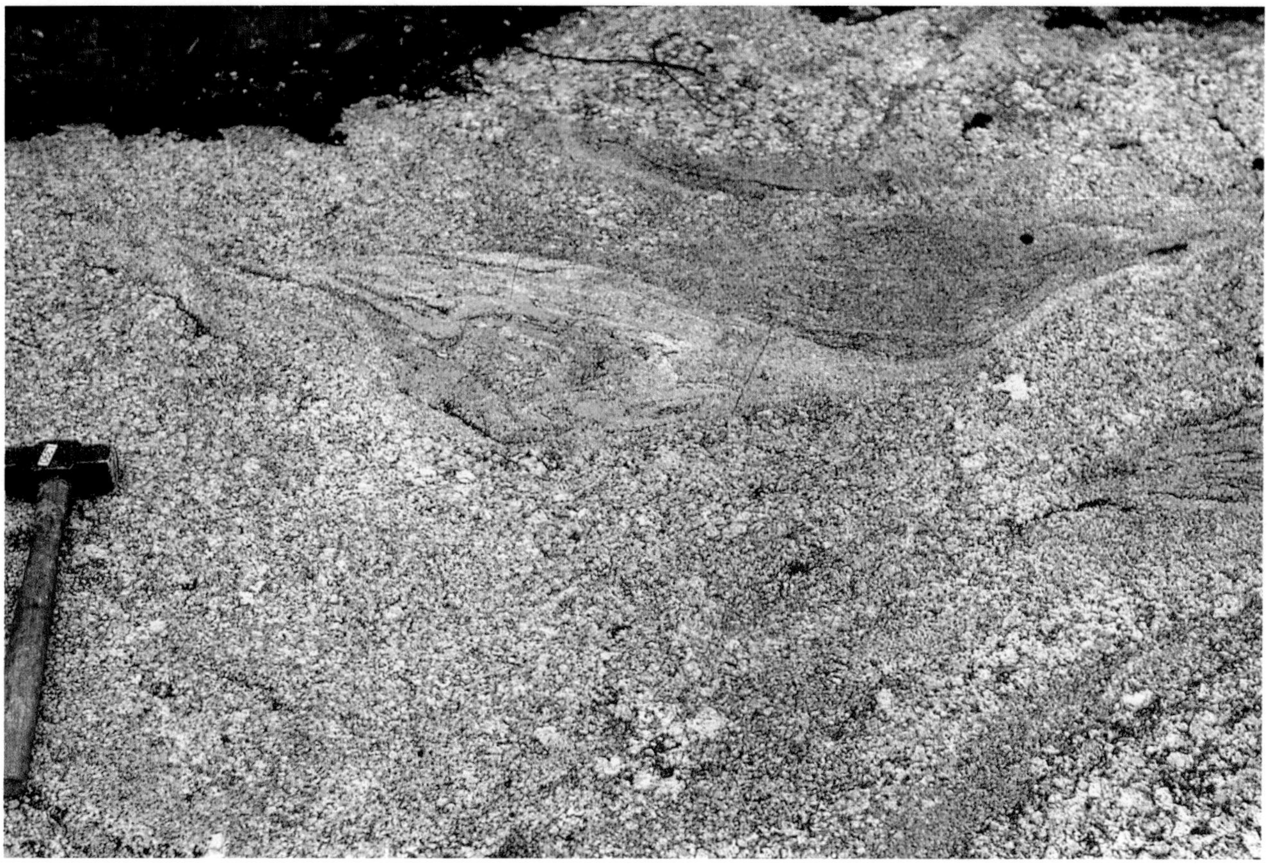

Fig. C10. Migmatites in the center of the anatectic region in the Opatica Subprovince do not contain contiguous structures that predate the partial melting; the structures they contain are essentially syn-anatectic. This example shows lens-shaped schollen of grey, foliated biotite–quartz–plagioclase tonalite in a pink, coarse-grained biotite–quartz–K-feldspar–plagioclase granite host. Some schollen are fragments of resister lithologies, but most have bulk compositions indicating that they are melt-depleted and hence residua. The microstructure in the leucocratic matrix varies considerably from place to place. In many places, as in this image, the microstructure is isotropic and granitic, but elsewhere, there are domains in which tabular crystals (plagioclase, K-feldspar, and biotite) are aligned, forming a magmatic foliation with or without biotite schlieren. In places, coarse-grained granitic veins have intruded finer-grained rocks, but elsewhere (e.g., just below the center of the photograph), coarse-grained rocks simply occur as diffuse patches in finer-grained ones. The leucocratic part of this migmatite contains some K-feldspar crystals that are euhedral, and others that have rational crystal faces at mutual contacts with quartz, textures that are indicative of crystallization from a melt. These observations suggest that in the center of the Opatica Subprovince, the fraction of melt was sufficiently high that the partially melted rocks were able to flow, and thus destroy the pre-anatexis structures they contained. The only pre-anatexis structures that remain are those preserved in the entrained schollen or rafts; hence, this is a diatexite migmatite and represents a more advanced stage of anatexis than that recorded in Fig. C9.

Location: Opatica Subprovince, Quebec, Canada. *Rock type*: diatexite migmatite; metatonalite protolith, anatexis under conditions of the upper amphibolite facies, T ca. 750°C, and P 5–7 kbar. *Scale*: the ruler is 15 cm long. *Image*: E.W. Sawyer.

Figure **C11**

Fig. C11. This migmatite is dominated by grey metasedimentary rocks and has a relatively minor proportion of leucosome. Since the streaks of leucosome are mostly located parallel to the bedding and foliation, the migmatite could be called stromatic. The mineralogy and whole-rock composition of the metasedimentary rocks in this migmatite are not consistent with an interpretation that they are paleosome, but indicate that they are the residuum left after the extraction of about 30 vol.% anatectic melt of granitic composition. This degree of melt loss far exceeds the volume of leucosome present in the outcrop; hence, anatexis and the formation of this migmatite were the result of an open-system process at the scale of the outcrop. The preservation of structures and layering that predate partial melting in the residuum after the loss of such a large volume of melt indicates that the melt was continuously drained from the residuum during anatexis. If the melt had accumulated in the rock up to a large fraction (>10%), then bulk flow would have occurred by one process or another, and that would have destroyed the layering. Because all the layers were fertile, and the melt-segregation process allowed only a small fraction of melt to exist in the rock at any one time, the resulting migmatite is essentially all neosome, but a residual one in which pre-anatectic structures are preserved. Large areas of fertile rocks undergoing this type of partial melting and segregation process could be the source of large volumes of granitic magma in the continental crust; for the Ashuanipi Subprovince, melt-depleted migmatites such as this are estimated to have produced 640 000 km^3 of granitic magma. In assigning this migmatite a further descriptive term, a decision has to be made as what characteristic needs to be emphasized; two possibilities are its morphology and its bulk composition.

Location: Ashuanipi Subprovince, Quebec, Canada. *Rock type*: residual metatexite migmatite; metaturbidite protolith, with partial melting at T 825–875°C, P 6–7 kbar. *Scale*: the ruler is 15 cm long. *Image*: E.W. Sawyer.

Figure C12

Fig. C12. In this migmatite, schollen (or rafts) of melt-depleted metasedimentary rocks identical to the residual parts of the metatexite migmatites (e.g., Fig. C11) occur in a coarser-grained, lighter-colored host; in this case, then, the rafts are the residual part of the neosome and not paleosome. However, in some outcrops, the schollen include ultramafic, mafic, and vein quartz resister lithologies, and these are all that remains of the paleosome. Microstructures that indicate the flow of magma (i.e., melt containing crystals) and crystallization of a melt are common in the matrix around the schollen. Examples of such microstructures include the tiling of tabular crystals of feldspar, the local development of euhedral crystals of feldspar and orthopyroxene, and of minerals with crystal faces against large interstitial crystals of equant, xenomorphic quartz. The most important characteristic of this migmatite, the one that identifies it as a diatexite migmatite, is the absence of structures that predate the partial melting in the matrix around the schollen. The matrix is neosome also, and the absence of discrete melanosome and leucosome in it indicates that the melt fraction and solid fraction did not become separated during anatexis. Rather, they flowed together as magma.

Location: Ashuanipi Subprovince, northern Quebec, Canada. *Rock type*: schollen diatexite migmatite; metaturbidite protolith partially melted at *T* 825–875°C, *P* 6–7 kbar. *Scale*: the ruler is 15 cm long. *Image*: E.W. Sawyer.

D.

SECOND-ORDER MORPHOLOGIES IN MIGMATITES

The second-order morphologies in migmatites can be related to the melt fraction present in the migmatite and, in a general way, to strain, or how the protolith responded to strain. With the photographs in this section, it will be possible first to examine the range of different morphologies that can be found in metatexite migmatites, where paleosome dominates, and in which structures that predate partial melting are preserved. Then, the morphologies found in diatexite migmatites, where neosome predominates, and in which coherent pre-anatectic structures are destroyed and replaced by structures formed during anatexis.

The start of partial melting [Figs. D1–D6]

An important part of mapping in metamorphic terranes is the determination of isograds. In high-grade metamorphic terranes, the "melt-in isograd" marks the beginning of the zone of anatexis. It is where migmatites begin to appear. However, noticing the first signs of partial melting in the field is not easy. The photographs in this section show examples of the incipient stages of anatexis. Because partial melting has barely started in these rocks, structures that pre-date partial melting, for example, bedding, compositional layering, foliations, and folds, remain prominent in the paleosome; hence, these are metatexite migmatites, albeit with a very small proportion of neosome.

Metatexite migmatites with a patch structure [Figs. D7–D10]

Patch migmatites are a type of metatexite migmatite in which the neosome occurs as patches in the paleosome. The patches of neosome can be sharply defined or they can be diffuse or nebulose. Similarly, the neosome may, or may not, have segregated into leucosome and melanosome. Patch migmatites are most commonly found in the lowest-grade parts of migmatite terranes, and are preserved in the parts of an anatectic terrane where the syn-anatectic strain was low (see also Fig. B1).

Metatexite migmatites with a nebulitic structure [Figs. D11, D12]

The term *nebulitic* has been used in two ways to describe migmatites. The term is most commonly used to describe domains of neosome that have an indistinct, diffuse or "fuzzy" boundary with a volumetrically dominant paleosome, so that the contact between them is typically interpreted to be gradational. This usage applies to metatexite migmatites. The second usage applies to the description of larger regions where faint, ghostlike relics of discontinuous, but not disrupted, paleosome can be seen through the "cloudiness" of the neosome. This use refers to diatexite migmatites. Nebulitic migmatites have a homogeneous microstructure (i.e., no foliation, or magmatic flow structure, and have not segregated into leucosome and melanosome). Consequently, they are interpreted to form in environments of very low syn-anatectic strain.

Metatexite migmatites with leucosome in dilatant structures [Figs. D13–D22]

Where anisotropic rocks are deformed, differences in the viscosity of the layers can result in strain-rate discontinuities along the layer boundaries that are accommodated by the formation of dilatant structures; boudinage and foliation boudinage are two well-known examples. If partial melt is present in the rock during deformation, then it will migrate to the dilatant sites, where the pressure in the rock is lowest. If the melt remains there and crystallizes, it will form leucosome. Leucosome located in dilatant sites is very common in migmatites and exhibits a very wide range of morphology that is principally controlled by the nature of the anisotropy. Leucosome in dilatant sites related to the development of folds will be dealt with later.

Metatexite migmatites with a net structure [Figs. D23–D30]

A net structure is formed in a metatexite migmatite where two or more sets of leucosome segregations intersect to create a net-like pattern of leucosome enclosing rhomb- or lozenge-shaped domains of paleosome or, in some cases, melanosome. A net structure is very common in metatexites, and its development appears to be a necessary step in the extraction of large volumes of melt from anatectic terranes and the creation of large areas of melt-depleted migmatites.

Metatexite migmatites with a layered or stromatic structure associated with low strain [Figs. D31–D34]

A stromatic or layered morphology results where the neosome in a migmatite occurs as parallel, laterally continuous bands in two dimensions and parallel sheets in the third. The bodies of neosome must also be parallel to the principal planar structure in the host. Most commonly, this is a foliation, but it can also be compositional layering, e.g., bedding in a sequence of thinly bedded pelites, semipelites, and psammites. The term stromatic usually is applied to the leucosome in a migmatite because it is the most prominent part of the neosome. However, there is no reason why a migmatite consisting of parallel bodies of melanosome, say after the loss of melt, cannot also be called stromatic. In this section, I show examples of migmatites with a stromatic morphology that is an initial, low-strain feature of the migmatite. In the next section, I will show examples with stromatic morphology arising from high strain during anatexis.

Metatexite migmatites with layered or stromatic structure due to transposition [Figs. D35–D40]

A feature of stromatic, or layered, migmatites from zones of high strain is the intensity and continuity of the layering, and the remarkable regularity in the width and spacing of the domains of leucosome. A microstructural examination of the leucosome is especially important. If the leucosome contains magmatic or submagmatic microstructures, then the term *layer* or *stromatic migmatite* is justified. However, if the microstructures all formed in the solid state, then the rocks may have acquired their layered structure after the melt in them had completely crystallized; in that case, the term migmatite, although possibly appropriate for the protolith of the highly strained rock, is not appropriate for the rock overall. It should be named accordingly (e.g., protomylonitic migmatite, mylonitized migmatite).

The transition from metatexite to diatexite migmatites [Figs. D41–D46]

Metatexite migmatites are characterized by the widespread preservation of the structures that existed before anatexis. In contrast, diatexite migmatites are characterized by the general absence of coherently preserved structures that predate the partial melting, i.e., diatexite migmatites are largely, or wholly, neosome. The destruction and overprinting of old structures in migmatites are progressive, and in many migmatite terranes, there is a transitional zone between metatexite and diatexite migmatites in which paleosome is sequentially disrupted and consumed into the neosome. One of the features of this transition is the preservation of fragments of the metatexite migmatite as abundant enclaves, which are called schollen, or rafts; hence, these transitional rocks are called schollen diatexites or raft diatexites. In other migmatite terranes, the change from metatexite to diatexite migmatite is abrupt. This first group of photographs shows examples of the early stages in the formation of diatexite migmatites.

Diatexite migmatites with schollen and with schlieren structures [Figs. D47–D52]

There is a systematic change in the morphology of diatexite migmatites in the direction of higher metamorphic grade and increased fraction of melt away from the metatexite to diatexite transition. The proportion of rafts, or schollen, decreases, and correspondingly, the proportion of leucocratic neosome increases. Schollen of metatexite material in the neosome tend to become progressively more lenticular in shape and are no longer angular, whereas schollen composed of resister lithologies, such as calc-silicates or vein quartz, are commonly rounded and equidimensional. Schlieren, the thin persistent layers of dark minerals found in diatexite migmatites (and in some granites), start to become prominent in the matrix and commonly define flow lines around the schollen.

Diatexite migmatites with schlieren structures [Figs. D53–D56]

The morphology of diatexite migmatites may appear to become less complex as the proportion of schollen decreases. However, this may well be a misconception, and many diatexite migmatites exhibit significant local variations in color, grain size, microstructure, and composition, most commonly in the form of discontinuous bands, or layers. The most widely known type of compositional variation are schlieren, the thin layers of dark minerals found in diatexite migmatites and in some granites. Commonly, schlieren are rich in biotite, but they can also contain other platy or rod-shaped minerals such as plagioclase, sillimanite, amphibole, and orthopyroxene. Schlieren generally show a considerable enrichment in accessory minerals. It is likely that a number of different processes can produce schlieren. They all require that the host was initially a magma, however. Mechanisms that have been suggested include, amongst others, flow sorting, filter pressing, and erosion of the rafts or schollen.

Diatexite migmatites [Figs. D57–D62]

The highest-grade parts of many migmatite terranes contain portions of outcrops, whole outcrops, or larger areas in which only syn-anatectic or younger structures are present. In these rocks, the proportion of neosome is very high, that of paleosome is correspondingly minor, and schollen and schlieren are few, or even absent. Consequently, changes in microstructure and composition that account for the morphological variations in these migmatites tend to be very subtle, but these are not homogeneous, or uniform, migmatites. Furthermore, many examples of diatexite migmatite contain leucosome and leucocratic veins of late anatectic age. There is no prefix term to describe the morphology of these rocks; they are the quintessential diatexite migmatites. Diatexite migmatites can be mesocratic, leucocratic, or melanocratic, and these variations in color have been found to correspond to the bulk composition of the protolith, the degree of enrichment in the melt fraction and in the residuum fraction, respectively. Consequently, a descriptive term used to describe the compositional, mineralogical, or microstructural variations within an outcrop of diatexite migmatite, or over a larger region, may be useful.

Diatexite migmatites at high strains [Figs. D63, D64]

Where the melt fraction in a diatexite migmatite is sufficiently high that crystals are in a dilute suspension, magma flow can occur without interaction between the solid crystals, and high strains are not fully recorded. As the melt fraction decreases, the crystals start to interact and to create a framework, and this can be deformed and preserve a record of the strain. Diatexite migmatites that had a framework of crystals enclosing melt-filled pockets, where deformed at high strains, can develop strongly layered morphologies comprising bands of the deformed crystal-framework material and leucosome bands consisting of the melt expelled from the framework as it collapsed.

Figure D1

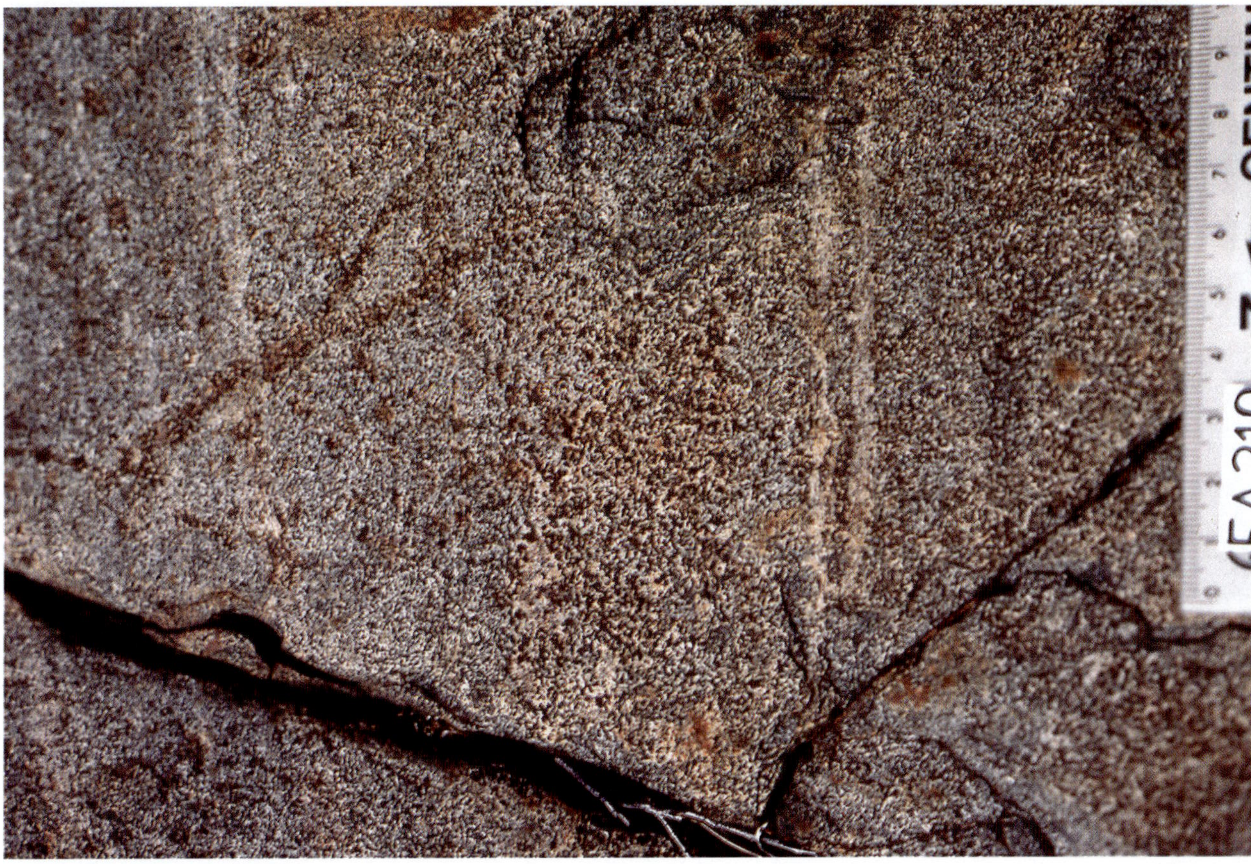

Fig. D1. Subtle evidence that partial melting has started in this granulite-facies metamafic rock is visible in the center of the image. Thin, white intergranular rims are developed around the solid product and reactant phases. The white rims consist of quartz and plagioclase, and represent crystallized films, and small pockets, of melt that formed on the grain boundaries between the reactant minerals. As the melt did not separate from the residuum, the neosome composition should represent the protolith composition. The melting reaction was hornblende + plagioclase + quartz = orthopyroxene + melt. The thin leucocratic bands parallel to the ruler are subsolidus segregations that predate anatexis. At the incipient stage of anatexis shown by this migmatite, the degree and extent of partial melting are very minor, so that at the outcrop scale, the migmatite preserves all of the features of its unmelted lower-grade equivalents; thus, it is a metatexite migmatite. As the edges of this neosome are very difficult to define precisely, and the neosome occupies a discrete, small volume in the paleosome, the migmatite could, if desired, be described as a nebulitic patch metatexite migmatite.

Location: Mount Hay, Arunta Inlier, Australia. *Rock type*: nebulitic patch metatexite migmatite; metamafic protolith, partial melting at T 825–875°C, P 6–7 kbar. *Scale*: the ruler is 15 cm long. *Image*: E.W. Sawyer.

Further reading: Harte, B., Pattison, D.R.M. & Linklater, C.M. (1991): Field relations and petrography of partially melted pelite and semi-pelitic rocks. *In* Equilibrium in Contact Metamorphism: the Ballachulish Igneous Complex and its Aureole (G. Voll, J. Topel, D.R.M. Pattison & F. Seifert, eds.). Springer-Verlag, Heidelberg, Germany (181–209).

Figure D2

Fig. D2. During incipient melting of this metamafic rock, the tiny fraction of melt that was generated migrated to the low-pressure sites adjacent to the competent crystals of garnet. The solid product of the incongruent breakdown of hornblende was clinopyroxene, and this mineral occurs in the groundmass, and hence remained at the site where the melt formed. The melting reaction was possibly hornblende + plagioclase + quartz = clinopyroxene + melt ± garnet. Most, if not all, of the garnet in this iron-rich metamafic rock was present before partial melting occurred; rocks in the lower amphibolite facies that have similar bulk compositions contain abundant porphyroblasts of garnet. The metamafic rock is Fe-rich because it crystallized from a melt that was the evolved product of the fractional crystallization of a more primitive parental basalt magma, one that had crystallized magnesian olivine and pyroxene.

Location: Abitibi Subprovince, south of Chibougamau, Quebec, Canada. *Rock type*: metatexite migmatite; metamafic protolith partially melted at conditions just below the "orthopyroxene-in" isograd, *T* 800–850°C, *P* 8–10 kbar. *Scale*: the ruler is 15 cm long. *Image*: E.W Sawyer.

Figure D3

Fig. D3. Incipient partial melting in this plagioclase–hornblende metamafic rock has produced very small, elongate patches of neosome. The neosome contains small subidioblastic crystals of clinopyroxene in a tonalitic, plagioclase + quartz groundmass. This assemblage is interpreted as indicating that the neosome represents the *in situ*, nonsegregated products of the incongruent breakdown of hornblende by a reaction such as hornblende + plagioclase + quartz = clinopyroxene + melt. The domains of neosome are 5–10 cm long, spaced about 75 cm apart, and are all elongate in the same direction. In this particular metamafic rock, there are no obvious structures associated with the orientation of the domains of neosome. However, the orientation of the domains of neosome in the metamafic layers does correspond to the orientation of the axial planes to small asymmetrical, Z-shaped folds developed in the adjacent metapelitic beds. Although partial melting was incipient in the metamafic layers, it had reached a far more advanced stage in the pelitic layers, which have lost significant amounts of anatectic melt and now have assemblages of residual minerals and appropriate melt-depleted bulk compositions (see Fig. B38).

Location: Saint-Fulgence, Grenville Province, Quebec, Canada. *Rock type*: patch metatexite migmatite; metamafic protolith, anatexis at conditions of the lower granulite facies, *T* 800–850°C, *P* 5–8 kbar. *Scale*: the ruler is 15 cm long. *Image*: E.W. Sawyer.

Figure D4

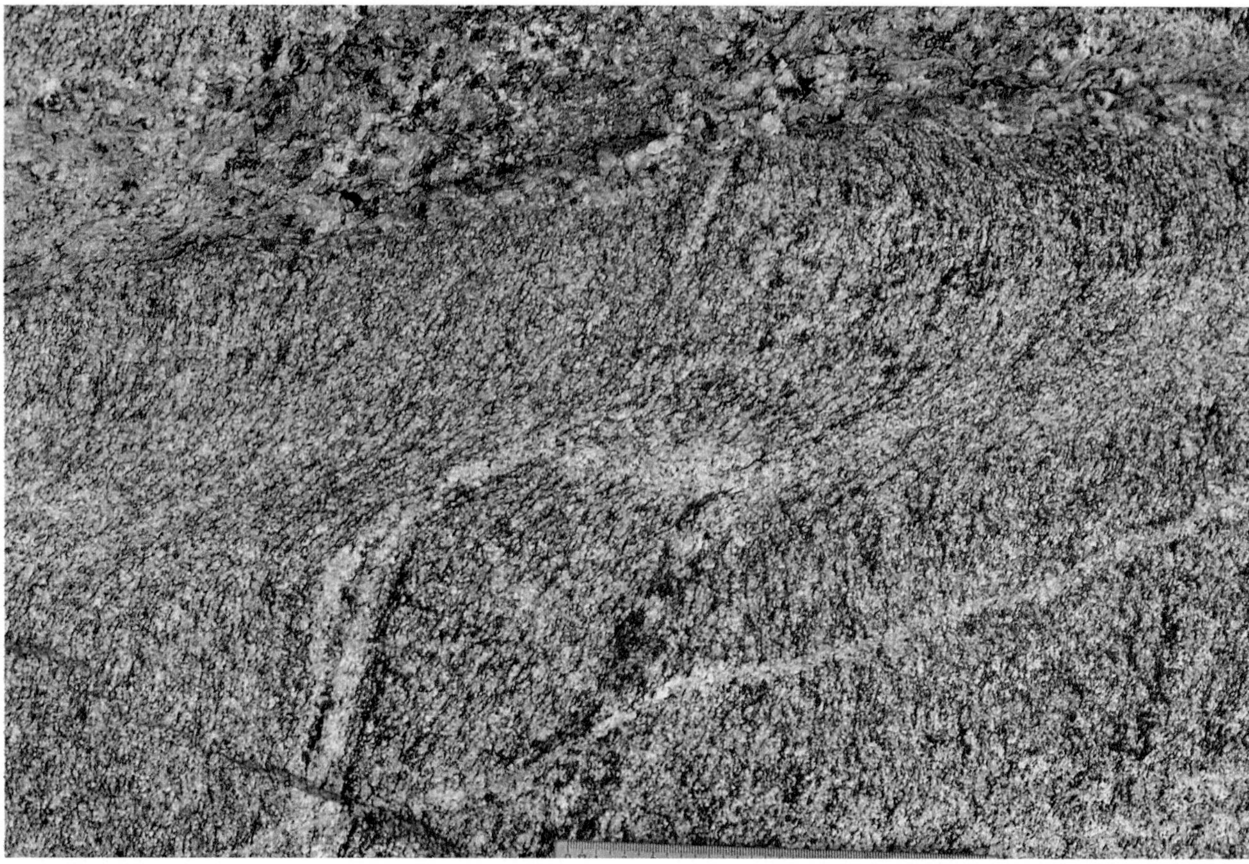

Fig. D4. The onset of partial melting in coarse-grained felsic rocks can be difficult to detect because the paleosome and neosome are very similar in color, mineralogy, and microstructure. This example from the "melt-in" isograd shows a small, diffuse leucocratic patch, or lens, located along a small, isolated shear zone, visible as a curvature of the pre-partial-melting foliation and compositional banding in the paleosome. The neosome is recognized by its slightly coarser grain-size, diffuse margins, and because it has essentially no foliation, characteristics that together indicate *in situ* formation. A thin vein of anatectic melt has intruded the paleosome just below the patch of *in situ* neosome. The preservation of pre-anatectic structures and the dominance of paleosome indicate this to be a metatexite migmatite. Furthermore, the form of the neosome means that the descriptive term "patch" could be applied also. The veins become more abundant, and the patches larger, in the direction of increasing metamorphic grade.

Location: Opatica Subprovince, Quebec, Canada. *Rock type*: patch metatexite migmatite; tonalite gneiss protolith, partial melting at conditions of the upper amphibolite facies, *T* ca. 750°C, *P* 5–7 kbar. *Scale*: the ruler is 15 cm long. *Image*: E.W. Sawyer.

Figure D5

Fig. D5. Partial melting in these pelitic migmatites occurred as the rocks were undergoing minor extension parallel to the layering. This migmatite is from just inside the "melt-in" isograd in the contact aureole of the Duluth Igneous Complex (DIC) and, therefore, shows a migmatite formed in a setting of very low degree of partial melting. The anatectic melt generated in the rocks has migrated a few millimeters from the grain boundaries where it formed into small extensional shear zones nearby, and created the network of millimeter-wide bands of leucosome. At the microscopic scale, the bands have diffuse margins suggestive of local segregation of the melt, rather than the injection of anatectic melt into fractures, and a granular microstructure, which may be the result of rapid cooling in the contact aureole. Because the bedding, a structure that existed before anatexis, is preserved in a coherent manner, this is a metatexite migmatite. The arrangement of the domains of leucosome, albeit very thin, in this migmatite has created a net-like pattern, and such morphology is very common in metatexite migmatites; this could also be described as a net-structured metatexite migmatite. The fine grain-size of this migmatite is characteristic of migmatites developed from low-grade metasedimentary protoliths in shallow contact-metamorphic aureoles (see also Fig. D6).

Location: Linwood Lake area 50 m from the western contact of the Duluth Igneous Complex, Minnesota, U.S.A. *Rock type*: metatexite migmatite; pelite and semipelite protolith, anatexis at *T* 700–750°C, *P* 2 kbar. *Scale*: the scale is 15 cm long. *Image*: E.W. Sawyer.

Further reading: Holness, M. & Clemens, J.D. (1999): Partial melting of the Appin quartzite driven by fracture-controlled H_2O infiltration in the aureole of the Ballachulish Igneous Complex, Scottish Highlands. *Contributions to Mineralogy and Petrology* **136**, 154–168.

Figure **D6**

Fig. D6. This migmatite records the early stages of partial melting in a metasedimentary enclave within the Kuna Crest Granodiorite in the western part of the Tuolumne Complex, Yosemite National Park, California. The metasedimentary protolith consisted of interlayered siliciclastic rocks ranging in composition from pelite to quartzite, but psammite predominates. The metasedimentary rocks were intruded by numerous thin granitic dikes, probably from the nearby El Capitan granite, some 15 Ma before they were incorporated into the younger Kuna Crest Granodiorite as an enclave and experienced minor partial melting. The photograph is dominated by paleosome that consists of grey, quartzofeldspathic metapsammitic schist that contains layers of pale buff quartzite 1–2 cm thick. Both are cut by thin, slightly discordant granitic (G) layers that have thin biotite-rich mafic selvedges. The evidence for partial melting in the enclave is the local presence of inconspicuous, thin, diffuse bodies of pale grey neosome (N) that have not segregated into leucosome and melanosome parts. The neosome has a slightly coarser grain-size than the grey metapsammite paleosome, and this may account for its slightly lighter color. The neosome appears to be located in shears developed on both the long and short limbs of asymmetrical crenulation folds prominent in the center of the photograph. Thus, they are oriented approximately parallel to the axial surfaces of the folds.

Location: May Lake area, Tuolumne Complex, Yosemite Park, U.S.A. *Rock type*: metatexite migmatite; metapsammite protolith containing felsic veins. *Scale*: the ruler is 15 cm long. *Image*: E.W. Sawyer.

Figure D7

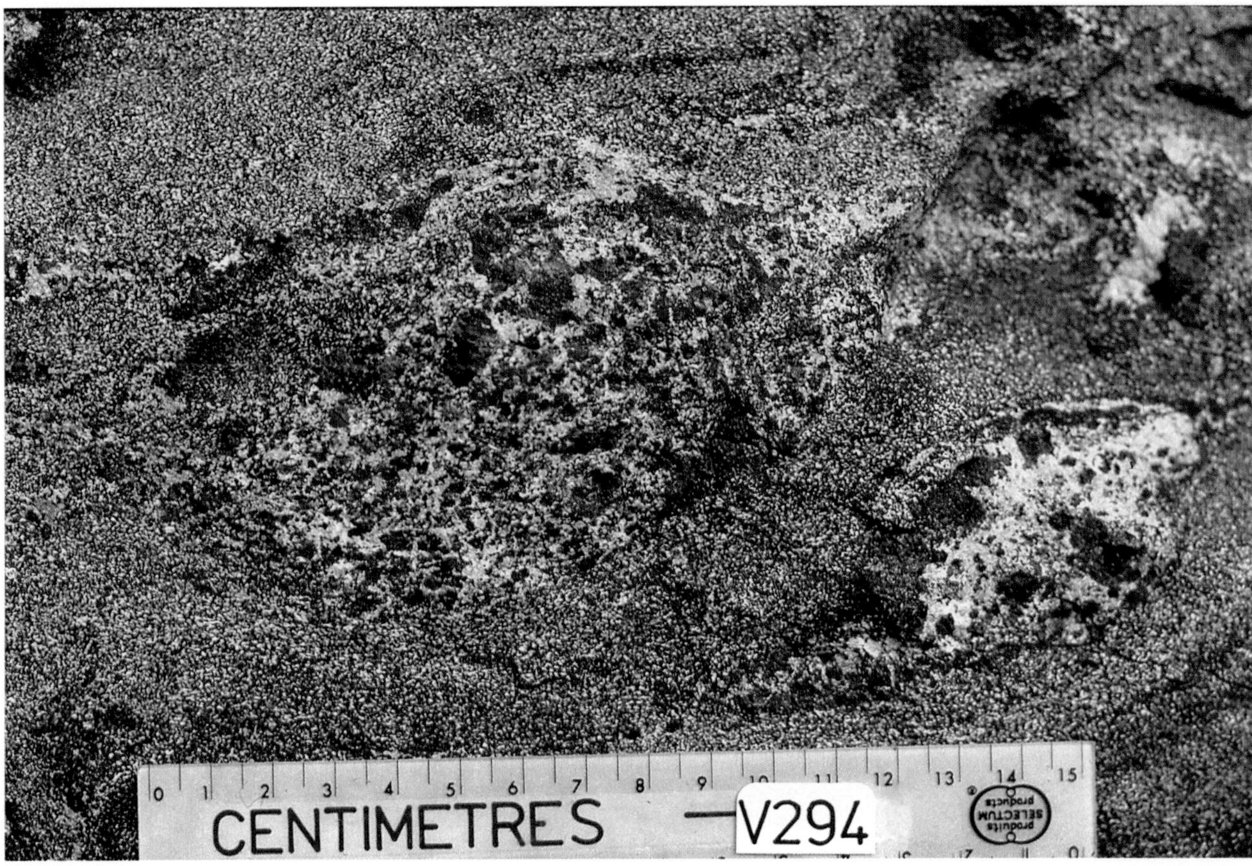

Fig. D7. This photograph shows the first migmatites that appear in an Archean prograde metamorphic sequence that runs from greenschist through amphibolite to the granulite facies. The neosome occurs as patches that are much coarser grained and more heterogeneous than the surrounding paleosome, a compositionally layered plagioclase–hornblende metamafic rock. The preservation of compositional layering that predates anatexis indicates that this is a metatexite migmatite. The patches of neosome consist of a plagioclase + quartz leucocratic portion and a clinopyroxene + hornblende + garnet melanocratic part. Clinopyroxene generally forms the largest crystals, and in many cases, it contains inclusions of hornblende. Thus, the melting reaction may have been similar to hornblende + plagioclase + quartz = garnet + clinopyroxene + melt. The largest patch of neosome has a uniform distribution of leucocratic and melanocratic constituents, and it has a diffuse border. However, the patch of neosome in the lower right has a very narrow, outer melanocratic rim, and a more leucocratic core, which contains large crystals of clinopyroxene. As the bulk geochemical composition of the patches is similar to that of the paleosome, the patches of neosome in this migmatite appear to have formed *in situ*. The differences in internal structure between adjacent patches of neosome must represent local differences in the degree to which the melt and residuum were able to separate.

Location: Abitibi Subprovince, south of Chibougamau, Quebec, Canada. *Rock type*: patch metatexite migmatite; metamafic protolith, partial melting at conditions close to the transition from the amphibolite to granulite facies, *T* 800–850°C, *P* 8–10 kbar. *Scale*: the ruler is 15 cm long. *Image*: E.W. Sawyer.

Figure D8

Fig. D8. The grey paleosome of this migmatite is a medium-grained garnet- and sillimanite-bearing biotite–quartz–plagioclase semipelitic schist. It contains several small, coarse-grained patches of neosome, each a few centimeters across, that contain the mineral assemblage plagioclase + K-feldspar + quartz + cordierite + biotite ± garnet. Biotite in the patches of neosome partially replaces cordierite or garnet (or both). The melting reaction was possibly biotite + quartz + sillimanite + plagioclase = melt + cordierite + K-feldspar ± garnet. Most of the patches of neosome have parts that are rich in ferromagnesian minerals and others that are rich in quartz, plagioclase, and K-feldspar. However, the distribution of the leucocratic and melanocratic parts within the neosome varies considerably from one patch to the next. The manner and extent to which the melt fraction separated from the solid fraction thus varied from neosome to neosome. Some patches of neosome are essentially equant in shape, but others, located in more micaceous layers, are elongate along the direction of the foliation. Anisotropy in the protolith thus influenced how individual patches of neosome grew. Both the grey paleosome and the patches of neosome are cut by planar, discordant leucocratic veins that are rich in quartz and K-feldspar. These veins are interpreted to be derived from a fractionated anatectic melt that intruded the migmatite when it was close to its solidus temperature.

Location: Lac Sainte-Marie, Central Metasedimentary Belt, Grenville Province, Quebec, Canada. *Rock type*: patch metatexite migmatite; metapelitic protolith, partial melting at T ca. 750–800°C, P 5–7 kbar. *Scale*: the ruler is 15 cm long. *Image*: E.W. Sawyer.

Figure D9

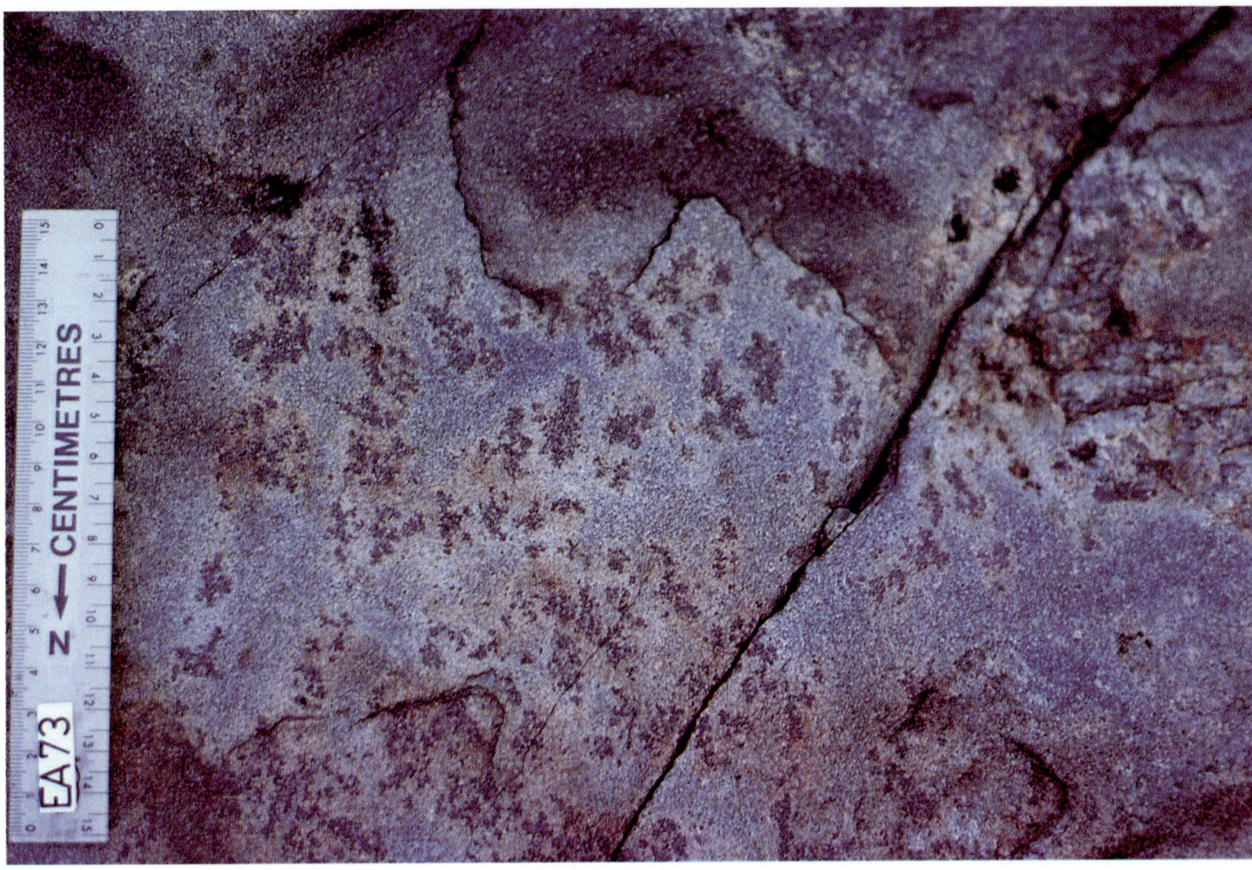

Fig. D9. Thick quartzofeldspathic layers, such as beds of metagreywacke, are more competent, and hence record lower internal strains than the more micaceous interbedded pelitic layers. Furthermore, because of their bulk composition, partial melting in the quartzofeldspathic layers typically starts at higher temperatures than in the pelitic ones. This patch migmatite formed in a feldspathic, Al-poor psammitic layer where partial melting was in its early stages. The migmatite consists of a light-grey biotite–quartz–plagioclase schist paleosome that hosts equant patches of neosome; these have a slight preferred orientation parallel to the local S_3 foliation (parallel to the scale). The patches of neosome are distinctive in that they have a dark inner part consisting of several orthopyroxene crystals, and a leucocratic outer part consisting of quartz, plagioclase, and K-feldspar; there is no biotite in the patches of neosome. Commonly, the orthopyroxene contains elongate inclusions of quartz. The inner part of the leucocratic rim, close to the orthopyroxene, is richer in K-feldspar than the outer part. The melting reaction in the psammitic layer was probably biotite + quartz + plagioclase = orthopyroxene + melt. Melting was more advanced, and the strain higher, in the adjacent pelitic layers; hence, migmatites with a stromatic, or layered, morphology developed there. The melting reaction in the pelitic layers probably included sillimanite, and may have been biotite + sillimanite + quartz + plagioclase = garnet + cordierite + K-feldspar + melt.

Location: Wuluma Hills, Arunta Inlier, Australia. *Rock type*: patch metatexite migmatite; Al-poor psammite protolith, with anatexis at *T* 825–875°C, *P* ca. 5 kbar. *Scale*: the ruler is 15 cm long. *Image*: E.W. Sawyer.

Figure D10

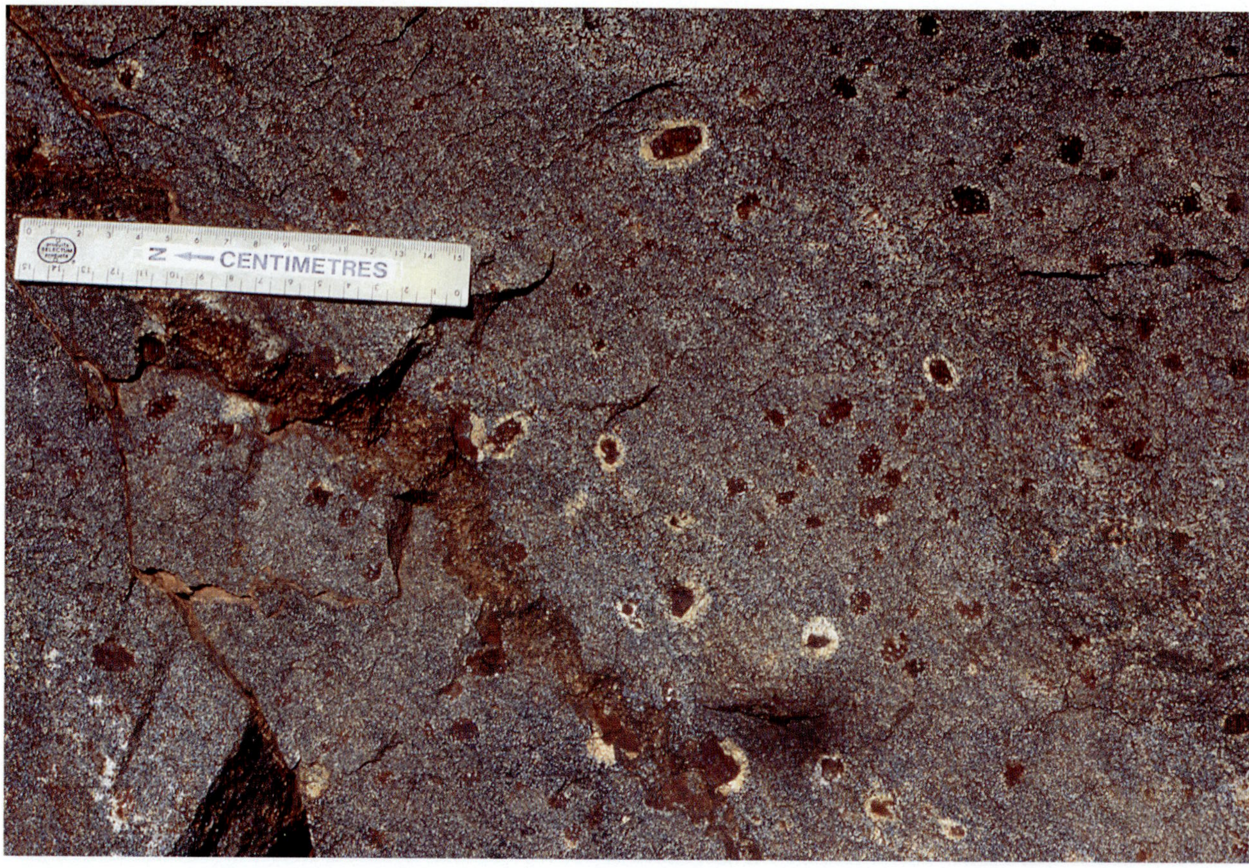

Fig. D10. This photomicrograph shows a variation in the morphological detail of the patch migmatites developed in quartzofeldspathic, Al-poor psammitic beds (cf. Fig. D9). The melanocratic cores of these patches of neosome consist of a single, large, idioblastic crystal of orthopyroxene that is surrounded by a narrow, leucocratic rim composed of quartz + plagioclase + K-feldspar. Some patches have a wider rim than others, and this may indicate the loss of melt from those patches that have a narrow rim. The peculiarities of local weathering have highlighted some of the leucocratic rims by making them pinkish white, but most are dull grey and inconspicuous. The mineral assemblage in the fine-grained paleosome around the patch is quartz + plagioclase + biotite. The absence of biotite from the patches of neosome suggests that they formed by a melting reaction involving the incongruent breakdown of biotite, such as biotite + plagioclase + quartz = orthopyroxene + melt. The patches of neosome in this photomicrograph are a maximum of 3 cm across, but similar patches can attain more than 30 cm across elsewhere in the Wuluma Hills area. Adjacent, slightly more aluminous beds contain patches of neosome that have a very similar appearance, but they contain cordierite, or cordierite + garnet, in addition to orthopyroxene (see Fig. F77).

Location: Wuluma Hills, Arunta Inlier, Australia. *Rock type*: patch metatexite migmatite; Al-poor psammite protolith, partially melted at T 825–875°C, P ca. 5 kbar. *Scale*: the ruler is 15 cm long. *Image*: E.W. Sawyer.

Figure **D11**

Fig. D11. The thin, fine-grained layers in various tones of pale grey are granodioritic, trondhjemitic, and intermediate orthogneisses that comprise the paleosome in this migmatite. Deformation, some of it previous to anatexis and some of it synchronous with it, has attenuated the felsic orthogneisses and generated the banding in them; the more competent and generally thicker intermediate layers have become boudinaged. Partial melting of the orthogneisses has generated small patches of neosome throughout this migmatite. The domains of neosome are coarser grained than the paleosome, and can be distinguished by the relatively large grains of ferromagnesian minerals they contain, and because they are neither foliated nor compositionally banded. Some domains of neosome are located in obvious low-pressure sites, such as at the ends of boudins or in shear bands. Others, however, occur as variously shaped patches (e.g., near the pocket knife) that do not appear to be related to any specific structure. Thus, partial melting is interpreted to have occurred during, and after, the penetrative deformation that affected the paleosome. Furthermore, the absence of discrete domains of melanosome in the migmatite indicates that the melt fraction did not separate from the residuum, which is consistent with the formation of the neosome after most of the penetrative deformation. These particular domains of neosome (labeled N) can be called nebulitic because they have do not have sharply defined borders, and in many places appear to be gradational from the paleosome (see also Fig. D1).

Location: Sand River Gneisses, Causeway locality, South Africa. *Rock type*: metatexite migmatite with nebulitic neosome; orthogneiss protolith, anatexis under granulite-facies conditions. *Scale*: the pocket knife is 11 cm long. *Image*: E.W. Sawyer.

Figure D12

Fig. D12. This metatexite migmatite is from just within the "melt-in" isograd; it formed by anatexis of a massive felsic protolith in a local environment where the syn-anatectic strain was relatively low. The neosome in this migmatite occurs as light, buff-colored, oval patches of granitic composition hosted by a grey, foliated leucotrondhjemitic paleosome. Because the patches of neosome (N) have a diffuse margin and do not have the foliation evident in the paleosome (P), they are somewhat nebulous in appearance; hence, this patch metatexite migmatite might also be called a nebulite. The neosome has a slightly coarser grain-size than the paleosome; the residual phases thus recrystallized to a larger grain-size while melt was present. Petrographic and geochemical data indicate that melt has not segregated from the residual solids during the formation of the patches of neosome; hence, no leucosome or melanosome was produced, i.e., the products of anatexis have remained strictly *in situ*.

Location: Opatica Subprovince, Quebec, Canada. *Rock type*: nebulitic patch metatexite migmatite, leucotrondhjemitic protolith, anatexis at conditions of the upper amphibolite facies, T ca. 750°C, P 5–7 kbar. *Scale*: width of the field of view is 45 cm. *Image*: E.W. Sawyer.

Further reading: Sawyer, E.W. (1998): Formation and evolution of granite magmas during crustal reworking: the significance of diatexites. *Journal of Petrology* **39**, 1147–1167.

Figure D13

Fig. D13. This photograph shows the morphology of metatexite migmatites formed in a pre-anatexis to syn-anatexis strike-slip shear zone in a metamafic protolith. The compositional layering remains prominent in the paleosome. The neosome consists of melanosome most commonly developed parallel to the compositional layering, and leucosome that occurs at a number of different structural sites: (1) along extensional (normal) shears forming a conjugate set oriented at about 30–40° to the layering, (2) in the space between boudins that have developed in the most competent layers (commonly layers of melanosome), and hence oriented approximately normal to the layering, and (3) in thin films of leucosome lying parallel to the compositional layering. Notice, however, that most of these are also closely related to the melanocratic layers (M). Thus, the leucosome is located in the dilatant sites that formed as the metamafic paleosome and then the layers of residuum (melanosome) underwent layer-parallel extension. The melt thus was mobile during synchronous partial melting and noncoaxial strain.

Location: Abitibi Subprovince, south of Chibougamau Quebec, Canada. *Rock type*: metatexite migmatite, metamafic protolith, anatexis at conditions near the transition from amphibolite to granulite facies, *T* 800–850°C, *P* 8–10 kbar. *Scale*: ruler is 15 cm long. *Image*: E.W. Sawyer.

Figure **D14**

Fig. D14. This migmatite consists of two parts. The melanocratic portion is dominant and consists of metamafic layers that have strongly melt-depleted bulk compositions relative to the likely metamafic protolith, and they are interpreted to be the residuum left after the extraction of anatectic melt. The mineral assemblage in the metamafic layers is hornblende + clinopyroxene + garnet + plagioclase; the color difference between layers is due to minor changes in the modal proportions of the same minerals. The migmatite contains a conspicuous set of regularly spaced domains of leucosome 15–25 cm long that are oriented either perpendicular to the layering, or between 30 and 45° to it. The leucosome occupies structures that formed as a result of extension parallel to the layering in the metamafic rock. All these bodies of leucosome are tonalitic in composition (plagioclase + quartz) and coarse-grained. Some of the bodies of leucosome contain angular, spalled-off fragments of wallrock. There is no discrete melanocratic rim around the leucosome, but this is not surprising given the residual nature of the entire host. Another set of domains of tonalitic leucosome occurs parallel to the layering of the metamafic rock. Although these are laterally persistent, they are thin (<5 mm) and inconspicuous.

Location: Abitibi Subprovince, south of Chibougamau, Quebec, Canada. *Rock type*: dilation-structured metatexite migmatite, metamafic protolith, anatexis at conditions close to the transition from amphibolite to granulite facies, T 800–850°C, P 8–10 kbar. *Scale*: the outcrop is 4 m across. *Image*: E.W. Sawyer.

Figure D15

Fig. D15. The paleosome to this migmatite consists of thick (up to 3 m), compositionally different layers of metamafic rock, which are interpreted to have originally been massive flows of basalt. A finer-grained mafic dike (upper left quadrant) crosses the layers of metabasalt. Thin bodies of leucosome are located parallel to the compositional layering and the foliation and in the pressure shadows around garnet crystals. Larger, irregularly shaped bodies of leucosome (e.g., near the dike) occur in dilatant sites and mark the position of incipient boudinage of the competent layer. These areas of leucosome have a melanocratic border, which indicates a local derivation of the melt that became the leucosome. The migmatite also contains a set of leucocratic veins, with exactly the same composition as the others, but located (e.g., just above the ruler) in planar-sided, dilatant fractures that systematically cross the outcrop at about 45° to the layering; these have no obvious rim of melanosome around them.

Location: Abitibi Subprovince, south of Chibougamau, Quebec, Canada. *Rock type*: dilation-structured metatexite migmatite, metabasalt protolith, anatexis at conditions close to the transition from amphibolite to granulite facies, *T* 800–850°C, *P* 8–10 kbar. *Scale*: the ruler is 15 cm long. *Image*: E.W. Sawyer.

Figure **D16**

Fig. D16. The paleosome in this migmatite consists of thin layers of metamafic rock in a hornblende–biotite granodiorite visible at the top of the photograph. Neosome occurs only in the metamafic layers of the migmatite, and it has a very distinctive and characteristic morphology. The neosome is segregated into well-defined leucosome and melanosome. The area of leucosome has a wide base that is rooted at the contact between the metamafic and the host granodiorite, and tapers to an apex that points toward the center of the layer of metamafic rock. In some cases, the triangular domains of leucosome on opposite walls of the mafic layer have grown sufficiently large that they meet, and leucosome bridges the entire layer of metamafic rock. The melanosome is a dark-colored halo around the domains of leucosome. It is depleted in quartz and feldspar, and consequently enriched in amphibole relative to the metamafic rock farther from the neosome. The metamafic rock is interpreted to have been much more competent than the host granodiorite. Thus, during deformation, brittle fractures developed in the metamafic rock as a result of strain partitioning. Because deformation and anatexis were synchronous in this migmatite, the growth of the brittle fractures provided a convenient low-pressure site into which the anatectic melt forming in the metamafic rock could migrate, and hence the domains of leucosome were formed. The melanosome halo, therefore, indicates the region in the metamafic rock from which the melt was derived.

Location: Mt. Odin, Monashee complex, British Columbia, Canada. *Rock type*: dilatant-structured metatexite migmatite; partial melting in the upper amphibolite facies. *Scale*: the pencil is 14 cm long. *Image and caption*: Paul McNeill.

Figure D17

Fig. D17. The dilatant structures into which melt can migrate, to form domains of leucosome, need not be large. In this example, the anatectic melt has collected in a series of dilatant sites produced as garnet crystals in the metamafic paleosome were pulled in a direction parallel to the foliation in the host. Consequently, many segments of leucosome are tabular in shape and have their long axes at a high angle to the foliation in the paleosome. However, the overall impression is that the domains of leucosome are oriented parallel to the layering.

Location: Georgian Bay, Grenville Province, Canada. *Rock type*: metatexite migmatite; the metamafic protolith partially melted at conditions of the upper amphibolite to granulite facies. *Scale*: the sunglasses are 14 cm across. *Image*: Olivier Vanderhaeghe.

Figure D18

Fig. D18. The main features of this outcrop are the layers of intermediate to mafic orthogneiss, which contain two sets of leucocratic veins of tonalite to granodiorite composition. A conspicuous set of leucocratic veins is located in dilatational structures, and the other set, much less obvious (and thinner), is located parallel to the compositional layering and the foliation. In the bottom half of the photograph, the dilatational sites are related to simple boudinage of melanocratic layers and to extension fractures that cross several layers, whereas in the upper part of the photograph, they are associated with a more highly disrupted mafic layer. Because most of the leucocratic veins have sharp contacts with their host, and do not have a melanocratic rim, it is not immediately evident that this is a migmatite; it could be a veined gneiss. However, the widest, dark grey layer in the top right contains a band of neosome (N) comprising small, irregularly shaped veins of tonalite that have small melanocratic patches rich in amphibole associated with them; this band is interpreted to have been derived from barely segregated anatectic melt and residuum. Hence, the outcrop is indeed a migmatite. Similar patches of unsegregated neosome occur in the lower left corner, in the melanocratic layer that has undergone boudinage. This melanocratic layer is, therefore, interpreted as residuum, and the leucocratic material in the boudin interpartitions, as leucosome. On the basis of the proportions of leucosome and melanosome in the outcrop, this migmatite probably contains more anatectic melt than was generated *in situ*. This must, therefore, be viewed as a site where anatectic melt was able to accumulate. The pressure gradients generated by the formation of the dilatant structures may have effectively sucked melt into this migmatite from nearby partially melted rocks.

Location: Thor–Odin dome, British Columbia, Canada. *Rock type*: metatexite migmatite; mafic and intermediate orthogneiss protolith, partially melted at conditions of the upper amphibolite facies. *Scale*: the coin in the center is 2.65 cm across. *Image*: Olivier Vanderhaeghe.

Figure D19

Fig. D19. The domains of leucosome in this metatexite migmatite are located in semiregularly spaced dilatant structures that have developed in a foliated metamafic paleosome. From the geometry of minor folds in the paleosome, the dilatancy appears to have occurred in several ways. In some cases, the small symmetrical folds suggest that regular boudinage occurred; however, elsewhere the curvature of the foliation suggests an asymmetrical foliation boudin, or flanking fold next to a zone of oblique shear. No matter how exactly the dilatancy occurred, the neosome that formed in and around it is very similar throughout the migmatite. Two parts, although not necessarily well separated, are evident in all the areas of neosome. The most prominent part is a coarse-grained leucosome, which contains the assemblage plagioclase + quartz + clinopyroxene. Typically, the leucosome occupies the dilatant structure itself, and has a partial envelope of coarse-grained melanosome with the mineral assemblage clinopyroxene + hornblende + garnet. In most cases, the melanosome is in the wallrock around the leucosome. This juxtaposition suggests the *in situ* formation of the neosome in and around the dilatant structure. However, in some cases, the melanosome occurs as fragments in the leucosome. This pattern indicates a further episode of dilatancy after some of the melanosome has developed. The paleosome is considerably finer grained than the neosome, and contains the assemblage hornblende + plagioclase + clinopyroxene + garnet + quartz. The large, idioblastic crystals of clinopyroxene present in the melanosome and parts of the leucosome are interpreted as the solid products of reaction that grew in the melt produced by a reaction that involved the incongruent breakdown of hornblende.

Location: Central Metasedimentary Belt, Grenville Province Quebec, Canada. *Rock type*: metatexite migmatite; metamafic protolith, granulite-facies anatexis *T* 800–850°C, *P* 5–8 kbar. *Scale*: the ruler is 15 cm long. *Image*: E.W. Sawyer.

Figure **D20**

Fig. D20. This migmatite contains well-developed compositional layering that can be traced across large (100 m) outcrops. The portion shown here comprises leucotonalite, mesocratic metamafic and melanocratic metamafic layers. The layers of metamafic rock on which the pocket knife rests contain conspicuous sublayers, some of which are particularly dark and interpreted to be the most residual in composition. Furthermore, the metamafic rocks were evidently the most competent ones in the migmatite, as they have undergone symmetrical and asymmetrical boudinage on the layer and the sublayer scales, and this has created a boudin-within-boudin structure. Small bodies of leucosome consisting of plagioclase + quartz occupy the interboudin partitions developed in the darkest sublayers. At a larger scale, the pattern is somewhat different, and segments of prominent layer-parallel domains of leucosome link with those in the large interboudin partitions between the layer-scale boudins. The melt-depleted melanocratic sublayers and their associated network of leucosome can be interpreted to have developed as the sequence became progressively melt-depleted during deformation. At some stage during melt depletion, the proportion of residual pyroxene and hornblende in the individual sublayers had increased sufficiently that the overall package of sublayers became stronger than the adjacent mesocratic layers. At this point, the assemblage started to boudinage as a whole and created the larger length-scale boudins. The site may then have ceased to lose melt and may have switched to become a zone of net accumulation of melt.

Location: Sand River Gneisses, Causeway locality, South Africa. *Rock type*: dilatant structured metatexite migmatite; felsic and mafic orthogneiss protolith, anatexis in the granulite facies, T 800–850°C, P 7–10 kbar. *Scale*: the pocket knife is 11 cm long. *Image*: E.W. Sawyer.

Figure D21

Fig. D21. The grey-weathering layers in this migmatite are Al-poor, orthopyroxene-bearing, biotite–plagioclase–quartz psammitic schists that have undergone relatively minor amounts of partial melting (see Figs. D9 and 10 for similar rocks from the same area). The darkest-weathering layers in the migmatite were originally semipelites; they now contain the mineral assemblage orthopyroxene + quartz + cordierite + biotite ± garnet, and are feldspar-absent. This assemblage of minerals and the whole-rock major- and trace-element compositions suggest that these layers are the severely depleted residuum left after the extraction of anatectic melt. In most circumstances, the Al-poor psammites are more competent than the semipelites. However, in this case, the extraction of substantial amounts of anatectic melt from the semipelitic layers has created a residuum with 70 modal % or more of orthopyroxene + quartz, and both are strong minerals. Consequently, as anatexis advanced and the semipelitic layers progressively lost melt, they systematically became more competent and started to develop extensional shears and to undergo boudinage in response to late layer-parallel extension. Anatectic melt was then able to migrate into these layers as the dilatant sites developed, to produce the array of small accumulations of leucosome visible in the most residual layers of this migmatite.

Location: Wuluma Hills, Arunta Inlier, Australia. *Rock type*: metatexite migmatite; semipelitic protolith, partially melted at *T* 825–875°C, *P* ca. 5 kbar. *Scale*: the ruler is graduated in cm. *Image*: E.W. Sawyer.

Figure **D22**

Fig. D22. This metatexite migmatite has a strong planar anisotropy. Within the grey paleosome, this anisotropy is due to variation in the composition of original beds, but it is also due to the segregation of melt parallel to the planar anisotropy in the protolith, creating parallel sheets of leucosome and mafic selvedge. Consequently, the principal morphology of the migmatite is stromatic, or layered. Late-stage extension, roughly parallel to the layering, resulted in the formation of two normal-sense shears and their associated flanking folds, in the slightly darker layer running across the center of the photograph. Melt then located in the layer migrated into the dilatant shears, and its crystallization products formed the leucosome. Both cross-cutting bodies of leucosome are diffuse at their tips, and the one on the right appears to have diffuse contacts with the layering it transects. The stromatic leucosome thus may not have been solid at the time the layer was extended.

Location: Port Navalo area, southern Brittany, France. *Rock type*: metatexite migmatite derived from quartzofeldspathic and aluminous gneisses, partial melting at T ca. 800°C, P ca. 8 kbar, then decompressed to 4 kbar, followed by cooling. *Scale*: the ruler is 15 cm long. *Image*: E.W. Sawyer.

Further reading: Jones, K.A. & Brown, M. (1990): High-temperature "clockwise" P–T paths and melting in the development of regional migmatites: an example from southern Brittany, France. *Journal of Metamorphic Geology* **8**, 551–578.

Marchildon, N. & Brown, M. (2003): Spatial distribution of melt-bearing structures in anatectic rocks from southern Brittany, France: implications for melt transfer at grain- to orogen-scale. *Tectonophysics* **364**, 215–235.

Figure D23

Fig. D23. The paleosome of this migmatite is a dark grey, well-foliated hornblende granodiorite; minor changes in color indicate only a weak compositional layering. Large equant to lens-shaped areas of neosome occupy the dilatant spaces created during the formation of both symmetrical and asymmetrical boudins. The absence of a strong compositional layering and the presence of a strong foliation suggest that these areas could be interpreted as examples of foliation boudinage. The neosome in these structures does not have a prominent melanocratic border around them, which could be taken to indicate that the melt in that case was externally derived. However, the areas of neosome have a diffuse margin, which is consistent with *in situ* partial melting. Furthermore, they contain scattered, large crystals of amphibole, which could be the solid products of the melting reaction. The presence of the amphibole also is consistent with *in situ* melting and very limited separation of the melt fraction from residuum. Hence, these bodies are better called neosome rather than leucosome. Thin sheets of leucosome with a sharp contact and very few mafic minerals occur parallel to the foliation; rarely, they have a narrow, discontinuous melanosome. The term leucosome is appropriate for these because the melt fraction and residual fractions did become segregated. Overall, this migmatite has two orientations of melt-bearing structures, and the intersection of these leads to the local development of an incipient net structure.

Location: Laag mountain area, Monashee complex, British Columbia, Canada. *Rock type*: dilation-structured metatexite migmatite; hornblende granodiorite protolith, partial melting in the upper amphibolite facies. *Scale*: the card is 8.5 cm long. *Image and caption*: Paul McNeill.

Figure D24

Fig. D24. This photograph is from the same area as Fig. D23. It shows the net-structured migmatite that developed where the proportion of melt in the migmatite was somewhat higher. The outcrop clearly shows two trends of melt-filled structures, one parallel to the foliation and the other discordant at 30–40° to the foliation, that together create the net-like morphology. The areas of leucosome that are parallel to the foliation have sharp borders and contain a very low proportion of mafic minerals. In contrast, the obliquely discordant structures contain a significantly higher proportion of mafic minerals, typically large euhedral crystals of amphibole, and have a diffuse border, which suggests an *in situ* origin. In most parts of the outcrop, the bodies of leucosome oriented parallel to the foliation are thin and volumetrically minor, which could indicate either that the bulk of the outcrop has only just begun to partially melt, or that these regions have already lost melt. However, part of the photograph shows several prominent, wide areas of leucosome that are parallel to the foliation, but fewer areas of leucosome accumulation in symmetrical and asymmetrical boudinage structures than elsewhere. The overall distribution of melt-filled structures in this migmatite has created a linked network of channels that could have drained melt from the migmatite, or aided melt in passing through the crustal level represented by this migmatite.

Location: Laag mountain area, Monashee complex, British Columbia, Canada. *Rock type*: dilation-structured metatexite migmatite; hornblende granodiorite protolith, partial melting in the upper amphibolite facies. *Scale*: the card is 8.5 cm long. *Image and caption*: Paul McNeill.

Figure **D25**

Fig. D25. The migmatite in this photograph contains two sets of leucosome domains that together create an incipient net-structured stromatic migmatite. One set is developed parallel to the strong tectonic foliation and attenuated compositional layering in the host rock; such leucosome can be described as stromatic. The other set consists of lens-shaped bodies of leucosome that are oblique to the foliation and occupy small, generally extensional shear zones; these bodies of leucosome can be described as discordant. The stromatic bodies of leucosome are generally thin, about 0.5 cm wide, and fine-grained, but many of them thicken and become coarser grained near the bodies of discordant leucosome. The contacts between the stromatic and discordant domains are gradational in many places, and interpreted to mean that both sets were interconnected and contained anatectic melt at the same time. However, there are places where the discordant areas of leucosome truncate the stromatic ones. This complex relationship is interpreted to represent a linked system in which melt generated in the host migrated along the tectonic foliation and accumulated in the shear zones; the gradational contacts preserve this linkage. The discordant bodies of leucosome truncate the stromatic ones where the shear zones that accommodated the melt propagated into regions of the host where the melt flux was much lower, or where it had ceased entirely. The truncated domains of stromatic leucosome in that region were no longer conduits for the transfer of melt.

Location: Tolstik Peninsula, Karelia, Russia. *Rock type*: incipient net-structured metatexite migmatite; mafic to intermediate protolith, anatexis in the upper amphibolite to lower granulite facies. *Scale*: the lens cap is 5.5 cm across. *Image and caption*: Mike Brown. Image previously published as fig. 5c in Brown (2007) and reproduced with the permission of the Geological Society, London.

Further reading: Brown, M. (1994): The generation, segregation, ascent and emplacement of granitic magma: the migmatite-to-crustally-derived granite connection in thickened orogens. *Earth Science Reviews* **36**, 83–130.

Brown, M. (2007): Crustal melting and melt extraction, ascent and emplacement in orogens: mechanisms and consequences. *Journal of the Geological Society of London*, **164**, 709–730.

Sawyer, E.W. (2001): Melt segregation in the continental crust: distribution and movement of melt in anatectic rocks. *Journal of Metamorphic Geology* **19**, 291–309.

Figure **D26**

Fig. D26. Bedding is well preserved in the lowest-grade part of the anatectic domain at Mount Stafford, Australia, and partial melting was considerably more advanced in the pelitic layers than in the psammitic, or semipelitic, ones. The scale rests on a psammitic bed, and above it is a crenulated and partially melted pelitic bed. Most of the leucosome in it, however, is not crenulated. There are two principal orientations for the bodies of leucosome, and consequently, the migmatite consists of a net-like pattern of leucosome that encloses rhombic portions of crenulated paleosome and melanosome. Hence, this metatexite migmatite has a net, or diktyonitic, structure. This particular net-structured migmatite is a little unusual in that neither of the orientations of leucosome is parallel to the foliation in the pelite. The more continuous set of leucosome is subparallel (i.e., crosses at a low angle) to the crenulated foliation in the pelite layer, whereas the other follows shear planes that are subparallel to the axial planes of the asymmetrical crenulations, or follow shears on either the short or long limbs of the folds.

Location: Mount Stafford, Australia. *Rock type*: net-structured metatexite migmatite; metapelitic protolith, Mt. Stafford metamorphic zone 2, *T* 650–675°C, *P* 3.2 kbar. *Scale*: width of the field of view is 75 cm. *Image*: E.W. Sawyer.

Further reading: White, R.W., Powell, R. & Clarke, G.L. (2003): Prograde metamorphic assemblage evolution during partial melting of metasedimentary rocks at low pressures: migmatites from Mt. Stafford, central Australia. *Journal of Petrology* **44**, 1937–1960.

Figure D27

Fig. D27. The most striking examples of net structure are developed in migmatites that have a paleosome with a strong planar anisotropy, e.g., pelitic and semi-pelitic metasedimentary material. In most cases in which net-structure migmatites are formed from felsic plutonic protoliths (e.g., granites, tonalites and trondhjemites), a planar anisotropy was created in the protolith before, or during, anatexis. Consequently, the paleosome in these migmatites shows structural evidence of high to very high strain. The paleosome in the migmatite shown in this photograph is a strongly foliated, leucocratic granodiorite and trondhjemite that contains boudins and lenses of more mafic rocks that are the relics of dikes disrupted during the strong, noncoaxial shearing event that generated the foliation. The paleosome is crossed by a network of neosome domains 1–3 cm wide that have a leucocratic margin and markedly more melanocratic center. The central part of each neosome contains large, euhedral crystals of orthopyroxene, which have later been partially altered. The ferromagnesian minerals in the center of the neosome are interpreted to be the solid products of the incongruent melting of biotite, and the leucocratic rim is interpreted to be derived from the anatectic melt. The photograph shows that the proportion of leucocratic to melanocratic materials in the neosome varies considerably from place to place. Some parts, such as to the right of the largest mafic boudin, have a higher fraction of melanosome. This can be explained if the array of neosome domains served as the system of channels through which anatectic melt flowed within and, ultimately, out of the migmatite. The parts of the network that are the most enriched in melanosome have lost more melt than the other parts. This variation reflects differences in the differential stress from one part of the outcrop to another.

Location: Sand River Gneisses, Causeway locality, South Africa. *Rock type*: net-structured metatexite migmatite; felsic orthogneiss protolith, granulite-facies anatexis. *Scale*: the pocket knife is 11 cm long. *Image*: E.W. Sawyer.

Figure **D28**

Fig. D28. A net-structured metatexite migmatite can develop from relatively homogeneous protoliths, but they are easily overlooked. The example in this photograph developed from a massive and compositionally uniform leucogranodioritic to tonalitic protolith. The areas of neosome that form the net-like array are 1–3 cm wide, and have a relatively uniform grain-size and distribution of minerals throughout (compare with the migmatite in Fig. D27 only a few tens of meters away). Overall, therefore, the neosome is not very different in either grain size or composition from the paleosome in this migmatite, and this makes it difficult to decide, from evidence at the outcrop alone, whether the neosome represents unsegregated melt and residuum, or anatectic melt (in which case it should be called leucosome). The absence of a strong planar anisotropy in the paleosome further contributes to the difficulty for the eye to pick out a net-like array of neosome domains in a migmatite where there is little contrast in color, grain size, or microstructure between neosome and paleosome. This is a common problem in migmatites derived from granites, granodiorites, tonalities, and trondhjemites (see also Fig. D12).

Location: Sand River Gneisses, Causeway locality, South Africa. *Rock type*: net-structured metatexite migmatite; felsic orthogneiss protolith, granulite-facies anatexis. *Scale*: the pocket knife is 11 cm long. *Image*: E.W. Sawyer.

Figure D29

Fig. D29. The distribution of net-structured migmatites is necessarily restricted to places where there are at least two sets of intersecting domains of leucosome. The paleosome in the migmatite depicted here is compositionally banded, intermediate gneiss that has been folded and has undergone partial melting. The resulting metatexite migmatite has a net structure in the hinge regions of the mesoscale folds. One set of leucosome domains is located parallel to the axial plane of the folds, specifically, in small shear zones on the flanks of minor folds located in the hinge region of the larger folds. The second set has developed along the compositional banding in the paleosome, but principally in the region of the fold hinges. The intersection of both sets of leucosome domains generates a net structure in the core of the mesoscale folds. The leucosome is continuous in both mineralogy and microstructure from one orientation to the other, suggesting that together the two sets of domains represent a connected network that contained a continuous anatectic melt. A partial melanocratic rim is developed around many of the leucosome domains, and its origin is an important matter to consider. If the rim is the residue from partial melting, then it indicates that a certain proportion of the melt was derived locally, and that is the critical evidence needed to identify this rock as a migmatite. However, sharp contacts locally between the leucocratic and mesocratic parts of the rock indicate the injection of some melt, but possibly this was simply redistribution of melt within the outcrop. Therefore, the opposing view is that all the felsic veins are from an external source, and that the dark rim is a mafic selvedge that results from rehydration of the wall-rocks following crystallization of the veins. In that case, the outcrop is an injected stockwork of felsic veins and not a migmatite. In either scenario, the hinge regions of the folds are the sites where the felsic melt moved through the system represented by the outcrop.

Location: Arunta Inlier, Australia. *Rock type*: net-structured metatexite migmatite; intermediate gneiss protolith, partially melted under conditions of the upper amphibolite facies. *Scale*: the pen is 15 cm long. *Image and caption*: Tony I.S. Kemp.

Figure **D30**

Fig. D30. One of the sets of leucosome domains that contributes to the net structure in this metatexite migmatite is oriented parallel to the foliation and compositional layering in the melanosome and paleosome. This set of leucosome can be called stromatic. The second set of leucosome domains is located in dilatant structures that are either interboudin partitions oriented perpendicular to the foliation (e.g., each side of the lens cap), or extensional shear planes oriented oblique to the foliation (e.g., top right). Petrographic continuity (i.e., the same modal proportions of quartz, plagioclase, and K-feldspar, the same grain-size, and the same microstructure) between the segments of leucosome located in the different structural sites in this migmatite implies that some proportion of the leucosome crystallized from a melt-bearing network that was continuous throughout the migmatite. This migmatite is located in the South Armorican Shear Zone, which had a dextral transcurrent sense of movement. Qualitative observations on the network of leucosome domains using adjacent, mutually perpendicular, flat outcrop surfaces reveals that the leucosome network is anisotropic relative to the subhorizontal elongation-induced lineation on the foliation planes. The anisotropic nature of the stromatic sets of leucosome suggests that the permeability in the plane of the foliation was larger parallel to the lineation than perpendicular to it. It may then be inferred that the melt in the migmatite flowed in the plane of the foliation and parallel to the lineation, and then flowed into the interboudin partitions and extensional shear surfaces in response to gradients in melt pressure that developed as these dilatant structures formed.

Location: Port Navalo, Brittany, France. *Rock type*: net-structured metatexite migmatite: pelitic and semipelitic protolith, granulite-facies anatexis at T ca. 800°C and P 9 kbar, with a stepped retrograde cooling path to 4 kbar. *Scale*: the lens cap is 6.5 cm across. *Image and caption*: Mike Brown. Image previously published as fig. 3b in Brown (2005) and reproduced with the permission of the Geological Society, London.

Further reading: Brown, M. (2004): The mechanism of melt extraction from lower continental crust of orogens. *Transactions of the Royal Society of Edinburgh: Earth Sciences* **95**, 35–48.

Brown, M. (2005): Synergistic effects from melting and deformation: an example from the Variscan Belt, western France. *In* Deformation Mechanism, Rheology and Tectonics: from Minerals to Lithosphere (D. Gapais, J.-P. Brun & P.R. Cobbold, eds.). *Geological Society, Special Publication* **243**, 205–226.

Jones, K.A. & Brown. M. (1990): High-temperature "clock-wise" P–T paths and melting in the development of regional migmatites: an example from southern Brittany, France. *Journal of Metamorphic Geology* **8**, 551–578.

Marchildon, N. & Brown, M. (2003): Spatial distribution of melt-bearing structures in anatectic rocks from southern Brittany, France: implications for melt transfer at grain- to orogen-scale. *Tectonophysics* **364**, 215–235.

Figure **D31**

Fig. D31. One of the first mechanisms to be proposed as an explanation for layered, or stromatic leucosome, is the repeated injection of melt parallel to bedding or foliation (i.e., *lit-par-lit* injection). If the injection is perfectly parallel to the layering, this mechanism can be difficult to detect. In this example of a stromatic migmatite, the very thin, fine-grained, leucocratic layers are an earlier generation of quartzofelspathic veins. The coarser-grained sheets of leucosome seem to be parallel to the layering, and hence are stromatic also. However, tracing individual sheets of "leucosome" across the outcrop reveals that many cross the layering at a very shallow angle (<5°); two places are indicated on the photograph. This discordance can be interpreted as evidence that they were derived from intrusions of anatectic melt, and did not form *in situ*. Consequently, they are either in-source leucosome or, more likely, leucocratic veins, but are not *in situ* leucosome.

Location: Wuluma Hills, Arunta Inlier, Australia. *Rock type*: stromatic metatexite migmatite; metapelitic protolith, anatexis at *T* 825–875°C, *P* ca. 5 kbar. *Scale*: the ruler is 15 cm long. *Image*: E.W. Sawyer.

Further reading: Sawyer, E.W., Dombrowski, C. & Collins, W.J. (1999): Movement of melt during synchronous regional deformation and granulite-facies anatexis, an example from the Wuluma Hills, central Australia. *In* Understanding Granites; Integrating New and Classical Techniques (A. Castro, C. Fernández & J.-L. Vigneresse, eds.). *Geological Society, Special Publication* **168**, 221–237.

Figure D32

Fig. D32. This stromatic migmatite consists of layers of leucosome (quartz + plagioclase + K-feldspar), bordered by biotite–muscovite–sillimanite melanosome and hosted by a paleosome consisting of a mesocratic, grey quartzofeldspathic schist. The leucosome contains pristine igneous microstructures in thin section, which suggests that the stromatic geometry was not acquired by high strain in a submagmatic or later state. The inferred melting reaction in the protolith is muscovite + plagioclase + quartz + K-feldspar + H_2O = melt, progressing to muscovite + plagioclase + quartz + H_2O = sillimanite + melt as K-feldspar became exhausted during progressive anatexis. Note the slight upward bulging of the leucosome across the compositional layering in its melanosome (main leucosome, center of the image); the other contact of the leucosome is planar. Such bulging structures have been termed "cauliflowers" and used to determine the "way up" in migmatites. The high proportion of leucosome relative to melanosome in this migmatite suggests that some anatectic melt from elsewhere has been added to the *in situ* melt that formed the initial stromatic leucosome.

Location: Glenelg River Complex, southeastern Australia. *Rock type*: stromatic metatexite migmatite; pelitic, semi-pelitic and psammitic schist protoliths, with anatexis in the upper amphibolite facies. *Scale*: the lens cap is 6.5 cm in diameter. *Image and caption*: Tony I.S. Kemp.

Further reading: Burg, J.-P. & Vanderhaeghe, O. (1993): Structures and way-up criteria in migmatites with applications to the Velay Dome (French Massif Central). *Journal of Structural Geology* **15**, 1293–1301.

Kemp, A.I.S. & Gray, C.M. (1999): Geological context of crustal anatexis and granitic magmatism in the northeastern Glenelg River Complex, western Victoria. *Australian Journal of Earth Sciences* **46**, 407–420.

Figure D33

Fig. D33. Use of the term stromatic has become broadened to include metatexite migmatites in which the bodies of leucosome are not laterally very extensive, but occur as numerous, high-aspect-ratio lenses oriented parallel to one another, and to either the foliation or the compositional layering in their host. This stromatic migmatite is one of these; it has many parallel, thin, high-aspect-ratio domains of leucosome that are sinuous and, locally, bifurcating. The leucosome is trondhjemitic (plagioclase + quartz) in composition, and surrounded by a metamafic rock with the assemblage hornblende + garnet + clinopyroxene + plagioclase that is locally coarser grained than some small pockets of metamafic material richer in hornblende and plagioclase (e.g., left of the coin). Thus, most of this migmatite is interpreted to be melanosome (i.e., residual in composition). Some portions of the leucosome contain large euhedral crystals of pyroxene, and other portions, garnet. These are the solid products of reaction of the incongruent breakdown of hornblende that were not separated from the anatectic melt, and may have grown to a large size because of the fluxing effects of the melt. The high modal proportion of garnet and pyroxene in the melanosome, and the relatively low proportion of leucosome, suggest that much of the anatectic melt generated in this migmatite has escaped.

Location: Kapuskasing Structural Zone, Ontario, Canada. *Rock type*: stromatic metatexite migmatite; mafic gneiss protolith partially melted at granulite-facies conditions, *T* ca. 850°C, *P* 11 kbar. *Scale*: the coin is 24 mm across. *Image*: D.R.M. Pattison. Previously published as fig. 2d in Hartel & Pattison (1996) and reproduced with the permission of Blackwell Publishing.

Further reading: Hartel, T.H.D. & Pattison, D.R.M. (1996): Genesis of the Kapuskasing (Ontario) migmatitic mafic granulites by dehydration melting of amphibolite: the importance of quartz to reaction progress. *Journal of Metamorphic Geology* **14**, 591–611.

Figure D34

Fig. D34. This photograph shows the contrasting appearance of stromatic leucosome developed from different metasedimentary protoliths. In the lower half of the photograph, sheets of medium-grained leucosome with the mineral assemblage plagioclase + quartz + K-feldspar ± garnet ± orthopyroxene are in a fine-grained host that contains the assemblage plagioclase + biotite + orthopyroxene + quartz + garnet. The host contains corroded plagioclase, quartz and biotite grains located in larger xenoblastic quartz and plagioclase crystals. This microstructure suggests that the melting reaction was perhaps biotite + quartz + plagioclase = orthopyroxene + garnet + melt. Therefore, the host is interpreted to be the residuum left after the removal of most of the melt from a plagioclase + quartz + biotite metagreywacke protolith, and the stromatic leucosome is the product of the segregated melt. The leucosome in the upper part of the photograph is much coarser grained and contains the mineral assemblage plagioclase + quartz + K-feldspar + garnet, in a host that contains the mineral assemblage plagioclase + quartz + K-feldspar + biotite + garnet + cordierite. This more aluminous assemblage is interpreted to be the melt-depleted residuum from a metapelitic protolith, and the leucosome in it is derived from the anatectic melt. The melting reaction in the metapelite may have been biotite + sillimanite + quartz + plagioclase = garnet + cordierite + K-feldspar + melt. If the garnet in the leucosome is the reaction product of the incongruent breakdown of biotite, then this is not strictly leucosome material; neosome would be a better term. Moreover, the position of the stromatic neosome could have been determined by the locus of garnet nucleation, rather than by the locus of collection of the melt fraction.

Location: Northern Ashuanipi Subprovince, Quebec, Canada. *Rock type*: stromatic metatexite migmatite; metapelitic and metagreywacke protoliths, anatexis at T ca. 850°C, P 6–7 kbar. *Scale*: the ruler is 15 cm long. *Image*: E.W. Sawyer.

Figure **D35**

Fig. D35. The stromatic migmatites at Port Navalo are located in the transcurrent South Armorican Shear Zone, and are typical of stromatic migmatites in zones of high strain. The subvertical compositional layering defined by the neosome (i.e., melanosome and leucosome) is parallel to the compositional layering in the paleosome, which in turn is parallel to the foliation defined by tabular minerals in both the paleosome and melanosome. The domains of leucosome are thin and laterally persistent, and are bordered by a thin mafic selvedge. Microstructures, such as rational faces on feldspar against quartz are common, and indicate that the stromatic structure was acquired with melt present as the rocks were sheared. The bodies of leucosome are rich in plagioclase. This fact, taken together with the very high proportion of leucosome in the outcrops, suggest that anatexis was an open-system process in this migmatite. A possible interpretation is that large volumes of anatectic melt were channeled upward through the South Armorican Shear Zone, and that the amount of leucosome in the outcrop presents the collapsed relic of the original conduit system made visible by the presence of the minerals (mostly plagioclase) that crystallized first from the magma and were left behind after the remaining evolved melt had migrated away.

Location: Port Navalo area, southern Brittany, France. *Rock type*: stromatic metatexite migmatite, quartzofeldspathic, and aluminous gneiss protolith, anatexis at T ca. 800°C, P ca. 8 kbar, then decompressed to 4 kbar and cooling. *Scale*: the ruler is 15 cm long. *Image*: E.W. Sawyer.

Further reading: Marchildon, N. & Brown, M. (2003): Spatial distribution of melt-bearing structures in anatectic rocks from southern Brittany, France: implications for melt transfer at grain- to orogen-scale. *Tectonophysics* **364**, 215–235.

Figure D36

Fig. D36. The effect of strain and transposition on the aspect ratio of bodies of leucosome can be seen in this exposure of migmatite that contains, in its state of lowest strain (center of the image near the lens cap), domains of leucosome with an aspect ratio up to 2. The leucosome in the adjacent zones of higher strain (top and bottom of the image) have an aspect ratio of up to 10, and are stromatic in the sense that they are elongate parallel to the foliation. The leucosome contains quartz, plagioclase, and K-feldspar. It is set in a melanocratic matrix that contains sillimanite, biotite, muscovite, plagioclase, garnet, and locally, K-feldspar that is believed to be the residuum after partial melting of siliciclastic turbidites via the reactions muscovite + plagioclase + quartz + H_2O = sillimanite + biotite and muscovite + plagioclase + quartz + H_2O = sillimanite + K-feldspar + biotite.

Location: western Maine, U.S.A. *Rock type*: incipient stromatic metatexite migmatite; pelitic and semipelitic schists as protolith, partially melted in the upper amphibolite facies; *T* 750–800°C, *P* 4–5 kbar. *Scale*: the lens cap is 6.5 cm across. *Image and caption*: Gary Solar. Image previously published as fig. 4e in Solar & Brown (2001) and reproduced with the permission of Oxford University Press.

Further reading: Solar, G.S. & Brown, M. (2001): Petrogenesis of migmatites in Maine, USA: possible source of peraluminous leucogranite in plutons? *Journal of Petrology* **42**, 789–823.

Figure **D37**

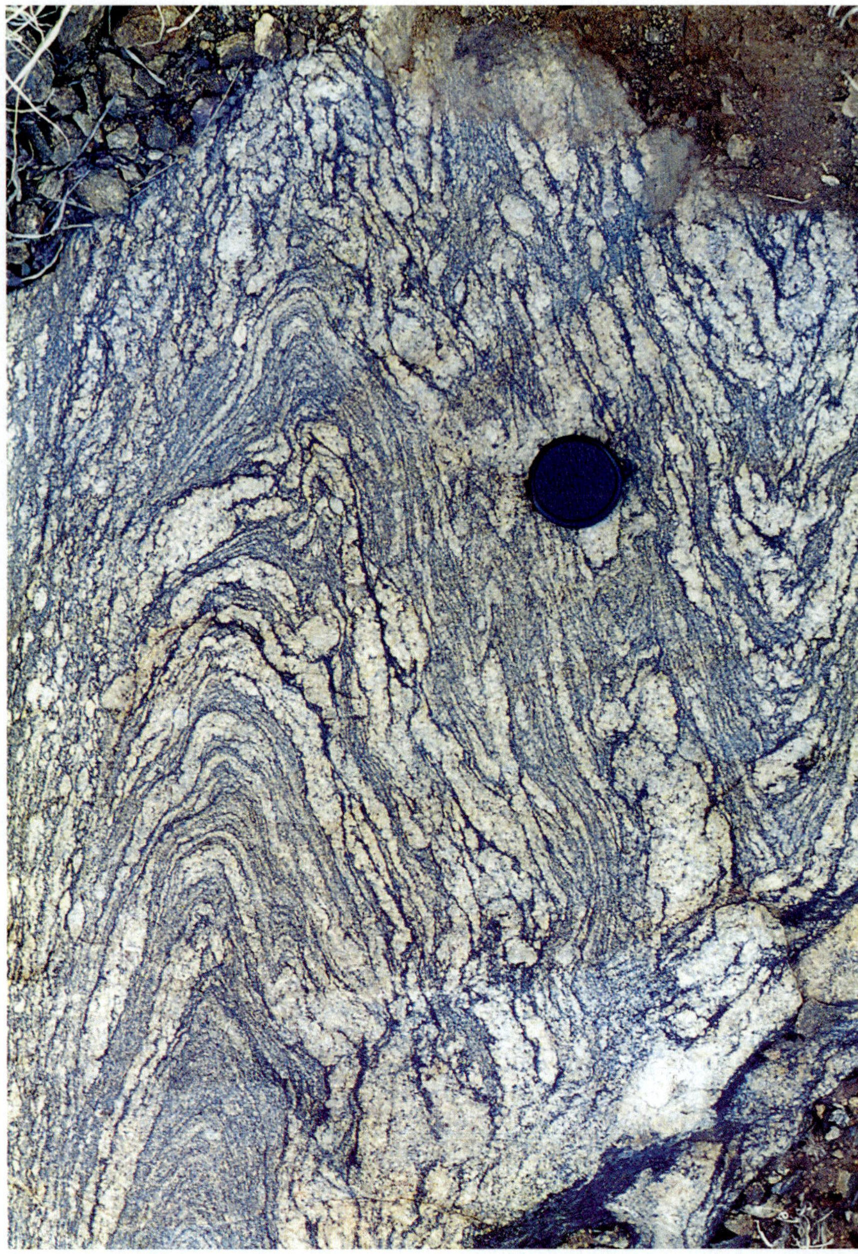

Fig. D37. The final stage of deformation on the lower flank of the fold in this stromatic migmatite occurred after the migmatite had crystallized, as it reduced the grain size and produced a protomylonite with feldspar porphyroclasts (bottom of the image). Elsewhere in this stromatic migmatite, both the formation of the layered structure of the leucosome and its subsequent folding occurred while melt was present (i.e., before the migmatite fully crystallized) and the migmatite was subject to high non-coaxial strain. As evidence for this interpretation, note that the bodies of quartz + plagioclase + K-feldspar stromatic leucosome are thickest in the fold hinges, where the melt fraction collected preferentially, and furthermore, note the presence of cross-cutting (center of the image), but compositionally similar, domains of leucosome oriented parallel, or subparallel, to the axial plane of the fold. A possible sequence of events is that, like at Port Navalo, this migmatite acquired its stromatic character in a shear zone, and as crystallization progressed, the flow regime was perturbed, and the stromatic layering became folded, moving the remaining melt into the hinge regions of the folds. The evidence that this is indeed a migmatite, and not just a rock made by multiple injections of felsic veins into a banded intermediate gneiss, is as follows: (1) the presence of patches of diffuse neosome elsewhere in the outcrop, and (2) mineral assemblages in the melanosome and leucosome that suggest that the melting reaction in the iron-rich protolith may have been biotite + plagioclase + quartz = magnetite + ilmenite + melt, which proceeded under conditions of relatively high fugacity of oxygen.

Location: Arunta Inlier, Australia. *Rock type*: stromatic metatexite migmatite; banded intermediate gneiss protolith, and anatexis at T ca. 850°C, P <5 kbar. *Scale*: the lens cap is 6.5 cm across. *Image and caption*: Tony I.S. Kemp.

Figure D38

Fig. D38. The morphology of the leucosome fraction in the psammitic layers of this migmatite is notably different from that in the pelitic layers; the layering is interpreted to be transposed bedding. The light grey layers (just to the left of the hammer and in the upper left) are foliated layers of psammite and have domains of leucosome with a relatively low aspect-ratio located in extensional shears and between boudins oriented at moderate to high angles to the foliation. The contacts between the leucosome and the psammite host are diffuse, which suggests *in situ* segregation of anatectic melt into dilatant sites during extension along the layering. The dark grey layers are strongly foliated pelites, and the majority of the domains of leucosome in them have a stromatic morphology with melanocratic borders that are rich in biotite. Many of the stromatic domains of leucosome display pinch-and-swell structures, which are indicative that they were stretched parallel to the layering as they crystallized. There are also a few domains of leucosome located in oblique extensional shears in the pelitic layers. However, the shear surfaces in the pelitic layers are oriented at lower angles to the foliation, and the domains of leucosome in them have a higher aspect-ratio than those found in the adjacent psammitic layers. The orientation of the extensional shear surfaces with respect to the foliation and the layering suggests that the psammite layers were more competent than the pelitic ones. Therefore, the greater tendency for stromatic morphology may be due to a partition of the strain into the weaker pelitic layers of the migmatite, which transposed the leucosome fraction there.

Location: Thor–Odin dome, British Columbia, Canada. *Rock type*: stromatic metatexite migmatite; pelitic and psammitic felsic protoliths, partially melted at conditions of the upper amphibolite facies. *Scale*: note the hammer. *Image*: Olivier Vanderhaeghe. Previously published as fig. 5a in Vanderhaeghe et al. (1999) and reproduced with the permission of the Canadian Journal of Earth Sciences.

Further reading: Vanderhaeghe, O., Teyssier, C. & Wysoczanski, R. (1999): Structural and geochronological constraints on the role of partial melting during the formation of the Shuswap metamorphic core complex at the latitude of the Thor–Odin dome, British Columbia. *Canadian Journal of Earth Sciences* **36**, 917–943.

Figure D39

Fig. D39. This photograph shows the same migmatites as in Figs. B23 and B24, but from an outcrop where a much higher degree of syn-anatectic strain (during D_2 in the local structural framework) has resulted in the development of a completely different morphology. The light-colored layers in this migmatite are neosome, not leucosome, as they contain conspicuous porphyroblasts of garnet and less conspicuous crystals of K-feldspar, which are the solid products of the melt-producing reaction, as well as leucocratic material that has crystallized from the anatectic melt itself. The layers of dark rock that alternate with the neosome are ilmenite–plagioclase–quartz–biotite–sillimanite–cordierite schist or gneiss. The schist is not properly described either as paleosome, because it has partially melted, nor as residuum, because it lacks the garnet. Perhaps, the most relevant term for it is melanocratic neosome. During non-coaxial shearing, the aggregates of garnet and K-feldspar in the neosome acted as competent bodies, and the anatectic melt developed in, or migrated to, the low-pressure sites around them. Thus, on this section, which is subparallel to the mineral lineation indicated by alignment of sillimanite needles, the morphology is quite clearly that of a stromatic migmatite.

Location: Round Hill, Broken Hill area, Australia. *Rock type*: stromatic metatexite migmatite; aluminous metapelite protolith partially melted at *T* 800°C and *P* 5–6 kbar. *Scale*: the ruler is 15 cm long. *Image*: E.W. Sawyer.

Further reading: White, R.W., Powell, R. & Halpin, J.A. (2004): Spatially-focussed melt formation in aluminous metapelites from Broken Hill, Australia. *Journal of Metamorphic Geology* **22**, 825–845.

Figure D40

Fig. D40. The Round Hill area in Australia has some astonishing migmatites; this example is remarkable for its morphological similarity to pressure-solution seams that are a common feature of low-grade pelitic schists. This migmatite has two principal constituents: a light-colored, coarse-grained neosome and a dark-colored schist that is best considered as a melanocratic neosome (see the caption to Fig. D39 for more explanation). The foliation in the schist part belongs to S_2 of the local structural sequence. The light-colored neosome can be divided into two parts: one contains garnet, K-feldspar, and quartz, and the other, a leucocratic granitic part without garnet. The garnet crystals define a prominent internal, layered structure in each area of neosome that is orthogonal to the external S_2. Moreover, the garnet-defined layering can be traced from one neosome to the next across the outcrop; it maintains the same orientation. This is compelling evidence that a penetrative fabric, or layering, existed in the migmatite before S_2 formed, and has been entirely erased from the schist (melanocratic neosome) layers. The garnet-rich portions form the widest parts of each neosome, whereas the garnet-absent granitic part occurs either as a wedge-shaped tail to, or as a narrower neck between, adjacent garnet-bearing parts. Since the domains of light-colored neosome are lenticular in form and elongate in the S_2 direction, the garnet-rich parts are interpreted to have been more competent than the leucocratic parts when the neosome was being deformed, dominantly by layer-normal shortening. This is consistent with the petrological interpretation of the neosome by White et al. (2004) in which the cores are the solid products of the incongruent breakdown reaction of biotite and sillimanite, and the granitic parts are the products of crystallization of the anatectic melt produced by the reaction. The preservation of crystal faces on feldspar in the light-colored neosome, and the general absence of plastic deformation features in the quartz and feldspar of the light-colored neosome, and in the minerals of the schist (melanocratic neosome) layers, indicate that deformation occurred while melt was present in the rock, i.e., during anatexis. Thus, the curvature of the foliation in the ilmenite–plagioclase–quartz–biotite–sillimanite–cordierite schist and the increase in the modal proportion of the mafic minerals visible in the schist around the competent garnet-rich portions of the neosome are consistent with the loss of volume from the schist layers. Thus, the overall structure of this migmatite is exactly analogous to the structure associated with pressure solution in low-grade metasedimentary rocks. In such rocks, the mica-rich microlithons are interpreted to have lost a mobile quartzofeldspathic component during dominantly layer-normal compression, and the competent, typically quartz-rich layers, to have gained some, or all, of the mobile material typically as overgrowths, or pressure-shadow fillings, around the most competent minerals. However, in this migmatite, the mobile material contained the components that made anatectic melt.

Location: Round Hill, Broken Hill area, Australia. *Rock type*: metatexite migmatite; aluminous metapelite protolith partially melted at T 800°C and P 5–6 kbar. *Scale*: the width of the field of view is 1 m. *Image*: E.W. Sawyer.

Further reading: Robin, P.-Y.F. (1979): Theory of metamorphic segregation and related processes. *Geochimica et Cosmochimica Acta* **43**, 1587–1600.

White, R.W., Powell, R. & Halpin, J.A. (2004): Spatially-focussed melt formation in aluminous metapelites from Broken Hill, Australia. *Journal of Metamorphic Geology* **22**, 825–845.

Figure D41

Fig. D41. The neosome in this migmatite shows very little segregation of the melt fraction from the residual fraction; hence, it is mesocratic, and medium-grained because partial melting has occurred in a contact aureole and not in a regional metamorphic setting. The local segregation of the melt from the residuum is indicated by the small leucocratic domains within the neosome. However, the most striking aspect of this migmatite is the large number of fragments of paleosome, called schollen, or rafts, which are present in the neosome. In the central part of the image, the schollen are small and tend to be elongate, but toward the top left and bottom right, the schollen are much larger. In this particular migmatite, the size and shape of the schollen reflect whether the protolith was thinly (small elongate schollen) or thickly (larger schollen) bedded. The migmatite shown in this photograph is largely neosome, and the paleosome occurs only as schollen in the neosome. Although some of the schollen have been rotated, others, most notably the elongate ones, have not, and the orientation of the bedding is more or less preserved through the neosome. There may not have been sufficient neosome developed to provide sufficient space for the longer schollen to have rotated. Hence, this migmatite can be considered as transitional from metatexite to diatexite migmatite. It is dominated by neosome formed from the fertile beds, but some continuity of older, pre-partial-melting structures is retained by the relics of the less fertile and sterile beds.

Location: Saint-Cast, Brittany, France. *Rock type*: schollen diatexite migmatite; semipelitic schist protolith, T ca. 700°C, P ca. 3 kbar. *Scale*: the ruler is 15 cm long. *Image*: E.W. Sawyer.

Figure **D42**

Fig. D42. This photograph shows the abrupt transition from a metatexite migmatite (top left) to a diatexite migmatite (bottom right). Schollen or rafts from a thick layer of metagreywacke appear to be detached from the metatexite and progressively transformed from tabular to rounded to lenticular shapes as they are incorporated into the diatexite migmatite. Smaller, tabular schollen are detached from the edges of a much larger one in the center of the photograph. The orientation and offset along melt-filled fractures in the schollen, the asymmetry of schollen shapes, and the trends of schlieren in the leucocratic matrix all indicate flow of the melt-rich part of the migmatite. Thus, this transition from metatexite and diatexite migmatites was dynamic, the diatexite magma flowed past the metatexite and in effect eroded and intruded it. Most of the melt in this diatexite migmatite was contributed by the metagreywacke layers, which underwent incongruent melting according to the reaction biotite + plagioclase + quartz = orthopyroxene + melt. The schollen contain the assemblage plagioclase + orthopyroxene + biotite + quartz, which is the residuum left after the extraction of the anatectic melt. They also contain abundant grain-scale microstructural evidence of partial melting (see Figs. F35–F38). The coarse-grained leucocratic matrix contains the assemblage plagioclase + quartz + K-feldspar + biotite + orthopyroxene, and represents the melt. The tiling of subhedral to euhedral crystals of plagioclase and K-feldspar, and the presence of rational faces on orthopyroxene, plagioclase and K-feldspar against quartz, both indicate a substantial fraction of melt in the leucocratic portion of the migmatite. However, local variations in the microstructure and whole-rock bulk composition indicate that some parts of the melt-rich material have a large component of cumulate plagioclase, whereas others have a large component of evolved melt.

Location: Ashuanipi Subprovince, Quebec, Canada. *Rock type*: schollen diatexite migmatite; metagreywacke protolith, anatexis in the granulite facies, *T* 825–875°C, *P* 6–7 kbar. *Scale*: the ruler is 15 cm long. *Image*: E.W. Sawyer.

Further reading: Guernina, S. & Sawyer, E.W. (2003): Large-scale melt-depletion in granulite terranes: an example from the Archaean Ashuanipi subprovince of Quebec. *Journal of Metamorphic Geology* **21**, 181–201.

Percival, J.A. (1991): Granulite-facies metamorphism and crustal magmatism in the Ashuanipi complex, Quebec–Labrador, Canada. *Journal of Petrology* **32**, 1261–1297.

Sawyer, E.W. (2001): Melt segregation in the continental crust: distribution and movement of melt in anatectic rocks. *Journal of Metamorphic Geology* **19**, 291–309.

Figure **D43**

Fig. D43. Large regions of diatexite migmatite developed from metamafic protoliths are very rare, probably because the metamorphic temperatures recorded from most of the exposed parts of the continental crust are not high enough to have caused a pervasive high degree of partial melting in such a protolith. However, the local-scale development of migmatites transitional from metatexite to diatexite is more commonly seen in metamafic protoliths. This photograph shows part of a small outcrop (ca. 15 m^2) in which the old, layered structure in a metatexite migmatite (upper part of the photograph) is progressively disrupted (lower part) and erased. The metatexite portion consists of foliated melanocratic layers that have been boudinaged, and of leucocratic material, which occurs parallel to the layering, in the spaces between boudins and in various fractures. The leucocratic material is tonalitic in composition and was largely derived from anatectic melt, but it contains a variable proportion of mafic material. In some places, the mafic material occurs sparsely as individual euhedral crystals of pyroxene and hornblende that could either be the solid product of the melting reaction, or else have crystallized from the anatectic melt. However, in many places, the mafic material is more abundant, and appears to consist of fragments derived from nearby melanocratic layers. These portions were derived from contaminated anatectic melt. The migmatite is substantially different in the lower part of the photograph, below the wide leucocratic vein, where there are no coherent melanocratic layers. Rather, the migmatite contains many small rounded, rotated, and partially disaggregated remnants (rafts or schollen) of melanocratic material engulfed in leucocratic material. The proportion of leucocratic material is greater in the diatexite portion than in the metatexite, and it is also petrographically different. It has a much higher proportion of mafic minerals, and these tend to occur as larger aggregates than in the metatexite. The tonalitic melt thus seems to have been contaminated with material derived from the disaggregating melanocratic schollen. The transition from metatexite to diatexite migmatite in this outcrop involved an increase in the fraction of melt, which assisted the breakup, rotation, and disaggregation of the melanosome and paleosome layers to yield a migmatite devoid of structures older than the anatexis.

Location: Beit Bridge Complex, Baklykraal Farm, South Africa. *Rock type*: transition from metatexite migmatite to schollen diatexite migmatite; metamafic protolith, subjected to granulite-facies anatexis. *Scale*: the pocket knife is 11 cm long. *Image*: E.W. Sawyer.

Figure **D44**

Fig. D44. This migmatite is quite complex, but most of it is neosome. The only part that has not partially melted, and hence can be called paleosome, is the rather uniform, medium-grey scholle (labeled P); it is a plagioclase + quartz + biotite Al-poor metagreywacke. All the other schollen, of various shades of grey, have undergone partial melting and coarsening of their grain size. In addition, most have lost some of their melt fraction; accordingly, they are called neosome (no loss of melt) or melanosome (loss of some melt). The darker and coarser-grained schollen contain the assemblage plagioclase + quartz + biotite + cordierite ± K-feldspar ± sillimanite, which is interpreted to be residual. The schollen have elongate, lenticular shapes, commonly with a narrow, biotite-rich rim that is interpreted to result from reaction between the melt around a scholle and the residual cordierite in it. All the schollen are enclosed in leucocratic material that contains the assemblage plagioclase + K-feldspar + quartz ± cordierite, and microstructures (e.g., crystal faces on feldspar crystals in contact with quartz, euhedral crystals of cordierite) that indicate crystallization from a granitic melt; hence, it can be called leucosome. Locally, the leucosome contains abundant curviplanar schlieren that are rich in biotite and accessory minerals, such as apatite and zircon. In places, groups of schlieren curve into, or are truncated by, other groups of schlieren, which is indicative of variations in flow rate within the leucosome magma. The elongation of the schollen and of the schlieren is parallel to the bedding in the nearby metatexite migmatites. The location of this diatexite migmatite was thus possibly controlled by the presence of particularly fertile beds in the protolith, or by the migration of anatectic melt into certain bedding-parallel sites. Two leucocratic veins in the migmatite are younger and cross the flow structure in the diatexite. A narrow, irregular band of leucosome (bottom right) is located in a small shear band and has diffuse margins. The diffuse contacts between the narrow leucocratic vein and the leucosome suggest that it is a late-stage segregation of melt expelled from the partially crystallized leucosome. A wide (>25 cm) coarse-grained leucocratic vein in the top left has sharp contacts and a very narrow mafic selvedge of biotite particularly where it is in contact with the schollen; this leucocratic vein was injected when the migmatite was near its solidus temperature.

Location: Quetico Subprovince, Ontario, Canada. *Rock type*: schollen diatexite migmatite; metapelitic protolith, partially melted in the upper amphibolite facies, *T* 700–800°C, *P* 3–4 kbar. *Scale*: the ruler is 15 cm long. *Image*: E.W. Sawyer.

Figure D45

Fig. D45. The Vaasa Migmatite Complex in western Finland contains a transition from a garnet–cordierite–K-feldspar–plagioclase–quartz–biotite gneiss to migmatite. This image shows the morphology of rocks that are transitional from metatexite to diatexite migmatite. The outcrop appears to be morphologically complex because it has numerous domains of different composition, grain size, and microstructure. There are several rounded, light greenish grey enclaves composed of calcareous resister lithologies (e.g., near the center); these are paleosome. However, all the rest of the outcrop is neosome of one sort or another. Dark-colored lens-shaped enclaves (e.g., near the top) are common throughout; these are fragments derived from the metatexite migmatite. The majority of these enclaves are from the mica-rich, residual portions of the metatexite that were left after the extraction of anatectic melt; they are melanosome. However, some enclaves also contain thin stromatic veins of leucosome. Much darker enclaves (higher contents of garnet) with a more granular microstructure (e.g., near the pen) appear to have been particularly prone to disaggregation and are surrounded by numerous small fragments. All the enclaves occur in a banded matrix in which layers of different grain size, microstructure, and modal mineralogy can be distinguished. Some of these bands are conspicuously leucocratic (e.g., left of the pen cap) and contain scattered crystals of garnet, but most are mesocratic, with a higher modal proportion of garnet, biotite, cordierite, and locally, orthopyroxene (e.g., left half of the image). However, all parts of the matrix contain domains where the platy or tabular minerals (micas or feldspars) define a foliation that is parallel to the compositional banding; both are attributed to flow of the matrix as magma. The presence throughout the matrix of scattered euhedral crystals of plagioclase, or plagioclase crystals with straight faces against quartz, is further evidence for the pervasive presence of melt in the matrix around the enclaves. Thus, the transition from metatexite to diatexite in the Vaasa Migmatite Complex can be interpreted as a change in the distribution of the melt fraction. Where the rock developed a pervasive fraction of melt between the grains, flow was able to occur and entrained enclaves able to rotate, so that the continuity of pre-anatectic structures was lost, and a diatexite migmatite formed. The compositional banding in the matrix between the enclaves is interpreted to be due to variations in the proportion of anatectic melt to residual crystals from band to band. As the proportion of enclaves and the small-scale variations in the proportion of melt to residuum decreased, the schollen diatexites in the Vaasa Migmatite Complex pass into diatexite migmatites (see Fig. D61).

Location: Kauhava area, Svecofennian domain, Finland. *Rock type*: schollen diatexite migmatite; metapelitic protolith, granulite-facies anatexis at T ca. 800°C, P 4–5 kbar. *Scale*: the pen is 15 cm long. *Image*: Hannu Mäkitie.

Further reading: Mäkitie, H. (2001): Eastern margin of the Vaasa Migmatite Complex, Kauhava, western Finland: preliminary petrography and geochemistry of the diatexites. *Bulletin of the Geological Society of Finland* **73**, 35–46.

Figure **D46**

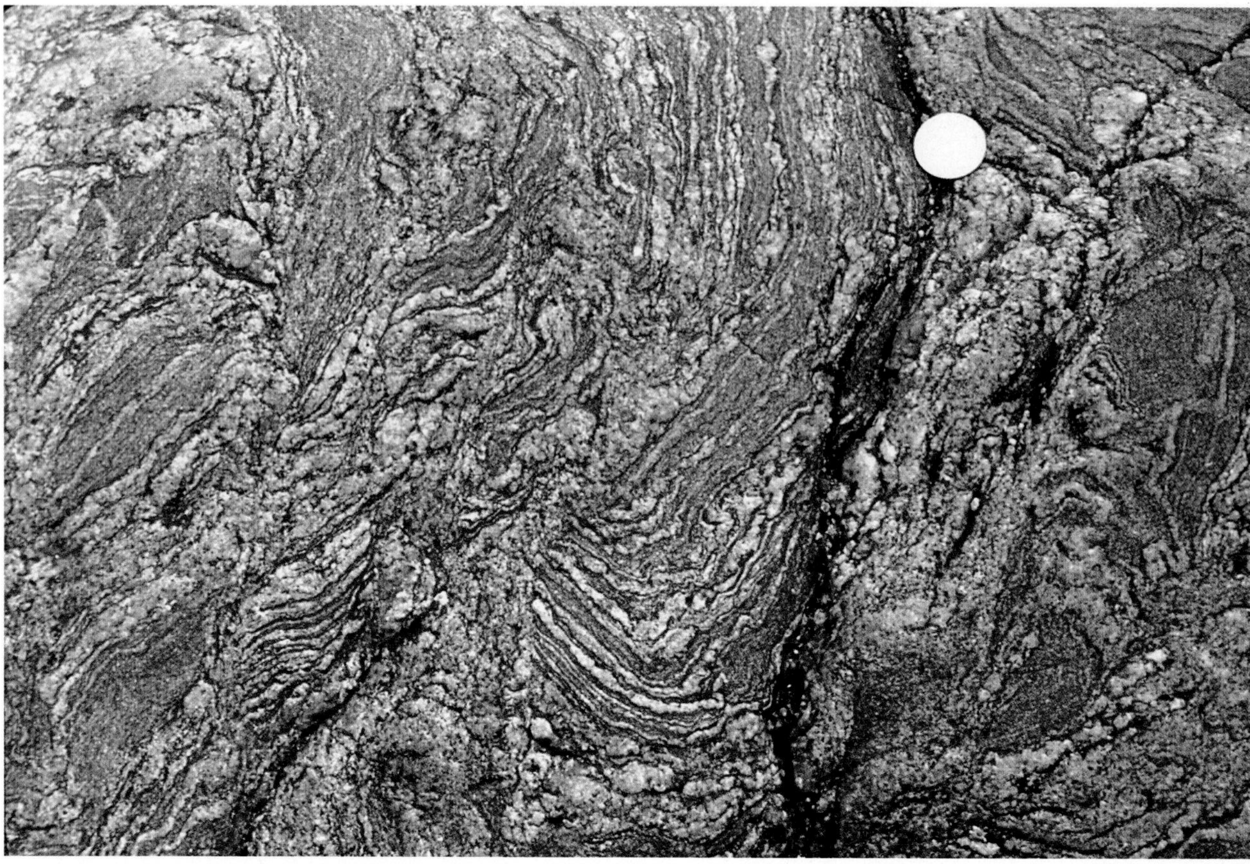

Fig. D46. This migmatite shows a transition from a stromatic metatexite to a diatexite migmatite. It contains a relatively high proportion of schollen (or rafts) set in a coarse-grained, granitic matrix. All of the schollen are of stromatic metatexite (i.e., neosome), which suggests an early stage of diatexite formation; in the more advanced stages of diatexite formation, most schollen of neosome have typically been assimilated, and paleosome (resister) lithologies predominate in the remaining schollen. The proportion of schollen is a little higher in the left half of the photograph, and the pre-existing structure outlined by the stromatic leucosome can be traced, more or less, from scholle to scholle. Some schollen are lenticular in shape, and their internal domains of stromatic leucosome have sigmoidal forms, which suggest that the schollen were separated by small shear zones. Other schollen contain tight to open folds that are truncated by the edges of the scholle. In the left half of the photograph, the matrix between the schollen is narrow, and locally contains discontinuous, biotite-rich schlieren that are generally parallel to the long sides of the schollen. However, at the ends of the schollen, the schlieren truncate the sigmoidal internal structure of the schollen and, consequently, appear to occupy shear surfaces between schollen. The migmatite is slightly different in the right half of the photograph; there, the proportion of schollen to matrix is lower, and there is more variation in the orientation of the internal structures between schollen, which suggests that the schollen were able to rotate in the matrix. Locally, biotite-rich schlieren in the matrix outline disharmonic flow-related folds. These relationships are interpreted to indicate that melt was able to migrate along small shear zones that developed in a stromatic metatexite migmatite during folding. Progressive infiltration of melt enabled some shear-bounded blocks of metatexite to detach and become schollen in a diatexite migmatite. As the melt infiltrated further into the metatexite migmatite, the schollen that were detached first were no longer at the relatively stagnant edge of the diatexite magma, and consequently were able to rotate in the flowing diatexite magma and to become dispersed.

Location: Saint-Malo, Brittany, France. *Rock type*: schollen diatexite migmatite; metasedimentary protolith. *Scale*: the coin is 2.0 cm across. *Image and caption*: Mike Brown.

Further reading: Brown, M. (1979): The petrogenesis of the St-Malo migmatite belt, Armorican Massif, France, with particular reference to the diatexites. *Neues Jahrbuch für Mineralogie, Abhandlungen* **135**, 48–74.

Figure D47

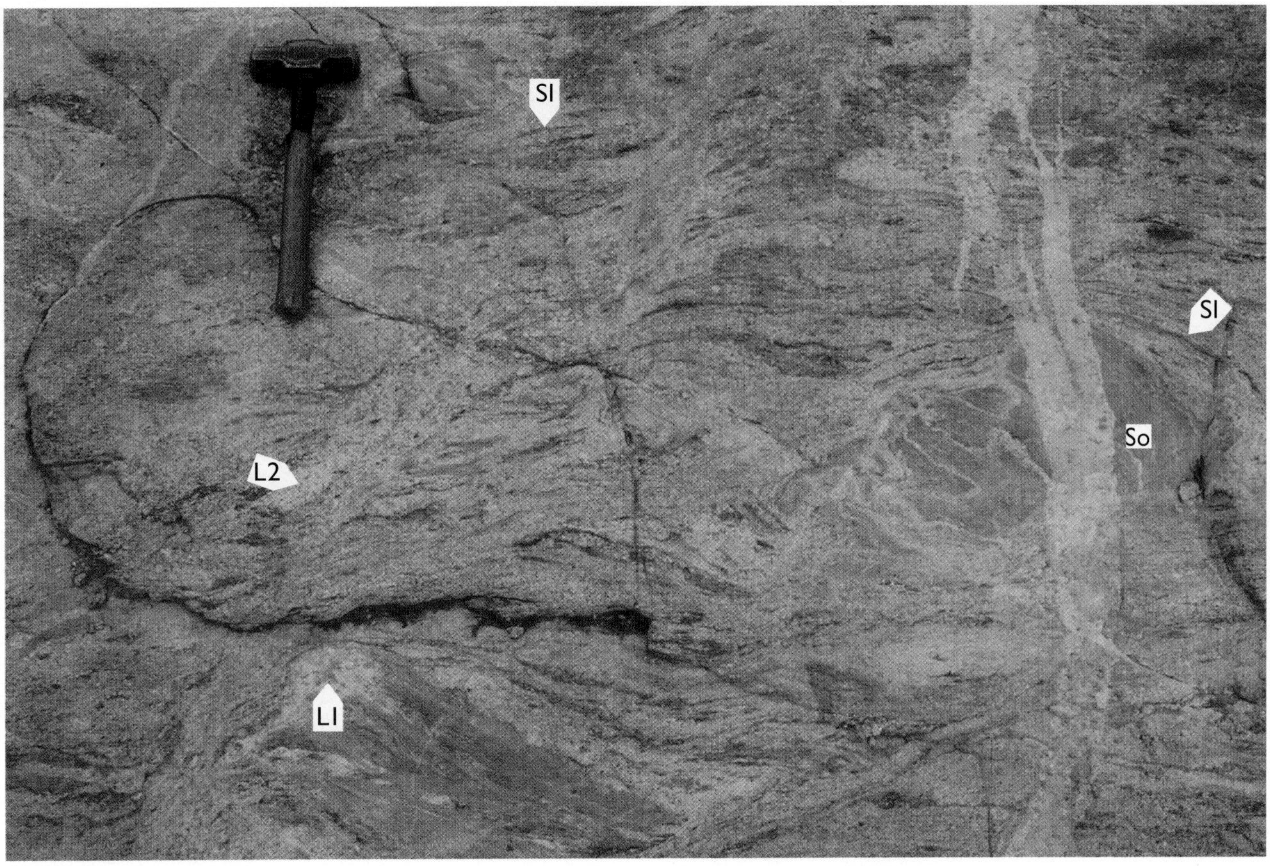

Fig. D47. This complex migmatite contains a high proportion of neosome, and there are no coherent or contiguous structures of pre-anatectic age; hence, this is a diatexite migmatite. Paleosome occurs only as several large, equant blocks called schollen (labeled So) of resister lithologies. There are many smaller melanocratic schollen, up to about 15 cm across in the neosome, and most of these contain sillimanite + biotite + garnet + quartz + plagioclase + muscovite and are interpreted as residuum. The leucocratic matrix contains the mineral assemblage quartz + plagioclase + muscovite + K-feldspar + biotite, and contains abundant schlieren (labeled Sl) rich in biotite and sillimanite. The schlieren define a foliation and flow banding that wraps around the larger schollen. The migmatite thus contained a sufficiently high fraction of melt that it could flow. This migmatite could be called either a schollen or a schlieren diatexite, since both are present. There is clear evidence of sequential segregation of melt from the melt-rich parts of the migmatite that host the schollen and the schlieren. Coarser-grained and generally more leucocratic material (Ll) occupy positions corresponding to pressure shadows at the corners of the schollen. Melt thus was segregated from the crystals during the main stage of magma flow. Furthermore, the sigmoidal curvature of the schlieren reveals a sinistral shear zone (just below and to the right of the hammer), which contains a diffuse, schlieren-poor leucosome (L2) along its center, and which truncates the curved schlieren. This is interpreted as a late-stage shear zone formed in the diatexite, which had already developed a rigid framework of crystals with interstitial residual melt. The melt was expelled as the crystal framework was deformed in the walls of the shear zone, and became concentrated in the dilatant center of the shear zone. There is evidence of later segregation of melt, which occurred when crystallization was nearly complete, i.e., the leucocratic vein to the right, which occupies a brittle fracture oriented orthogonal to the layering and perpendicular to the direction of elongation in the outcrop.

Location: Tumbledown Anatectic Domain, western Maine, U.S.A. *Rock type*: schollen diatexite migmatite with schlieren; pelitic and semipelitic metasedimentary protolith, anatexis in the upper amphibolite facies; *T* 750–800°C, *P* 4–5 kbar. *Scale*: the hammer handle is 40 cm long. *Image and caption*: Gary Solar.

Further reading: Solar, G.S. & Brown, M. (2001): Petrogenesis of migmatites in Maine, USA: possible source of peraluminous leucogranite in plutons? *Journal of Petrology* **42**, 789–823.

Figure **D48**

Fig. D48. This photograph shows a diatexite migmatite with both schollen and schlieren. The two prominent schollen are metamafic resister lithologies that contain the assemblage orthopyroxene + plagioclase + ilmenite. They were probably mafic dikes in the metapelitic protolith, i.e., they are paleosome. However, the main point of interest in this migmatite is the strong morphological variation present within different parts of the neosome. The neosome material between the two schollen (1) is leucocratic and coarse-grained, (2) contains the assemblage plagioclase + quartz + K-feldspar + biotite + cordierite + garnet, (3) generally lacks a prominent foliation, although it has some diffuse, darker wisps that are a little richer in biotite and, locally, domains where the feldspar crystals form a tile microstructure, and (4) contains large regions where euhedral crystals of plagioclase form a framework with interstitial quartz. Together, these characteristics indicate that this neosome crystallized from a granitic melt that contained very few residual minerals and records a relatively low level of syn-anatectic shear strain. This region of neosome and the two paleosome schollen constitute a single structural domain within the migmatite that is enveloped by a different neosome that appears to have flowed around it. The enveloping neosome (1) is medium-grained, (2) contains the assemblage plagioclase + quartz + biotite + garnet + cordierite + sillimanite ± K-feldspar, (3) has a strong foliation due to the alignment of idioblastic and euhedral crystals of plagioclase, biotite, sillimanite, and locally, cordierite, and (4) contains a compositional banding due to the alternation of leucocratic bands with mafic schlieren of biotite + sillimanite + garnet composition and diffuse, elongate lenses of fine-grained quartz + biotite + sillimanite + garnet + cordierite + plagioclase melanosome. Together, these characteristics indicate that this neosome was derived from a magma rich in residual crystals, contained schollen of melanosome, and experienced high shear strain while in magmatic and submagmatic states. Thus, the morphological differences between areas of neosome in this migmatite are a direct function of their structural position in the flowing diatexite magma. High shear strains in the partially crystallized, residuum-contaminated diatexite magma drove the segregation of fractionated anatectic melt out of the banded neosome, and the melt collected into the low-pressure site between the two paleosome schollen, where it became the coarse-grained and residuum-poor neosome.

Location: Lac Kenogami, Grenville Province, Quebec, Canada. *Rock type*: schollen diatexite migmatite; metapelitic protolith, anatexis at T 800–850°C, P 4–7 kbar. *Scale*: the ruler is 15 cm long. *Image*: E.W. Sawyer.

Figure D49

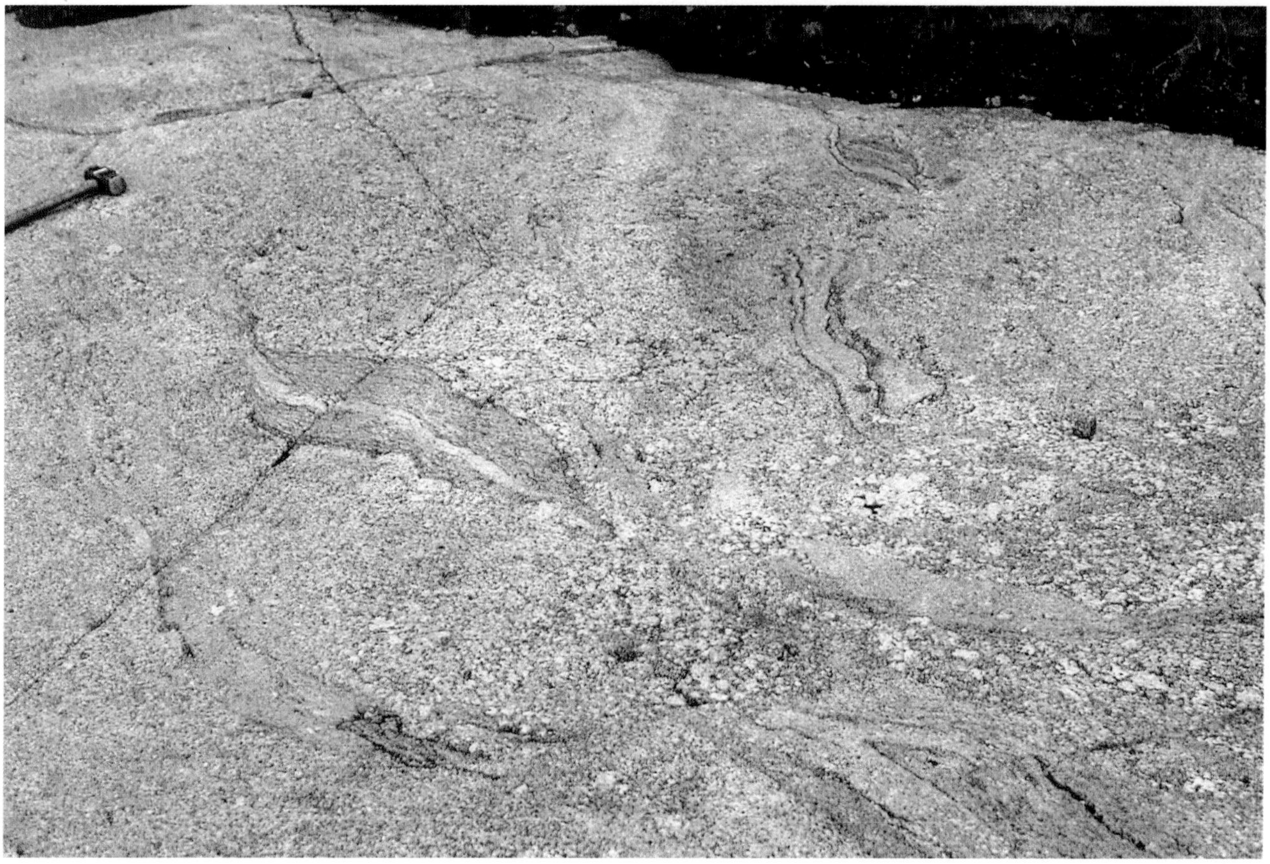

Fig. D49. Schollen diatexite migmatites are best known from metasedimentary protoliths, but they also form from felsic igneous protoliths. This example, derived from a leucotonalite and leucotrondhjemite protolith, contained a higher fraction of melt than the previous examples shown, and the schollen are fewer and have a more lenticular shape. Some of the schollen have a very high aspect-ratio, and their tapered ends pass laterally into biotite-rich schlieren. Many of the schollen have melt-depleted (i.e., residual) bulk compositions that, because of the nature of the protolith, are dominated by plagioclase, the principal mineral present in excess, but they are also enriched in the accessory minerals. The compositions and microstructures of other schollen suggest they are layers of plagioclase accumulated from the crystallizing anatectic melt that have become disrupted and incorporated into the flowing magma. A few schollen are resister lithologies. The pink host to the schollen is a leucogranite in composition, with the assemblage K-feldspar + plagioclase + quartz + biotite. However, the leucogranite host contains a wide variety of microstructures; some parts contain a strong magmatic, or submagmatic, foliation defined by the orientation of tabular K-feldspar and plagioclase, whereas other parts are coarser grained and isotropic and occur as layers, patches, or irregular veinlike bodies. Some of the different microstructural types of leucogranite that occur as layers, or as wispy bands, are truncated by other more continuous bands, or by biotite-rich schlieren. Such variations are typical of many diatexite migmatites; the banding and schlieren are evidence for the flow of a compositionally heterogeneous diatexite magma, and the leucogranite patches and veins provide evidence of the late segregation of the melt at a time when the diatexite magma was in an advanced stage of crystallization. Fractionated melt in the intercrystalline pores was expelled as the framework of euhedral grains of feldspar began to compact in response to shearing; the melt that was expelled collected into brittle fractures and other dilatant structures, to produce the late veins and patches of coarse-grained, isotropic leucogranite present in the outcrop.

Location: Opatica Subprovince, Quebec, Canada. *Rock type*: schollen diatexite migmatite; leucotonalite and leucotrondhjemite protolith, anatexis at conditions of the upper amphibolite facies, *T* ca. 750°C, and *P* 5–7 kbar. *Scale*: the hammer is 45 cm long. *Image*: E.W. Sawyer.

Figure **D50**

Fig. D50. This migmatite has a very high proportion of leucocratic neosome, and its morphology is typical of the regions in anatectic terranes where melt has accumulated. The upper part of the outcrop contains mesocratic schollen of metagreywacke resister layers with the simple assemblage plagioclase + quartz + biotite and schlieren. These are layers that did not melt and were disrupted and incorporated into the layer of anatectic melt. The central and lower parts of the outcrop contain melanocratic lenticular schollen that contain the mineral assemblage plagioclase + quartz + cordierite + biotite + K-feldspar + sillimanite. It represents the residuum left after the melting reaction: biotite + sillimanite + quartz + plagioclase = cordierite + K-feldspar + melt. Very thin schlieren of biotite are present in the leucocratic layers, and some schlieren (labeled Sl) seem to link some of the lens-shaped melanocratic schollen. The asymmetrical shape of most of the residual schollen and the curvature of the schlieren define the shear sense of flow (dextral) in the migmatite. The presence of euhedral crystals of cordierite and feldspar, and of equant domains of strain-free quartz in the leucocratic parts of the diatexite migmatite, indicates that flow occurred in the magmatic and submagmatic states. Several periods of melt segregation and emplacement can be recognized. A syn-anatectic shear zone separates the domain of resister schollen from that of residual schollen, and contains a slightly coarser-grained, more leucocratic center, which probably represents fractionated melt expelled from the walls of the shear zone. The lower edge of the domain that contains the resister schollen and associated schlieren is truncated (indicated by the arrows a and b) by the syn-anatectic shear zone. A later leucocratic vein has intruded the migmatite just above the scale; the irregular contacts of this vein suggest that it was injected before the host had cooled below its solidus. The last stage is represented by a coarse-grained, leucocratic vein oriented orthogonally to the layering in the resister schollen domain.

Location: Quetico Subprovince, Ontario, Canada. *Rock type*: schlieric diatexite migmatite; metapelitic and metagreywacke protolith, upper-amphibolite-facies melting, *T* 700–800°C, *P* 3–4 kbar. *Scale*: the ruler is 15 cm long. *Image*: E.W. Sawyer.

Further reading: Sawyer, E.W. (1987): The role of partial melting and fractional crystallization in determining discordant migmatite leucosome compositions. *Journal of Petrology* **28**, 445–473.

Figure D51

Fig. D51. This migmatite is essentially all neosome, but it exhibits a considerable range in microstructure, modal mineralogy, and color. Although there are no layers, rafts, or schollen of paleosome, there are ghostlike schollen of mesocratic neosome. The lightest-colored parts can be called leucosome. The melanocratic (darkest) parts of the migmatite have two forms. The most prominent occur as selvedges adhering to quartz–feldspar pods around which the flow structure in the host is deflected. These selvedges are interpreted as fragments of early, coarse-grained leucosome, which together with their melanocratic border were disrupted as partial melting advanced and as the flow of the resulting diatexite magma occurred. Also present are thin biotite-rich streaks without attached leucosome; these are biotite schlieren. Furthermore, there are many larger mesocratic domains (below and to the right of the lens cap) that contain a layering and appear to be attenuated, from which one can infer that they were weak rather than strong. These domains could represent layers of partially melted protolith that were either less fertile than the other layers and, therefore, did not partially melt as much, or they are layers in which the melt did not segregate much from the residuum. The result is that these layers were not sufficiently competent to form well-defined rafts or schollen. Rather, they form darker wispy layers and streaks in the migmatite. Hence, this outcrop could be viewed as representing an advanced stage in the transition from a metatexite to diatexite migmatite in which the mesocratic parts have partially melted, and record the last vestiges of the original compositional layering in the protolith.

Location: Glenelg River Complex, Australia. *Rock type*: schlieric diatexite migmatite; quartzofeldspathic protolith, partial melting in the upper amphibolite facies. *Scale*: the lens cap is 6.5 cm in diameter. *Image and caption*: Tony I.S. Kemp.

Further reading: Kemp, A.I.S. & Gray, C.M. (1999): Geological context of crustal anatexis and granitic magmatism in the northeastern Glenelg River Complex, western Victoria. *Australian Journal of Earth Sciences* **46**, 407–420.

Figure **D52**

Fig. D52. This migmatite consists entirely of neosome. The schollen are fragments of coarse-grained, stromatic metatexite that have a prominent internal layering due to the alternation of thin bands of leucosome with melanosome. Domains of leucosome contain the assemblage plagioclase + quartz + K-feldspar + muscovite, and the melanosome consists of biotite + plagioclase + quartz. The schollen are lenticular in shape and strongly aligned. Internal layering is parallel to the long axis of the schollen, except in the scholle with the lowest aspect-ratio, where the internal layering is sigmoidal. The matrix that encloses the schollen is mesocratic and broadly granitic, with the mineral assemblage plagioclase + quartz + K-feldspar + muscovite + biotite, but it has both melanocratic and leucocratic parts. Thin, biotite-rich domains define schlieren (Sl) that are parallel to the elongation of the schollen, but the biotite schlieren are much less prominent and more irregular where there are no schollen present, e.g., in the lower right. Coarser-grained, leucocratic domains in the mesocratic matrix occur as (1) lenses oriented parallel (L1) to the biotite schlieren and to the direction of elongation of the schollen and (2) small domains (L2) located at the ends of low-aspect-ratio schollen, a position that is analogous to pressure shadows around competent porphyroblasts. Microstructures in the mesocratic matrix, such as the alignment of large crystals of feldspar and biotite that define a foliation, equant quartz grains with only minor evidence for grain-boundary migration, and large crystals of plagioclase that have rational faces against quartz, suggest that the matrix crystallized from a melt, and that the foliation is magmatic. The parallel orientation of the schollen, biotite schlieren, and magmatic foliation may, therefore, have developed in the diatexite magma (granitic melt with fragments of metatexite migmatite and crystals of biotite and feldspar) as a result of gradients in the flow rate (i.e., shear-strain gradients). The leucocratic domains in the matrix may represent either a fractionated leucocratic melt, or a melt devoid of entrained crystals of residual biotite that was able to migrate to low-pressure sites as the framework of crystals in the partially crystallized diatexite deformed toward the end of its crystallization.

Location: Saint-Malo, Brittany, France. *Rock type*: schollen diatexite migmatite; pelitic to semipelitic protolith, partial melting in the upper amphibolite facies, at T ca. 750°C, P 4–7 kbar. *Scale*: the coin is 2.2 cm in diameter. *Image*: Mike Brown.

Further reading: Brown, M. (1979): The petrogenesis of the St-Malo migmatite belt, Armorican Massif, France, with particular reference to the diatexites. *Neues Jahrbuch für Mineralogie, Abhandlungen* **135**, 48–74.

Figure D53

Fig. D53. Most of the neosome in this migmatite is mesocratic and contains the assemblage cordierite + biotite + K-feldspar + quartz + plagioclase; sillimanite occurs as inclusions in the cordierite. The bulk composition of the neosome is the same as that of the likely protolith, collected from just below the "melt-in" isograd. Moreover, the mesocratic neosome contains prismatic cordierite and K-feldspar, has tiled, unstrained crystals of plagioclase, and has a grain size that is 5–10 times that of the scattered schollen of paleosome (nonmelted plagioclase + quartz + biotite schist). Therefore, this migmatite is largely partially melted protolith in which significant separation of the melt from its residuum did not occur, at least not on the scale of the 10-kg sample used for determining its whole-rock composition. The melting reaction was probably biotite + sillimanite + quartz + plagioclase = cordierite + K-feldspar + melt. The principal compositional banding in this migmatite results from the concentration of biotite to various degrees during flow as a magma. Hence, some schlieren are darker than others. As there are no structures preserved that predate anatexis, and as biotite schlieren are the dominant morphological element, this is a schlieric diatexite migmatite. Scattered, elongate schollen contribute to the banding. The migmatite also contains some K-feldspar crystals that are up to 5 cm across. A diffuse patch of coarse-grained diatexite (just to the right of the scale) lacks the strong compositional banding typical of the rest of the diatexite and results from the late injection of a mesocratic magma that contained large crystals of K-feldspar into the already banded diatexite magma. A leucocratic K-feldspar-bearing magma was also injected into the diatexite after it had become compositionally banded, and this formed the prominent coarse-grained granitic vein that is oriented subparallel to the compositional banding and crosses the center of the photograph. Note the abundance of features such as the asymmetrical shape of the schollen, the asymmetrical shape of the leucosome-filled shadows around schollen and other competent objects, sigmoidal schlieren and truncated flow-banding, which can be used to determine the shear sense at the stage at which this diatexite migmatite was in a magmatic state.

Location: Quetico Subprovince, Ontario, Canada. *Rock type*: schlieric diatexite derived from pelitic protoliths; amphibolite-facies conditions, *T* 750–800°C, *P* 3–4 kbar. *Scale*: the ruler is 15 cm long. *Image*: E.W. Sawyer.

Further reading: Milord, I. & Sawyer, E.W. (2003): Schlieren formation in diatexite migmatite: examples from the St. Malo migmatite terrane, France. *Journal of Metamorphic Geology* **21**, 341–362.

Milord, I., Sawyer, E.W. & Brown, M. (2001): Formation of diatexite migmatite and granite magma during anatexis of semi-pelitic metasedimentary rocks: an example from St. Malo, France. *Journal of Petrology* **42**, 487–505.

Figure D54

Fig. D54. This diatexite migmatite has no schollen, but it has abundant, subparallel, biotite-rich schlieren. The schlieren are sinuous and comparatively short (<25 cm), and they range in width from <1 to about 7 mm. The narrowest schlieren are chains of imbricate crystals of biotite, but the biotite in the core of the thickest schlieren commonly form decussate aggregates. At the margins, however, the biotite crystals are either subparallel or imbricate. Some individual crystals of biotite are kinked, and this may indicate that they were derived from the disaggregation of coarse-grained schollen that contained melanosome. The matrix containing the schlieren is a biotite–muscovite–K-feldspar–quartz–plagioclase leucogranite. Tabular, subhedral to euhedral crystals of plagioclase in the leucogranite are generally aligned (locally imbricate or tiled) and define a foliation that is essentially parallel to the schlieren; the widespread presence of straight crystal faces against quartz indicates that the foliation is magmatic. The grain size is generally 4–5 mm, but markedly coarser (7–8 mm) in small, scattered, equant domains (e.g., just below the coin). Feldspar crystals in the coarse-grained domains are generally not aligned. This randomness may indicate that these domains were the last parts of the diatexite migmatite to have crystallized.

Location: Saint-Malo, Brittany, France. *Rock type*: schlieren diatexite migmatite; pelitic and semipelitic protolith, partial melting at T ca. 750°C, P 4–7 kbar. *Scale*: the coin is 2.0 cm across. *Image*: Mike Brown.

Figure D55

Fig. D55. This diatexite migmatite contains only a few small, elliptical, and ghostlike remains of schollen. Alternating biotite-rich and quartzofeldspathic layers in the schollen closely resemble the melanosome and leucosome found in the neosome of metatexite migmatites elsewhere in the Saint-Malo terrane. The leucocratic host contains the assemblage plagioclase + quartz + K-feldspar + muscovite + biotite and, locally, domains where tabular crystals of plagioclase are imbricated (i.e., tile microstructure) and have rational faces against interstitial quartz. Taken together, these features indicate that the leucocratic part of this migmatite crystallized from a magma of granitic composition as it was sheared during flow. The most prominent features of this migmatite are the discontinuous biotite-rich bands that it contains. Some of these could be called schlieren; the possibility that the shorter, low-aspect ones could be pieces of the biotite-rich melanosome derived from disaggregated schollen should be considered, however.

Location: Saint-Malo migmatite terrane, France. *Rock type*: schlieren diatexite migmatite; semipelitic protolith, partial melting at T 750–800°C, P 4–5 kbar. *Scale*: the ruler is 15 cm long. *Image*: E.W. Sawyer.

Further reading: Milord, I. & Sawyer, E.W. (2003): Schlieren formation in diatexite migmatite: examples from the St-Malo migmatite terrane, France. *Journal of Metamorphic Geology* **21**, 341–362.

Figure **D56**

Fig. D56. There are two main components to this diatexite migmatite. The first contains schlieren and irregularly shaped melanocratic patches (e.g., in the lower right near the ruler), and the second is distinguished by its large crystals of cordierite (up to 4 cm across) and a lack of schlieren. Schlieren are up to 20 cm long and composed of biotite and sillimanite, with minor amounts of cordierite and ilmenite. The melanocratic patches consist of the assemblage biotite + cordierite + sillimanite + ilmenite + plagioclase, and are interpreted to be residuum left after the incongruent melting of biotite; they are melanosome. The melting reaction was probably biotite + sillimanite + quartz = melt + cordierite + ilmenite, and the protolith was pelitic. The leucocratic matrix that contains the melanocratic patches and schlieren consists of the assemblage plagioclase + quartz + K-feldspar + biotite + cordierite (i.e., a granitic to granodioritic bulk composition) and is interpreted to have been derived from the anatectic melt generated in making the melanosome. A layering due to the alternation of leucocratic and melanocratic bands is locally folded. The second component has a granitic bulk composition, with an assemblage of plagioclase + quartz + K-feldspar + cordierite + biotite, which suggests derivation from an anatectic melt extracted from a metapelitic source. In places, the second component contains enclaves of the schlieren-bearing component, indicating that the latter is older, and was more coherent. Elsewhere (e.g., above the scale), faint traces of schlieren persist into the cordierite-rich component, and result in a diffuse, or nebulitic, contact between the two. The presence of schlieren and melanosome implies that the parent magma contained crystals and fragments of residual material, but the magma that generated the cordierite-rich component did not. Folding of the schlieren-bearing portion could begin once crystallization reached the rigid percolation threshold (ca. 55% crystals) and could have redistributed, or expelled, some of the melt residing in the framework of crystals. Places that lost sufficient melt by compaction to reach the particle-locking threshold (ca. 72% crystals) behaved as if they were solid, but other places in the migmatite still contained as much as 50 vol.% melt. The redistributed melt was depleted in the residuum component, and hence very leucocratic. Whether the magma that produced the cordierite-bearing component of the diatexite was simply redistributed, or was introduced from a more distant source, it incorporated the parts of the schlieren-bearing component with >72% crystals as rigid enclaves. However, where substantial amounts of melt remained, the contact between the two was not sharp.

Location: Peña Negra, Spain. *Rock type*: schlieren diatexite migmatite; pelitic protolith, partial melting at T ca. 750°C, P 4–7 kbar. *Scale*: the ruler is 15 cm long. *Image*: E.W. Sawyer.

Figure D57

Fig. D57. The Ashuanipi Subprovince contains a number of large bodies of diatexite that show a transition from metatexite migmatites and schollen diatexites at their borders, to more uniform interiors. This example illustrates a mesocratic orthopyroxene–biotite–K-feldspar–quartz–plagioclase diatexite migmatite from the interior of a kilometer-size body of diatexite migmatite. Small lenticular aggregates of biotite in the rock define a weak foliation in which some tabular crystals of feldspar also are aligned. The larger, aligned crystals of feldspar are plagioclase, whereas K-feldspar, quartz, the smaller crystals of plagioclase, and orthopyroxene are interstitial; this rock has a bulk composition close to that of the protolith. However, in other diatexite migmatites from the same body, the plagioclase, and some of the K-feldspar, form larger tabular crystals that are aligned and define a magmatic foliation; the whole-rock composition of these rocks is generally closer to that of a granitic, anatectic melt. The orthopyroxene in diatexite migmatites that have a granitic bulk composition generally shows a much greater degree of replacement by biotite than the orthopyroxene in the diatexite migmatites that have compositions more like that of the protolith.

Location: Ashuanipi Subprovince, Quebec, Canada. *Rock type*: mesocratic diatexite migmatite; metagreywacke protolith, anatexis at *T* 825–875°C, *P* 6–7 kbar. *Scale*: the ruler is 15 cm long. *Image*: E.W. Sawyer.

Figure **D58**

Fig. D58. This diatexite migmatite contains small, scattered schollen of residual metasedimentary material consisting of the assemblage plagioclase + biotite + orthopyroxene + quartz, and a few mafic schlieren, within a leucocratic, weakly foliated matrix of granodioritic bulk composition. The principal reaction responsible for partial melting in the protolith was biotite + quartz + plagioclase = orthopyroxene + melt ± ilmenite. The matrix is coarse-grained, and the microstructure in much of it consists of a framework composed mostly of weakly aligned euhedral crystals of plagioclase and a small proportion of euhedral crystals of K-feldspar. Anhedral quartz and K-feldspar, together with a minor amount of biotite, fill the angular interstitial spaces in the feldspar framework. This microstructure is interpreted to have formed during the crystallization of a granitic to granodioritic anatectic magma under relatively static conditions. However, there are some local variations in the microstructure of the migmatite. One of the more common types consists of small, round patches in the diatexite (on the bottom edge, just left of center) in which the grain size is noticeably coarser, the rock is more leucocratic, and the large euhedral crystals of feldspar are predominantly K-feldspar. These patches are interpreted as the sites where the last stages of crystallization occurred from a magma that had a more evolved (i.e., fractionated) bulk composition than the initial magma from which rest of the diatexite crystallized. The diatexite magma had, therefore, undergone fractional crystallization and had previously crystallized a major amount of plagioclase.

Location: Ashuanipi Subprovince, Quebec, Canada. *Rock type*: mesocratic diatexite migmatite; metagreywacke protolith, anatexis at T 825–875°C, P 6–7 kbar. *Scale*: the sticker (top left) is 3 cm long; width of the photograph corresponds to 115 cm. *Image*: E.W. Sawyer.

Figure **D59**

Fig. D59. Partial melting of semipelitic rocks in the Saint-Malo terrane involved the breakdown of muscovite through reactions such as: muscovite + plagioclase + quartz + H_2O = melt, and muscovite + plagioclase + quartz = melt + K-feldspar + sillimanite + biotite. Biotite was not generally a reactant and remained stable. Therefore, changes in the proportion of biotite can be used to infer which parts of the diatexite migmatites are residual (biotite content increased) or melt-enriched (biotite content diluted) relative to the protolith. Under these circumstances, mapping the diatexites as melanocratic, mesocratic, or leucocratic is a useful field technique for tracing regions of melt loss and melt gain. This particular example of diatexite migmatite contains large crystals of feldspar, some of which have rational faces against quartz, indicating crystallization in a melt. However, the diatexite is melanocratic, has the mineral assemblage plagioclase + quartz + K-feldspar + biotite, and a bulk composition that indicates considerable enrichment in the residual components and biotite relative to an anatectic granite derived from a pelite. Euhedral crystals of plagioclase and biotite are oriented and define the foliation. Some crystals of plagioclase have rational faces against quartz, and the domains of quartz are equant. Thus, the foliation is interpreted to be magmatic; however, the microstructure has a higher proportion of straight grain-boundaries and lacks the interlocking grains typical of granites. Schlieren of biotite are oriented parallel to the foliation. A key problem to resolve for residual diatexite migmatites is how the residual composition was attained, and whether this happened before, or during, the bulk flow, or magma-flow processes that destroyed the pre-anatectic structures in the protolith and paleosome.

Location: Saint-Malo migmatite terrane, Brittany, France. *Rock type*: melanocratic diatexite or, alternatively, residual diatexite; pelitic and semipelitic protolith, anatexis at *T* ca. 750°C, *P* 4–7 kbar. *Scale*: the ruler is 15 cm long. *Image*: E.W. Sawyer.

Further reading: Milord, I., Sawyer, E.W. & Brown, M. (2001): Formation of diatexite migmatite and granite magma during anatexis of semi-pelitic metasedimentary rocks: an example from St. Malo, France. *Journal of Petrology* **42**, 487–505.

Figure **D60**

Fig. D60. This diatexite migmatite contains scattered small schollen and could, therefore, be called a schollen diatexite. Most of the schollen are small fragments of biotite-rich melanosome and are probably the remnants of larger rafts derived from stromatic migmatite, such as that just left of the lens cap, which have disintegrated. Other rafts are of resister lithologies, notably calc-silicate rocks and quartz veins. However, the main characteristic of this migmatite is the uniformity of the matrix, which appears to have neither mafic schlieren, leucocratic segregations nor a flow banding in it. Although the bulk composition is leucogranitic and the mineral assemblage is plagioclase + quartz + K-feldspar + muscovite + biotite, the microstructure is not granite-like in that many groundmass grains are polygonal rather than interlocking. Euhedral and subhedral crystals of plagioclase and microcline (up to 5 mm in length) are generally oriented and define a foliation, together with the larger crystals of biotite present; the quartz domains and grains are equant and generally devoid of strain features. Therefore, the foliation is interpreted to be of magmatic origin. Although the magma that produced this diatexite migmatite was subject to gradients in shear strain, which produced the mineral alignments, no segregation of the melt fraction from the solid appears to have occurred. Either the shear stresses were very weak, or they ended before about 50% crystallization and the rigid percolation threshold was reached. Muscovite, however, is not oriented and appears to have crystallized late. This diatexite migmatite is considerably more leucocratic than that shown in Fig. D59, also from the Saint-Malo migmatite terrane; the difference indicates that diatexite migmatites can be open systems. The higher proportion of residuum in the migmatite in Fig. D59 indicates the loss of some melt from it, whereas this migmatite contained a much higher fraction of anatectic melt, either because the residuum had already been separated or because it was diluted by the addition of anatectic melt. If the latter is correct, then one can infer that this migmatite occupies a structural site where anatectic melt could accumulate.

Location: Saint-Jacut-de-la-Mer, Saint-Malo migmatite terrane, Brittany, France. *Rock type*: leucocratic diatexite or, alternatively, leucocratic schollen diatexite; pelitic and semipelitic protolith, anatexis at T ca. 750°C, P 4–7 kbar. *Scale*: the lens cap is 6.5 cm across. *Image*: Mike Brown. Previously published as fig. 3c in Brown (1995) and reproduced with the permission of the Société géologique de France.

Figure D61

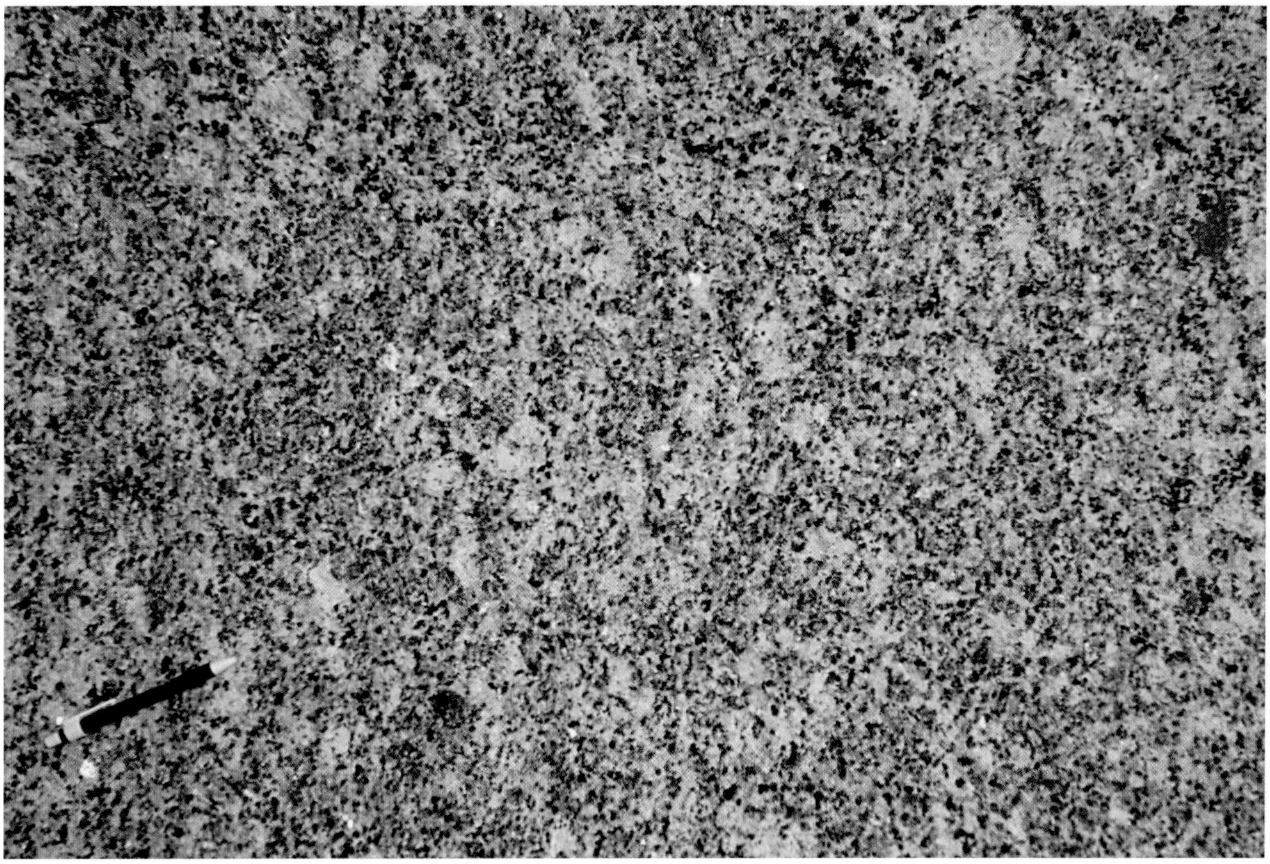

Fig. D61. This very uniform, coarse-grained K-feldspar–biotite–quartz–plagioclase diatexite migmatite lacks both schollen and schlieren, although it does have a weak foliation. The principal variation in the diatexite visible in this outcrop-scale photograph results from scattered porphyroblasts or phenocrysts of K-feldspar. Plagioclase has crystal faces against quartz, and this microstructure is interpreted to indicate that the plagioclase crystallized from a melt. Some of the biotite in the diatexite is intergrown with quartz or plagioclase (or both), and this microstructure suggests that the biotite + quartz may have replaced residual orthopyroxene (see, for example, Figs. F78–F81). Orthopyroxene occurs commonly, although not abundantly, in other, compositionally less evolved diatexite migmatites in the Kauhava area.

Location: Kauhava area, Svecofennian domain, Finland. *Rock type*: diatexite migmatite; metapelitic protolith, granulite-facies anatexis at T ca. 800°C, P 4–5 kbar. *Scale*: the pencil is 14 cm long. *Image*: Hannu Mäkitie. Previously published as fig. 2a in Mäkitie (2001), and reproduced with the permission of the Geological Society of Finland.

Further reading: Mäkitie, H. (2001): Eastern margin of the Vaasa Migmatite Complex, Kauhava, western Finland: preliminary petrography and geochemistry of the diatexites. *Bulletin of the Geological Society of Finland* **73**, 35–46.

Figure **D62**

Fig. D62. This close-up photograph shows a plagioclase + quartz + K-feldspar + muscovite + biotite diatexite migmatite of granitic bulk composition in which the microstructure and the distribution of minerals are almost isotropic. The principal variations in the diatexite migmatite are (1) scattered small phenocrysts of K-feldspar that are only slightly coarser-grained than the matrix, (2) a very weak foliation due to the alignment of some of the biotite, and (3) several small (ca. 10 mm) enclaves consisting of crenulated aggregates of biotite, sillimanite, and muscovite. The thin, dark, fine-grained layers oriented roughly parallel to the scale and located near the center of the photograph are mylonite seams that formed as a result of sinistral strike–slip shearing after the diatexite migmatite had crystallized.

Location: foreshore at Saint-Malo, Saint-Malo migmatite terrane, France. *Rock type*: diatexite migmatite; pelitic and semipelitic protolith, anatexis at *T* ca. 750°C, *P* 4–7 kbar. *Scale*: the ruler is 15 cm long. *Image*: E.W. Sawyer.

Figure D63

Fig. D63. This photograph shows a parallel layered structure that developed in a 4-m-wide shear zone crossing an otherwise rather uniform diatexite migmatite that has a magmatic foliation. The host diatexite has been transformed by the shear zone into alternating bands of (1) coarse-grained leucosome in which magmatic textures, such as feldspar with crystal faces against quartz, are present and (2) finer-grained layers with a higher proportion of mafic minerals (biotite + orthopyroxene) and a microstructure that resembles a cumulate-like arrangement of tabular crystals of plagioclase, a very small proportion of which show fractures healed by plagioclase of slightly more sodic composition. This layered structure is interpreted to have formed at the late stages of crystallization of the host diatexite body. During cooling, the synmagmatic shear strain responsible for the foliation in the host diatexite became progressively focused into the narrowing melt-bearing parts of the diatexite body. The existing framework of crystals in these narrow melt-bearing domains was progressively deformed, its grain size reduced, and the crystals fractured. The principal effect was to compact the framework by reducing the intracrystalline pore-space and, in so doing, to drive out the melt trapped there, to form adjacent melt-rich and mafic-mineral-poor leucocratic layers. The truncations of the layering present are exactly analogous to foliation cut-offs observed in mylonites.

Location: Ashuanipi Subprovince, Quebec, Canada. *Rock type*: high-strain layered diatexite migmatite; *T* 825–875°C, *P* 6–7 kbar. *Scale*: the ruler is 15 cm long. *Image*: E.W. Sawyer.

Figure D64

Fig. D64. Diatexite migmatite with a strong layered appearance due to alternations of thick bands of leucosome and thin ribbons of mesocratic paleosome. Some of the dispersed megacrysts of alkali feldspar and fragments of quartz in the diatexite migmatite have an augen-like appearance and an asymmetrical shape, which indicates a dextral sense of shear. The megacrysts and fragments of quartz may be the remains of an older, disrupted vein of granitic pegmatite. The presence of finer-grained trails around some of the fragments of quartz and of K-feldspar megacrysts suggests the presence of high strain in a submagmatic state. However, the paradox is that features of solid-state strain are not present in the diatexite; this would suggest that the shearing was related to magmatic flow. Thus, this migmatite highlights one of the key problems that needs to be resolved if microstructures in migmatites are to be fully evaluated; to what extent are the features of intercrystalline deformation that are acquired by minerals as the crystal framework deforms erased by slow cooling at a high temperature, perhaps with melt or melt-derived aqueous fluid present? Put another way, can the microstructural history of migmatites be reset?

Location: Glenelg River Complex, southeastern Australia. *Rock type*: schlieric diatexite migmatite; quartzofeldspathic protolith, anatexis in the upper amphibolite facies. *Scale*: the pencil is 15 cm long. *Image and caption*: Tony I.S. Kemp.

Further reading: Kemp, A.I.S. & Gray, C.M. (1999): Geological context of crustal anatexis and granitic magmatism in the northeastern Glenelg River Complex, western Victoria. *Australian Journal of Earth Sciences* **46**, 407–420.

E.

OTHER MORPHOLOGIES OF MIGMATITE

Some morphologies are not specific to either metatexite or diatexite migmatites. The photographs in this section illustrate the two most common of these morphologies.

Syn-anatectic folding: fold structures in migmatites [Figs. E1–E8]

It is very important to distinguish migmatites that were folded while they contained anatectic melt (i.e., migmatites with fold structures) from migmatites that were folded after they had completely crystallized (i.e., folded migmatites). The presence of melt in a rock creates weak domains that can control the geometry of the folds (and other structures such as shear zones) that subsequently develop, and thus can affect the final morphology of a migmatite. Once the melt fraction in a migmatite starts to crystallize and the fraction of melt in the leucosome part, or in a diatexite migmatite, declines below the rigid percolation threshold (less than ca. 45% melt), a framework of crystals will have formed, and an increase in strength will occur. When crystallization progresses further to the particle-locking threshold (<28% melt left), leucosome and diatexites both have attained a significant fraction of their solid-state strengths, and may at that point become the strong layers in a migmatite.

Migmatites with a vein structure [Figs. E9–E14]

Migmatites commonly become veined by the intrusion of felsic magma in anatectic or orogenic events long after the one in which they were formed. However, it is veining during the anatectic event that formed the migmatites that is of interest here. Even with that restriction, veining could occur at any stage between incipient melting, when the fraction of melt is low, through to peak temperatures, when migmatites contain a larger fraction of melt, back to near-solidus conditions when the fraction of melt is, again, very small.

Figure E1

Fig. E1. This migmatite has a well-developed layering consisting of (1) a mesocratic plagioclase + hornblende paleosome, (2) a melanocratic hornblende + plagioclase residuum from partial melting, and (3) a plagioclase + quartz leucosome that represents the product of the former anatectic melt; hence, this is a stromatic metatexite migmatite. Some features in this outcrop are typical of metatexite migmatites that were folded while they contained melt. Many layers that contained melt show a slight thickening in the fold-hinge region relative to the fold limbs. For example, the neosome layer (N) at the lower left has leucosome and melanosome components that are intimately mixed in the fold core, but which tend to become separated and thinned on the fold limbs. Another, more complex layer of neosome (on which the scale rests) is a little different; it has well-developed layers of melanosome that may have influenced the distribution of melt during the folding. The leucosome layer associated with it is thickest on the long limb of an asymmetrical synform, and not in the fold core. However, the leucosome on the fold limb contains several thin screens of boudinaged melanosome, suggestive of dilation within the competent layer of hornblende-rich melanosome as it was folded. This created local space into which the anatectic melt could migrate. The core of the largest fold contains several short, thin streaks of leucosome (L) that are oriented parallel, or subparallel, to the axial plane of the fold; the streaks have roots in the layer-parallel leucosome. This pattern is consistent with the retention of some melt in the fold-hinge regions during folding. If the melanosome marks the place where anatectic melt was formed, then the best evidence for the overall movement of anatectic melt away from the fold limbs is the presence of melanosome there without any associated leucosome. The fraction of melt preserved in this migmatite as leucosome is much lower than the degree of partial melting estimated by petrological means, i.e., this migmatite had lost a significant proportion of its melt before it crystallized. Therefore, one can suppose that the melt produced on the limbs of the fold has migrated out of the migmatite through the core of folds.

Location: Abitibi Subprovince, south of Chibougamau, Quebec, Canada. *Rock type*: stromatic metatexite migmatite with fold structures; metamafic protolith, *T* 800–850°C, *P* 8–10 kbar. *Scale*: the ruler is 15 cm long. *Image*: E.W. Sawyer.

Figure E2

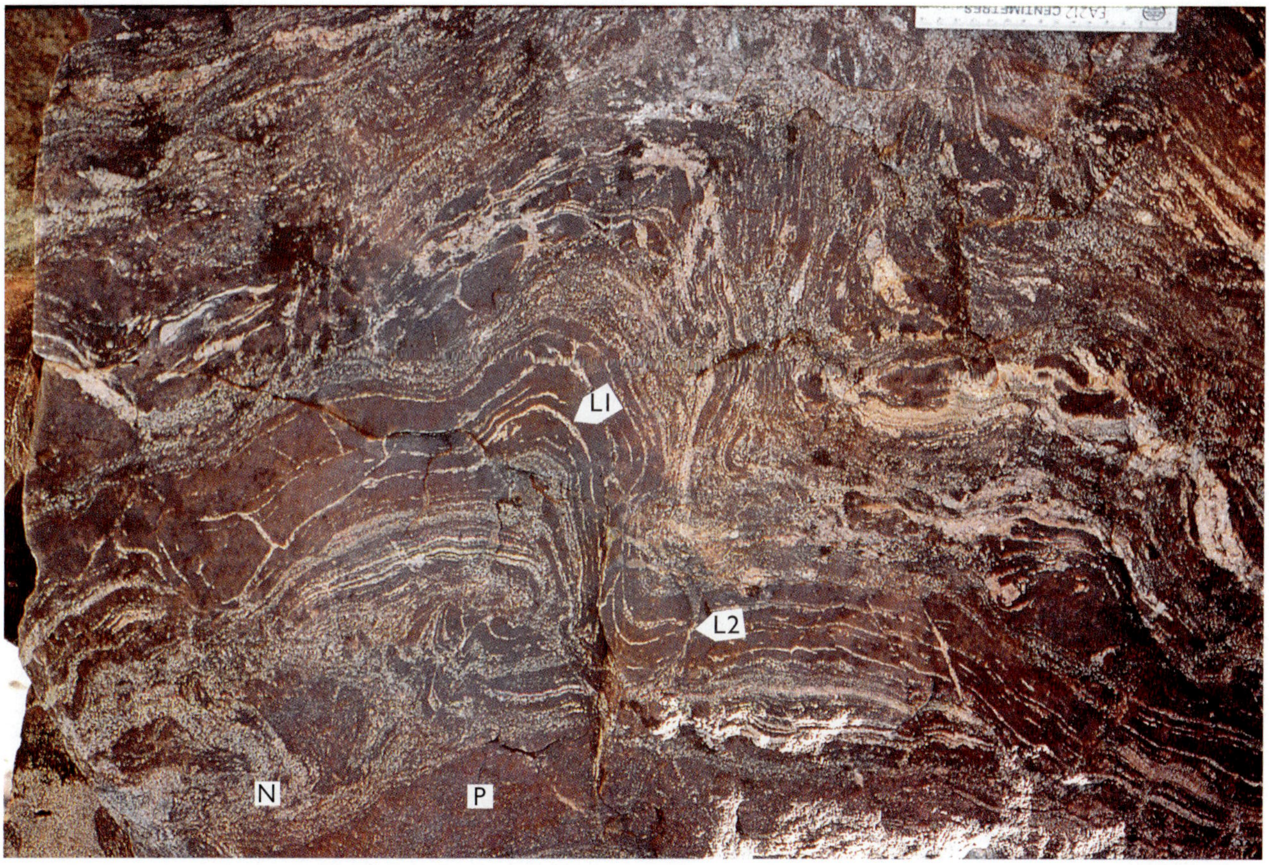

Fig. E2. This complex metatexite migmatite had a higher fraction of melt present during folding than the previous example. The dark, competent layer (P) at the base of the outcrop is only gently folded, and the melt-rich layer of neosome (N) above it has acted as a plane of décollement above which the layers were more tightly folded and deformed. The folds are asymmetrical; flat-lying long limbs are separated by steep, thinned short limbs. The competent, dark layers are thicker in the antiformal hinge than in the synformal one. Most melt present during the folding was located parallel to the layering, and occurs either as thin bodies of stromatic leucosome in the competent layers (L1) or as thicker layers of neosome (unsegregated melt + residuum mixtures) between the competent ones. The layers of leucosome (e.g., L1) and neosome are generally thickest in the fold hinges, both antiformal and synformal, and thinnest on the steep limbs of folds, indicating that melt moved from the fold limbs to the hinges during folding. Variously oriented fragments of the dark-colored competent layers in the neosome provide further evidence for magma flow in the hinges during the folding. A small number of high-aspect-ratio domains of leucosome (L2) are located in the synformal hinge and parallel to the axial plane of folds; these areas of leucosome contain very little residual material and are indicative of the late-stage segregation of melt into the axial planes. Domains of leucosome are also located in other sites on the fold limbs, such as symmetrical and asymmetrical boudins and conjugate, normal-sense extensional shears. Some domains of leucosome are located along reverse-sense shears, however.

Location: Mt. Hay, Arunta Inlier, Australia. *Rock type*: metatexite migmatite with fold structures; metapelitic and semipelitic protolith, partial melting in the granulite facies, *T* 825–875°C, *P* 6–7 kbar. *Scale*: the ruler is 15 cm long. *Image*: Marianne Bonnay.

Further reading: Collins, W.J. & Sawyer, E.W. (1996): Pervasive granitoid magma transfer through the lower-middle crust during non-coaxial compressional deformation. *Journal of Metamorphic Geology* **14**, 565–579.

Sawyer, E.W., Dombrowski, C. & Collins, W.J. (1999): Movement of melt during synchronous regional deformation and granulite-facies anatexis, an example from the Wuluma Hills, central Australia. *In* Understanding Granites; Integrating New and Classical Techniques (A. Castro, C. Fernández & J.-L. Vigneresse, eds.). *Geological Society, Special Publication* **168**, 221–237.

Figure E3

Fig. E3. This metatexite migmatite had an intermediate gneissic protolith and was folded when anatectic melt was present; the view is down plunge. Most bodies of leucosome in the core of the fold (near the hammer) are parallel to the compositional layering in the paleosome, and are notably thicker in the fold hinges compared to the limbs. Few of the leucosome domains in the core area have a melanocratic border. The melt thus migrated there from elsewhere. There are far fewer domains of leucosome in the layer outside the fold core, but they are oriented parallel to the axial plane of the fold and have a conspicuous melanosome around them, indicating that the neosome probably formed *in situ*. The domains of leucosome in the top right are interesting; most are planar (stromatic) and do not have a melanosome. The one that has melanosome appears to be made up of a series of short, closely spaced segments that are parallel to the axial plane of the fold. Thus, it is not stromatic in the same sense as the others. Some shearing parallel to the axial plane has occurred in the core of the fold, and has resulted in solid-state recrystallization in the affected domains of leucosome. Deformation thus outlasted anatexis at this locality.

Location: Arunta Inlier, Australia. *Rock type*: metatexite migmatite with fold structure; intermediate orthogneiss protolith, anatexis in the upper amphibolite to lower granulite facies. *Scale*: the hammer is 60 cm long. *Image and caption*: Tony I.S. Kemp.

Figure E4

Fig. E4. This metatexite migmatite developed in interlayered mafic (Ma), psammitic (Ps, grey and brown colors), and pelitic (Pe) rocks crossed by a few felsic dikes. It is an example of how layering can influence the position, morphology, and composition of the leucosome. The pelitic layers were incompetent, and flowed during the folding. Thus, they are thin, or even absent, on the limbs of the fold and thickest in the fold hinges. In contrast, the psammitic and metamafic layers were competent and produced parallel folds, and underwent boudinage on the long limbs of these folds. The psammitic layer was the least affected by partial melting. It contains very sparse areas of leucosome or melanosome. The metamafic layers experienced more extensive partial melting and developed an internal layering in which the most melt-depleted parts are the darkest. Partial melting was most extensive in pelitic layers, and these now have a (biotite + garnet)-rich residual composition. During folding, the partial melt generated near the fold hinges in the pelitic layers migrated along these layers and formed triangular bodies of granitic leucosome (e.g., L1) in the hinges of both the antiformal and the synformal folds. The melt also migrated along the pelite layers on the long limbs of the folds, and occupied the space between the boudins in the adjacent mafic layer, where it crystallized and formed triangular bodies of leucosome (L2). Both types of leucosome contain large, distinctive poikiloblasts of garnet, which, as the solid products of the incongruent breakdown of biotite, are evidence that such leucosome was derived from the pelitic layers. Furthermore, these bodies of leucosome have a magmatic microstructure; hence, they crystallized after the folding and the boudinage. Very thin, stromatic bodies of leucosome also occur in the pelitic layers, and these contain large crystals of garnet and plagioclase that have pressure shadows and finer-grained tails. These domains of leucosome thus were partially crystalline during the folding, and underwent grain-size reduction. The competent layers contain leucosome in similar structural positions; stromatic bodies of leucosome lying parallel to the layering are slightly thicker in the fold hinges, and are boudinaged on the fold limbs. Melt thus moved into the space created as planes parallel to the layering opened, and the leucosome was partially solid during the folding, probably through loss of melt. Small, equant bodies of leucosome occur in the dilatant spaces between boudins in the competent layers. The leucosome in the metamafic layers is tonalitic in composition and does not contain garnet, features consistent with a derivation from the metamafic protolith. Hence, anatectic melt migrated principally within, and not across, the various layers in this migmatite. Lastly, a complexly shaped area of leucosome (L3) occurs on the outside of the disrupted layer of psammite. It is interpreted to occupy the low-pressure site created during folding and breakup of the psammite layer. This leucosome contains garnet where it is in contact with metapelite, but is garnet-free where it is in contact with the metamafic rock, again indicating local derivation of the melt in this migmatite.

Location: Sand River Gneisses, Causeway locality, South Africa. *Rock type*: metatexite migmatite with fold structures; interlayered protolith, anatexis in the granulite facies. *Scale*: the pocket knife is 11 cm long. *Image*: E.W. Sawyer.

Figure E5

Fig. E5. For some geologists, the image of a migmatite with fold structure is that shown in this photograph. However, migmatites of this kind are uncommon and generally restricted to small (meter-scale) domains in which the strain history is complex. Common examples illustrate progressive deformation in which structures are rotated and overprinted during a single event of deformation, such as in shear zones, or in places where later folding is superimposed on earlier structures, producing interference. The domains of leucosome in this particular migmatite do not systematically have a melanosome. In places, the melanosome is absent, but elsewhere it is thin, and locally, such as on the inside of fold closures, it is wide. The domains of leucosome are characteristically thin, have a granular microstructure, and a grain size of 1–2 mm. The folds are disharmonic and convolute. The fold hinges show a considerable range in orientation, both in plunge and azimuth. However, in some similar examples, the orientation of fold hinges is systematic, and dome or basin structures are the result of sections cut through the noses of sheath folds.

Location: Glenelg River Complex, southeastern Australia. *Rock type*: migmatite with fold structure in semipelitic schists; upper amphibolite facies. *Scale*: the lens cap is 6.6 cm across. *Image and caption*: Tony I.S. Kemp.

Figure E6

Fig. E6. There are three principal lithologies in this metatexite migmatite. A grey layer of biotite–plagioclase–quartz psammite extends across the center of the photograph, and in the bottom right, there is a darker grey layer of hornblende gneiss. Both lithologies are unaffected by anatexis and could be considered paleosome in this migmatite. The third lithology is pelitic, but during anatexis, the pelitic layers were transformed into neosome with well-defined leucocratic and melanocratic portions. Tight asymmetric folding with a wavelength of about 25 cm has affected all three lithologies, but in the pelitic layer, the folds are notably convolute and polyclinal. Two different leucocratic components can be identified in the pelitic layers: (1) the thin, fine-grained leucocratic layers (L1) have sharp borders and a thin dark rim interpreted to be a mafic selvedge, and (2) the coarser-grained, light-grey, biotite–plagioclase–K-feldspar–quartz domains of leucosome (L2) have more diffuse edges and a wider garnet–sillimanite–biotite melanosome. Some of the fine-grained leucocratic layers are isoclinally folded and form intrafolial folds in the layering; these have been reoriented by the tight asymmetrical folds. The domains of coarser-grained leucosome occur as lenses, some of which are located along shear planes on the limbs of the tight asymmetrical folds. From these observations, the fine-grained leucocratic layers are interpreted to be granitic veins intruded into the pelite; these layers subsequently developed a mafic selvedge by reaction with the host, then in the solid state. The veins and host were isoclinally folded and later refolded during anatexis. The bodies of coarser-grained leucosome may have formed *in situ*, by partial melting of the pelite, synchronously with the asymmetrical folding. In addition to a convolute and polyclinal geometry, there are also many shear zones and discontinuities across which the orientation of the stromatic layering changes abruptly. Further complexity results from the formation of shear-bounded lobes of stromatic neosome that resemble escape structures; these have intruded across adjacent layers and, in one case, into the psammite layer. This suite of structures suggests that there were some very weak domains in the pelite layer, possibly because they contained partial melt. The overall style of folding in this metatexite migmatite is typical of that in many high-grade migmatites, particularly where all the layers were relatively weak (contrast with Figs. E2 and E4, in which some layers were strong and resulted in the formation of parallel folds).

Location: Thor–Odin dome, British Columbia, Canada. *Rock type*: stromatic metatexite migmatite with fold structures; pelitic and psammitic felsic protoliths, partial melting at T ca. 800°C, P 4–7 kbar. *Scale*: hammer for scale. *Image*: Olivier Vanderhaeghe.

Further reading: Vanderhaeghe, O., Teyssier, C. & Wysoczanski, R. (1999): Structural and geochronological constraints on the role of partial melting during the formation of the Shuswap metamorphic core complex at the latitude of the Thor–Odin dome, British Columbia. *Canadian Journal of Earth Sciences* **36**, 917–943.

Figure E7

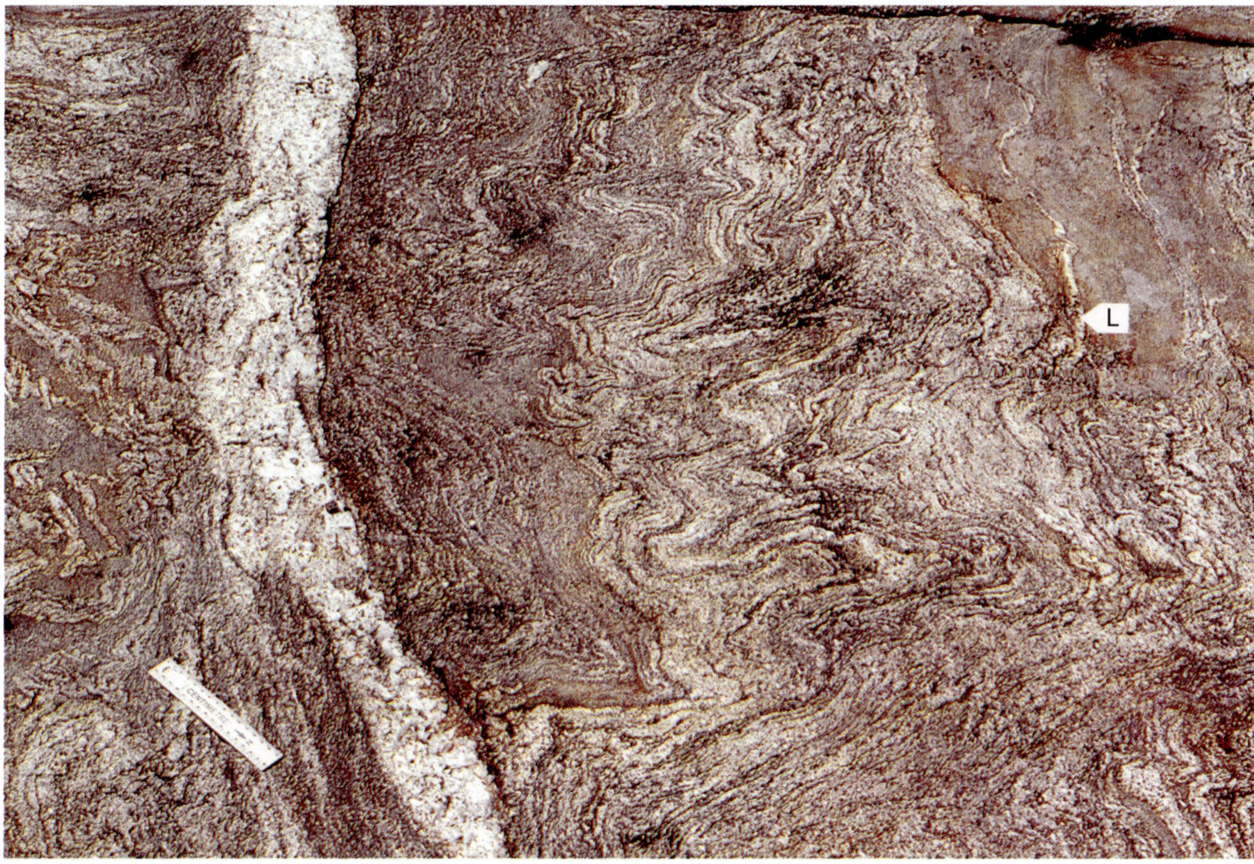

Fig. E7. Neosome predominates in this migmatite, and paleosome occurs only in relics, possibly boudins, in a fine-grained biotite–quartz–plagioclase psammitic schist. The neosome is stromatic, but contains two types of stromatic leucosome, although they have the same plagioclase + quartz + K-feldspar assemblage. One type is in contact with mesocratic garnet–cordierite–biotite–plagioclase–quartz pelitic schist; the schist could be residual in composition, and the leucosome may have formed by *in situ* partial melting. The second, less common type is distinguished by the presence of a narrow, biotite-rich mafic selvedge; in places, such leucosome (L) and its mafic selvedge are discordant and can be traced from the pelitic layers across into the remnants of psammite. This stromatic leucosome is interpreted to be have been injected, and its mafic selvedge, to be the result of reaction with the host rock. The outcrop displays complex folding. The style of folding of some bodies of leucosome (e.g., middle of the left edge of the photograph) could be called ptygmatic. In general, however, the folds in the layered neosome are nonplanar, and the style is convolute; there is a considerable range in the orientation of traces of the axial plane of folds, but some of this may be due to local refolding. Many bodies of leucosome are thicker in the fold hinges, and in places, there are thin, diffuse domains of leucosome oriented approximately parallel to the local direction of the axial plane. The migmatite may, therefore, have been folded when it was close to its solidus, and only a very small fraction of melt was left that could be moved, or segregated, during the folding. The 10-cm-wide leucocratic vein on the left cuts across the stromatic layering in this portion of the migmatite. It can, however, be traced some 25 m across the outcrop. In one direction, it dies out in a sharply defined discordant brittle fracture, but in the other, its borders become increasingly diffuse against the adjacent bodies of stromatic leucosome in the host, and it is rooted in a large layer-parallel lens of leucosome located in the core of a mesoscale (10-m) antiform. The discordant leucocratic vein is, therefore, an example of local, structurally controlled accumulation of melt and its subsequent escape along a propagating fracture. This migmatite is difficult to classify. The neosome can be described as stromatic or layered, but at the scale of the photograph, there are no coherent pre-anatectic structures visible (although at a larger scale, individual, thick units of psammite could be mapped). Consequently, the migmatite is not adequately described as a metatexite. The absence of coherent pre-anatectic structures is a distinguishing feature in diatexite migmatites, but this migmatite lacks other characteristics, such as magmatic flow structures. It is thus not properly described as a diatexite migmatite either. It could be described as a transitional stage between metatexite and diatexite migmatite. If all the stromatic leucosome with a mafic selvedge was in fact intruded into the migmatite, which at the time contained only a small proportion of *in situ* leucosome, then this would be a vein-structured migmatite derived from a metatexite migmatite.

Location: Nemiscau Subprovince, Canada. *Rock type*: transitional metatexite to diatexite migmatite with fold structures; metasedimentary protolith, anatexis in the upper amphibolite facies. *Scale*: the ruler is 15 cm long. *Image*: E.W. Sawyer.

Figure E8

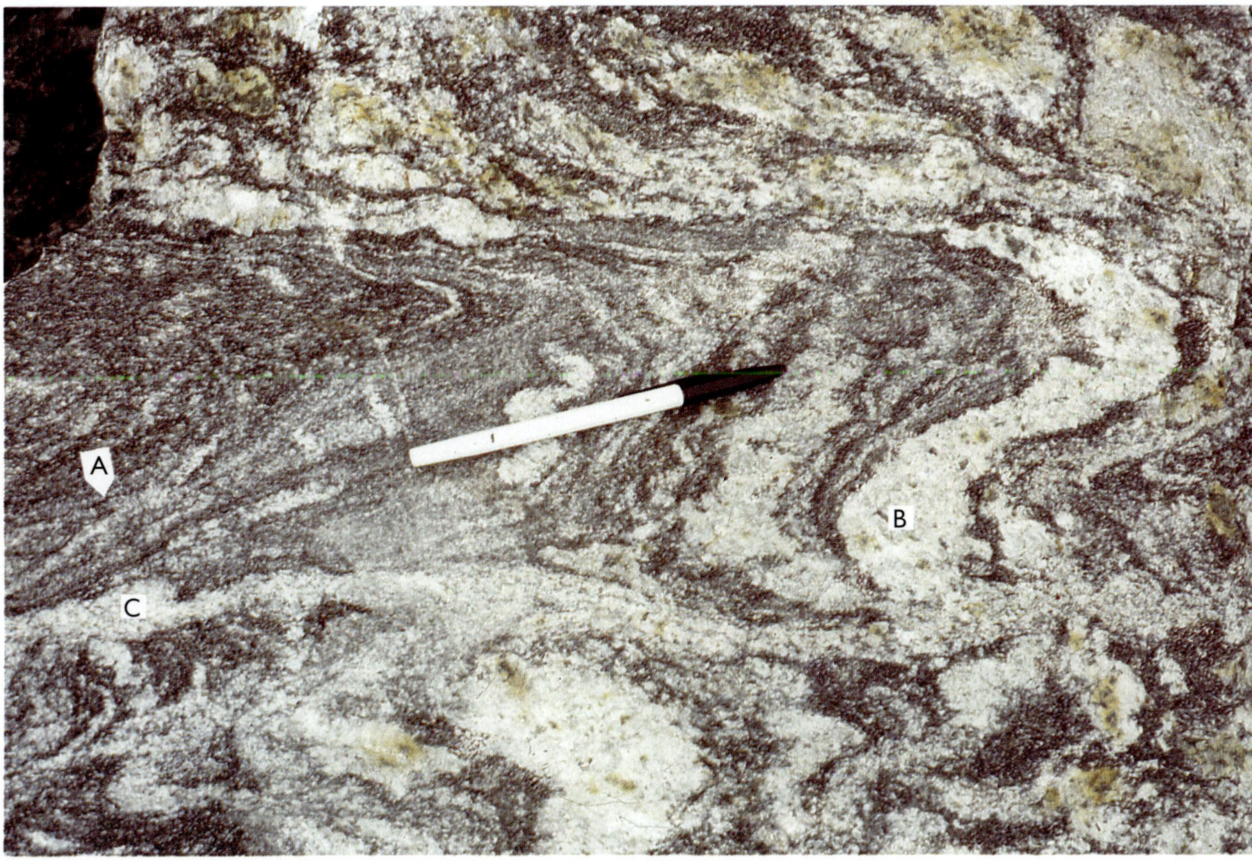

Fig. E8. Thin bodies of leucosome (<1 cm) formed *in situ* in this migmatite and have associated melanosome. They (A) are parallel to the compositional banding, believed to have been bedding, in the mesocratic neosome. However, some areas of "leucosome" (B) are much wider (about 5 cm) and seem to cut the compositional layering at a low angle; these domains of "leucosome" generally have a mafic selvedge, but there are some contacts without them. The wider bodies of leucosome may have crystallized from a melt that was injected, or alternatively, they may result from *in situ* partial melting fluxed by the ingress of H_2O along fracture planes. Together, the thin and thicker bodies of leucosome give the migmatite a layered or stromatic morphology, and both were tightly folded in the presence of partial melt. A grey leucocratic vein (C) with a different microstructure crosses the outcrop just below the pen. This vein has thin mafic selvedges where it crosses the metagreywacke, but tends to have diffuse contacts with the thicker, folded bodies of leucosome. The vein is oriented subparallel to the axial planes of the folds and appears to truncate the folded layers of leucosome; it represents an injection of another felsic melt along a sheared limb. However, the local presence of diffuse margins (e.g., around the splay in the vein) and contacts with the wider domain of leucosome indicate that it might have been injected while the host still contained some interstitial melt. Note that bends in the leucocratic vein coincide with adjacent, large domains of leucosome in the host. The curvature may, therefore, be related to late compression across the axial plane of the fold, at a stage when the two large domains of leucosome contained less than 25% melt, had passed the particle-locking threshold, and were the more competent layers. The morphology of this fold-structured migmatite is noticeably different from that in Figs. E1, E2, and E4; the bodies of leucosome generally have diffuse contacts, and there are no dilatant structures into which melt has migrated. There may well have been a lower contrast in viscosity between the metagreywacke layers and the leucosome in this migmatite.

Location: Abaukoma, Yaoundé, Cameroon. *Rock type*: migmatite transitional from metatexite to diatexite with fold structures; metagreywacke protolith, anatexis at T 800–850°C, P 10–12 kbar. *Scale*: the pen is 15 cm long. *Image*: Pierre Barbey.

Further reading: Barbey, P., Macaudière, J. & Nzenti, J.P. (1990): High-pressure dehydration melting of metapelites: evidence from migmatites of Yaoundé (Cameroon). *Journal of Petrology* **31**, 401–427.

Nzenti, J.P., Barbey, P., Macaudière, J. & Soba, J. (1988): Origin and evolution of the late Precambrian high-grade Yaoundé gneisses. *Precambrian Research* **38**, 91–109.

Figure E9

Fig. E9. The ruler rests on a slightly rusty-colored magnetite–K-feldspar–biotite–cordierite–quartz–plagioclase metapelitic layer that contains a few domains of K-feldspar–plagioclase–quartz leucosome. The cordierite in the metapelite contains rounded inclusions of sillimanite, quartz and biotite that are not in mutual contact. Therefore, the rusty layers are interpreted as the residuum from the partial melting of a pelite by a reaction such as biotite + sillimanite + quartz + plagioclase = melt + cordierite + K-feldspar, and subsequent segregation of most of the anatectic melt fraction. These assemblages provide the evidence that this outcrop is indeed a migmatite. The more numerous grey layers are biotite–quartz–plagioclase metapsammite, and because they show no evidence of having partially melted, they are considered paleosome. The outcrop contains prominent discordant veins of leucogranite that have a narrow, biotite-rich mafic selvedge. Locally, the mafic selvedge crosses the layering (bedding) and is developed at contacts with the pelitic and the psammitic layers. Hence, the selvedge is considered to be the result of a reaction between the granitic melt and the host rock. The leucogranitic veins are peraluminous, and their whole-rock major- and trace-element composition is compatible with derivation from an anatectic melt from a pelitic protolith, one similar to that present in the outcrop. However, the veins are depleted in the heavy rare-earth elements, which suggests that they formed from a melt that had garnet as a residual phase. Moreover, the major- and trace-element contents suggest they were derived from a higher degree of partial melting than occurred in the pelitic layers present in this outcrop. As cordierite, and not garnet, is residual in the pelitic layers of this migmatite, the melt from which the discordant leucogranitic veins were derived was generated deeper in the crust, where garnet was stable, and the degree of partial melting higher. Thus, a case can be made that the magma from which the leucocratic veins crystallized belongs to the same anatectic event that formed their migmatite host; hence, this is a vein-structured migmatite and not a veined migmatite.

Location: Quetico Subprovince, Ontario, Canada. *Rock type*: vein-structured metatexite migmatite; metapelitic protolith, anatexis at T 700–800°C, P 3–4 kbar. *Scale*: the ruler is 15 cm long. *Image*: E.W. Sawyer.

Figure E10

Fig. E10. At first glance, this looks like a stromatic migmatite with prominent subparallel veins of leucosome in a grey, medium-grained semipelitic host. Close inspection of the "leucosome" fraction reveals that they cross the foliation and compositional layering in the host by a few degrees and have narrow, biotite-rich mafic selvedges and not melanosome. Some domains (e.g., the lowermost one in the photograph) have a pegmatitic microstructure. All these features are inconsistent with formation of leucosome *in situ*. Examination of the grey host rock shows that most of it is a uniform biotite–quartz–plagioclase psammitic schist that does not appear to be residual in composition and, therefore, not to have melted. There are a few barely visible diffuse, nebulitic quartzofeldspathic patches developed in the outcrop in which the grain size is larger. These features indicate than an incipient stage of partial melting was in fact reached in a few scattered places in the host. The rock thus is properly called a migmatite. However, the stromatic morphology is the result of the injection of granitic melt, some of it quite evolved in composition, into a host that had just barely melted. Thus, the mesocratic host is mostly paleosome, and the "leucosome" fraction is correctly considered as leucocratic veins or dikes.

Location: Wuluma Hills, Arunta Inlier, Australia. *Rock type*: vein-structured metatexite migmatite; psammitic protolith, partially melted at *T* 800–850°C, *P* ca. 5 kbar. *Scale*: the ruler is 15 cm long. *Image*: E.W. Sawyer.

Figure E11

Fig. E11. Veins of peraluminous granite are intruded subparallel to the foliation and compositional layering in a stromatic migmatite formed from a pelitic protolith by the dehydration-induced melting of a muscovite-bearing assemblage. The stromatic migmatite has a plagioclase + quartz + muscovite (trondhjemitic) leucosome and biotite + quartz + sillimanite + garnet melanosome. The granitic veins contain K-feldspar and have compositions consistent with derivation from pelitic schist similar to the protolith of the stromatic migmatite. However, the stromatic leucosome consists of the accumulated plagioclase crystallized from the anatectic melts, whereas the granitic sheets and veins have a slightly more evolved composition, but have still lost evolved fractionated melt. That melt is believed to have risen to higher levels in the crust and to have formed granite plutons.

Location: western Maine, U.S.A. *Rock type*: vein-structured stromatic metatexite migmatite; upper amphibolite facies, T ca. 750°C, P 4 kbar. *Scale*: the lens cap is 6.5 cm across. *Image and caption*: Gary Solar.

Further reading: Solar, G.S. & Brown, M. (2001): Petrogenesis of migmatites in Maine, USA: possible source of peraluminous leucogranite in plutons? *Journal of Petrology* **42**, 789–823.

Figure E12

Fig. E12. The veined migmatite in this outcrop was created when a series of thin, subparallel veins were injected into a metatexite migmatite host. The metatexite migmatite is compositionally layered at a scale of 10–50 cm. The layers were probably bedding, and contain the mineral assemblage biotite + sillimanite + garnet + quartz, which is interpreted to be the residuum after partial melting of a pelitic protolith. The domains of quartzofeldspathic leucosome oriented parallel to the layering are interpreted to represent part of the anatectic melt derived from the pelite; these domains of leucosome have a mafic selvedge. The steeply dipping structure in the metatexite migmatite is cut by a large number of thin, less steeply dipping plagioclase + quartz veins, most of which do not have a mafic selvedge. The local presence of mafic selvedges may indicate that the veins were injected while their host was still hot, and close to the solidus. Some of the leucocratic veins are of uniform thickness, but many consist of trains of lenticular segments. The low-temperature retrogression of the assemblages of residual minerals in the migmatite, caused by an influx of aqueous fluids, can be seen as the greenish blue alteration associated with an array of fractures that traverse the outcrop.

Location: western Maine U.S.A. *Rock type*: vein-structured metatexite migmatite; upper amphibolite facies, T ca. 750°C, P 4 kbar. *Scale*: note the person at the top right. *Image and caption*: Gary Solar.

Further reading: see Fig. E11.

Figure E13

Fig. E13. This diatexite migmatite has a very diffuse compositional banding, and a rather uniform, mesocratic appearance because very little segregation of the melt from its residuum has occurred. Some of the compositional banding is the result of the presence of quartz-rich, subsolidus segregations into veins (A) and their biotite-rich selvedges in the protolith; these lower-grade relics are, therefore, paleosome in this migmatite. An asymmetrical lens-shaped raft of paleosome has a biotite-rich mafic selvedge at its edge closest to the pen, and contains an internal compositional layering due to thin quartz-rich veins. A very leucocratic melt has segregated into the pressure-shadow region to the right of the enclave. The position and shape of this segregation are consistent with the overall dextral sense of shear determined from folding of the main banding and the sigmoidal curvature of the leucocratic bands. Most of the leucocratic bands in this migmatite have no associated melanosome. As some are clearly oblique to the weak modal banding in the mesocratic neosome, they are interpreted to result from the injection of leucocratic veins into the diatexite. Some of the veins (B) have diffuse margins, and thus probably were injected into the diatexite while it still contained anatectic melt. However, other veins have sharp contacts. Melting in these rocks occurred through the breakdown of muscovite: muscovite + plagioclase + quartz + K-feldspar + H_2O = melt, followed by muscovite + plagioclase + quartz + H_2O = melt + sillimanite. As biotite was not involved, changes in biotite concentration act, therefore, as a good tracer to identify the location and the processes involved in melt segregation.

Location: Glenelg River Complex, southeastern Australia. *Rock type*: vein-structured diatexite migmatite; quartzofeldspathic gneiss protolith, anatexis in the upper amphibolite facies, *T* ca. 700°C, *P* ca. 5 kbar. *Scale*: the pen in 14 cm long. *Image and caption*: Tony I.S. Kemp.

Further reading: Kemp, A.I.S. & Gray, C.M. (1999): Geological context of crustal anatexis and granitic magmatism in the northeastern Glenelg River Complex, western Victoria. *Australian Journal of Earth Sciences* **46**, 407–420.

Figure E14

Fig. E14. The schollen in this diatexite migmatite are orthopyroxene–quartz–biotite–plagioclase schists, and represent the residuum left after partial melting of a biotite–quartz–plagioclase psammitic schist protolith. Some partial melt may have remained in the residuum to form the fraction of stromatic leucosome present in the schollen, although some of this material represents injected granitic veins, not generated *in situ*. The leucocratic matrix around the schollen is granodioritic in composition, and has the mineral assemblage orthopyroxene + biotite + K-feldspar + quartz + plagioclase. Tabular crystals (euhedral orthopyroxene and feldspar) in the matrix are aligned and define a magmatic foliation. Furthermore, trains of platy biotite define mafic schlieren in the matrix that are parallel to the magmatic foliation. The schollen diatexite is crossed by a number of thin, generally planar leucocratic veins, or dikes (e.g., upper left and upper right). These veins are leucogranitic in composition (i.e., more K-feldspar and less orthopyroxene than the matrix) and do not have melanosome or, in general, a mafic selvedge associated with them. In most places, the contact between the veins and the granodioritic matrix is gradational, but in the places where the schlieren are truncated, it appears to be sharp. The leucocratic veins and the granodioritic matrix were thus partially molten at the same time, i.e., the leucocratic veins are part of the anatectic event in this migmatite. The whole-rock composition of the leucocratic veins and the matrix indicates that they are petrogenetically related; the veins are the evolved residual melt derived by fractional crystallization of anatectic melt, and the granodioritic matrix is the remaining plagioclase-enriched cumulate.

Location: Ashuanipi Subprovince, Quebec, Canada. *Rock type*: vein-structured schollen diatexite migmatite; biotite–quartz–plagioclase protolith, anatexis at *T* 825–875°C, *P* 6–7 kbar. *Scale*: the ruler is 15 cm long. *Image*: E.W. Sawyer.

F.

MICROSTRUCTURES CHARACTERISTIC OF MIGMATITES

Investigators in recent years have identified certain microstructures that can be used to infer the presence of partial-melting reactions, the former presence of the melt itself, or the crystallization from a melt. In this section, I illustrate some of those microstructures currently used in the study of migmatites. The first few figures establish a continuity of microstructures found in rapidly quenched partial-melting experiments with those found in nature in progressively more slowly cooled natural melts. Then, in the next series of figures, I look at microstructures related specifically to dehydration melting and the trapping of melt.

Results from quenched deformation-melting experiments: a starting point [Figs. F1–F4]

Many of the microstructures preserved in rapidly quenched experiments in which rocks have been deformed while partially molten closely resemble those found in some migmatites. In recent years, experimental techniques have advanced greatly, and it is now possible to deform rocks while they are partially molten in order to ascertain how partially molten rocks deform in nature, and how partial melt is extracted. This group of figures illustrates some of the microstructures produced in four deformation – partial-melting experiments. In all the back-scattered electron (BSE) images, the reaction products (RP) are the brightest grey and very fine-grained, and typically occur in a matrix of melt, which is an intermediate grey. Biotite (Bt) grains also appear light grey, similar to the reaction products, but are very coarse-grained and show the (001) cleavage planes. Plagioclase (Pl) and quartz (Qtz) are much darker and have a similar appearance, although quartz is slightly darker. The black areas in the images are cracks that result from decompression of the sample at the end of the experiment. The foliation in the gneiss is oriented parallel to the bottom of each image.

Subsurface contact-aureoles: the Glenmore plug, Scotland [Figs. F5, F6]

The partial melting that has occurred around near-surface intrusions, or lava flows, has experienced a unique combination of melting at very low pressures and very rapid rates of cooling. The most rapidly cooled of the partially melted rocks are glassy.

Erupted, partially melted xenoliths: El Joyazo, Spain [Figs. F7–F10]

The rapidly quenched microstructures found in partially melted xenoliths and enclaves brought up from the middle crust by volcanic eruptions are perhaps the closest link between crustal anatexis and partial-melting experiments.

Subsurface contact-aureoles: the Rum Igneous Complex, Scotland [Figs. F11, F12]

The melt in rocks from subsurface contact-aureoles that experienced somewhat slower rates of cooling does not quench to glass, but crystallizes to a granophyric microstructure. These rocks may also show evidence that the initial microstructure has been modified by recrystallization.

Shallow contact-aureoles: the Traigh Bhàn na Sgùrra sill on Mull, Scotland [Figs. F13–F16]

The Traigh Bhàn na Sgùrra sill on the island of Mull, Scotland, was a major conduit for magma flow during the Tertiary. Most of the sill experienced instantaneous injection of basaltic magma, which then solidified *in situ*, whereas locally, magma continued to flow for up to 5 months. The mafic rocks in the regions of long-lived flow have no chilled margins, and are surrounded by a wide contact-aureole in which significant degrees of partial melting occurred in the psammitic country-rocks. Melting was generally H_2O-absent and due to the breakdown of muscovite or due to dry melting in the system quartz–feldspar. The volume increase associated with muscovite-dehydration melting resulted in the formation of abundant microcracks filled with melt. The pressure of metamorphism was only about 0.6 kbar. The four images are of partially melted psammites from this contact aureole.

Shallow- to medium-depth contact-aureoles: the Duluth Igneous Complex [Figs. F17–F28]

Anatexis in the aureole of the Duluth Igneous Complex occurred at a depth equivalent of 1.5–2 kbar. The cooling rate was sufficiently slow that the melt did not quench to glass, but was able to crystallize completely in most cases, albeit to a fine grain-size. However, there is granophyre locally. The first five images are from residual rocks in the contact aureole that show microstructures formed during partial melting. The next five images are from melt-rich parts of rocks (i.e., diatexite migmatites and leucosome) in the aureole, and show microstructures related to the crystallization of the anatectic melt.

Deeper contact-aureoles: the Ballachulish Igneous Complex, Scotland [Figs. F29–F32]

The rate of cooling recorded in partially melted rocks from contact aureoles decreases steadily with depth in the Earth's crust. Contact aureoles such as that around the Ballachulish Igneous Complex, in Scotland, contain microstructures that have been significantly modified by recrystallization during the slower cooling. Consequently, these migmatites provide an important intermediate step in understanding the sequence of microstructures that develop in partially melted rocks, from those produced in quenched experiments and rapidly cooled shallow intrusive bodies on the one hand, to those found in very slowly cooled regional migmatite terranes on the other. The first two images are from partially melted rocks of pelitic and semipelitic bulk compositions, and in these, dehydration melting was important. The following images are of a partially melted psammite, the Appin Quartzite, from the high-temperature inner part of the aureole, and partial melting in this rock was driven by the fracture-controlled infiltration of H_2O.

Regional migmatite terranes: the Ashuanipi Subprovince [Figs. F33–F36]

The Ashuanipi Subprovince of northern Quebec contains 90 000 km^2 of anatectic rocks, a large proportion of which are siliciclastic metasedimentary units that have melt-depleted bulk compositions. Microstructures that indicate the former presence of anatectic melt are especially common, although not abundant, in the psammitic (metagreywacke) rocks. However, microstructures in migmatites from regional anatectic terranes show a much greater degree of modification by recrystallization and the slow growth of minerals than the microstructures in migmatites from contact aureoles.

Regional migmatite terranes: the Opatica Subprovince [Figs. F37, F38]

Partial melting in the Opatica Subprovince occurred in a typical protolith of Archean tonalite–trondhjemite–granite (TTG) suite that had been strongly foliated well before anatexis. Most of the rocks of the TTG suite are leucocratic with very low biotite or biotite + hornblende contents. The microstructures all show considerable textural modification due to slow cooling. Nevertheless, the similarities with those found in contact aureoles are evident.

Microstructures in residual rocks [Figs. F39–F46]

Most residual rocks are enriched in ferromagnesian minerals that are the solid products of the incongruent melting of biotite, or hornblende, and hence they are melanocratic. However, residual rocks can also be enriched in the minerals that were present in excess with respect to the melting reaction, such as quartz or plagioclase, and this can result in residual rocks that are not noticeably melanocratic. The most familiar form of residual, i.e., melt-depleted rock is the melanosome adjacent to a leucosome. However, residual rocks can also occur as layers, bands, patches or as quite extensive regions in migmatites, and in some cases, without domains of leucosome nearby. Residual rocks are volumetrically the largest component in many migmatite terranes, the anatectic melt fraction having migrated to higher levels in the crust, and there formed granite plutons.

Crystallization-induced microstructures in the melt-derived parts of migmatites: leucosome and leucocratic veins [Figs. F47–F58]

Domains of leucosome are formed in the structures in which melt collected, or flowed through, in migmatites. However, there may have been a substantial fraction of residual material (refractory minerals plus the solid products of incongruent melting, i.e., peritectic phases) entrained with it. The figures in this section illustrate some of the microstructures found in leucosome, and show, not surprisingly, that there is considerable similarity with microstructures found in plutonic igneous rocks, most particularly cumulus microstructures.

Crystallization-induced microstructures in the melt-rich parts of migmatites: diatexite migmatites [Figs. F59–F65]

Diatexite migmatites at one stage contained a substantial amount of melt, although they likely also had entrained solid material that could consist of refractory minerals, the solid products of incongruent melting, or the early products of crystallization. The fraction of melt was commonly dominant (i.e., >50 vol.%), but not in every case, as flow can start at quite low fractions of melt, depending on the flow mechanism involved. The figures in this section illustrate some of the microstructures characteristic of diatexite migmatites; readers will note some similarity with microstructures in plutonic igneous rocks.

Microstructures formed by flow in diatexite migmatites [Figs. F66–F69]

A characteristic feature of diatexite is the presence of structures that indicate flow of magma, such as flow-banding, schlieren, oriented elongate enclaves, and mineral alignments. Because of the size of these features and the coarse grain-size of many regionally developed diatexite migmatites, it is commonly difficult to appreciate the structures resulting from magmatic flow in thin sections. Hence, this section starts with two examples of fine-grained diatexite migmatites from the contact aureole at Duluth.

Inclusions of melt quenched to glass in minerals [Figs. F70–F73]

Surface irregularities on minerals either growing from or growing into a melt can lead to the entrapment of small pockets of melt in the crystal as it grows. If the crystal and its melt-filled inclusion are then rapidly cooled, the melt will quench to glass. With slower cooling, the melt in inclusions has time to interact with its host and to crystallize. Melt inclusions are important in the study of migmatites, because analyzing them is one way to directly determine the composition of the anatectic melt. Glass inclusions found in the minerals of partially melted enclaves brought to the surface in volcanic eruptions are, perhaps, the best way of sampling anatectic melts produced at middle and deep levels of the crust.

Cordierite–, garnet–, and orthopyroxene–quartz intergrowth microstructures [Figs. F74–F77]

The high-grade parts of many anatectic terranes consist of migmatites derived from pelitic or semipelitic protoliths that contain domains of neosome with a distinctive intergrowth microstructure. The neosome is, overall, lighter colored and coarser grained than its host, and has two parts. One consists of large crystals of a ferromagnesian mineral intergrown with quartz, whereas the other part is essentially quartzofeldspathic and leucocratic. Such neosome can occur in different morphologies, but typically, it is vein- or patch-like in shape, and a diagnostic feature is that it does not have a melanocratic border, or rim. The ferromagnesian mineral is most commonly garnet, but exactly analogous microstructures with cordierite or orthopyroxene also occur. The mineral assemblage in the host corresponds to the reactant assemblage, and that in the neosome, to the product assemblage of the melt-producing reaction. This type of neosome and its characteristic intergrowths of minerals appear to be diagnostic of biotite-dehydration-melting reactions. Mineralogical and whole-rock compositions of the neosome indicate that most have lost melt, and typically a substantial fraction of the melt. Analogous microstructures might be expected during dehydration melting of hornblende in a mafic protolith.

Biotite–quartz and biotite–plagioclase symplectitic intergrowth microstructures [Figs. F78–F81]

Symplectitic intergrowth microstructures involving biotite and quartz or plagioclase can be found in many migmatites, although they are rarely abundant. Typically, they partially replace some of the ferromagnesian residual minerals such as garnet, cordierite, and orthopyroxene. The replacement also involves a K-bearing phase, commonly believed to be the last-remaining anatectic melt in the migmatite, although some studies have suggested that it is K-feldspar. Hence, biotite–quartz and biotite–plagioclase symplectites are generally regarded as late anatectic microstructures, formed while a small amount of melt still remained in the migmatite.

Biotite–sillimanite and biotite aggregates replacing garnet or cordierite [Fig. F82]

Coarse-grained aggregates of biotite, or of biotite + sillimanite, are common in some migmatites developed from metapelitic (i.e., aluminous) protoliths. These aggregates are generally regarded as a late-stage alteration of garnet or cordierite that formed after much of the anatectic melt had already migrated away, and as the last remaining melt crystallized.

Plagioclase [Fig. F83]

The plagioclase crystals reported from most migmatites derived from crustal rocks show very little compositional variation within crystals, and the optical zoning that they contain is generally rather subtle. However, it is becoming evident that isolated crystals of plagioclase with a complexly zoned calcic core are a feature of anatectic terranes in which there has been an input of mantle-derived magma. Therefore, the zoning patterns in the plagioclase crystals found in migmatites should be noted.

Contact between the leucosome and melanosome in metatexite migmatites [Figs. F84–F96]

This group of photomicrographs shows some details of the contact and spatial relationships found between the leucosome and melanosome in neosome from metatexite migmatites.

Microstructure of schlieren in diatexite migmatites [Figs. F97–F100]

Schlieren are thin, planar aggregates of platy, tabular, or acicular minerals in diatexite migmatites and some granites. Biotite is the mineral most commonly associated with schlieren, but examples have been reported that are dominantly amphibole, orthopyroxene, sillimanite, or plagioclase. Many schlieren are also enriched in the accessory phases, commonly because they occur as inclusions in the major phases, most notably biotite, but in some schlieren, most of the grains of the accessory phases occur outside the major minerals. The microstructure of schlieren have not been widely studied or well understood, but there appears to be considerable variation. Two principal origins have been proposed to explain the development of schlieren in migmatites: (1) accumulations of biotite resulting from the impingement of grains rotating in a magma and (2) attenuated fragments of residual material derived from disaggregating schollen.

Microstructure of biotite-rich selvedges in migmatites [Figs. F101–F104]

Many veins and dikes of felsic composition have a narrow selvedge of mafic minerals separating them from the host rock. The mafic selvedges that are developed on leucocratic veins injected into the paleosome lithologies in migmatites are probably the easiest to notice in the field, and all the examples shown are of this type. In some cases, a few of the quartz and feldspar crystals located at the edge of the leucocratic veins have undergone an increment of late growth out into the mafic selvedge, to produce somewhat bulbous projections. The microstructures present in mafic selvedges, like those in schlieren, appear to be quite diverse, but they are, in general, pooly known.

Figure F1

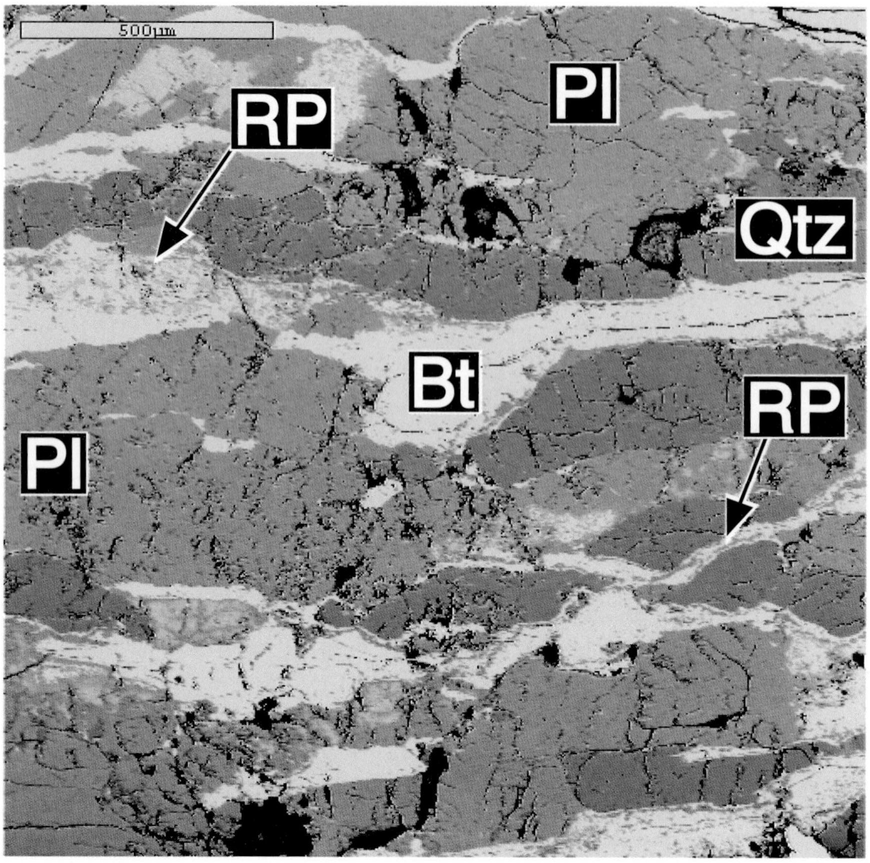

Fig. F1. This experiment involving a low fraction of melt was conducted at 900°C and contained less than 1% partial melt. The mica flakes between the two grains of plagioclase shown have partially reacted, and their grain size has been reduced. The adjacent quartz and plagioclase grains appear to have flowed along with the fine-grained products of reaction that formed at the edge of the biotite grains. The reaction products (RP) are too small to distinguish individually, but are generally light-colored and located close to biotite.

Location: Deformation and melting experiment BPQ–D38. *Rock type*: biotite–quartz–plagioclase gneiss, deformed at T = 900°C, P 10 kbar, strain rate $1 \times 10^{-5}\ s^{-1}$, run duration 36 h. *Scale*: bar is 500 µm long. BSE image. *Image and caption*: Caleb Holyoke III. This image was published as fig. 8a in Holyoke & Rushmer (2002), and is reproduced with the permission of Blackwell Publishing.

Further reading: Holyoke, C.W., III & Rushmer, T. (2002): An experimental study of grain scale melt segregation mechanisms in two common crustal rock types. *Journal of Metamorphic Geology* **20**, 493–512.

Figure F2

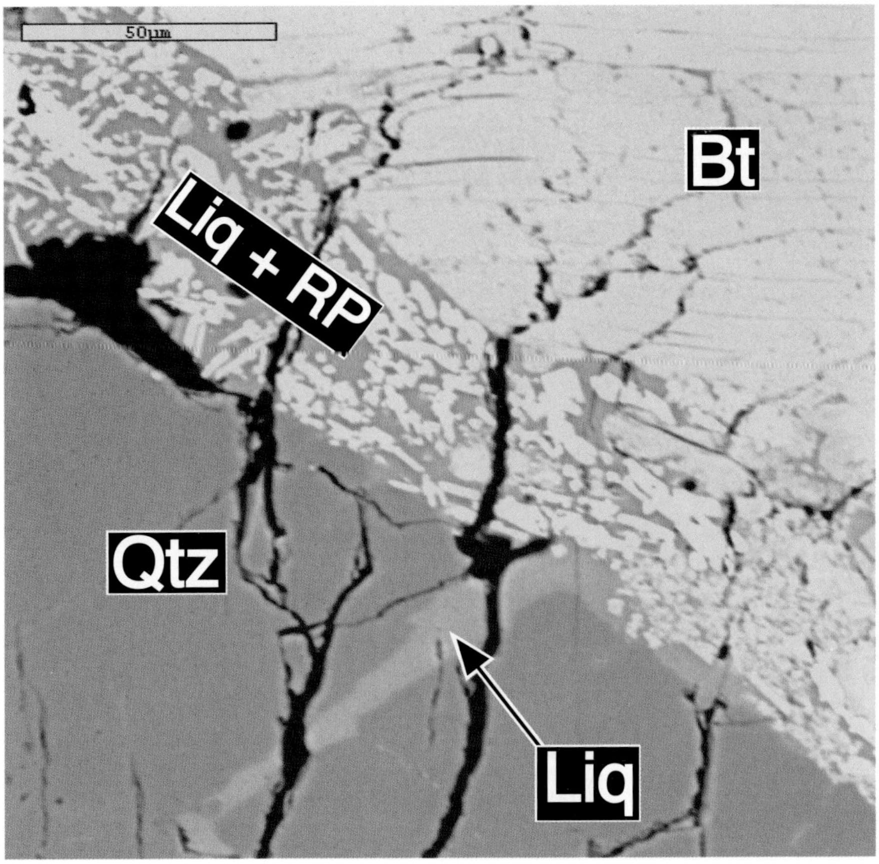

Fig. F2. This experiment also contained a low fraction of melt; it was performed at 920°C and contained about 1% melt. This BSE image is a close-up of the boundary between biotite (Bt) and quartz (Qtz). Solid reaction products (RP) and glass (Liq), which is quenched partial melt, occur along the boundary between the quartz and biotite grains in the gneiss. Notice the strong alignment of the solid products of reaction (which shows as light grey) parallel to the biotite–quartz boundary. This relationship suggests that the zone of melt plus reaction products that developed on the grain boundary was weak and that shear strain was concentrated there and rotated the grains in the melt. In some places, the glass (Liq) occurs within the quartz crystal, and indicates that the anatectic melt entered pre-existing fractures in the quartz, which was a strong mineral in this experiment.

Location: Deformation and melting experiment BPQ–D40. *Rock type*: biotite–quartz–plagioclase gneiss, deformed at T = 920°C, P 10 kbar, strain rate $1 \times 10^{-5}\ s^{-1}$, run duration 6 h. *Scale*: bar is 50 µm long. BSE image. *Image and caption*: Caleb Holyoke III. This image was published as fig. 8b in Holyoke & Rushmer (2002), and is reproduced with the permission of Blackwell Publishing.

Further reading: see Fig. F1.

Figure F3

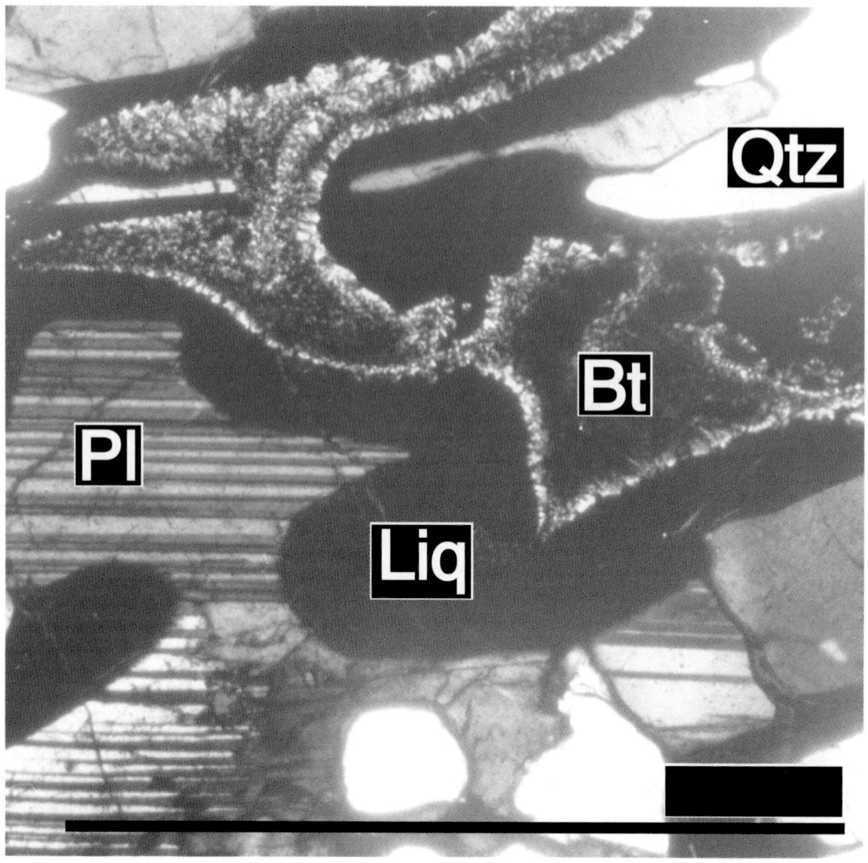

Fig. F3. In this experiment the fraction of melt was high; partial melting at 950°C produced 20 vol.% melt. At this content of melt, the sample failed by cataclasis, or by melt-assisted granular flow, and most of the strain is partitioned into the zones of disaggregation formed during sample failure. The partial melt in the sample migrated to these disaggregated zones as it moved to the edges of the sample, where it accumulated between the sample and its jacket. Thus, deformation in the sample is heterogeneous, and some regions are left relatively undeformed. This photomicrograph shows the microstructure that develops in the relatively unstrained regions; note the rounded and corroded edges of the reactant quartz (Qtz), plagioclase (Pl) and biotite (Bt) grains. Pools of glass (Liq), which is the quenched partial melt, occur around much of the biotite, and along many grain boundaries between quartz and plagioclase. The light grey rims around the biotite, and elsewhere in the sample, are small accumulations of solid products of reaction.

Location: Deformation and melting experiment BPQ–D45. *Rock type*: biotite–quartz–plagioclase gneiss, deformed at T = 920°C, P 10 kbar, strain rate $1 \times 10^{-5}\ s^{-1}$, run duration 36 h. *Scale*: bar is 1 mm long. Optical microscope, crossed polarizers. *Image and caption*: Caleb Holyoke III. This image was published as fig. 8d in Holyoke & Rushmer (2002), and is reproduced with the permission of Blackwell Publishing.

Further reading: see Fig. F1.

Figure F4

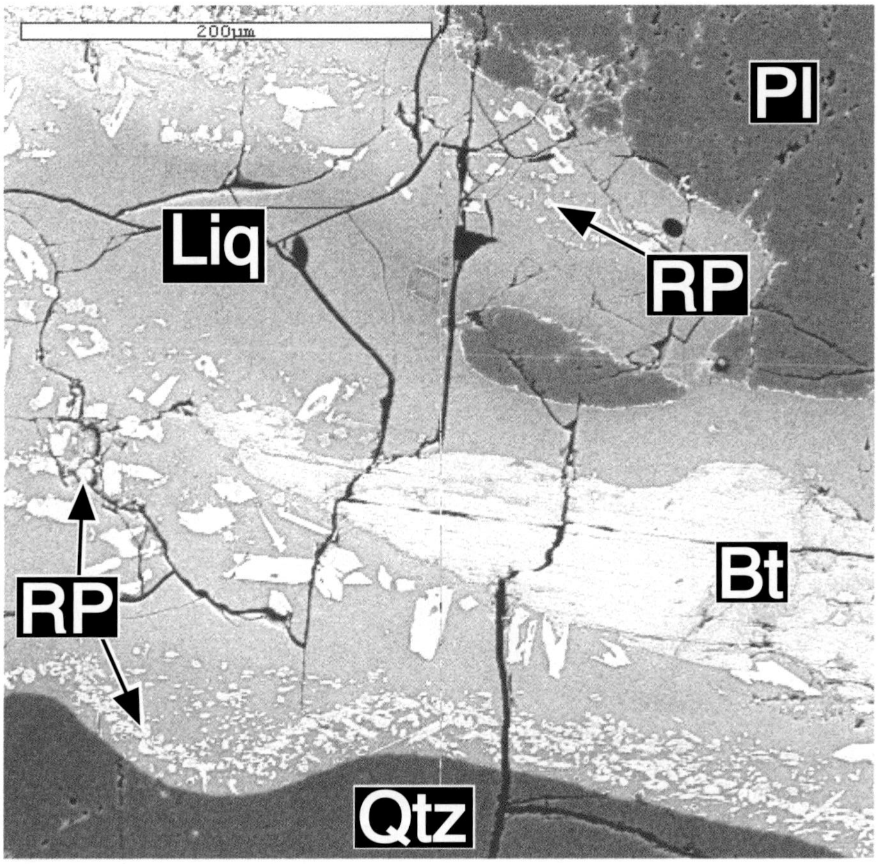

Fig. F4. Results of an experiment that contained a high fraction of melt (35%) run at 950°C. The image shows a large crystal of biotite (Bt) surrounded by glass (Liq) that contains many small crystals of various reaction-products (RP). The biotite crystal was originally located between large grains of quartz (Qtz) and plagioclase (Pl). The biotite, quartz, and plagioclase all have a rounded outline, indicating that they have partially reacted. The incongruent breakdown of biotite produced melt (now quenched to glass) and the numerous small crystals of orthopyroxene and spinel that are in the glass. Crystals of the reaction products are largest in the center of the glass domains, close to the corroded biotite, and they do not have a strong preferred orientation. However, smaller crystals of the reaction products occur abundantly in a band located close to, and parallel with, the edge of the quartz grain. A large proportion of these small crystals are aligned, and this is due to flow of the crystal-bearing melt as the sample was shortened. Therefore, this band of crystals is analogous to the schlieren seen in some migmatites. Some of the reaction products in the glass at the left-hand edge of the image appear as skeletal crystals; such crystals are a feature also seen in the reaction products in some anatectic rocks from low-pressure contact-aureoles, e.g., the Duluth Igneous Complex.

Location: Deformation and melting experiment BPQ–D32. *Rock type*: biotite–quartz–plagioclase gneiss, deformed at T = 950°C, P 10 kbar, strain rate $1 \times 10^{-5}\ s^{-1}$, run duration 36 h. *Scale*: bar is 200 μm long. Optical microscope, crossed polarizers. *Image and caption*: Caleb Holyoke III. This image was published as fig. 9d in Holyoke & Rushmer (2002), and is reproduced with the permission of Blackwell Publishing.

Further reading: see Fig. F1.

Figure F5

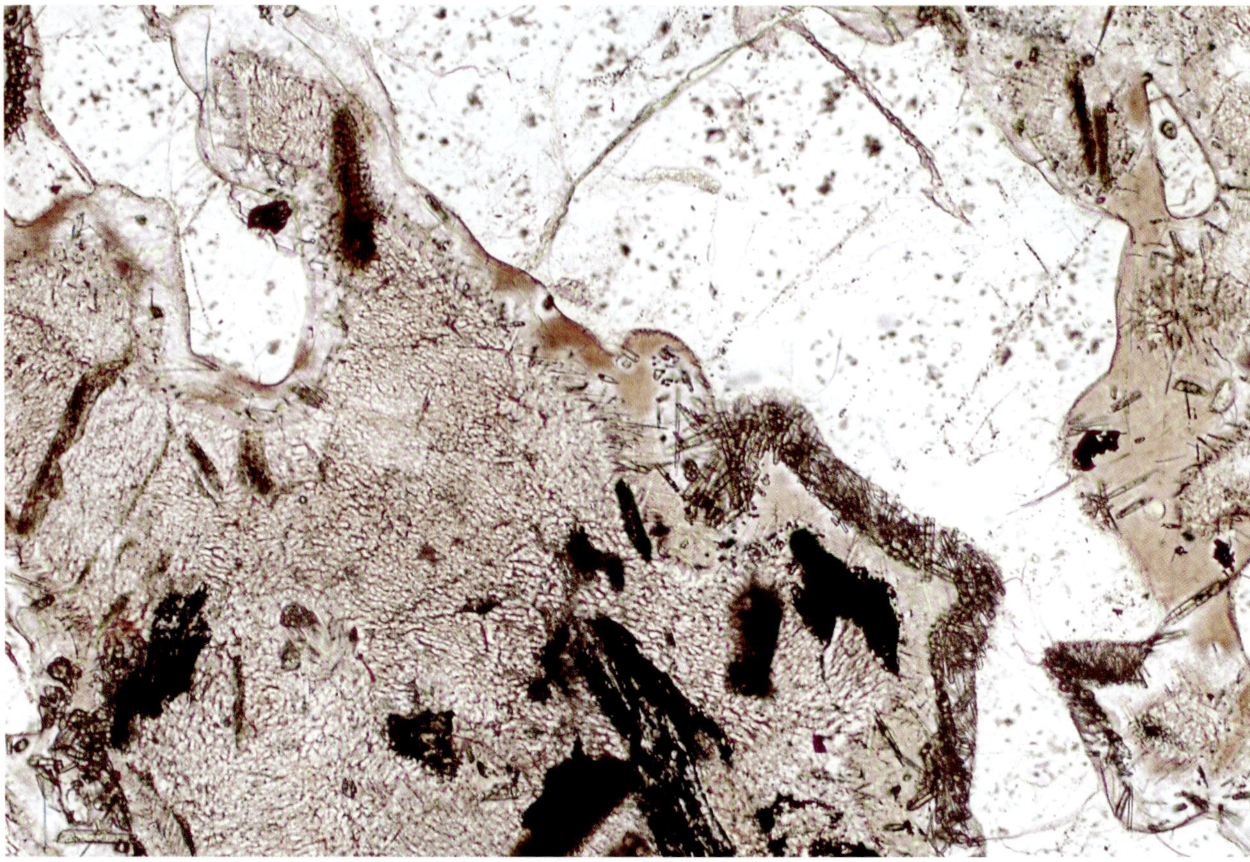

Fig. F5. Glassy rocks extend almost 1 m from the contact of a dolerite plug 30 m in diameter at Glenmore, Ardnamurchan intrusive complex, Scotland. That extent of melted rocks around such a tiny intrusive body indicates that the plug represents a long-lived conduit of magma. This first photomicrograph is of partially melted arkose that underwent a relatively low degree of partial melting in the outer part of the contact aureole. Relict quartz is clear and unclouded, with a corroded outline. The relict feldspar is now sieve-textured and turbid, or cloudy-looking; the dark elongate patches are former grains of biotite. Brown glass forms a rim around the relict quartz and occupies the site of the original grain-boundary between the quartz and adjacent feldspar. The width of the glass rim indicates the extent to which quartz and feldspar have been consumed in generating melt. The small elongate needles that have nucleated on the quartz, which locally form a fringe, are orthopyroxene.

Location: Glenmore dolerite plug, Ardnamurchan, Scotland. *Rock type*: partially melted arkose. *Scale*: the field of view is 0.9 mm. Plane-polarized light. *Image and caption*: Marian Holness.

Further reading: Butler, B.C.M. (1961): Metamorphism and metasomatism of the Moine Series by a dolerite plug at Glenmore, Ardnamurchan. *Mineralogical Magazine* **32**, 866–897.

Holness, M.B., Dane, K., Sides, R., Richardson, C. & Caddick, M. (2005): Melting and melt segregation in the aureole of the Glenmore Plug, Ardnamurchan. *Journal of Metamorphic Geology* **25**, 29–44.

Figure F6

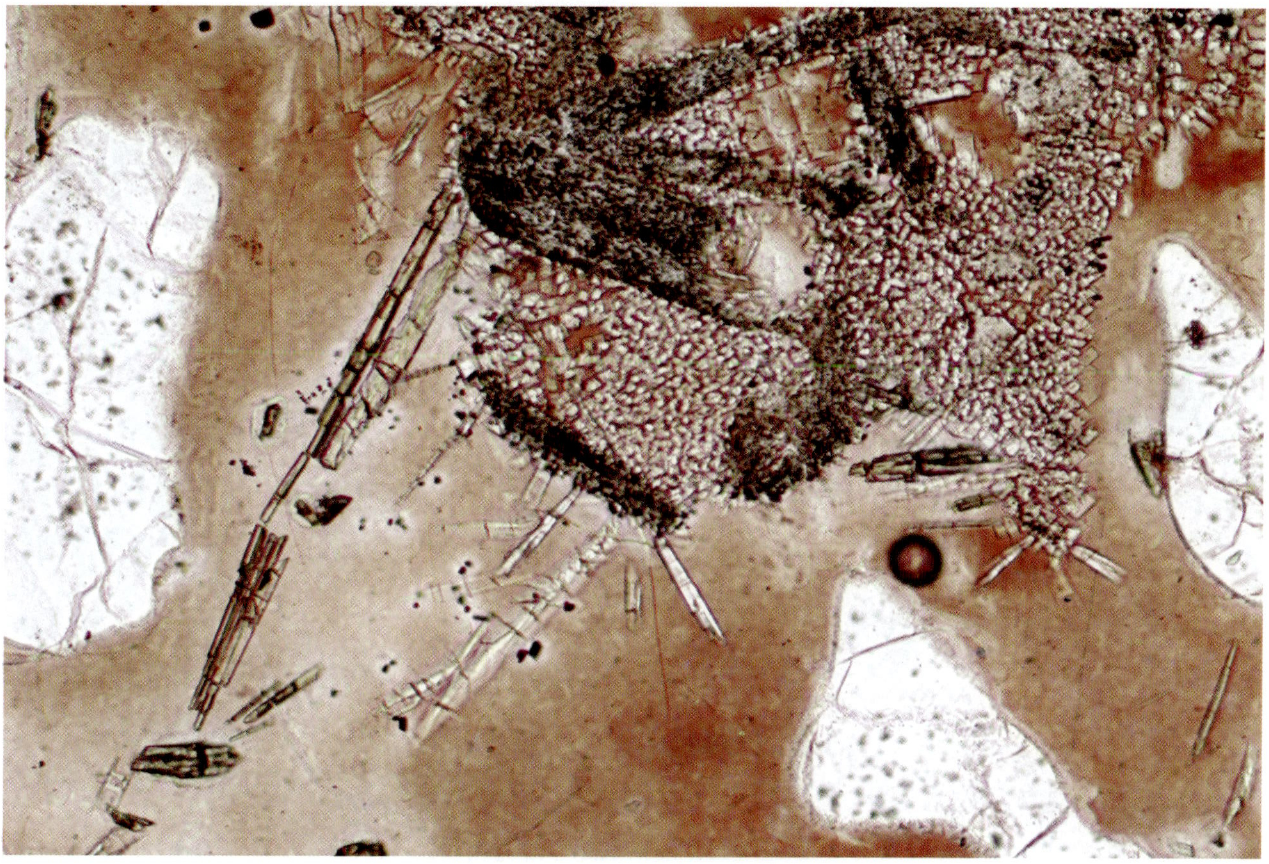

Fig. F6. The photomicrograph shows a partially melted arkose from Glenmore in which there was a high degree of partial melting, and the rock contained a larger fraction of melt. The fused arkose in this sample consists largely of brown glass. The glass contains corroded relics of clear quartz, sieve-textured feldspar (i.e., the areas of fine-grained crystals with brown glass), and biotite (the dark elongate patches near the center of the photograph). Elongate, needlelike crystals in the glass are orthopyroxene, and a product of the partial-melting reaction.

Location: Glenmore dolerite plug, Ardnamurchan, Scotland. *Rock type*: partially melted arkose. *Scale*: the field of view is 0.9 mm. Plane-polarized light. *Image and caption*: Marian Holness.

Figure F7

Fig. F7. Close-up view of a partially melted residual enclave erupted in a dacite lava. The enclave is full of bluish cordierite, black biotite, white sillimanite, and euhedral crystals of red garnet. In thin section, all these minerals are intergrown with glass and contain glass inclusions. The glass, of course, represents quenched anatectic melt. The detailed images in the three following figures are from enclaves such as this.

Location: El Joyazo, southeastern Spain. *Rock type*: residual pelite enclave in dacite volcanic rock; partial melting at T 850 ± 50°C, P >5 kbar, but cooled rapidly by eruption. *Scale*: the field of view is 4.0 cm. *Image and caption*: Bernardo Cesare.

Further reading: Cesare, B. (2000): Incongruent melting of biotite to spinel in a quartz-free restite at El Joyazo (SE Spain): textures and reaction characterization. *Contributions to Mineralogy and Petrology* **139**, 273–284.

Cesare, B. & Maineri, C. (1999): Fluid-present anatexis of metapelites at El Joyazo (SE Spain): constraints from Raman spectroscopy of graphite. *Contributions to Mineralogy and Petrology* **135**, 41–52.

Cesare, B., Salvioli Mariani, E. & Venturelli, G. (1997): Crustal anatexis and melt extraction in restitic xenoliths at El Joyazo (SE Spain). *Mineralogical Magazine* **61**, 15–27.

Figure **F8**

Fig. F8. Microstructures involving biotite from a domain that corresponds to partial melting in a quartz-present and Al_2SiO_5-free system. The biotite crystals show a thin reaction-rim consisting of orthopyroxene, ilmenite, and glass. The reaction-rim is discontinuous, and appears to be developed mainly at contacts between biotite and quartz (top and right). In contrast, the boundaries between biotite and cordierite (bottom left) seem to be in chemical equilibrium. The inferred melting reaction, biotite + quartz = orthopyroxene + ilmenite + melt, could not be balanced with the chemical compositions determined from the minerals that comprise this microstructure.

Location: El Joyazo, southeastern Spain. *Rock type*: biotite from an enclave of residual pelite in a dacite, partial melting at T 850 ± 50°C, P >5 kbar, but cooled rapidly by eruption. *Scale*: the field of view is 0.65 mm wide. Plane-polarized light. *Image and caption*: Bernardo Cesare.

Further reading: as for Fig. F7.

Figure F9

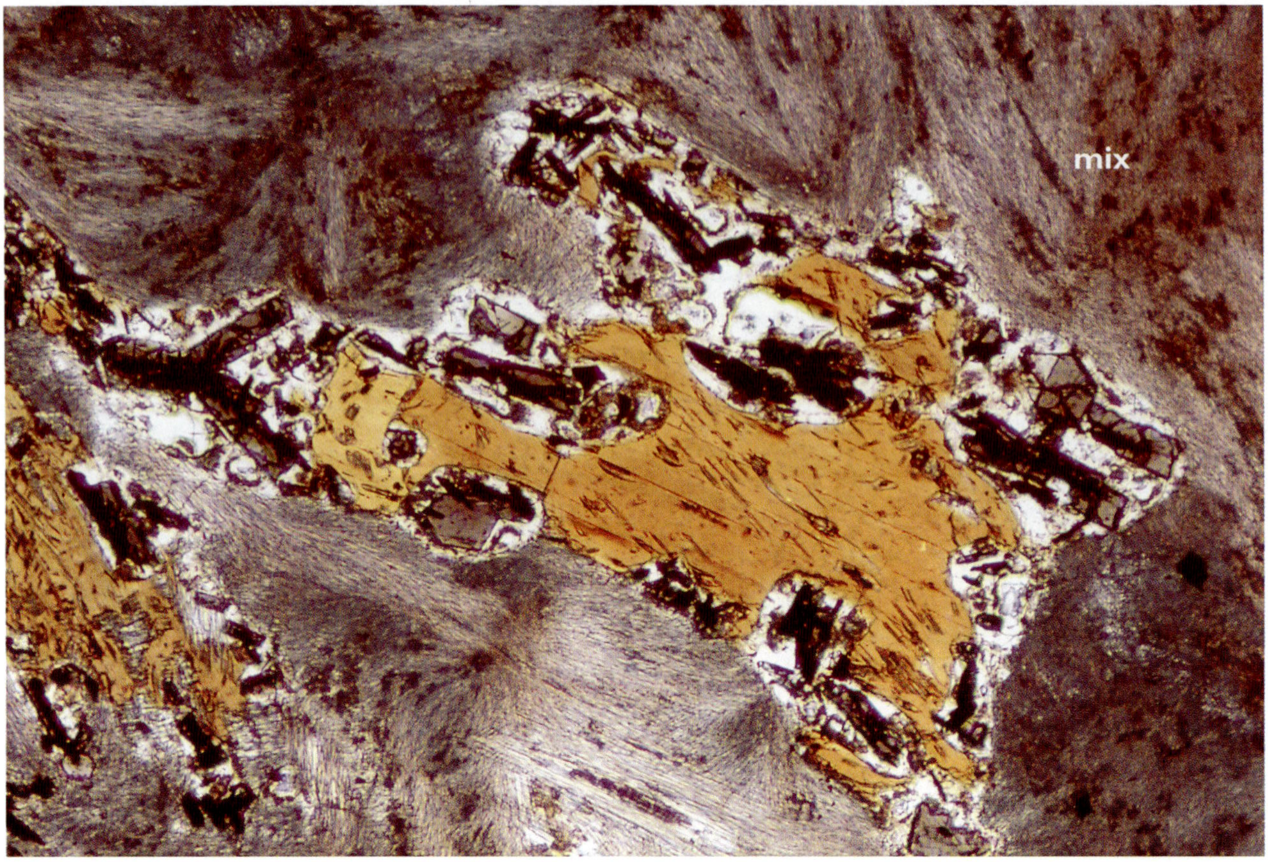

Fig. F9. Microstructures involving biotite from a domain that corresponds to partial melting in a quartz-free system. The biotite crystals are surrounded by a grey, fine-grained mixture (mix) of fibrolitic sillimanite and glass (former melt). All the biotite crystals are irregular in shape, and have deep embayments, which indicates that they have been corroded. Furthermore, the crystals display well-developed, clear reaction-rims. Each rim consists of brownish green hercynitic spinel and laths of opaque (i.e., black) ilmenite in a colorless (i.e., white) glass. Notice the euhedral, or subhedral, form of the spinel where it is in contact with the glass, which represents the former melt. On the basis of compositions of the minerals in this microstructure and in its surroundings, Cesare (2000) has balanced the possible biotite-melting reaction as $biotite_1$ + $plagioclase_1$ + mix + garnet = ilmenite + melt + $biotite_2$ + hercynite + $plagioclase_2$.

Location: El Joyazo, southeastern Spain. *Rock type*: biotite from an enclave of residual pelite in dacite, partial melting at T 850 ± 50°C, P >5 kbar, but cooled rapidly by eruption. *Scale*: the field of view is 0.65 mm wide. Plane-polarized light. *Image and caption*: Bernardo Cesare.

Further reading: as for Fig. F7.

Figure F10

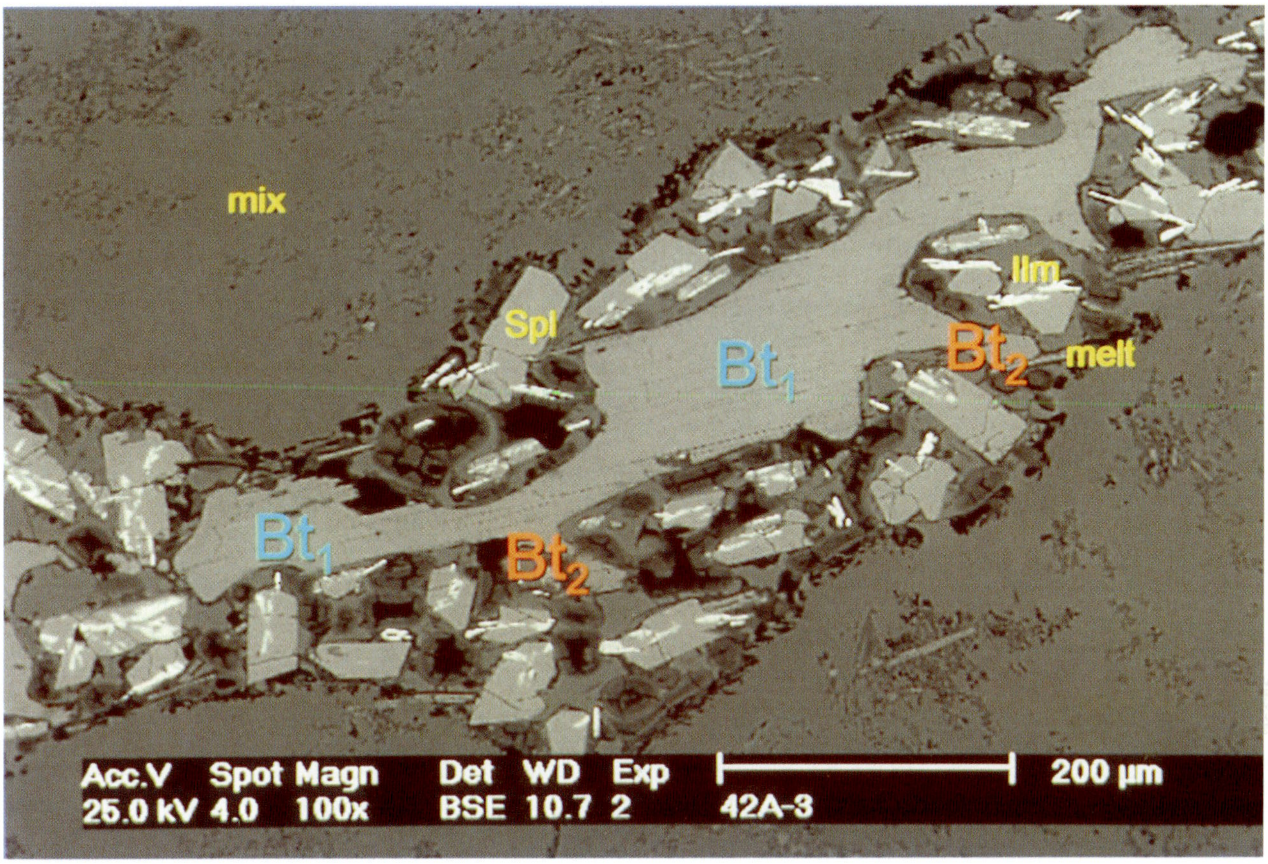

Fig. F10. Back-scattered electron image (BSE) of a corroded grain of biotite from a quartz-free domain in an enclave of residual pelite. The biotite is hosted by an aggregate of finely intergrown, fibrolitic sillimanite and glass (labeled "mix"). The outline of the biotite has deep embayments, which indicates that it has been corroded, and hence was unstable. A well-developed reaction-rim occurs around the biotite and contains crystals of hercynitic spinel (Spl) and laths of ilmenite (Ilm) in a glass (melt); the ilmenite appears white in this BSE image. Crystals of both spinel and ilmenite are euhedral or subhedral, and are interpreted as products grown in the melt formed by the incongruent breakdown of biotite. The composition of the outermost rim of the biotite is richer in both MgO and TiO_2 (Bt_2) than the core (Bt_1) of the crystal. The possible biotite-melting reaction can be balanced as $biotite_1$ + $plagioclase_1$ + mix + garnet = melt + ilmenite + $biotite_2$ + hercynite + $plagioclase_2$.

Location: El Joyazo, southeastern Spain. *Rock type*: biotite from an enclave of residual pelite erupted in dacite lava, partial melting at *T* 850 ± 50°C, *P* >5 kbar, but cooled rapidly by eruption. *Scale*: the field of view is 0.65 mm wide. Plane-polarized light. *Image and caption*: Bernardo Cesare.

Further reading: as for Fig. F7.

Figure F11

Fig. F11. Partially melted Lewisian gneiss containing quartz and plagioclase, with pyroxene aggregates after hornblende (not all are obvious from the photograph). This amphibole-bearing gneiss has been contact-metamorphosed at low pressure by the Rum Igneous Complex. Partial melting occurred locally in the gneiss where it was in close proximity to the mafic intrusive body. The example shown in this photomicrograph contains a thin rim of granophyre developed along the boundary between quartz and plagioclase crystals; the orientation of this slide is such that most of the quartz crystals in the photomicrograph are white. This granophyre is of special interest in that much of the quartz in it forms elongate plates; these are paramorphic after tridymite. Generally, quartz in granophyre from partially melted rocks occurs as rounded and elongate blebs.

Location: Proimh Lochs, Isle of Rum, Scotland. *Rock type*: partially melted Lewisian gneiss. *Scale*: the field of view is 4.25 mm. Cross-polarized light. *Image and caption*: Marian Holness.

Further reading: Holness, M.B. & Isherwood, C.E. (2003): The aureole of the Rum Tertiary Igneous Complex, Inner Hebrides, Scotland. *Journal of the Geological Society of London* **160**, 15–27.

Figure F12

Fig. F12. The center of the photomicrograph is occupied by a large grain of feldspar that is full of inclusions. Nearby are two large, rounded grains of quartz. The boundary between these quartz and feldspar grains is decorated by a "string of beads" microstructure consisting of small grains of feldspar; these are interpreted to have crystallized from a thin rim of anatectic melt that separated the larger feldspar and quartz grains. This microstructure is interpreted as a slowly cooled version of the rather more common granophyric rims that develop on grain boundaries during local *in situ* partial melting followed by more rapid cooling.

Location: Rum Igneous Complex, Scotland. *Rock type*: partially melted Lewisian gneiss, *P* ca. 150 bar. *Scale*: the field of view is 0.9 mm. Cross-polarized light. *Image and caption*: Marian Holness.

Further reading: as for Fig. F11.

Figure F13

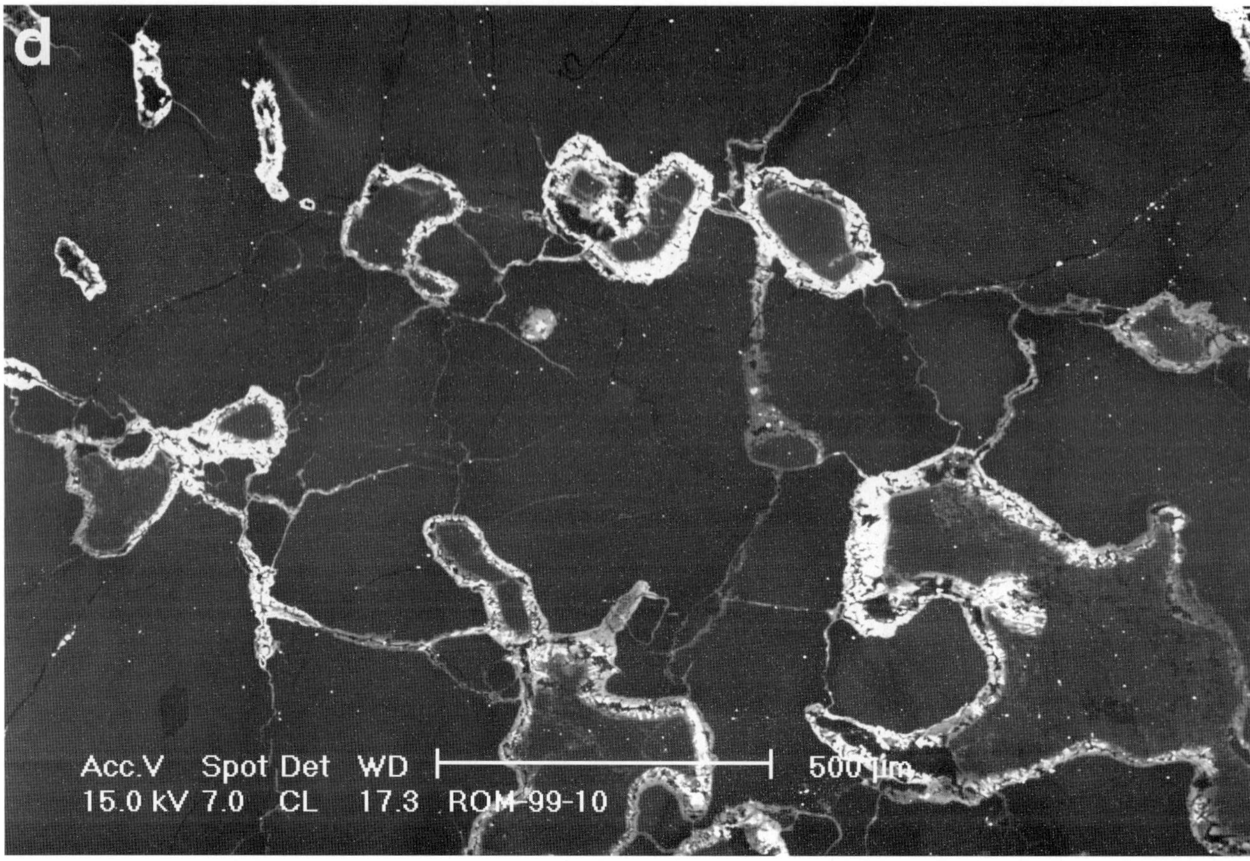

Fig. F13. This sample of psammite did not contain much mica, and the melt present at the peak temperature was derived predominantly from the quartz and feldspar in the rock. In this cathodoluminescence (CL) image, each feldspar grain is surrounded by a brightly luminescent rim of granophyre from which cracks emanate; they are inferred to have been filled with melt, and now form a completely interconnected network. Note that the final generation of cracks (black in this image) cut across the features related to partial melting and anatectic melt.

Location: Isle of Mull, Scotland. *Rock type*: partially melted psammite 44 cm from the contact with the dolerite sill. *Scale*: indicated on the image. *Image and caption*: Marian Holness. This image was published as fig. 13d in Holness & Watt (2002), and is reproduced with the permission of Oxford University Press.

Further reading: Holness, M.B. & Humphreys, M.C.S. (2003): The Traigh Bhàn na Sgùrra sill, Isle of Mull: flow localisation in a major magma conduit. *Journal of Petrology* **44**, 1961–1976.

Holness, M.B. & Watt, G.R. (2002): The aureole of the Traigh Bhàn na Sgùrra sill, Isle of Mull: reaction driven micro-cracking during pyrometamorphism. *Journal of Petrology* **43**, 511–534.

Figure F14

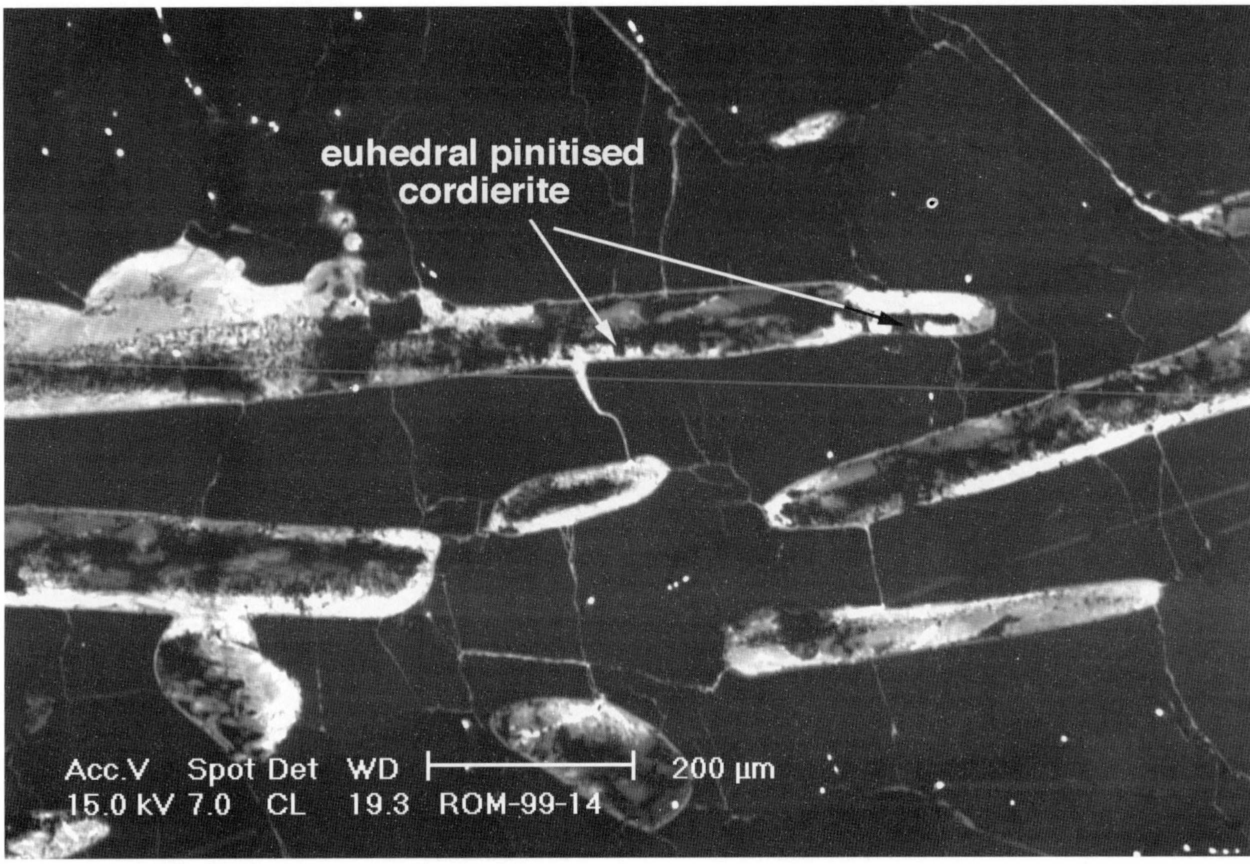

Fig. F14. This cathodoluminescence image of a psammite shows elongate bright regions that are the sites of reacted muscovite grains. Note also the dark, rectangular grains of retrograded, euhedral cordierite within domains representing the former grains of muscovite. The muscovite was set in quartz, which is dark in this image. The image shows an abundance of bright, linear features oriented at a high angle to the foliation, which in the sample is defined by the orientation of the muscovite. These bright linear features are cracks generated by the volume change associated with the breakdown of muscovite. These cracks have been filled with anatectic melt; they are exactly analogous to those formed experimentally by Connolly et al. (1997).

Location: Isle of Mull, Scotland. *Rock type*: partially melted psammite 80 cm from contact with the sill. *Scale*: indicated on the image. *Image and caption*: Marian Holness. This image was published as fig. 9 in Holness & Watt (2002), and is reproduced with the permission of Oxford University Press.

Further reading: Connolly, J.A.D., Holness, M.B., Rubie, D.C. & Rushmer, T. (1997): Reaction-induced micro-cracking: an experimental investigation of a mechanism for enhancing anatectic melt extraction. *Geology* **25**, 591–594.

Figure F15

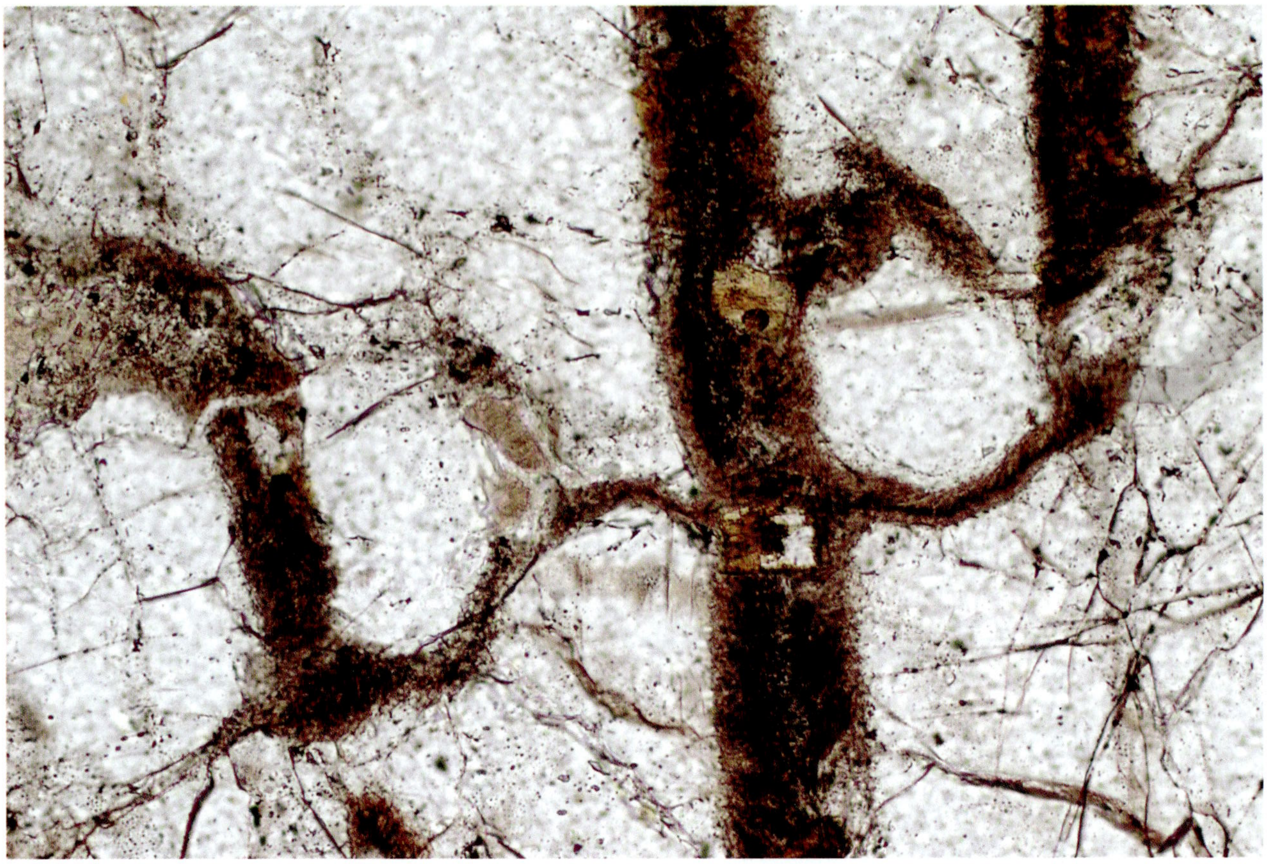

Fig. F15. This photomicrograph is a close-up of reacted grains of muscovite in a psammite. The muscovite is now replaced by mullite, biotite, K-feldspar, spinel, and some prismatic grains of cordierite. The cordierite is now entirely replaced by a yellow product of alteration. The sites of the reacted muscovite (parallel to the short side of the photomicrograph) were linked by melt-filled fractures, which are now filled with brownish granophyre. Most of the fractures in this image contained anatectic melt, but this is only really apparent on back-scattered electron or cathodoluminescence images. The large colorless grains are quartz.

Location: Isle of Mull, Scotland. *Rock type*: partially melted psammite 80 cm from contact with the sill. *Scale*: the field of view is 0.9 mm wide. Plane-polarized light. *Image and caption*: Marian Holness.

Further reading: see Fig. F13.

Figure F16

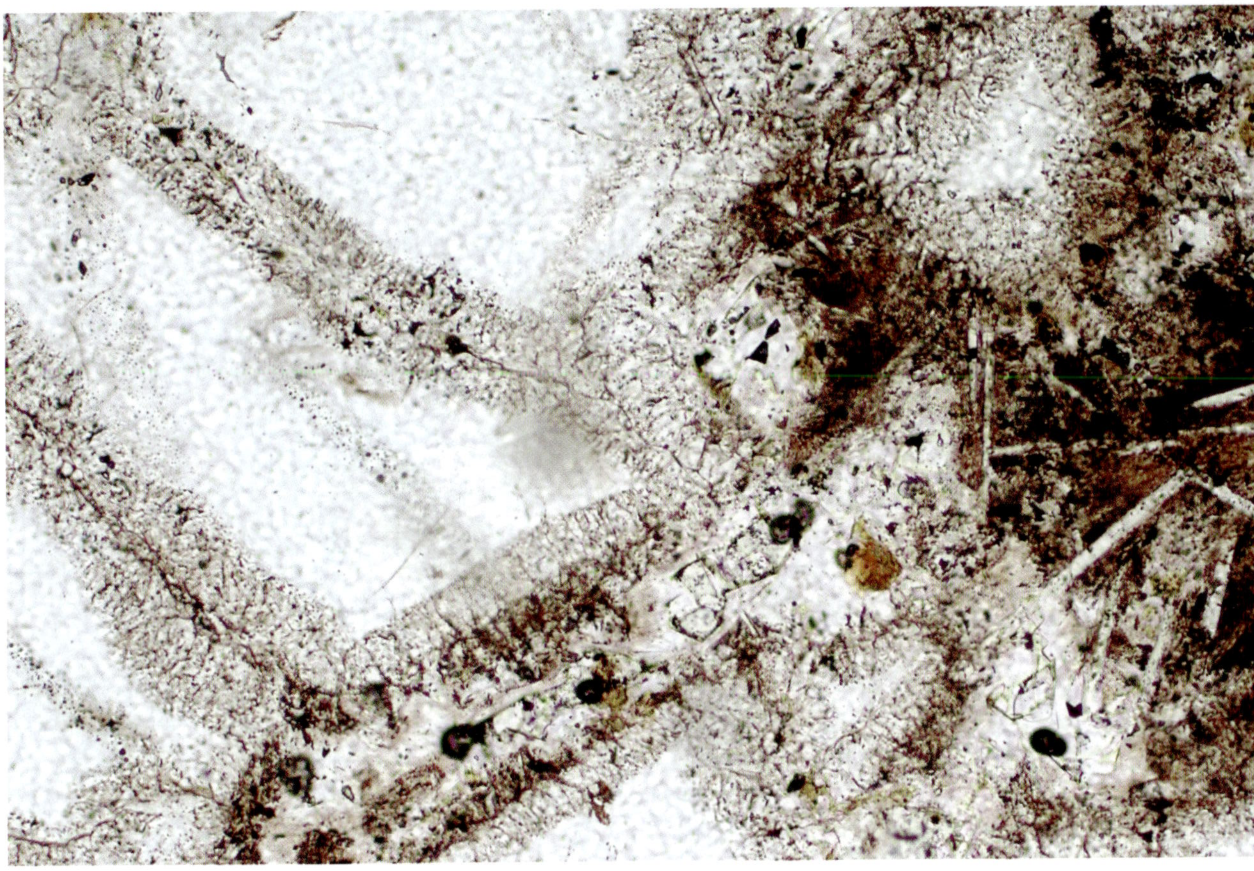

Fig. F16. This partially melted psammite is highly fractured and consists principally of large, colorless grains of rounded quartz (e.g., left side of the photomicrograph). The fractures are quite wide and show the growth of quartz on their walls. Possibly, the melt remained static in the fractures for some time, and the fracture walls moved back (retreated) as the quartz dissolved into the melt. When cooling began and the melt started to crystallize, the quartz that grew out of the melt formed euhedral overgrowths of quartz on the walls, trapping melt in small inclusions. The dark, ovoid domains on the right are the sites of reacted grains of muscovite; note the colorless plates of tridymite that are now replaced by quartz, associated with the reacted muscovite.

Location: Isle of Mull, Scotland. *Rock type*: partially melted psammite from the contact with the sill. *Scale*: the field of view is 0.9 mm wide. Plane-polarized light. *Image and caption*: Marian Holness.

Figure **F17**

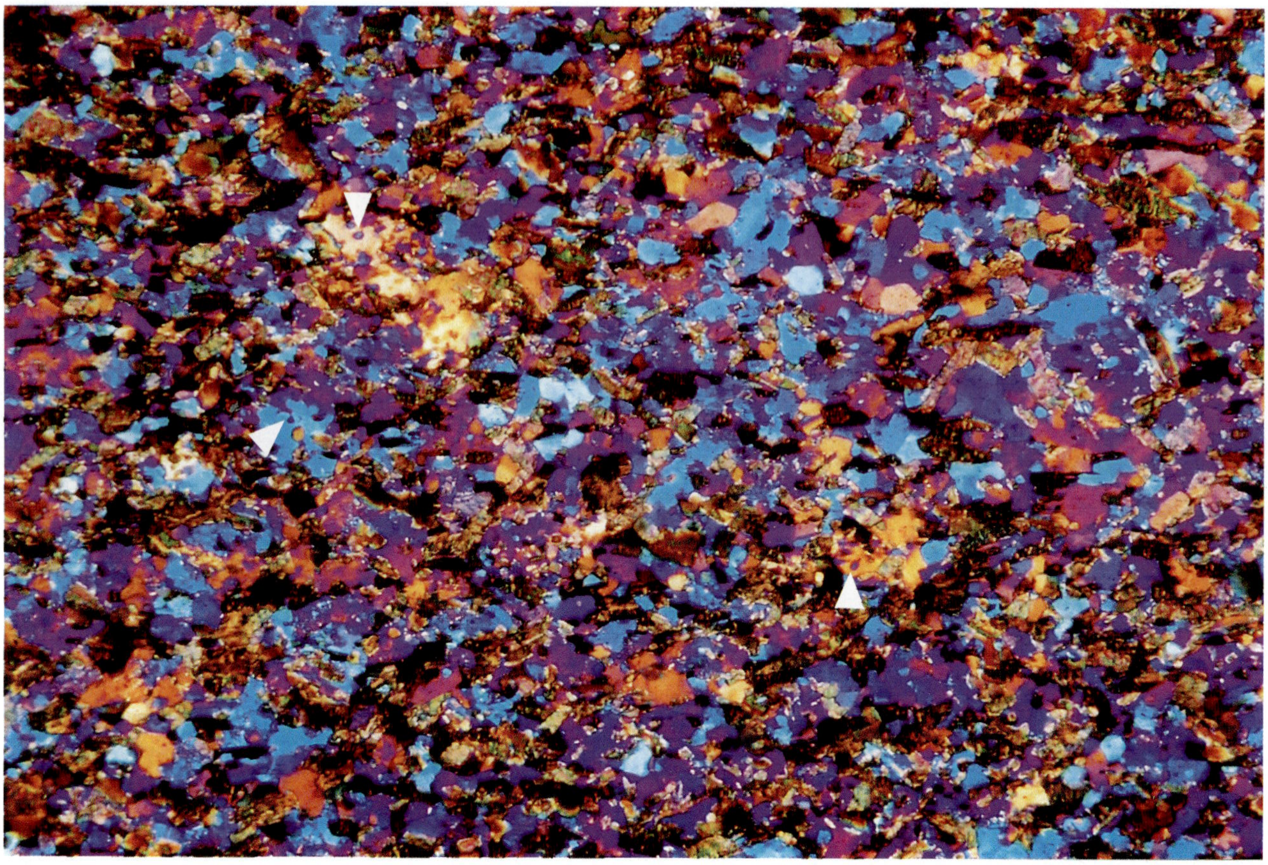

Fig. F17. This pelitic rock was collected from near the "melt-in" isograd >140 m below the contact in the footwall of the Duluth Igneous Complex. This image was taken using a one-wave, quartz accessory plate. Depending on the orientation of grains, this either adds, or subtracts, about 570 nm to the color value. Thus, grains that would otherwise be first-order grey are changed to between first-order red and second-order blue, and consequently, individual grains in a matrix become easier to distinguish, as they are no longer all grey. Locally, the rock contains small irregular patches of quartz or feldspar (three are indicated with arrows) that contain rounded inclusions of quartz, plagioclase, and rarely, biotite; these patches are interpreted as former pockets of anatectic melt and can be thought of as *in situ* microleucosome. This rock contains a high modal percentage of biotite, no orthopyroxene and only minor amounts of cordierite, which indicates that partial melting was at an incipient stage. The presence of biotite along many of the grain boundaries makes it difficult to find cuspate extensions of quartz or feldspar between the corroded grains of the matrix.

Location: Duluth Igneous Complex, Minnesota, U.S.A. *Rock type*: metatexite migmatite; Virginia Formation metapelite protolith, partial melting at the "melt-in" isograd, *T* ca. 700–725°C. *Scale*: the width of the photograph is 2.5 mm. Cross-polarized light and quartz plate inserted. *Image*: E.W. Sawyer.

Figure F18

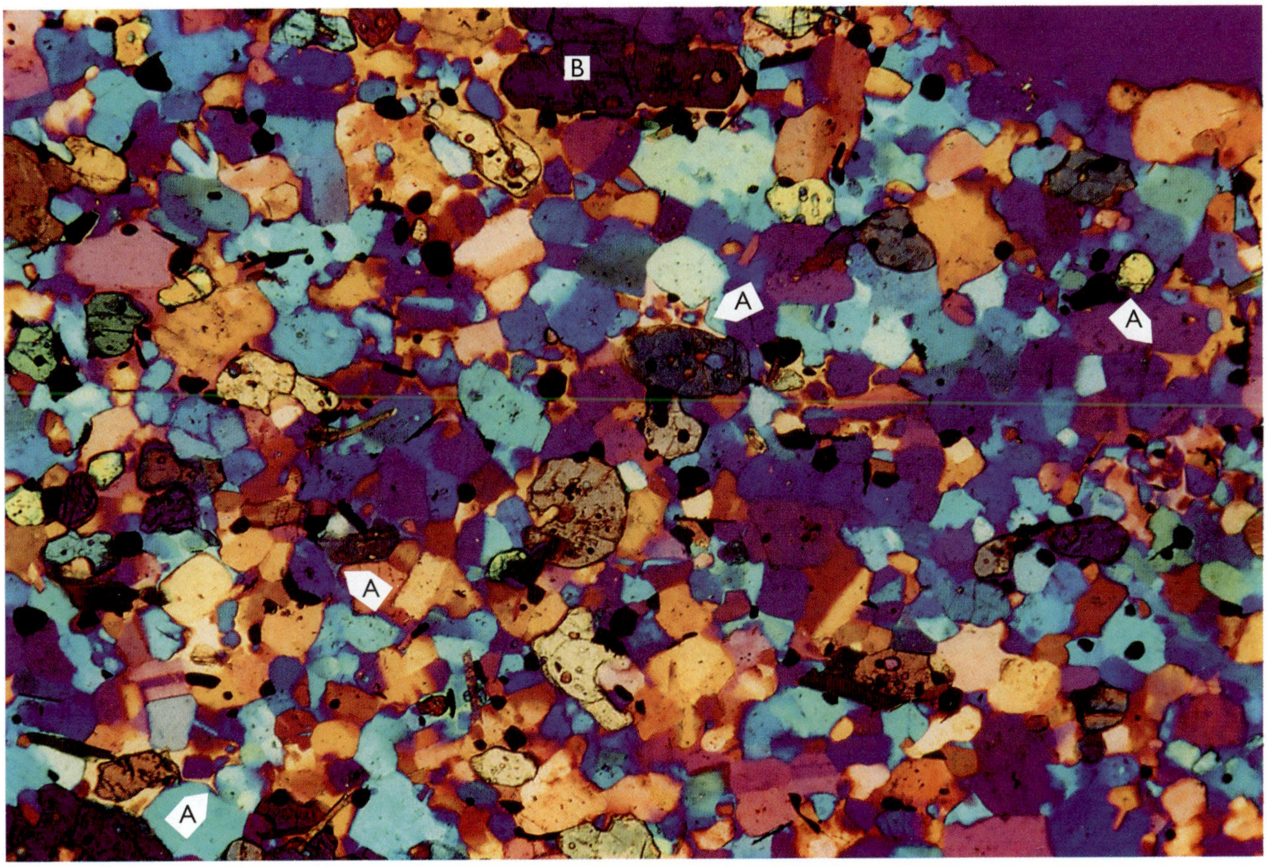

Fig. F18. Pelitic rocks with residual bulk compositions occur in metasedimentary xenoliths of the Virginia Formation from the norite of the Duluth Igneous Complex. This image was taken using a one-wave quartz accessory plate so that some of the former pools of melt appear yellow and are, therefore, easier to see (others are blue, red, and purple, but harder to see). These former pockets of melt have cuspate shapes (A) with thin, tapering extensions penetrating along adjacent grain-boundaries between the reactant minerals. Moreover, the matrix minerals in contact with the yellow domains are rounded, and thus interpreted as evidence that they are corroded. The former pockets of melt contain small, rounded grains of quartz, plagioclase, and biotite. Most of the orthopyroxene crystals (high relief) are subidioblastic, but a large one (B) grown into a melt pool has rational faces and is, therefore, idioblastic. The microstructures indicate that the melting reaction in this residual rock was biotite + plagioclase + quartz = orthopyroxene + melt + ilmenite. The melting was well advanced, and little of the reactant biotite and quartz remains; most of the biotite present in the rock forms large bladed porphyroblasts and grew later from the melt.

Location: Duluth Igneous Complex, Minnesota, U.S.A. *Rock type*: xenolith of Virginia Formation metapelite located about 25 m into the intrusive body, anatexis at $T > 900°C$. *Scale*: the width of the photograph is 2.5 mm. Cross-polarized light and quartz plate inserted. *Image*: E.W. Sawyer.

Figure F19

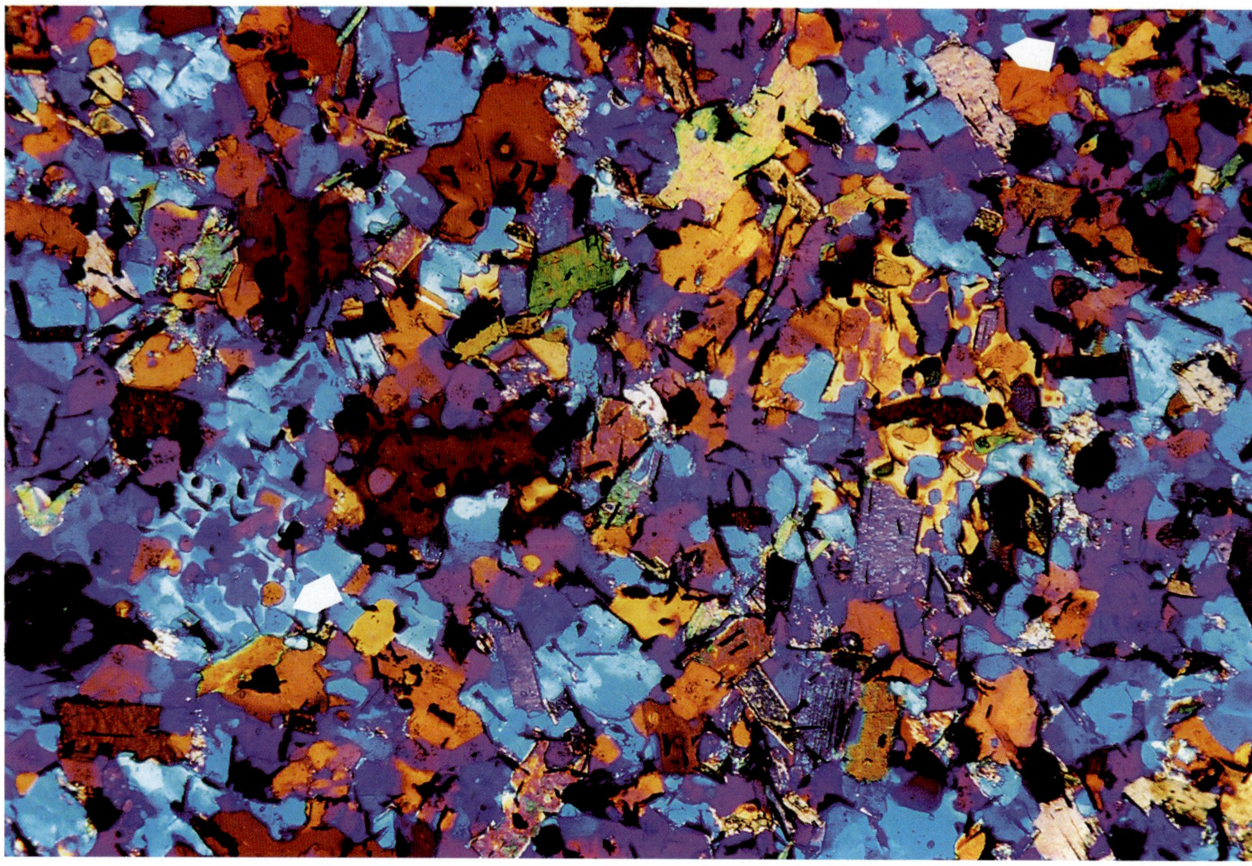

Fig. F19. Microstructures indicative of partial melting can also be found in many of the other lithologies around the Duluth Igneous Complex. Irregular domains of quartz and feldspar containing rounded inclusions of quartz, biotite, plagioclase, and cordierite mark the sites of partial melting in this graphite + Fe-oxide-bearing metapelite. There are several such domains in this photomicrograph, and the most prominent one is the yellow area just right of the center. Others are indicated with arrows. Note the basal sections of large, irregularly shaped crystals of biotite in the rock; these are interpreted to have grown either from the anatectic melt, or from a reaction between the melt and minerals such as K-feldspar or cordierite. The bulk composition of the metapelite, and the low pressure of metamorphism in the aureole of the Duluth Igneous Complex, mean that K-feldspar and cordierite are abundant in many of the pelitic hornfelses, and the earliest generation of both minerals formed before partial melting began.

Location: Duluth Igneous Complex, Minnesota, U.S.A. *Rock type*: metatexite migmatite; graphite + Fe-oxide-bearing metapelite protolith, anatexis at T ca. 750–800°C. *Scale*: the width of the photograph is 2.5 mm. Crossed polarizers and quartz plate. *Image*: from a thin section provided by Steve Hauck.

Figure F20

Fig. F20. The onset of partial melting in sulfide-rich (>10 modal %) pelitic rocks produces microstructures in the silicate portion that are very similar to those in normal pelitic rocks. The ferromagnesian silicate minerals are far more magnesian, however, and have compositions very close to the Mg end-members. Equant domains of anhedral quartz and plagioclase several times larger than the grains in the matrix occur throughout the rock (three are indicated by arrows) and are interpreted to be crystallized pockets, or pools, of anatectic melt. Many of the large crystals of quartz and feldspar that comprise the domains contain small, rounded inclusions of quartz, plagioclase, and biotite. These inclusions are interpreted to be the corroded remnants of the reactant minerals. Most of the former pools of melt have cuspate forms; some thin, tapered offshoots extend along the boundaries of rounded matrix grains.

Location: Duluth Igneous Complex, Minnesota, U.S.A. *Rock type*: metatexite migmatite; the protolith was the bedded pyrrhotite unit, Virginia Formation, anatexis at *T* ca. 800°C. *Scale*: the width of the photograph is 2.5 mm. Crossed polarizers and quartz plate inserted. *Image*: from a thin section provided by Steve Hauck.

Figure F21

Fig. F21. Partial melting has also affected the (clinopyroxene + magnetite)-bearing quartz-rich rocks of the Biwabik Formation, which occur in the footwall of the Duluth Igneous Complex. This photograph shows an example where incipient partial melting has led to the corrosion of the quartz at grain junctions and to the development of films of anatectic melt on some grain boundaries. Upon freezing, the films of melt on the grain boundaries have crystallized to feldspar and quartz. Locally, some larger pools of melt have developed (A), and they crystallized to a characteristic micrographic intergrowth (granophyre) of quartz and feldspar (A). Melt that occupied smaller pores (B) crystallized only plagioclase. The anatectic melt in rapidly cooled migmatites quenches to glass (see Figs. F5 and F6), whereas in very slowly cooled migmatites such as those formed by regional metamorphism, it crystallizes to produce normal igneous microstructures. However, in the regime of intermediate rates of cooling, typical of contact aureoles in the 1–3-kbar range, the anatectic melt commonly crystallizes to granophyre, as in this example (see also Fig. F11).

Location: Duluth Igneous Complex, Minnesota, U.S.A. *Rock type*: metatexite migmatite; magnetite-bearing quartzite protolith, anatexis at *T* ca. 900°C. *Scale*: the width of the photograph is 6 mm. Cross-polarized light. *Image*: E.W. Sawyer.

Figure F22

Fig. F22. In this partially melted impure quartzite from the Biwabik Formation, some of the quartz grains (Qr) are rounded owing to dissolution into the anatectic melt, and they contain abundant inclusions and cracks; these are interpreted to be grains of residual quartz. However, other quartz grains have straight edges, and this quartz is interpreted to have grown from the anatectic melt, possibly as an overgrowth on a grain of residual quartz. Some crystals of quartz (Qm) are rectangular and may indicate that the quartz grew as beta quartz. Large grains of feldspar (F), some of which display wispy films of sodic plagioclase of exsolution origin, occur between the quartz crystals. The large, grey crystal of feldspar near the center of the photograph contains both rounded and rectangular crystals of quartz. The boundary of this grain is highly irregular; locally, it extends between quartz crystals and terminates in cuspate projections at quartz triple junctions (arrow). Feldspar crystals in the upper right and lower left are in extinction position and, therefore, appear black, but they too extend between quartz grains and have cuspate terminations. This microstructure is interpreted to indicate that the feldspar in this migmatite crystallized from the anatectic melt that was interstitial to the quartz grains.

Location: Duluth Igneous Complex, Minnesota, U.S.A. *Rock type*: metatexite migmatite; impure quartzite protolith, anatexis at *T* ca. 900°C. *Scale*: the width of the photograph is 2.7 mm. Cross-polarized light. *Image*: E.W. Sawyer.

Figure **F23**

Fig. F23. In the melt-rich parts of migmatites, the former patches of melt have a microstructure dominated by crystallization of the melt, and not by the dissolution of reactant minerals. This photomicrograph from a diatexite migmatite shows a region of melt that contained small, euhedral (rectangular) grains of plagioclase, equant euhedral grains of cordierite, some with a hexagonal shape showing sector-twinning (Crd), euhedral (rectangular) grains of untwinned K-feldspar, and crystals of blade-shaped biotite; there is also a prominent and somewhat skeletal grain of orthopyroxene that is partially replaced by biotite and other products of alteration. Most of the melt that contained these crystals crystallized to large subhedral grains of quartz and K-feldspar. This particular photomicrograph shows mostly the quartz. The abundance of rectangular feldspar and cordierite crystals gives this domain of former melt a distinctive blocky outline, which contrasts with the cuspate outline of pockets of melt formed during partial melting.

Location: Duluth Igneous Complex, Minnesota, U.S.A. *Rock type*: diatexite migmatite; pelitic protolith, 130 m from the footwall contact; *T* 850–900°C. *Scale*: the width of the photograph is 2.5 mm. Cross-polarized light. *Image*: E.W. Sawyer.

Figure F24

Fig. F24. The last parts of the granitic melt to crystallize are dominated by quartz and K-feldspar. In this part of a melt pocket, it is the quartz that crystallized last, and it formed large irregularly shaped grains that enclose plagioclase (Pl), K-feldspar (Kfs), cordierite (Crd), and biotite (Bt), i.e., all the other minerals. K-feldspar, which occurs as several large, euhedral crystals (e.g., the grain at the center of the photomicrograph), and biotite both appear to have crystallized relatively late. The quartz, K-feldspar, and biotite all contain small, euhedral inclusions of plagioclase and cordierite, although the inclusions are far more abundant in the quartz. The common occurrence of rational faces on the crystals and the euhedral habit of most of the crystals together indicate that this microstructure formed as the anatectic melt crystallized.

Location: Duluth Igneous Complex, Minnesota, U.S.A. *Rock type*: diatexite migmatite; Virginia Formation pelite protolith, 103 m from the contact, *T* ca. 850–900°C. *Scale*: the width of the photograph is 2.5 mm. Cross-polarized light. *Image*: E.W. Sawyer.

Figure F25

Fig. F25. The two prominent, rectangular crystals in this photomicrograph are euhedral phenocrysts of K-feldspar, and both contain abundant inclusions. The distribution of the inclusions in the light grey crystal at the lower right is zoned. An inner region contains an abundance of small, round inclusions of reddish brown biotite, and these are interpreted to be partially dissolved reactant biotite. An adjacent zone contains several larger inclusions of subhedral biotite that possibly grew in the anatectic melt, and rounded reactant quartz. Both of these zones are surrounded by a region with just a few small inclusions of subhedral crystals of cordierite, interpreted to have grown from the melt. The outermost zone is best developed at the ends of the crystal, and contains abundant euhedral and subhedral crystals of cordierite and plagioclase, plus some rounded inclusions of quartz and iron oxide. The other phenocryst of K-feldspar has a similar outer zone in which euhedral inclusions of cordierite are concentric to the crystal outline. However, the interior of this K-feldspar is different; it contains large subhedral crystals of biotite with quartz, which resembles the groundmass outside the phenocryst; possibly, the K-feldspar was hollow and contained melt, which crystallized later. The groundmass around the K-feldspar crystals comprises large, anhedral crystals of quartz that contain abundant inclusions of euhedral plagioclase and cordierite, and larger subhedral inclusions of biotite. The zone in the K-feldspar phenocryst that contains the rounded inclusions of biotite may have grown as a solid product of the melting reaction, and may have been the substrate upon which the K-feldspar that grew from the anatectic melt nucleated, and then trapped euhedral crystals of cordierite as it grew. The margins of both crystals of K-feldspar partially envelop much larger euhedral crystals of plagioclase, cordierite, and biotite that are shared with the large subhedral grains of quartz that are adjacent. This relationship may indicate that the last increment of K-feldspar growth occurred as quartz crystallized from the last melt in the migmatite.

Location: Duluth Igneous Complex, Minnesota, U.S.A. *Rock type*: diatexite migmatite; Virginia Formation pelite protolith, 103 m from the contact; *T* ca. 850–900°C. *Scale*: the width of the photograph is 3.5 mm. Cross-polarized light. *Image*: E.W. Sawyer.

Figure F26

Fig. F26. As an H_2O-undersaturated granitic melt progressively crystallizes, the H_2O content in the remaining melt fraction increases if most of the minerals that crystallize first, such as plagioclase, are anhydrous. This photomicrograph shows a domain where the crystallization of the last increments of now H_2O-rich melt has resulted in the growth of blades of muscovite (Ms, yellow and blue interference colors), and also to the replacement of earlier-formed orthopyroxene (Opx) by small bladed crystals of biotite and muscovite. The extent of the former pocket of melt is indicated by the quartz (white); the melt also contained abundant, small euhedral crystals of plagioclase, K-feldspar, and cordierite, together with some larger, bladed crystals of biotite. Note (see Figs. F17–F28) that the crystals with the largest grain-size in the migmatites at Duluth are generally the minerals that were the last to crystallize, typically quartz, plagioclase, and K-feldspar.

Location: Duluth Igneous Complex, Minnesota, U.S.A. *Rock type*: diatexite migmatite; Virginia Formation pelite protolith, 97 m from the contact; T ca. 850–900°C. *Scale*: the width of the photograph is 2.5 mm. Cross-polarized light. *Image*: E.W. Sawyer.

Figure **F27**

Fig. F27. Some of the migmatites in the contact aureole at the Duluth Igneous Complex (DIC) contain crystals that have a skeletal, or hollow, form; skeletal crystals are a common feature in many rapidly quenched partial-melting experiments. This photomicrograph, of a diatexite migmatite derived from a pelitic protolith, shows two different morphologies of cordierite. Most of the cordierite grains (Crd1) occurs as very small (<0.25 mm across), equant, euhedral crystals, some of which show a well-developed sector twinning. Similar small crystals of cordierite also occur abundantly as inclusions in plagioclase, K-feldspar, and quartz. Consequently, these small crystals are interpreted to have crystallized in the anatectic melt. The center of the photograph shows a rare, large (ca. 1.5 mm), isolated phenocryst of cordierite (Crd2) that has grown in the most leucocratic and coarsest-grained part of the diatexite migmatite. Because this leucocratic domain consists predominantly of large anhedral crystals of quartz, plagioclase, and K-feldspar, it is interpreted to be one of the last places in the diatexite to have crystallized. The phenocryst is skeletal, it has angular embayments, and its core contains quartz, plagioclase, K-feldspar, and biotite that resemble the groundmass outside the phenocryst. Interestingly, there are some small fan-shaped clusters of muscovite (Ms) inside the cordierite, which suggests that the minerals in the core of the cordierite crystallized from an evolved melt that was partially enclosed by the cordierite. This fact, and the presence of straight crystal faces against both quartz and feldspar, lead to the interpretation that the cordierite grew from the melt, and that its skeletal, or embayed, morphology is a growth feature, and not due to its dissolution. In their most melanocratic parts, some diatexite migmatites in the DIC aureole contain cordierite that has a third morphology. Such cordierite is generally not twinned and forms subidioblastic crystals up to 1 mm across that contain equant inclusions of biotite and quartz. These crystals are interpreted to be residual cordierite that formed by metamorphic reactions prior to partial melting.

Location: Duluth Igneous Complex, Minnesota, U.S.A. *Rock type*: diatexite migmatite; Virginia Formation pelite protolith, 103 m from the contact; *T* ca. 850–900°C. *Scale*: the width of the photograph is 3.5 mm. Cross-polarized light. *Image*: E.W. Sawyer.

Figure F28

Fig. F28. This photomicrograph shows elongate crystals of orthopyroxene and plagioclase that are strongly aligned in a narrow zone oriented parallel to the bottom-left-to-top-right diagonal of the photograph. The strong preferred orientation of the elongate grains occurs in the center of this incipient in-source leucosome, and is due to flow of magma that contained the elongate crystals, i.e., this is an example of the flow of a crystal-rich magma. Smaller, equant crystals of subhedral plagioclase, K-feldspar, and cordierite occur between the larger, better-oriented crystals, and were not aligned because of their equant shapes. Large, irregularly shaped regions of quartz, K-feldspar, and minor plagioclase occur around and between the elongate (aligned) and equant (not aligned) subhedral crystals of plagioclase and orthopyroxene, and are interpreted to be the places where the last of the interstitial melt crystallized after flow of the magma had ceased. Some patches of interstitial melt were surrounded by euhedral crystals that had grown from the melt, and their characteristic blocky outline is preserved (two patches are indicated by arrows). The orthopyroxene in this rock is a solid product of the incongruent melting of biotite (some authors use the term "peritectic phase" for minerals formed in this way), and because they grew in the presence of an anatectic melt, some crystals have well-developed faces. In contrast, most of the plagioclase present crystallized from the melt and is, therefore, considered to be a melt-crystallization product, i.e., a liquidus phase. The leucosome is considered to be incipient because the domain does not have clearly defined edges, and is not evident in the hand sample.

Location: Duluth Igneous Complex, Minnesota, U.S.A. *Rock type*: incipient leucosome in a residual metatexite migmatite; Virginia Formation pelite protolith, xenolith $T > 900°C$. *Scale*: the width of the photograph is 2.5 mm. Cross-polarized light. *Image*: E.W. Sawyer.

Figure **F29**

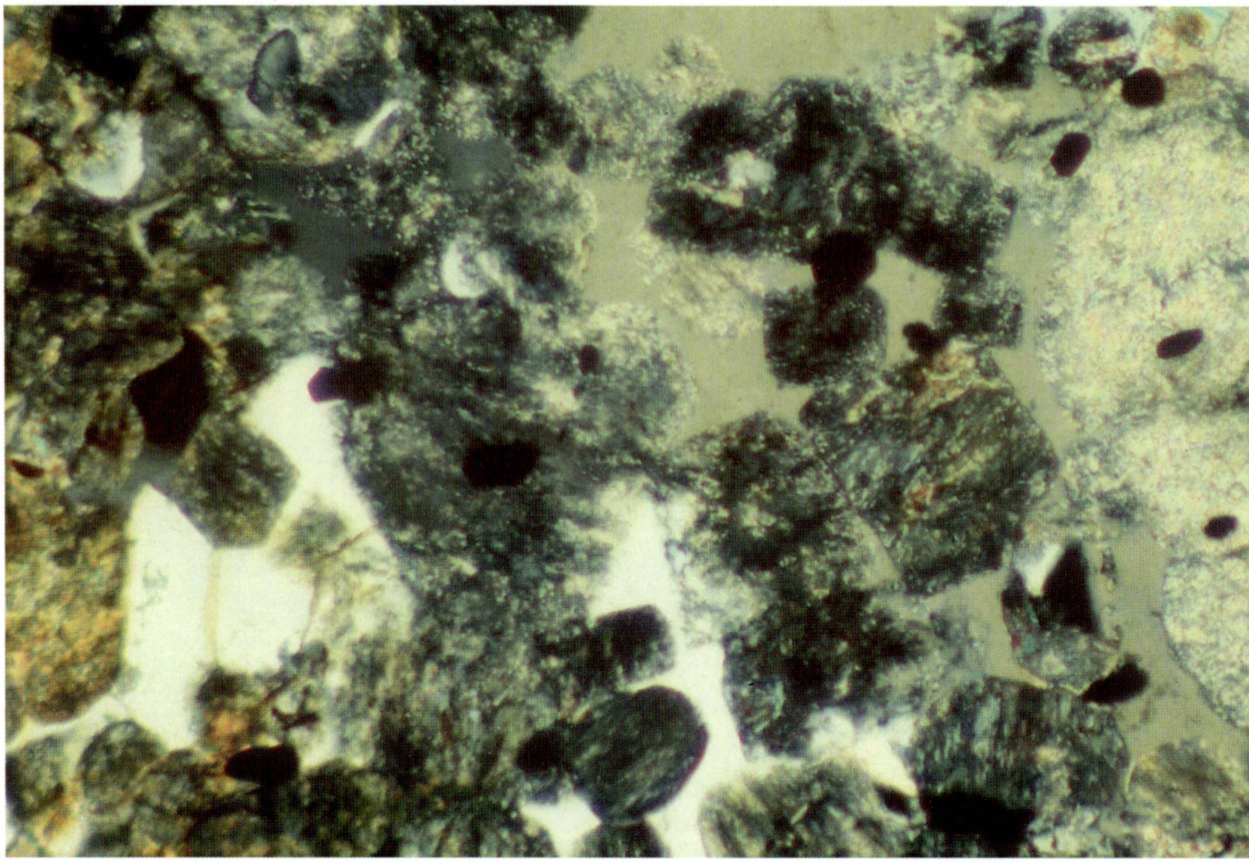

Fig. F29. This photomicrograph is from a thin section taken from a nebulitic metatexite migmatite derived from a semi-pelitic protolith. The quartz has an interstitial to poikilitic morphology and surrounds subhedral to euhedral grains of K-feldspar and, to a lesser extent, cordierite. The K-feldspar is highly altered, and so appears patchily dark, and somewhat streaked. The cordierite grain on the extreme right is altered largely to a shimmer aggregate. The outline of several K-feldspar grains shows clear evidence of a definite shape of the crystals. The quartz, which is clear (white) to yellowish brown in color, has an interstitial, pore-filling form between the grains of K-feldspar and cordierite, but it is poikilitic overall and in the field of view consists largely of two crystals: one in the lower left and center has white birefringence-colors, and the other has yellowish colors. The K-feldspar is considered to have crystallized from the anatectic melt earlier than the quartz, which crystallized around the K-feldspar (and cordierite) crystals as it filled the remaining "pore space"; hence, it has both a poikilitic and an interstitial habit.

Location: Ballachulish Igneous Complex, Scotland. *Rock type*: nebulitic metatexite migmatite; semipelitic protolith. *Scale*: the width of the photograph is 1.5 mm. Cross-polarized light. *Image and caption*: Ben Harte and Claire M. Linklater.

Further reading: Harte, B., Pattison, D.R.M. & Linklater, C.M. (1991): Field relations and petrography of partially melted pelite and semi-pelitic rocks. *In* Equilibrium in Contact Metamorphism: the Ballachulish Igneous Complex and its Aureole (G. Voll, J. Topel, D.R.M. Pattison & F. Seifert, eds.). Springer-Verlag, Heidelberg, Germany (181–209).

Figure F30

Fig. F30. This photomicrograph is from a sample of nebulitic metatexite migmatite developed from a semipelitic protolith. It shows quartz grains in a matrix of K-feldspar. The K-feldspar is highly altered; it appears dark and streaky. The shapes of the quartz grains range from roundish to subhedral, and there are small segments of the grains that are possibly crystal faces. The quartz is interpreted to be principally rounded, or corroded, residual grains formed by the partial melting of quartz-rich rocks. The flat-faced segments on the quartz crystals are considered to be portions that grew later as the melt crystallized. The original crystal of K-feldspar had a morphology interpreted as interstitial, and is considered to have crystallized largely from the melt, which formerly occupied the pore space.

Location: Ballachulish Igneous Complex, Scotland. *Rock type*: nebulitic metatexite migmatite; semipelitic protolith. *Scale*: the width of the photograph is 1.5 mm. Cross-polarized light. *Image and caption*: Ben Harte and Claire M. Linklater.

Further reading: same as Fig. F29.

Figure F31

Fig. F31. This image shows microstructures typical of the Appin Quartzite (which actually contains abundant feldspar (Fsp), and so may be better described as an arkose) from close to the intrusive body, which has melted almost completely. The quartz grains are recognized by their clarity, or transparency, and are roundish, whereas in contrast, the feldspar grains are slightly cloudy, or turbid, and have highly cuspate forms; the feldspar–quartz dihedral angles are low. The feldspar is typical of that grown in melt pockets, and its shape suggests that it has replaced the melt.

Location: Ballachulish Igneous Complex, Scotland. *Rock type*: partially melted Appin Quartzite one meter from the contact, *T* ca. 775°C, *P* 3 kbar. *Scale*: the field of view is 0.9 mm wide. Cross-polarized light. *Image and caption*: Marian Holness.

Further reading: Holness, M.B. & Clemens, J.D. (1999): Partial melting of the Appin quartzite driven by fracture-controlled H_2O infiltration in the aureole of the Ballachulish Igneous Complex, Scottish Highlands. *Contributions to Mineralogy and Petrology* **136**, 154–168.

Holness, M.B. & Watt G.R. (2001): Quartz recrystallisation and fluid flow during contact metamorphism: a cathodoluminescence study. *Geofluids* **1**, 215–228.

Figure F32

Fig. F32. The quartz grains in this photomicrograph have undergone significant growth. The tiny, triangular grains of feldspar (Fsp) that used to be located at three-grain junctions are now entirely enclosed within a single grain of quartz (Qtz). The feldspar (turbid) is highly cuspate in shape, and has some elongate, tapering extensions along the boundaries between the adjacent grains.

Location: Ballachulish Igneous Complex, Scotland. *Rock type*: partially melted Appin Quartzite 1 m from the contact, *T* ca. 775°C, *P* 3 kbar. *Scale*: the field of view is 0.9 mm wide. Cross-polarized light. *Image and caption*: Marian Holness.

Further reading: see Fig. F31.

Figure F33

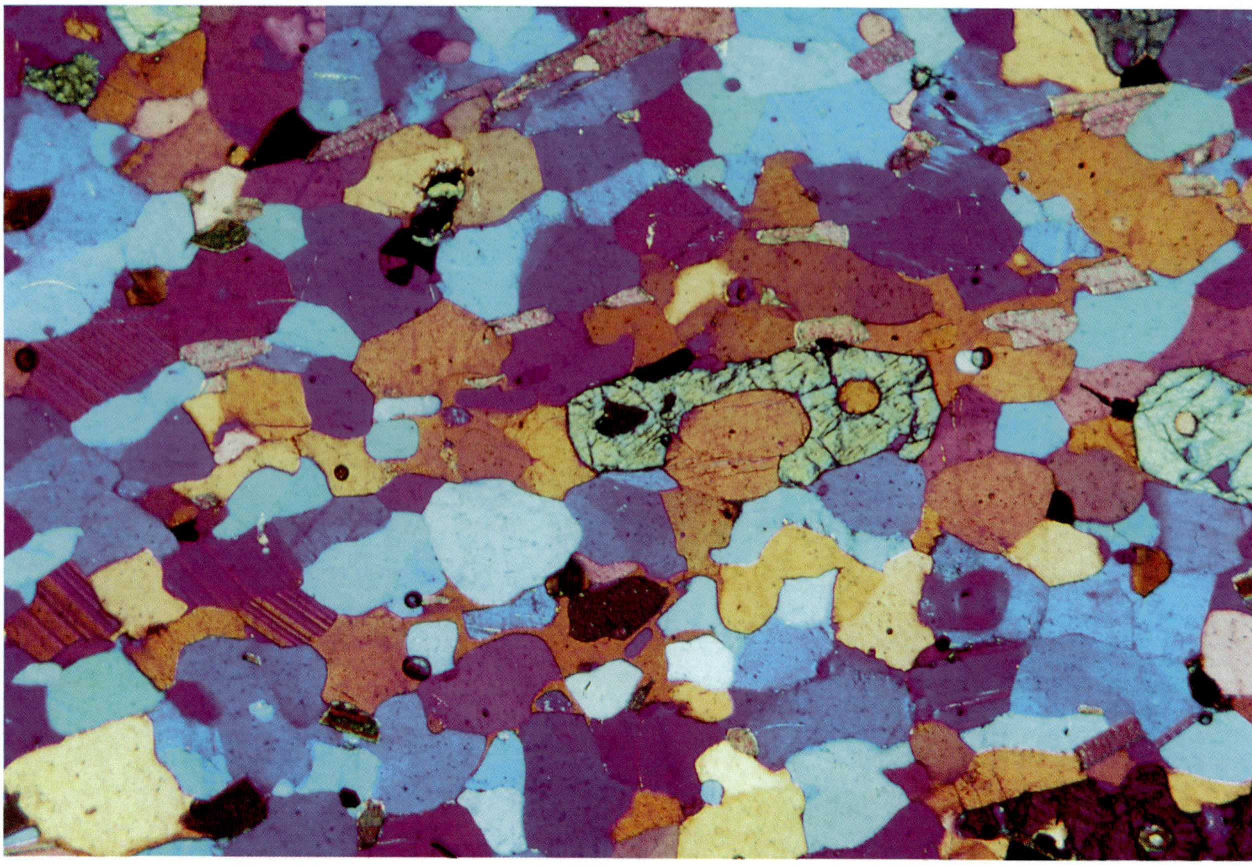

Fig. F33. This sample contains irregularly shaped domains of K-feldspar (domains of quartz and plagioclase are rarer) that are up to four times the size of the equant, polygonal-shaped groundmass minerals, quartz, and plagioclase and bladed biotite. The largest domains are shown in orange for clarity, but there are others of different colors as well. Typically, the irregularly shaped domains have cuspate outlines, because the adjacent minerals are rounded owing to dissolution; Mehnert et al. (1973) produced very similar microstructures in partial-melting experiments. In places, thin, tapered extensions from the domains follow the grain boundaries between adjacent minerals. The microstructure present at the time of partial melting thus is preserved and did not undergo textural re-equilibration either when melt was still present or subsequently in the subsolidus state (e.g., Holness et al. 2005, see Further reading below). Most of the cuspate domains contain rounded inclusions of quartz, plagioclase, and biotite. An orthopyroxene crystal located in the center of the photomicrograph has well-developed, rational faces where it is in contact with a K-feldspar domain. This microstructure is interpreted to indicate a site where partial melting occurred by the reaction: biotite + quartz + plagioclase = melt + orthopyroxene + Fe-oxide. The distribution of melt in this rock was, therefore, controlled by where the melting reaction occurred, i.e., by the location of biotite in the rock. Consequently, the former pools of melt are, in general, oriented parallel to the foliation. In other places, thin films of minerals representing former melt occur along grain boundaries and at grain junctions. From point counting and a mass-balance calculation, this particular rock is interpreted to have a residual bulk composition resulting from the loss of 15.8% melt, but 2.1% melt was retained in the pools, grain boundaries, and grain junctions. The microstructures that indicate the former presence of melt in this rock are unusually large (up to 1 mm, or four times the grain size) for migmatites formed in a regional metamorphic terrane; more typical examples are shown in Figs. F35 and F36.

Location: Ashuanipi Subprovince, Quebec, Canada. *Rock type*: residual part of a metatexite migmatite; metagreywacke (psammite) protolith, anatexis in the granulite facies at *T* 825–875°C, *P* 6–7 kbar. *Scale*: the long side of the photograph corresponds to 3.5 mm. Cross-polarized light and quartz accessory plate inserted. *Image*: E.W. Sawyer.

Further reading: Holness, M.B., Cheadle, M.J. & McKenzie, D. (2005): On the use of changes in dihedral angle to decode late-stage textural evolution in cumulates. *Journal of Petrology* **46**, 1565–1583.

Sawyer, E.W. (2001): Melt segregation in the continental crust: distribution and movement of melt in anatectic rocks. *Journal of Metamorphic Geology* **19**, 291–309.

Figure F34

Fig. F34. Same image as shown in Fig. F33, but in cross-polarized light without an accessory plate. The microstructures indicative of the former presence of melt are much harder to see without the aid of the quartz accessory plate. Using the accessory plate confers a considerable advantage in distinguishing, and hence identifying, microstructures that contain differently oriented grains of minerals that all have low, first-order interference-colors, such as quartz, plagioclase, and K-feldspar.

Location: Ashuanipi Subprovince, Quebec, Canada. *Rock type*: residual part of a metatexite migmatite; metagreywacke (psammite) protolith, anatexis in the granulite facies at T 825–875°C, P 6–7 kbar. *Scale*: the long side of the photograph corresponds to 3.5 mm. Cross-polarized light. *Image*: E.W. Sawyer.

Figure F35

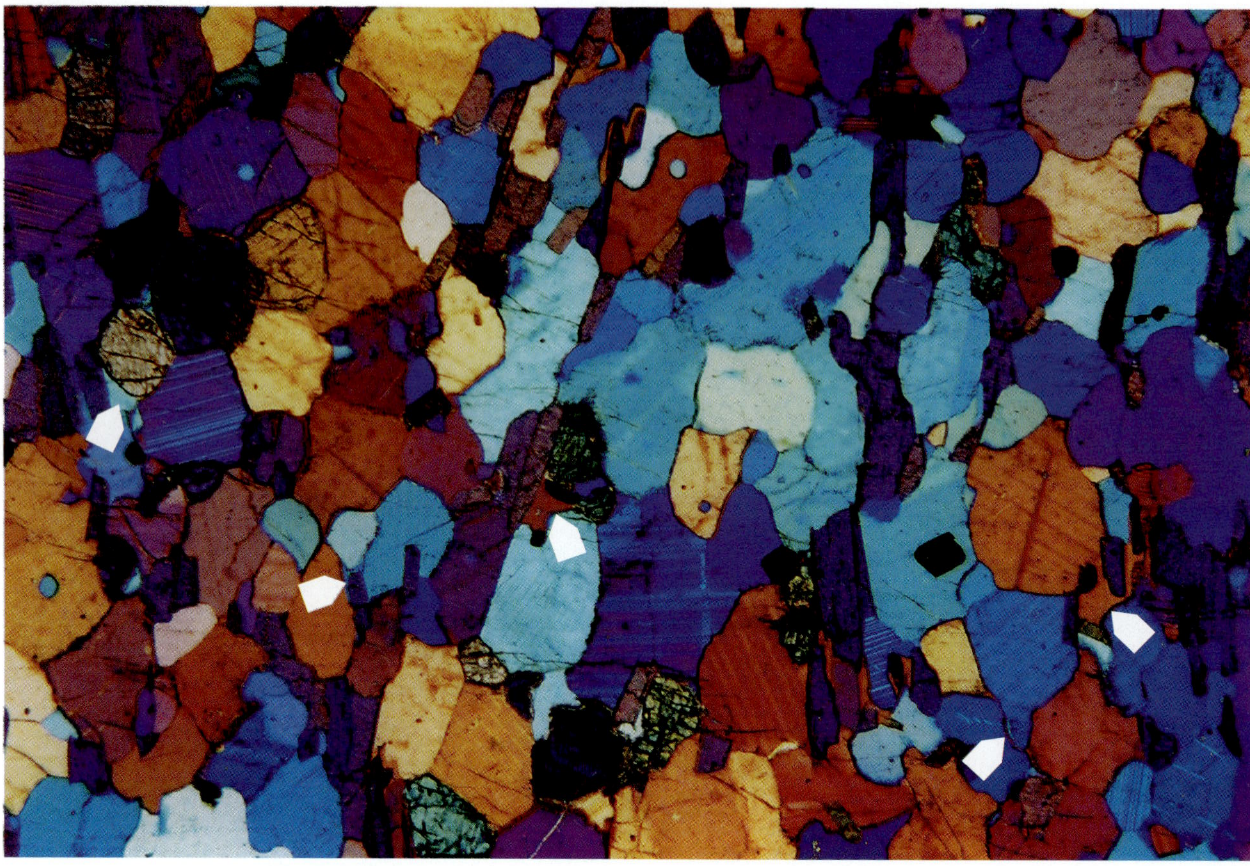

Fig. F35. This residual metagreywacke contains a few scattered irregularly shaped domains of K-feldspar, quartz and plagioclase in contact with rounded quartz, plagioclase, and especially biotite. The domains are very small (<0.15 mm), comparable to the grain size, and not easy to find. The ones that are most obvious in this image are orange in color, but others are blue or violet; some are indicated by arrows. The domains are interpreted as the sites where partial melting occurred via the reaction: biotite + quartz + plagioclase = melt + orthopyroxene + Fe-oxide. The domains of former melt in this rock are much smaller, relative to the grain size of the matrix, than those in Figs. F33 and F34, and have a geometry that is more suggestive of films of melt located along grain boundaries, rather than discrete pools of melt in the matrix. The distribution of the former films of melt is strongly linked to the location of biotite, the hydrous reactant mineral present in the rock; hence, melt formed preferentially in the plane of the foliation.

Location: Ashuanipi Subprovince, Quebec, Canada. *Rock type*: residual part of a metatexite migmatite; metagreywacke (psammite) protolith, anatexis in the granulite facies at T 825–875°C, P 6–7 kbar. *Scale*: the width of the photograph corresponds to 4.0 mm. Cross-polarized light plus quartz accessory plate. *Image*: E.W. Sawyer.

Further reading: Guernina, S. & Sawyer, E.W. (2003): Large-scale melt-depletion in granulite terranes: an example from the Archaean Ashuanipi subprovince of Quebec. *Journal of Metamorphic Geology* **21**, 181–201.

Figure F36

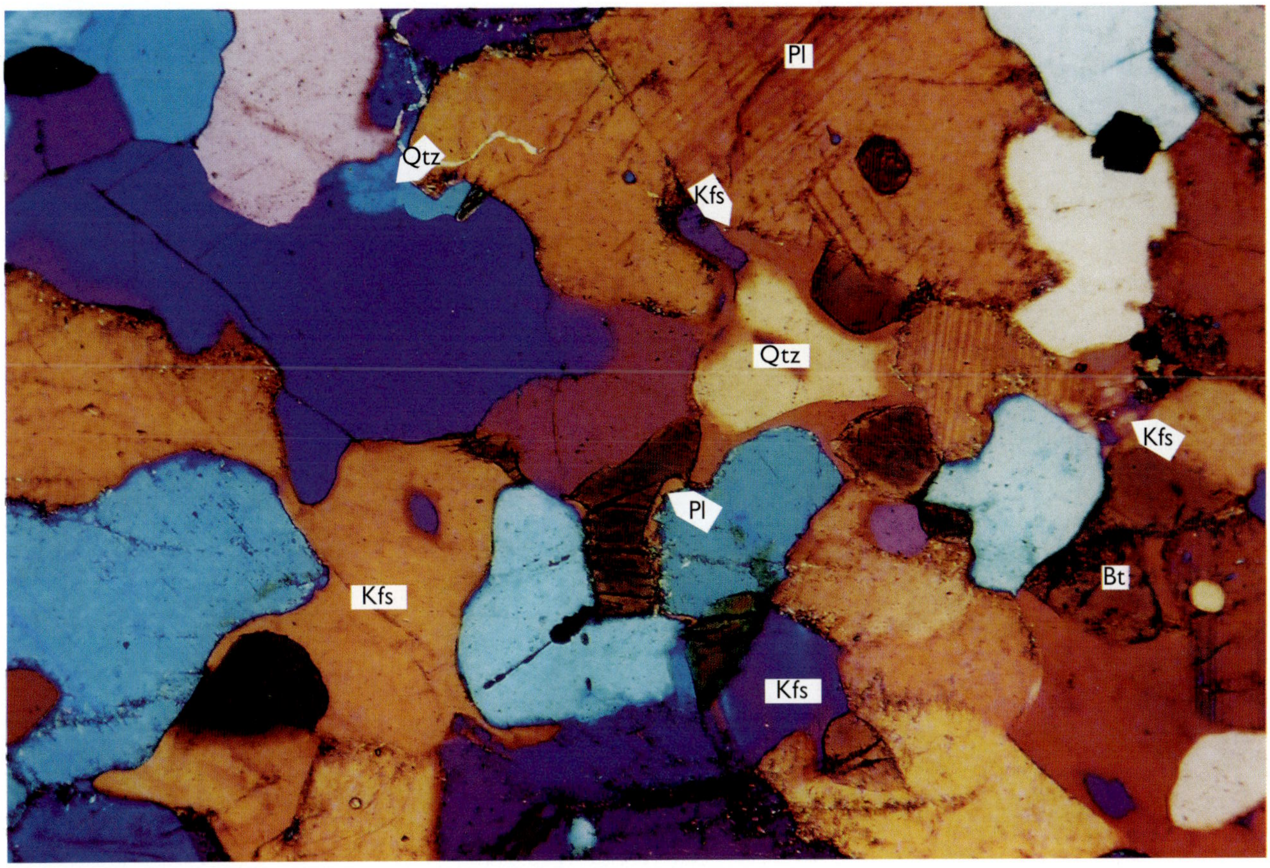

Fig. F36. The thin section shown in this photomicrograph has been cut parallel to the foliation plane to show the shape of the former pools of melt in the plane of the foliation. The center of this image shows several irregularly shaped, cuspate minerals between the rounded matrix grains of biotite (Bt), quartz (Qtz), and plagioclase (Pl). There is no orthopyroxene in this photomicrograph, but it occurs elsewhere in the thin section. Interstitial films of K-feldspar, plagioclase, and quartz are identified by arrows. The rounded outline of the quartz, biotite, and plagioclase grains near the K-feldspar is interpreted to be the result of dissolution during the melt-producing metamorphic reaction. On the basis of this interpretation and the presence of orthopyroxene elsewhere in the thin section, partial melting is inferred to have occurred via the reaction: biotite + quartz + plagioclase = melt + orthopyroxene + Fe oxide. K-feldspar does not occur on the mutual boundaries between adjacent grains of plagioclase or between adjacent grains of quartz; it occurs adjacent to biotite. Consequently, the irregularly shaped grains are interpreted to have crystallized from anatectic melt that formed at triple-grain junctions, or along grain boundaries adjacent to the grains of reactant biotite. The distribution of melt in the foliation plane of this rock is thus controlled by where the melting reaction occurred. The K-feldspar has a cuspate outline and thin tapered projections along the grain boundaries between adjacent residual minerals. The microstructure that developed at the time of partial melting thus has not been modified to any significant degree subsequently. The presence of intergranular K-feldspar, quartz, and plagioclase suggests that the anatectic melt that crystallized might have had a granitic bulk composition, and this is broadly confirmed by determining the mineral modes of the intergranular patches by point counting.

Location: Ashuanipi Subprovince, Canada. *Rock type*: residual part of a metatexite migmatite; metagreywacke (psammite) protolith, anatexis in the granulite facies, *T* 825–875°C, *P* 6–7 kbar. *Scale*: the width of the photograph corresponds to 2.5 mm. Cross-polarized light plus quartz accessory plate. *Image*: E.W. Sawyer.

Further reading: see Figs. F33 and F35.

Figure **F37**

Fig. F37. This is a photomicrograph of a biotite–quartz–plagioclase leucotonalite that underwent partial melting and lost most of the melt fraction that was generated. In the upper left quadrant, a large purple, rounded grain of quartz lies in a group of large crystals of plagioclase; the quartz is enclosed by a thin film of K-feldspar, which separates it from direct contact with the neighboring grains of plagioclase. The K-feldspar in this rock occurs only as films such as this between quartz and plagioclase; it does not form large subhedral crystals like quartz and plagioclase. Furthermore, the distribution of K-feldspar is not restricted to the vicinity of biotite. The lower right quadrant of this image shows rounded (i.e., corroded) crystals of quartz (yellowish) and plagioclase (blue) that are separated by a thin film of twinned plagioclase (orange–yellow). This microstructure is very much like that shown in many quenched partial-melting experiments in which a thin film of glass separates the corroded reactant phases. In this case, the plane of the thin section cuts through the part of the melt pocket that crystallized to plagioclase. Biotite does not appear to be corroded in this rock, or to be spatially associated with the K-feldspar films, which suggests that the melting reaction may not have included biotite as a reactant. Melting possibly occurred by a fluid-present reaction such as quartz + plagioclase + K-feldspar + H_2O = melt, and progressed until all the reactant K-feldspar was consumed.

Location: Opatica Subprovince, Quebec, Canada. *Rock type*: melt-depleted residuum; leucotonalite protolith, anatexis at *T* ca. 750°C, *P* 5–7 kbar. *Scale*: the width of the photograph corresponds to 4 mm. Cross-polarized light plus quartz accessory plate. *Image*: E.W. Sawyer.

Further reading: Sawyer, E.W. (1998): Formation and evolution of granite magmas during crustal reworking: the significance of diatexites. *Journal of Petrology* **39**, 1147–1167.

Sawyer, E.W. (2001): Melt segregation in the continental crust: distribution and movement of melt in anatectic rocks. *Journal of Metamorphic Geology* **19**, 291–309.

Figure F38

Fig. F38. This photomicrograph shows a thin, three-branched film of K-feldspar located along the grain boundary between rounded (corroded) grains of quartz and plagioclase in a biotite–quartz–plagioclase leucotonalite. To the left, a smaller film of K-feldspar encloses a rounded grain of plagioclase that is still in optical continuity with a larger grain of plagioclase nearby. This microstructure resembles those found in quenched partial-melting experiments as well, and is interpreted as evidence for partial melting. Considering the whole thin section, the films of K-feldspar occur between matrix grains of plagioclase and quartz, and not with biotite. Hence, the melting may have occurred by a reaction such as quartz + plagioclase + K-feldspar + H_2O = melt. Films of plagioclase and quartz are rare; in general, the grain boundaries between the crystals of residual quartz and plagioclase contain small rounded grains of quartz (indicated with an arrow) and plagioclase. These small grains are interpreted to be the remains of melt films that have subsequently recrystallized. Thus, the distribution of films of K-feldspar and the small recrystallized grains of quartz and plagioclase can be used to infer the former distribution of melt in the rock (e.g., Sawyer 2001).

Location: Opatica Subprovince, Canada. *Rock type*: melt-depleted residuum, leucotonalite protolith, *T* ca. 750°C, *P* 5–7 kbar. *Scale*: the image is 3 mm wide. Cross-polarized light plus the quartz accessory plate. *Image*: E.W. Sawyer.

Figure **F39**

Fig. F39. This photomicrograph shows some of the features typically seen in the residual rocks adjacent to areas of leucosome. At the outcrop scale, this migmatite contains narrow (<1 cm), parallel bands of leucosome that, together with their adjacent domains of melanosome, are oriented parallel to the bedding, resulting in a stromatic metatexite migmatite. The image shows the full width of the melanosome; the leucosome (plagioclase + quartz + K-feldspar) is just visible at right edge of the image, and the plagioclase + quartz + biotite + garnet paleosome at the other. The modal proportions of quartz and plagioclase (colorless minerals) decrease in a systematic fashion throughout the melanosome toward the leucosome, and concomitant with this, there is an increase in the modal proportion of biotite. Moreover, the average grain-size of biotite in the melanosome increases toward the leucosome (i.e., from left to right across the photomicrograph). In contrast, the grain size of quartz and plagioclase remains approximately constant, except for the domains where biotite is especially abundant; there, quartz and plagioclase crystals are smaller. Garnet is uniformly distributed in this particular melanosome, although the largest grain is located closest to the leucosome. Garnet grains close to the leucosome contain small, rounded inclusions of biotite, a feature absent from garnet in the paleosome. Many grains of garnet, especially those nearest the leucosome, are partially enclosed by coarse-grained biotite, which suggests that biotite has partially replaced garnet. The only occurrence of cordierite (slightly yellow because of alteration) in the melanosome is in, and around, a shallow embayment in the largest crystal of garnet close to the leucosome; the significance of this relationship is unclear. There is no sillimanite, or K-feldspar, in the melanosome; the partial-melting reaction may have been biotite + sillimanite + quartz + plagioclase + H_2O = melt + garnet. Biotite and cordierite could both result from interaction between the residual crystals and the remaining anatectic melt in the melanosome following the migration of the melt to the leucosome.

Location: Nemiscau Subprovince, Quebec, Canada. *Rock type*: melanosome from a stromatic metatexite migmatite; semipelitic protolith, upper amphibolite or lower granulite facies (no orthopyroxene). *Scale*: the long side of the photomicrograph corresponds to 12 mm. Plane-polarized light. *Image*: E.W. Sawyer.

Further reading: White, R.W. & Powell, R. (2002): Melt loss and the preservation of granulite facies mineral assemblages. *Journal of Metamorphic Geology* **20**, 621–632.

Figure F40

Fig. F40. This photomicrograph shows the same field of view as Fig. F39, but is taken in cross-polarized light to better show the variation in the grain size of quartz and plagioclase.

Location: Nemiscau Subprovince, Quebec, Canada. *Rock type*: melanosome in a stromatic metatexite migmatite; semipelitic protolith, upper amphibolite or lower granulite facies (no orthopyroxene). *Scale*: the long side of the photomicrograph corresponds to 12 mm. Cross-polarized light. *Image*: E.W. Sawyer.

Figure **F41**

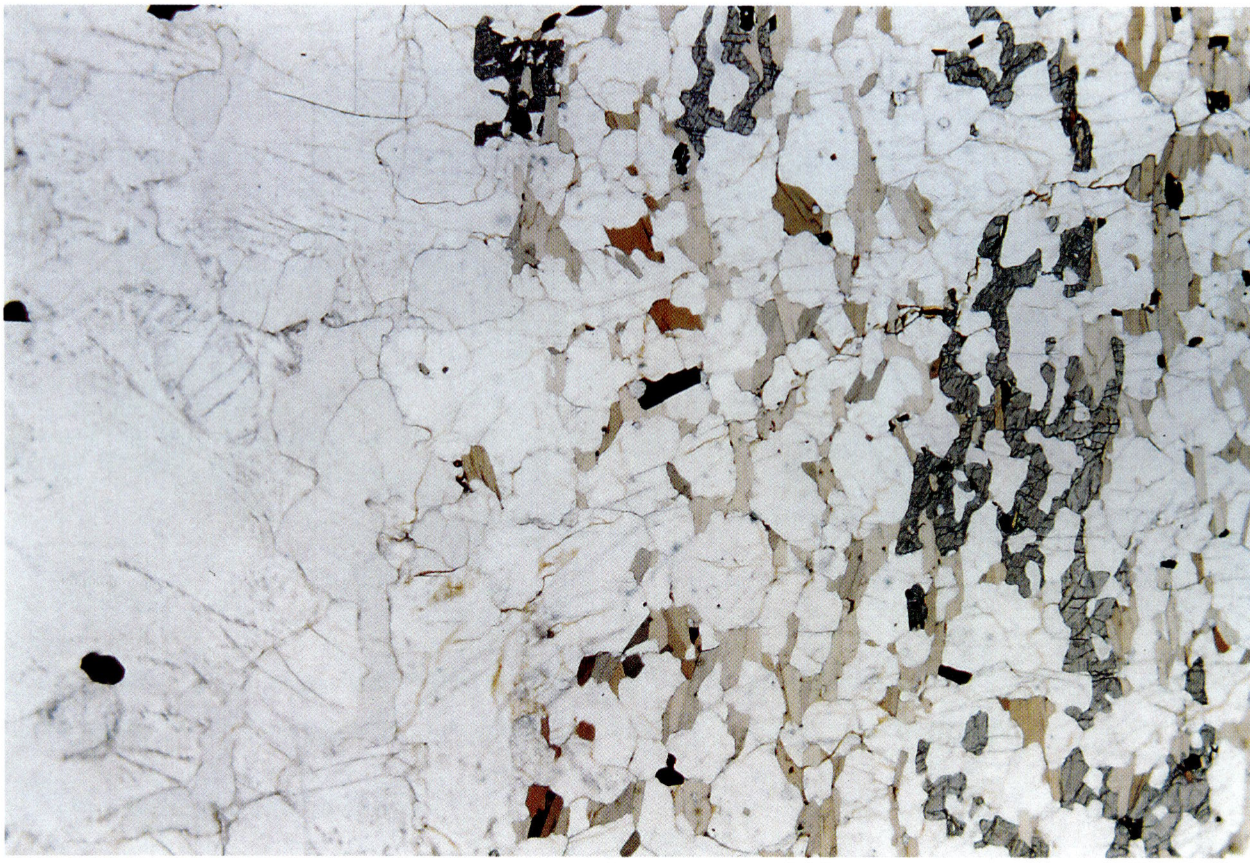

Fig. F41. The edge of the leucosome in this photomicrograph is clearly defined by the abrupt change in the modal abundance of ferromagnesian minerals and by the change in the grain size of quartz and feldspar; the contact is comparatively planar. In outcrop, the migmatite consists of only two parts, the leucosome representing the products derived from anatectic melt and the residuum; there is no paleosome. The areas of leucosome are not bordered by a conspicuous melanosome; rather, the melt-depleted region complementary to the leucosome is the entire layer hosting the leucosome, and such layers can be as much as 50 cm wide. The melt-producing reaction in this migmatite was biotite + quartz + plagioclase = melt + orthopyroxene + ilmenite (or magnetite). These layers are remarkably uniform in chemical composition, mineralogy, and microstructure. They have melt-depleted major- and trace-element compositions, consist of the assemblage plagioclase + biotite + orthopyroxene + quartz + ilmenite, and locally contain microstructures indicative of the former presence of melt (see Figs. F33–F36). The leucosome in the photomicrograph is not bordered by melanosome that becomes progressively more enriched in ferromagnesian minerals toward it. However, the photomicrograph (see Fig. F42 for the same thin section in cross-polarized light) shows that there is a very slight increase in the grain size of plagioclase immediately adjacent to the leucosome and that the modal content of quartz is lowest next to the leucosome. The grain size of the plagioclase + quartz + K-feldspar leucosome is smaller at its margin compared to the interior, possibly because the plagioclase crystallizing from the melt at the margins of the leucosome was able to nucleate on plagioclase crystals in the wallrock. Hence, there are more, but smaller, crystals of plagioclase at the margins of the leucosome. The neosome shown in this photomicrograph is typical of the leucosome–residuum relationship in the metatexite migmatites derived from aluminum-poor metagreywackes in the Ashuanipi Subprovince; orthopyroxene is the greenish grey to pinkish mineral with high relief.

Location: Ashuanipi Subprovince, Quebec, Canada. *Rock type*: metatexite migmatite; aluminum-poor metagreywacke protolith, granulite facies, T 825–875°C, P 6–7 kbar. *Scale*: the width of photograph corresponds to 13.5 mm. Plane-polarized light. *Image*: from a thin section lent by Robert Thériault.

Further reading: Guernina, S. & Sawyer, E.W. (2003): Large-scale melt-depletion in granulite terranes: an example from the Archaean Ashuanipi subprovince of Quebec. *Journal of Metamorphic Geology* **21**, 181–201.

Sawyer, E.W. (2001): Melt segregation in the continental crust: distribution and movement of melt in anatectic rocks. *Journal of Metamorphic Geology* **19**, 291–309.

Figure **F42**

Fig. F42. The photomicrograph shows the same field of view as Fig. F41, but in cross-polarized light to show the microstructures indicative of recovery and recrystallization in the quartz. Grain-boundary readjustments in the quartz and feldspar during and after deformation have modified some of the primary grain-boundary relationships (e.g., straight faces on plagioclase against quartz) formed when this leucosome crystallized from anatectic melt.

Location: Ashuanipi Subprovince, Quebec, Canada. *Rock type*: metatexite migmatite; aluminum-poor metagreywacke protolith, granulite facies, *T* 825–875°C, *P* 6–7 kbar. *Scale*: the width of the photograph corresponds to 13.5 mm. Cross-polarized light. *Image*: From a thin section lent by Robert Thériault.

Figure F43

Fig. F43. This photomicrograph is from a meter-wide layer of Archean pelitic schist. The pelite layer contains a few small domains of leucosome that are located in small shear bands and in irregular dilatant sites that are oriented approximately orthogonal to the layering. The borders of the leucosome domains are diffuse, and they do not have a conspicuous melanosome around them. If compared to the average of 30 lower-grade pelitic schists from the same sequence, the whole-rock major- and trace-element composition of the pelitic schist can be interpreted as being residual after the loss of some anatectic melt. The textural relationships among the minerals in this rock confirm its residual nature. The mineral assemblage is plagioclase (Pl) + biotite + quartz + cordierite + K-feldspar + ilmenite; sillimanite is present, but is not part of the paragenesis. Cordierite (Crd) forms porphyroblasts up to 7 mm long that are elongate in the plane of the foliation; most of the porphyroblasts contain acicular inclusions of sillimanite and small rounded inclusions of quartz and of biotite. However, the biotite, quartz, and sillimanite inclusions are never in mutual contact; they are separated by cordierite. Some inclusions of quartz contain small needles of sillimanite. Sillimanite does not occur in the schist outside the cordierite poikiloblasts. This spatial relationship among the minerals suggests that the melt-producing reaction was biotite + sillimanite + quartz + plagioclase = melt + cordierite + K-feldspar + ilmenite. Most of the cordierite is fresh, with only very minor alteration to chlorite at the ends of some porphyroblasts; the removal of the anatectic melt and the H_2O it contained effectively dehydrated the residuum so that very little rehydration of the cordierite could occur as metamorphic temperatures declined beyond the solidus. Minor amounts of texturally late muscovite (Ms) occur in the schist; it may be the product of late, near-solidus reaction of cordierite + K-feldspar with the last remaining, interstitial anatectic melt.

Location: Quetico Subprovince, Ontario, Canada. *Rock type*: residuum in a metatexite migmatite; pelitic protolith, *T* 700–800°C, *P* 3–4 kbar. *Scale*: the long side of the photomicrograph corresponds to 2.5 mm. Cross-polarized light. *Image*: E.W. Sawyer.

Further reading: Sawyer, E.W. (1987): The role of partial melting and fractional crystallization in determining discordant migmatite leucosome compositions. *Journal of Petrology* **28**, 445–473.

Figure F44

Fig. F44. In outcrop, the cordierite + K-feldspar + biotite + quartz pelitic schist from which this sample was collected contains scattered domains of K-feldspar + quartz + plagioclase leucosome oriented parallel to the foliation. Smaller but compositionally similar domains occur in extensional shear bands. Because the leucosome does not amount to more than 10% of the outcrop, the two orientations create a very open, net-like array of leucosome in outcrop. This photomicrograph is from a sample of pelitic schist collected from between the net-like array of leucosome domains. Both cordierite (Crd) and K-feldspar (Kfs) form abundant (20 modal % each) rounded poikiloblasts that contain abundant inclusions of biotite and round, corroded quartz. The foliation in the schist is defined by the preferred orientation of biotite, and wraps around the cordierite + K-feldspar domains, creating an augen-like microstructure; some penetrative deformation thus occurred in this migmatite after the poikiloblasts had formed. Sillimanite is not present in this particular sample, but in the adjacent schist, a small proportion of the cordierite poikiloblasts contains aggregates of sillimanite needles that outline small crenulation-type folds, but the sillimanite inclusions are not in contact with quartz and biotite, also present as inclusions in the cordierite. Hence, a possible melt-producing reaction for the migmatite in the photomicrograph was biotite + sillimanite + quartz = melt + cordierite + K-feldspar. Because cordierite is widely preserved in this residual rock and not replaced by biotite, muscovite, or sillimanite, H_2O generated from the dehydration melting of reactant biotite must have left the residuum in the anatectic melt. Hence, from its mineralogy and the lack of retrograde rehydration, this sample of migmatite is interpreted to be melt-depleted residuum.

Location: Central Zone of the Damara Belt, Namibia. *Rock type*: residuum in a metatexite migmatite; plagioclase-poor, aluminous pelite protolith, anatexis in the lower granulite facies, T ca. 800°C, P 3–4 kbar. *Scale*: the long side of the image corresponds to 5.5 mm. Cross-polarized light. *Image*: E.W. Sawyer.

Further reading: Pickering, J.M. & Johnston, A.D. (1998): Fluid-absent melting behavior of a two-mica metapelites: experimental constraints on the origin of Black Hills granite. *Journal of Petrology* **39**, 1787–1804.

White, R.W. & Powell, R. (2002): Melt loss and the preservation of granulite facies mineral assemblages. *Journal of Metamorphic Geology* **20**, 621–632.

White, R.W., Powell, R. & Holland, T.B J. (2001): Calculation of partial melt equilibria in the system Na_2O–CaO–K_2O–FeO–MgO–Al_2O_3–SiO_2–H_2O (NCKFMASH). *Journal of Metamorphic Geology* **19**, 139–153.

Figure **F45**

Fig. F45. This photomicrograph shows a highly residual, granulite-facies rock that in outcrop occurs as thick, light-colored layers and as isolated boudins in syn-anatectic shear zones. Neither mode of occurrence has many spatially associated domains of leucosome. This sample was collected >20 cm from the nearest domain of leucosome. The mineral assemblage is cordierite + orthopyroxene + biotite + sillimanite + spinel. Cordierite is easily identified by its prominent, yellow pleochroic halos, and the orthopyroxene, by its high relief. A few tiny, isolated, equant grains of quartz also are present in some of the biotite aggregates, but quartz represents no more than trace amounts. Biotite occurs as large individual grains, or as aggregates generally interstitial to the cordierite and partially enclosing some orthopyroxene grains. Very small round inclusions of biotite occur in the cordierite; in contrast, the inclusions of biotite in the orthopyroxene are larger and more irregular in shape. However, both types are interpreted as relics of reactant biotite. The photomicrograph shows that some cordierite–cordierite and cordierite–orthopyroxene boundaries are almost black. This coloration reflects the presence of various very fine-grained minerals. Green hercynite occurs as a dusting of tiny crystals along some orthopyroxene–cordierite contacts (A) and, rarely, as tiny aggregates in cordierite; the spinel is possibly a product of quartz-absent prograde reactions. Aggregates of very small elongate crystals of biotite occur along some cordierite and orthopyroxene grain boundaries; these are commonly oriented at a high angle to the grain edges and might be a late alteration of cordierite and orthopyroxene. There are also cordierite–cordierite boundaries that are dark brown (B); this is due to a very fine intergrowth of biotite and sillimanite in which the orientation of the tiny acicular crystals is controlled by the crystallographic orientation of the cordierite, and not the orientation of the grain boundary. Melting involving biotite, quartz, sillimanite, and plagioclase is interpreted to have proceeded to the extent that quartz, sillimanite, and plagioclase were all consumed, leaving a residuum dominated by cordierite + orthopyroxene. Some biotite may be residual, but the aggregates of large grains of biotite associated with the quartz blebs, the aggregates of small grains of biotite, and the biotite–sillimanite intergrowths are all interpreted as the result of a reaction between the residual minerals cordierite and orthopyroxene, and a small amount of intergranular melt left on the grain boundaries after virtually all of the melt generated by anatexis had migrated away and contributed to the leucosome elsewhere in the migmatite.

Location: Wuluma Hills, Arunta Inlier, Australia. *Rock type*: residuum in a metatexite migmatite; aluminous metapelite protolith, granulite-facies anatexis, T 825–875°C, P ca. 5 kbar. *Scale*: the long side of the image corresponds to 7 mm. Plane-polarized light. *Image*: E.W. Sawyer.

Figure F46

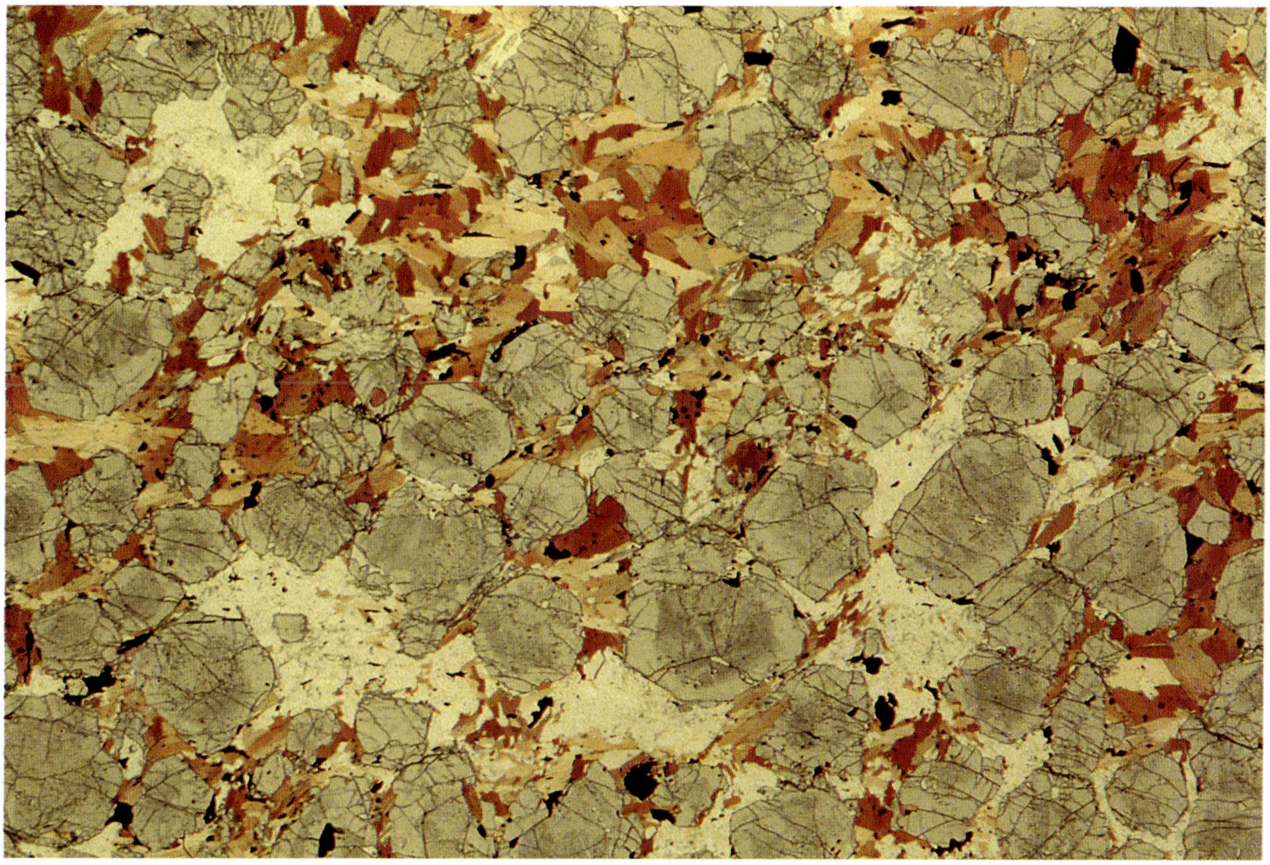

Fig. F46. Rocks that have undergone a high degree of partial melting and loss of melt are distinctive because of their very high modal proportion of residual minerals. The extremely melt-depleted nature of the pelitic migmatite in this photomicrograph is indicated by the abundance of large, zoned crystals of garnet. The colorless (white) interstitial areas between the garnet crystals consist of quartz, K-feldspar, and plagioclase (An_{20-40}), and are interpreted to be derived from the crystallization of anatectic melt that was interstitial to the garnet crystals. Large crystals of biotite partially replace the garnet, and they may have formed by reaction between the residual garnet and the small amount of anatectic melt trapped between the garnet crystals. The rock also has a high modal abundance of accessory minerals; this might also be consistent with a highly residual nature. Geochemical mass-balance calculations based on the whole-rock composition of this rock and a plausible pelitic protolith are consistent with the garnet-rich rock being the melt-depleted residuum from a pelitic protolith that has undergone extensive partial melting, and has lost at least 50 vol.% anatectic melt. However, it is not clear whether such extensive partial melting occurred by dehydration melting of biotite at metamorphic temperatures in excess of 850°C, or whether it occurred at somewhat lower temperatures (750–800°C) through H_2O-fluxed partial melting. The highly unusual field-relations of this particular rock are examined in Fig. B33.

Location: Abaukoma, Yaoundé, Cameroon. *Rock type*: residuum in a metatexite migmatite; metapelitic protolith, anatexis at *T* ca. 850°C, *P* 10–12 kbar. *Scale*: the long side of the photograph corresponds to 25 mm. Plane-polarized light. *Image*: Pierre Barbey.

Further reading: Barbey, P., Macaudière, J. & Nzenti, J.P. (1990): High-pressure dehydration melting of metapelites: evidence from migmatites of Yaoundé (Cameroon). *Journal of Petrology* **31**, 401–427.

Nzenti, J.P., Barbey, P., Macaudière, J. & Soba, J. (1988): Origin and evolution of the late Precambrian high-grade Yaoundé gneisses. *Precambrian Research* **38**, 91–109.

Figure **F47**

Fig. F47. Mount Stafford is located in the Proterozoic Arunta Inlier of central Australia. The area around Mount Stafford consists largely of Paleoproterozoic metasedimentary rocks in which a sequence of five metamorphic zones that range from the greenschist to the granulite facies have been mapped. The area is intriguing because granulite-facies anatexis has occurred at shallow levels in the crust, corresponding to pressures of about 3.2 kbar, and has many features that suggest contact metamorphism. For example, partial melting in the interbedded metapsammites, metasiltstones (protolith for cordierite granofels) and aluminous metapelites has, in places (zones 3 and 4), occurred without segregation of the melt fraction from the solid fraction and generated some unusual migmatites (see Figs. A1 and B17) that have essentially isotropic neosome, i.e., without foliation or magmatic flow-induced structures. However, leucosome is developed in some migmatites. An optically continuous quartz crystal (light grey) encloses euhedral crystals of K-feldspar within a leucosome; this microstructure is typical of the crystallization of a granitic magma. The K-feldspar has perfectly developed crystal faces against the interstitial quartz; contacts between K-feldspar crystals are less well developed; however, they are, in general, also planar. Numerous small inclusions of quartz occur in the core of many crystals of K-feldspar. Most probably the interior of the K-feldspar grains formed during the subsolidus breakdown of coexisting muscovite and quartz. The absence of strain features in the quartz, such as undulose extinction, or evidence for the mobility of grain boundaries, is very unusual for a regional metamorphic terrane, and illustrates that the migmatites at Mt. Stafford crystallized in an environment of low strain.

Location: Mt. Stafford, Arunta Inlier, Australia. *Rock type*: leucosome; semipelitic protolith, anatexis at T 780–810°C, P 3.5–4 kbar. *Scale*: the long side of the image corresponds to 3 mm. Cross-polarized light. *Image and caption*: Richard White. This image was published as fig. 2g in White et al. (2003) and is reproduced with the permission of Oxford University Press.

Further reading: Greenfield, J.E., Clarke, G.L., Bland, M. & Clarke, D.L. (1996): *In situ* migmatite and hybrid diatexite at Mt. Stafford, central Australia. *Journal of Metamorphic Geology* **14**, 413–426.

Vernon, R.H., Clarke, G.L. & Collins, W.J. (1990): Local, mid-crustal granulite facies metamorphism and melting: an example from the Mount Stafford area, central Australia. *In* High Temperature Metamorphism and Crustal Anatexis (J.R. Ashworth & M. Brown, eds.). Mineralogical Society, Series **2**. Unwin Hyman, London, U.K. (272–319).

White, R.W., Powell, R. & Clarke, G.L. (2003): Prograde metamorphic assemblage evolution during partial melting of metasedimentary rocks at low pressures: migmatites from Mt. Stafford, central Australia. *Journal of Petrology* **44**, 1937–1960.

Figure F48

Fig. F48. An optically continuous crystal of K-feldspar (light grey, faintly striped mineral) encloses euhedral crystals of quartz (hexagonal outlines) in a leucosome from metamorphic zone 4 at Mt. Stafford. Some of the quartz crystals appear black because they are viewed along the *c* crystallographic axis and seem, therefore, isotropic. Biotite crystals are bladed in shape, and have some straight faces (i.e., they are subhedral) where in contact with the K-feldspar. This microstructure is interpreted to indicate that both biotite and quartz crystallized from the anatectic melt forming the leucosome, and before the K-feldspar crystallized from the remaining melt.

Location: Mt. Stafford, Arunta Inlier, Australia. *Rock type*: leucosome; semipelite protolith, anatexis at *T* 780–810°C, *P* 3.5–4 kbar. *Scale*: the long side of the image corresponds to 2.2 mm. Cross-polarized light. *Image and caption*: Richard White. This image was published as fig. 2h in White et al. (2003) and is reproduced with the permission of Oxford University Press.

Further reading: as for Fig. F47, Greenfield, J.E., Clarke, G.L. & White, R.W. (1998): A sequence of partial melting reactions at Mt. Stafford, central Australia. *Journal of Metamorphic Geology* **16**, 363–378.

Figure **F49**

Figure **F49**

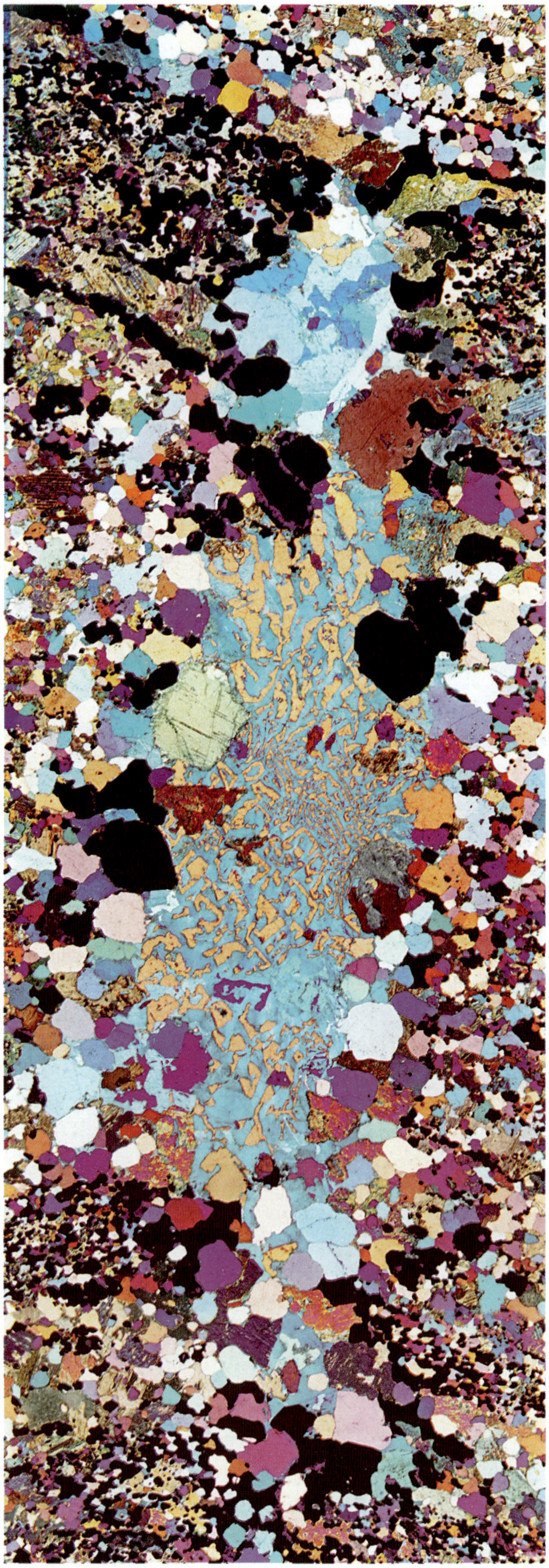

Fig. F49. This mosaic shows an *in situ* lens of leucosome (about 20 mm long) located between two boudins in a metasedimentary rock of the Biwabik Formation, from the footwall of the Duluth Igneous Complex. The host is thinly bedded iron-formation in which the bedding is manifested by changes in the modal proportions and grain size of magnetite, plagioclase, quartz, and clinopyroxene. The bedding is oriented subparallel to the short side of the figures. The leucosome has very irregular margins at the grain scale, a feature that is common in bodies of *in situ* leucosome. The central part of the leucosome is characterized by large crystals of plagioclase that contain conspicuous graphic intergrowths of quartz. However, in a few plagioclase crystals, the graphic intergrowth is with K-feldspar (microcline) and not quartz. Most of the quartz and plagioclase at the ends of the lens of leucosome occurs as separate anhedral, equant grains; a graphic intergrowth of quartz in plagioclase there is very rare, and is also very poorly formed. Because quartz is the more abundant mineral at the ends of the leucosome domain, and plagioclase more abundant in the center, there is a compositional gradient along the leucosome. Nevertheless, overall it has a granodioritic bulk composition. Some of the magnetite and clinopyroxene crystals that project into the leucosome from the wallrocks have crystal faces against the plagioclase in the leucosome. These minerals are thus interpreted to have grown into a space filled with melt. In contrast, the opposite ends of these same crystals have curved and interlocking grain boundaries, consistent with growth in a solid, i.e., the wallrock. The grain size of the crystals in the wallrock shows a systematic increase toward the leucosome. This is most readily seen around the widest part of the leucosome, where all the minerals show an increase in grain size by a factor of 5–10 toward the leucosome. The increase in grain size may occur because there was anatectic melt along a significant proportion of the grain boundaries close to the leucosome, and this enabled a more rapid grain-boundary migration, and hence grain coarsening to occur there. Magnetite crystals that are completely enclosed within the leucosome are either euhedral or skeletal in form, whereas those in the wallrocks have irregular or rounded grain-boundaries. The presence of a granophyric microstructure, and of crystals with skeletal shapes in the leucosome, are both features that indicate a rapid to moderately rapid rate of cooling for the anatectic melt in the contact aureole of the Duluth Igneous Complex.

Location: Duluth Igneous Complex, Minnesota, U.S.A. *Rock type*: *in situ* leucosome in a metatexite migmatite; Fe-oxide-rich metasedimentary protolith, T ca. 900°C, P 1.5–2 kbar. *Scale*: the width of the photograph is 10 mm. Left: cross-polarized light. Right: cross-polarized light plus the quartz accessory plate. *Image*: E.W. Sawyer.

Figure F50

Fig. F50. One of the first macroscopic signs of partial melting in the contact aureole around the Duluth Igneous Complex (DIC) is the development of small (<10 mm) patches of neosome in the pelitic hornfelses. This small *in situ* patch of neosome is typical of those in the lowest-grade migmatites. It has a large inner part that consists of a quartz + plagioclase + K-feldspar + muscovite leucosome in which the muscovite (Ms) is commonly arranged in small aggregates of radiating crystals. In contrast, neosome, leucosome, and leucocratic veins in the higher-temperature inner part of the contact aureole rarely contain muscovite. The quartz and feldspar in the leucosome tend to form small anhedral, interlocking crystals, which results in a microstructure that is substantially different from that in the leucosome from the highest-grade parts of the aureole (e.g., Figs. F53 and F54); in particular, the grain size is so small that no grains extend from the margins to the center of the leucosome. The outer part of the neosome is a narrow annulus (<0.5 mm) of melanosome that is rich in biotite. However, cordierite does not appear to be enriched in the melanosome compared to the metapelite host farther away. Melting in the outer parts of the aureole thus may not have involved the incongruent breakdown of biotite, but could have involved the H_2O-fluxed melting of quartz, plagioclase, and K-feldspar.

Location: Duluth Igneous Complex, Minnesota, U.S.A. *Rock type*: *in situ* leucosome in a patch metatexite migmatite; Virginia Formation metapelite, *T* ca. 750°C, *P* 1.5–2 kbar. *Scale*: the width of the photograph is 9 mm. Cross-polarized light. *Image*: E.W. Sawyer.

Figure F51

Fig. F51. This in-source leucosome, from about halfway between the onset of partial melting and the appearance of diatexite migmatites, is typical of the larger bodies of leucosome from the outer, lower-grade part of the contact aureole of the Duluth Igneous Complex. Like most of the leucosome in the outer part of the aureole, this one is in an extensional shear zone (see Fig. F88 for an example) and, therefore, is slightly discordant to the foliation in the fine-grained host, a quartz–biotite–plagioclase–K-feldspar–cordierite pelitic schist. Comparing this leucosome with that in Fig. F53 shows that there are significant petrological differences between the outer and inner aureoles. The main difference is the widespread occurrence of bladed muscovite (Ms) in the leucosome of the outer aureole, and the narrow, dark-colored rim around the leucosome (most clearly evident on the left side) due to alteration of cordierite in the adjacent wallrock; the dark rim is not melanosome. Both of these features indicate that the melt from which the leucosome crystallized in the outer aureole was richer in H_2O than the melt in the inner aureole. The bulk composition of the leucosome suggests that it is not the product of fractional crystallization. Therefore, the higher H_2O contents may be due to the occurrence of H_2O-fluxed melting in the outer aureole, whereas dehydration melting predominated in the inner aureole. Another difference is that the domains of leucosome in the outer aureole are significantly finer-grained relative to their width than those in the inner aureole. Consequently, they have a different microstructure dominated by small anhedral grains that have irregular grain-boundaries.

Location: Wyman Creek, Duluth Igneous Complex, Minnesota, U.S.A. *Rock type*: in-source leucosome, metatexite migmatite; pelitic protolith, pyroxene hornfels facies, *T* ca. 750°C, *P* 1.5–2 kbar. *Scale*: the long side of the photomicrograph corresponds to 9 mm. Cross-polarized light. *Image*: E.W. Sawyer.

Figure F52

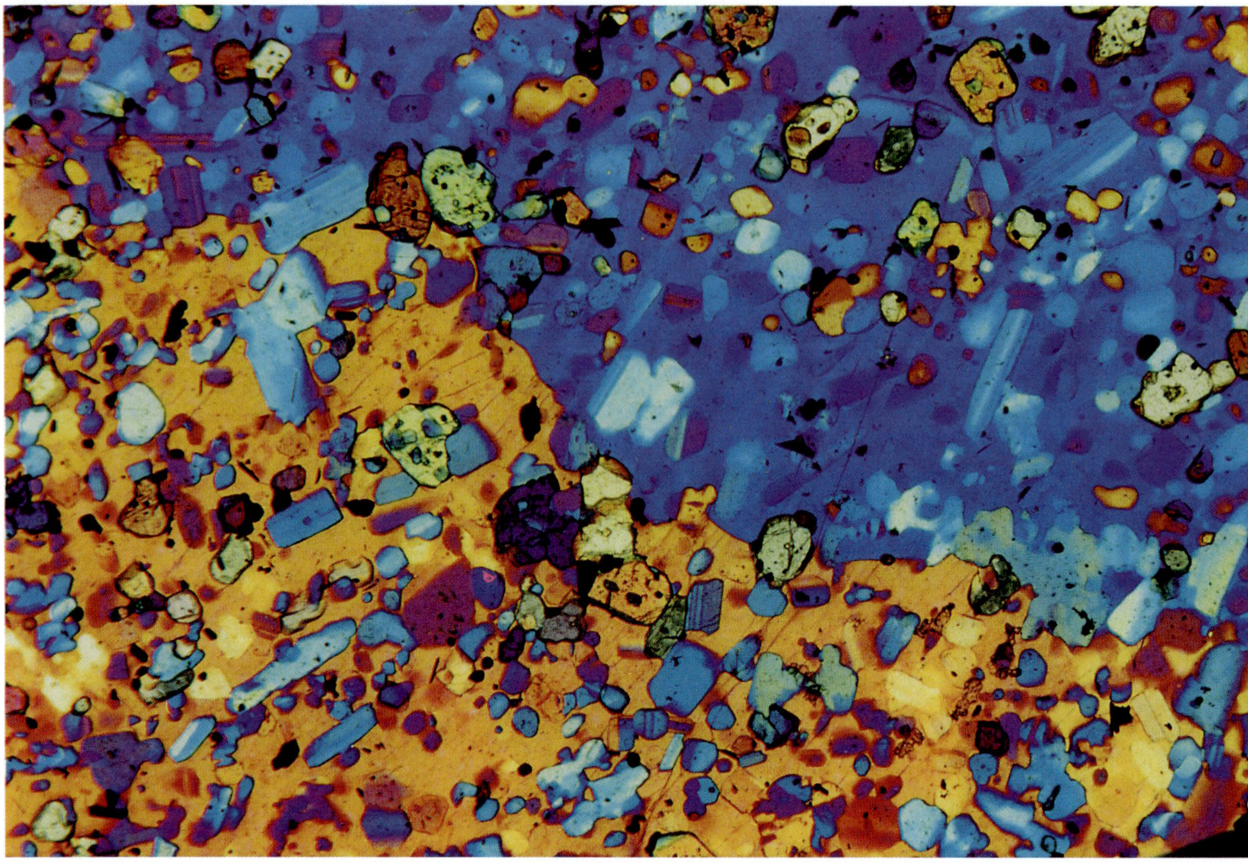

Fig. F52. This photomicrograph shows part of a small (2 cm long) elongate patch of *in situ* leucosome in a residual metapelite that was collected from a xenolith located close to the contact with the Duluth Igneous Complex. Consequently, this occurrence of leucosome is from the very highest grade part of the contact aureole. The patch of leucosome contains evidence that the melt in it flowed, and hence that its elongate shape was acquired before the leucosome solidified. The melt in the leucosome contained a few scattered, equant crystals of orthopyroxene (high relief), but it also contained many small, elongate, euhedral crystals of plagioclase. The plagioclase crystals locally show a strong alignment that is parallel to the long axis of the patch of leucosome. This alignment is interpreted to have occurred when the small lens of magma in the pelite was deformed and became elongated. The last part of the microstructure to form developed when the melt containing the aligned minerals crystallized and formed the large xenomorphic grains of quartz, K-feldspar, and plagioclase that now host the aligned smaller crystals. The lack of any optical evidence for crystal-plastic deformation in the last-crystallized minerals, such as undulose extinction or the formation of polygonal subgrains, indicates that the final stages of crystallization of the melt occurred after the deformation that elongated the leucosome. The photomicrograph shows the contact between large, late-crystallized crystals of quartz (blue) and K-feldspar (yellow).

Location: Duluth Igneous Complex, Minnesota, U.S.A. *Rock type*: *in situ* leucosome in a residual metatexite migmatite; Virginia Formation pelite protolith, T > 900°C, P 1.5–2 kbar. *Scale*: the width of the photograph is 2.5 mm. Cross-polarized light and the quartz plate inserted. *Image*: E.W. Sawyer.

Figure F53

Fig. F53. This example of leucosome is from the transition from metatexite to diatexite migmatite. The photomicrograph shows a fine-grained pelitic hornfels (upper right) that has a foliation defined by the orientation of biotite. The foliation is crossed at an angle of about 20° by a 4-mm-wide, discordant domain of leucosome (lower left). There is no melanosome next to this segment of the leucosome, which suggests that the leucosome is not *in situ*. Consequently, it should be called a leucocratic vein. However, as the leucosome can be traced back a few centimeters to where it becomes parallel to the foliation and has a melanosome that is enriched in cordierite, at least that part of the leucosome can be considered to be *in situ*. Thus, the term leucosome does seem appropriate. The microstructure of the leucosome is dominated by large, anhedral crystals of quartz (Qtz), which extend from the margins of the leucosome to its interior (cf. Fig. F51). Close examination shows that the contact of the leucosome appears to be somewhat diffuse, and quartz crystals in the leucosome can be traced to the wallrock contact, and locally beyond it into the wallrocks while maintaining optical continuity. The melt from which the leucosome crystallized contained abundant small crystals, and these are now preserved as inclusions in the large crystals of quartz. Abundant large, dark bluish grey or black (in extinction position) equant inclusions in the quartz are K-feldspar (Kfs), and these are interpreted to have crystallized from the melt. Small, dark, hexagonal crystals with sector twinning are euhedral cordierite (Crd), and this too probably grew from the melt. Plagioclase is abundant and occurs mostly as small anhedral grains interstitial to the large crystals of quartz and K-feldspar. Some plagioclase exhibits lamellar twins. Biotite occurs as small rounded inclusions in quartz and feldspar and as large, randomly oriented bladed crystals; the former are interpreted to be relics of residual biotite, and the latter, as late biotite grown from the melt. There is no muscovite. The photomicrograph is not representative of the leucosome as a whole, as it is somewhat richer in quartz; this part was selected because it shows a greater range in interference colors, making it easier to distinguish the individual crystals.

Location: Wetlegs, Duluth Igneous Complex, Minnesota, U.S.A. *Rock type*: leucosome in a metatexite migmatite; pelitic protolith, pyroxene hornfels facies, T ca. 800°C, P 1.5–2 kbar. *Scale*: the long side of the photomicrograph corresponds to 4.5 mm. Cross-polarized light. *Image*: E.W. Sawyer.

Figure F54

Fig. F54. This photomicrograph shows a K-feldspar-rich leucocratic vein 4 mm wide in a partially melted, sulfide-bearing quartz + phlogopite + plagioclase + K-feldspar + cordierite aluminous metasediment. The contacts between the leucocratic vein and its host are irregular, possibly because the host was itself partially molten when the vein was injected. The vein has the mineral assemblage K-feldspar + quartz + plagioclase, with minor amounts of muscovite and tourmaline. The microstructure in the leucocratic vein is dominated by equant shapes of crystals, and is different from that in leucosome from lower-temperature migmatites (cf. Figs. F50 and F51). Small crystals of quartz and plagioclase are prevalent at the border, whereas large (2 mm) crystals of K-feldspar are preponderant in the center. However, in places (e.g., at the lower left), the center of the vein contains large crystals of muscovite that are associated with an intergrowth of tourmaline (light grey) and K-feldspar (dark blue–grey). Some of the large crystals of K-feldspar in the center of the vein have straight contacts against quartz, a microstructure generally interpreted as evidence for crystallization from a melt. Leucosome and leucocratic veins of syenogranitic composition are relatively common in the migmatites around the Duluth Igneous Complex. Some of these are undoubtedly primary anatectic melts derived from plagioclase-poor pelitic protoliths (see Further reading). However, most are the result of the fractional crystallization of the anatectic granitic melts. From the presence of the tourmaline and muscovite, one can infer that this leucocratic vein crystallized from an anatectic melt that had an evolved bulk composition, possibly due to the crystallization and fractionation of plagioclase.

Location: Wetlegs, Duluth Igneous Complex, Minnesota, U.S.A. *Rock type*: in-source leucosome, sulfide-bearing metatexite migmatite; pelitic protolith, pyroxene hornfels facies, T 800–850°C, P 1.5–2 kbar. *Scale*: the long side of the photomicrograph corresponds to 9 mm. Cross-polarized light. *Image*: E.W. Sawyer.

Further reading: Grant, J.A. (2004): Liquid compositions from low-pressure experimental melting of pelitic rock from the Morton Pass, Wyoming, USA. *Journal of Metamorphic Geology* **22**, 65–78.

Figure F55

Fig. F55. Although the mineral assemblage in this leucocratic vein, K-feldspar + quartz + plagioclase + muscovite + tourmaline, is the same as that in Fig. F54, its microstructure is very different. The constituent minerals of this vein are generally not equant, and have a far greater tendency to show irregular and embayed shapes. In places, however, crystal faces are developed where K-feldspar or plagioclase are against quartz. Tourmaline (white mineral near the center of the image) is rare and occurs as irregularly shaped crystals intergrown with K-feldspar and quartz. Muscovite occurs as large grains, locally intergrown with quartz, and as a minor replacement of K-feldspar. The difference in microstructure may be due to the greater width of this vein (15 mm), or to its location farther out in the aureole, and therefore, its formation in lower-temperature migmatites. The presence of tourmaline and muscovite indicates that this leucocratic vein crystallized from a melt that may have had an evolved, or fractionated, composition.

Location: Wetlegs, Duluth Igneous Complex, Minnesota, U.S.A. *Rock type*: leucocratic vein in sulfide-bearing metatexite migmatite; pelitic protolith, pyroxene hornfels facies, *T* ca. 750–800°C, *P* 1.5–2 kbar. *Scale*: the long side of the photomicrograph corresponds to 9 mm. Cross-polarized light. *Image*: E.W. Sawyer.

Figure **F56**

Fig. F56. This photomicrograph, from a domain of leucosome 15 mm wide in a migmatite derived from a plagioclase-rich pelite protolith (similar to that in Fig. F43), shows features common to leucosome from many regional migmatite terranes. Equant, subhedral to euhedral crystals of plagioclase form a framework of touching crystals. The interstices in the framework of feldspar crystals are filled with quartz, as in this photograph, or in other cases, by quartz + K-feldspar. Some of the plagioclase crystals have straight, rational faces against the quartz, a feature that is commonly taken to indicate crystallization from a melt. However, some of the other plagioclase crystals in the framework have lobate contacts with quartz. Straight plagioclase–plagioclase contacts are commonly the result of a planar, rational face on one of the crystals and simple impingement by the other, but many plagioclase–plagioclase contacts are irregular. The bulk composition of a plagioclase-rich leucosome, such as this one, can commonly be modeled as a plagioclase cumulate derived by the fractional crystallization of an anatectic melt of granitic composition. Evidently, if an evolved melt has been expelled in making a cumulate leucosome, such as this one, then other bodies of leucosome (see Fig. F57) or veins somewhere in the migmatite should be derived from this evolved, expelled melt and have appropriate mineral assemblages, microstructures, and bulk compositions (see Further reading).

Location: Quetico Subprovince, Ontario, Canada. *Rock type*: in-source leucosome with plagioclase cumulate microstructure and composition from a metatexite migmatite; pelitic protolith, T 700–800°C, P 3–4 kbar. *Scale*: the long side of the photomicrograph corresponds to 5.5 mm. Cross-polarized light. *Image*: E.W. Sawyer.

Further reading: Sawyer, E.W. (1987): The role of partial melting and fractional crystallization in determining discordant migmatite leucosome compositions. *Journal of Petrology* **28**, 445–473.

Figure F57

Fig. F57. Not all feldspar frameworks in leucosome consist of plagioclase. This photomicrograph shows another example of leucosome, 25 mm wide, from the Quetico Subprovince, but in this instance, the feldspar framework is constructed of subhedral and euhedral crystals of microcline. The interstices in the microcline framework are filled with quartz, and in places, the microcline has straight crystal faces against the quartz, evidence that its precursor (a disordered feldspar) crystallized from a melt. This leucosome also contains some subhedral and euhedral crystals of plagioclase (noticeably cloudy owing to alteration to small crystals of white mica) which, although smaller than the microcline crystals, are nevertheless also part of the feldspar framework in the leucosome. Plagioclase crystals in the feldspar framework rarely contain inclusions of quartz or K-feldspar, whereas the K-feldspar crystals in the framework commonly contain inclusions of both quartz and plagioclase. Bodies of K-feldspar-rich leucosome in metatexite migmatites derived from aluminum-poor psammitic and plagioclase-rich pelitic protoliths can commonly be modeled as crystallizing from the fractionated, or evolved, melts generated by the fractional crystallization of plagioclase from an anatectic melt of granitic composition. Thus, this leucosome could have crystallized from the evolved melt generated by a cumulate process that made a plagioclase-rich leucosome of the type shown in Fig. F56.

Location: Quetico Subprovince, Ontario, Canada. *Rock type*: stromatic, in-source leucosome in a metatexite migmatite; pelitic protolith, anatexis at *T* 700–800°C, *P* 3–4 kbar. *Scale*: the long side of the photomicrograph corresponds to 9 mm. Cross-polarized light. *Image*: E.W. Sawyer.

Figure **F58**

Fig. F58. This leucocratic vein is hosted by a biotite–quartz–plagioclase psammitic schist (greywacke layer) that has not undergone partial melting, and hence it is interpreted to have been injected. The overall bulk composition of the leucocratic vein is granitic to syenogranitic, which suggests that it crystallized from a melt that had an evolved bulk composition. Compared to the leucosome in Fig. F57, which has a similarly high content of microcline, this example is richer in biotite; possibly the biotite is residual. The feldspar crystals tend to be rectangular, and locally, they have straight boundaries against quartz, which is interpreted to indicate that they crystallized from a melt.

Location: Quetico Subprovince, Ontario, Canada. *Rock type*: stromatic, leucocratic vein in a metatexite migmatite; pelitic protolith, anatexis at T 700–800°C, P 3–4 kbar. *Scale*: the long side of the photomicrograph corresponds to 14 mm. Cross-polarized light. *Image*: E.W. Sawyer.

Figure F59

Fig. F59. The equant, euhedral to subhedral crystals of plagioclase in this diatexite migmatite form a very open framework in which the crystals touch, almost exclusively at their corners. The interstitial space is close to the theoretical maximum for a framework, and is filled mostly with quartz. There is a minor amount of titaniferous biotite and trace amounts of K-feldspar also. Almost all of the plagioclase grains have crystal faces against quartz, which is interpreted as evidence for growth in a melt. The preservation of this delicate framework and the equant but irregular shape of the interstitial quartz suggest that there was little synmagmatic or postmagmatic strain. However, the undulose extinction in some of the quartz indicates minor post-solidification strain. The plagioclase is unlikely to be residual in origin, because in the residual, melt-depleted metagreywackes, it has a grain size of about 0.5 mm, whereas in the diatexite migmatites, the grains are 1–5 mm across. Crystals of residual plagioclase in the anatectic melt may have provided sites for the growth of plagioclase from the melt.

Location: Ashuanipi Subprovince, Quebec, Canada. *Rock type*: diatexite migmatite; plagioclase + quartz + biotite metagreywacke protolith, anatexis at *T* 825–850°C, *P* 6–7 kbar. *Scale*: the long side of the photomicrograph corresponds to 12.5 mm. Cross-polarized light. *Image*: E.W. Sawyer.

Figure **F60**

Fig. F60. This photomicrograph shows a diatexite migmatite that has a very dense framework structure comprised of equant subhedral to euhedral crystals of plagioclase. Virtually all the plagioclase crystals are in contact along their faces rather than just at their corners; consequently, there is very little interstitial space. That interstitial space is filled with quartz, K-feldspar, and biotite. Locally, the crystals of plagioclase have straight, rational faces against the interstitial quartz, suggesting that they crystallized in a melt. The pore-space is partially lined with a thin film of K-feldspar in some places, but it is primarily filled with quartz. The irregular form of biotite and its location at grain junctions suggest that it crystallized from the melt in the interstitial pore-space. In outcrop, this diatexite migmatite contains biotite schlieren and oriented, elongate rafts, or schollen, of residual metagreywacke, which together indicate a high shear strain in the diatexite magma. The plagioclase in the framework is not strongly oriented, nor do the plagioclase crystals appear to be deformed; hence, the flow must have occurred while the diatexite contained a significant fraction of melt. Subsequently, a large proportion of the fractionated melt generated by the crystallization of the plagioclase was expelled. The result is a plagioclase-dominated diatexite with a bulk composition that can be modeled by the accumulation of plagioclase.

Location: Ashuanipi Subprovince, Quebec, Canada. *Rock type*: plagioclase-rich diatexite migmatite; plagioclase + quartz + biotite metagreywacke protolith, partial melting at T 825–850°C, P 6–7 kbar. Scale: the long side of the photomicrograph corresponds to 12.5 mm. Cross-polarized light. Image: E.W. Sawyer.

Figure F61

Fig. F61. The plagioclase framework in this diatexite migmatite is very open, like that in Fig. F59, with most plagioclase grains in contact at the corners only. About half of the interstitial space has been filled by a large, single crystal of K-feldspar (now microcline); quartz and biotite fill the rest of the interstitial space. The bulk composition of this rock is granitic. Some interstices show that the space was first lined with a thin film of K-feldspar, and then filled with quartz. Overall, this microstructure indicates that plagioclase crystallized as euhedral and subhedral grains first, and formed the framework structure before K-feldspar and then quartz began to crystallize and fill in the pore space. A few crystals of plagioclase have rational faces against either K-feldspar or quartz, indicative of crystallization from a melt. The overall microstructure can be interpreted as resulting from the crystallization and accumulation of plagioclase before the other phases formed.

Location: Ashuanipi Subprovince, Quebec, Canada. *Rock type*: diatexite migmatite; plagioclase + quartz + biotite metagreywacke protolith, anatexis at *T* 825–850°C, *P* 6–7 kbar. *Scale*: the long side of the photomicrograph corresponds to 7 mm. Cross-polarized light. *Image*: E.W. Sawyer.

Figure **F62**

Fig. F62. The open framework made up of subhedral to euhedral crystals of feldspar in this diatexite migmatite includes of both K-feldspar (now microcline), which predominates, and plagioclase. The bulk composition of this migmatite is monzogranite to syenogranite. As the protolith was a plagioclase + quartz + biotite schist, this bulk composition suggests derivation from an evolved, or fractionated, anatectic melt that had already crystallized some plagioclase. Such a melt could have been expelled from a diatexite migmatite like that shown in Figs. F59 or F60. Most of the space between the feldspar grains in the framework is filled by large crystals of quartz, but in places, it is filled with an aggregate of small, equant, anhedral crystals of sodic plagioclase, microcline, and quartz. Elsewhere, irregularly shaped crystals of biotite, or aggregates of biotite, fill the space between the framework of feldspar grains. The overall microstructure in this diatexite migmatite indicates the crystallization and accumulation of K-feldspar and plagioclase from a melt, or a magma. At the outcrop scale, the migmatite contains schlieren and oriented schollen (rafts) of residual metagreywacke, and both these features indicate flow of the diatexite in the magmatic state. The crystals of feldspar that form the open framework structure in this diatexite migmatite are not strongly aligned; the orientation of the schollen and the formation of the schlieren thus must predate completion of the framework.

Location: Ashuanipi Subprovince, Quebec, Canada. *Rock type*: diatexite migmatite; plagioclase + quartz + biotite metagreywacke protolith, anatexis at T 825–875°C, P 6–7 kbar. *Scale*: the long side of the photomicrograph corresponds to 13.5 mm. Cross-polarized light. *Image*: E.W. Sawyer.

Figure F63

Fig. F63. Crystal frameworks are created where sufficient crystals are present in a magma that they begin to impinge on one another. Plagioclase is typically the first major phase to crystallize abundantly in magmas of granitic composition, and hence it commonly forms the framework. However, other minerals can also be part of a framework structure. This photomicrograph shows a biotite–orthopyroxene–quartz–plagioclase diatexite migmatite in which the crystal framework consists mostly of euhedral plagioclase, but also contains euhedral crystals of orthopyroxene (Opx) that are of a similar size and shape to the plagioclase. The interstices of the framework are filled by quartz and biotite. Both the plagioclase and the orthopyroxene have crystal faces against the interstitial quartz, which can be interpreted as evidence of growth in, or from, a magma. Grains of orthopyroxene in the residual rocks associated with this diatexite migmatite are small (<0.5 mm) and xenoblastic; therefore, the 4-mm-long euhedral crystals in this photomicrograph are interpreted as a cotectic phase, and not as inherited residual grains. The orthopyroxene shows only minor replacement by biotite along the edges and corners. This finding, and the general absence of K-feldspar in this rock, indicate that the crystallizing magma had lost most of the evolved melt fraction, and with it, H_2O. Had the evolved melt and its H_2O remained, then toward the final crystallization of the magma, the exsolved H_2O would have led to the replacement of the orthopyroxene by hydrous phases, such as biotite. Indeed, the orthopyroxene in diatexite migmatites that have a more evolved bulk composition and a crystal framework containing both K-feldspar and plagioclase is generally extensively altered to biotite, or to biotite + quartz symplectite, as shown in Figs. F80 and F81.

Location: Ashuanipi Subprovince, Quebec, Canada. *Rock type*: diatexite migmatite; plagioclase + quartz + biotite metagreywacke protolith, *T* 825–875°C, *P* 6–7 kbar. *Scale*: the long side of the photomicrograph corresponds to 13.5 mm. Cross-polarized light. *Image*: from a thin section lent by Robert Thériault.

Figure **F64**

Fig. F64. Many diatexite migmatites did not develop an obvious framework of crystals. This example contains scattered, euhedral crystals of plagioclase, which locally form small aggregates, or clumps of crystals. The individual crystals in some of the aggregates are arranged in an imbricate or tiled pattern that is interpreted to result from the impingement of crystals that were rotating in a magma undergoing shearing. In much of this thin section, there is no framework of large euhedral or subhedral crystals, indicating that the diatexite magma was not carrying many large crystals. The microstructure of this diatexite consists mostly of equant, but commonly irregularly shaped crystals of quartz, plagioclase, K-feldspar, muscovite, and biotite. Partial melting at Saint-Malo principally progressed through the breakdown of muscovite; biotite remained stable in most of the migmatites there. Consequently, much of the biotite in the diatexite migmatite is residual and did not crystallize from the anatectic melt. Furthermore, melting reactions involving the breakdown of muscovite generated melts with relatively high contents of H_2O. In the late stages of crystallization, muscovite was a liquidus phase. In this diatexite migmatite, a small amount of K-feldspar had crystallized before muscovite appeared on the liquidus. However, some migmatites at Saint-Malo have substantial amounts of K-feldspar, whereas in others, muscovite formed instead of K-feldspar.

Location: Saint-Malo migmatite terrane, Brittany, France. *Rock type*: diatexite migmatite; semipelitic protolith, anatexis at T ca. 750°C, P 4–7 kbar. *Scale*: the long side of the photomicrograph corresponds to 11 mm. Cross-polarized light. *Image*: E.W. Sawyer.

Further reading: Brown, M. (1979): The petrogenesis of the St-Malo migmatite belt, Armorican Massif, France, with particular reference to the diatexites. *Neues Jahrbuch für Mineralogie, Abhandlungen* **135**, 48–74.

Milord, I., Sawyer, E.W. & Brown, M. (2001): Formation of diatexite migmatite and granite magma during anatexis of semi-pelitic metasedimentary rocks: an example from St. Malo, France. *Journal of Petrology* **42**, 487–505.

Weber, C., Barbey, P., Cuney, M. & Martin, H. (1985): Trace element behaviour during migmatization. Evidence for a complex melt–residuum–fluid interaction in the St. Malo migmatitic dome (France). *Contributions to Mineralogy and Petrology* **90**, 52–62.

Figure F65

Fig. F65. Depending on the earlier history of crystallization, some diatexite migmatites formed from crystal-poor anatectic melts that had previously crystallized much plagioclase. These diatexite migmatites have fractionated, or evolved, bulk compositions. This photograph is of such a biotite–plagioclase–quartz–K-feldspar diatexite migmatite of syenogranite to granite composition from the Opatica Subprovince. In outcrop, it is a pink-colored, coarse-grained, leucocratic rock dominated by K-feldspar, a proportion of which forms euhedral crystals up to 10 mm long. Although the rock contains biotite schlieren and elongate schollen (or rafts) of paleosome and residual material, it does not have a foliation due to aligned feldspar, or aligned biotite. Either the microstructure has been reset or the flow indicated by the schlieren and oriented schollen occurred when the magma contained few crystals. The photomicrograph shows some large, subhedral crystals of plagioclase (composition An_{18-20}) with a cloudy interior, but most of the feldspar is clear microcline. K-feldspar occurs as large euhedral to subhedral crystals that with plagioclase form an open framework structure, as large crystals that fill some of the interstices in the framework structure, and as smaller irregularly shaped grains that together with quartz, biotite, and a vermicular intergrowth (myrmekite) of quartz and sodic plagioclase (An_{8-10}), fill the remaining interstitial space. The plagioclase in the K-feldspar-rich diatexites of the Opatica Subprovince is consistently more sodic (An_{15-18}) than the plagioclase (An_{22-25}) that comprises the feldspar framework in the K-feldspar-poor, plagioclase-rich diatexite migmatites. The anatectic melt in the Opatica Subprovince thus underwent fractional crystallization as the diatexite migmatites were formed.

Location: Opatica Subprovince, Quebec, Canada. *Rock type*: diatexite migmatite; plagioclase + quartz + K-feldspar + biotite leucotonalite or trondhjemite protolith, anatexis in the upper amphibolite facies, T ca. 750°C, P 5–7 kbar. *Scale*: the long side of the photomicrograph corresponds to 11.5 mm. Cross-polarized light. *Image*: E.W. Sawyer.

Further reading: Sawyer, E.W. (1998): Formation and evolution of granite magmas during crustal reworking: the significance of diatexites. *Journal of Petrology* **39**, 1147–1167.

Sawyer, E.W. (2001): Melt segregation in the continental crust: distribution and movement of melt in anatectic rocks. *Journal of Metamorphic Geology* **19**, 291–309.

Figure **F66**

Fig. F66. The tabular, subhedral crystals of biotite (0.7 mm) in this diatexite migmatite have a very strong preferred orientation that defines the foliation in the rock. The matrix around the biotite consists of much smaller (ca. 0.2 mm), interlocking, equant crystals of plagioclase, quartz, and K-feldspar, with still smaller crystals of euhedral cordierite scattered throughout. Locally, the matrix contains somewhat larger, euhedral, lath-shaped crystals of plagioclase (e.g., in the bottom right corner), and the long axes of these are oriented parallel to the foliation defined by the large crystals of biotite. The diatexite magma consisted of abundant large flakes of biotite, scattered large laths of plagioclase, and small, euhedral crystals of cordierite (the product of the incongruent breakdown of biotite) in an anatectic melt of granitic composition. The orientation of the biotite and similar-sized plagioclase crystals is interpreted to have resulted from rotation during shearing and flow when the diatexite was a magma. The cordierite crystals are small (0.07 mm) and have a low aspect-ratio, and so any preferred orientation they may have is not evident. Because most of the crystallization occurs near the solidus in a melt of granitic composition, the late-crystallizing minerals cannot grow freely and tend to develop interlocking shapes. The magmatic-flow stage of the diatexite may have ended abruptly owing to a sharp increase in viscosity as the magma neared solidification.

Location: Duluth Igneous Complex, Minnesota, U.S.A. *Rock type*: diatexite migmatite; pelite protolith, anatexis in the pyroxene hornfels facies, T 800–850°C, P 1.5–2 kbar. *Scale*: the long side of the photomicrograph corresponds to 5.5 mm. Cross-polarized light. *Image*: E.W. Sawyer.

Figure F67

Fig. F67. This cordierite–biotite–K-feldspar–quartz–plagioclase diatexite migmatite illustrates the main criteria for identifying a magmatic foliation. The large bladed crystals of biotite (0.5–0.7 mm long) are aligned and define a foliation that is approximately parallel to the long side of the photomicrograph. The largest of the plagioclase crystals (about 0.5 mm long) in the diatexite also are tabular in shape, and their long axes are subparallel to the orientation of the large grains of biotite. Most of the small crystals (0.15 mm) in the groundmass (cordierite, plagioclase, K-feldspar, and quartz) are equant and, therefore, do not appear to contribute to the foliation. The large and somewhat embayed porphyroblast of cordierite in the center of the photomicrograph is slightly elongate in a direction subparallel to the biotite, although the cordierite could also be considered as a pair of better-oriented, elongate grains that have become joined. In contrast, the large, equant but highly irregular crystal of quartz (white in the center of the picture) that envelops the porphyroblast of cordierite shows no tendency to be elongate and, moreover, displays a uniform pattern of extinction, indicating that it has relatively little internal strain. The microstructure in this diatexite migmatite is interpreted to indicate that the elongate crystals of biotite, plagioclase, and cordierite became oriented as a result of flow gradients in the diatexite magma. They developed in response to an applied shear stress. The large intergranular crystal of quartz crystallized from one of the last pockets of melt remaining after the tabular crystals had become oriented to create the foliation. The absence of undulose extinction in the quartz means that no significant penetrative deformation occurred after this diatexite had fully crystallized.

Location: Duluth Igneous Complex, Minnesota, U.S.A. *Rock type*: diatexite migmatite; pelite protolith, anatexis in the pyroxene hornfels facies, T 850–900°C, P 1.5–2 kbar. *Scale*: the long side of the photomicrograph corresponds to 2.7 mm. Cross-polarized light. *Image*: E.W. Sawyer.

Figure **F68**

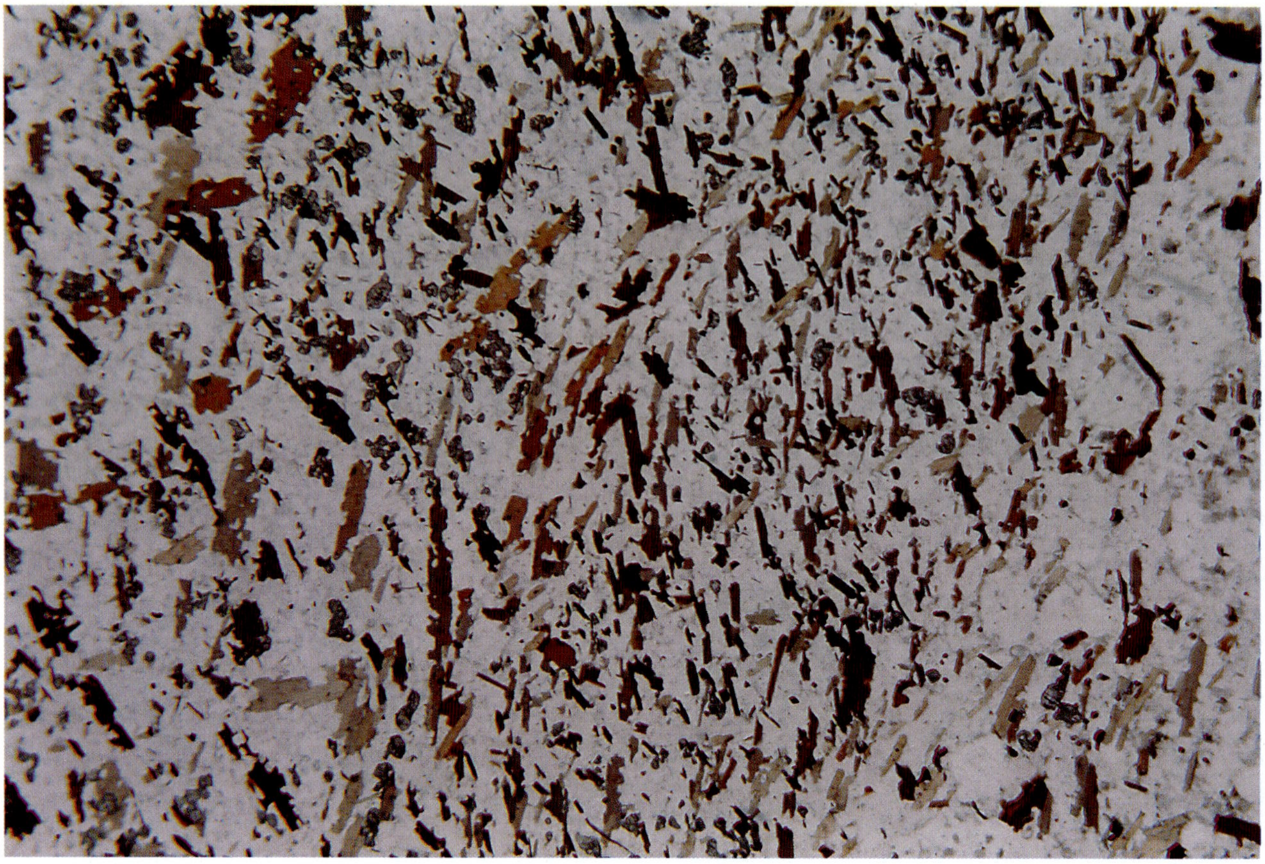

Fig. F68. This photomicrograph shows three layers in a diatexite migmatite; they have the same biotite + quartz + plagioclase + K-feldspar + orthopyroxene + cordierite assemblage, but the modal proportions and grain size differ among the layers. The cordierite-rich layer in the lower right is coarse grained, and because it has the lowest content of biotite, it appears the lightest in color. A coarse-grained orthopyroxene-rich (the high-relief mineral) layer on the left side of the photomicrograph has a slightly higher content of biotite, whereas the fine-grained, orthopyroxene-rich layer that separates them has the highest content of biotite and appears the darkest. These layers probably reflect an original compositional layering (i.e., bedding) in the metasedimentary protolith. The biotite in this migmatite displays two principal morphologies: (1) small bladed crystals that commonly define a foliation and (2) large, subhedral, randomly oriented (basal sections are common) crystals that commonly have an irregular outline (several are in the top left corner). The first type is interpreted to be residual biotite, and the second type, which is present in virtually all the diatexite migmatites at Duluth, is interpreted to have grown during crystallization of the diatexite magma. The coarser-grained layers mostly contain the late generation of large, irregularly oriented biotite, but this biotite became oriented close to the contact with the fine-grained layer. In contrast, the fine-grained layer contains mainly the small residual crystals of biotite, and they define an S-shaped, sigmoidal foliation. The fine-grained layer is interpreted to be derived from a less fertile layer of the protolith, and so generated less anatectic melt than adjacent layers. This layer recorded strain throughout its history. The fertile layers next to it consumed their reactant biotite and progressed to the magmatic state, and thus had the record of their early strain history erased when they became a magma. They carried the less-melted layer as a scholle, and during transport, it was ablated at its margins (magmatic erosion). When cooling started, the large crystals of biotite began to grow from the diatexite magma, and because of the gradient in shear strain close to the scholle, those large crystals of biotite near the scholle became oriented and recorded the sense of the late magmatic shear around the scholle. Most of the diatexite crystallized after the magma was sheared and did not develop a magmatic foliation.

Location: Duluth Igneous Complex, Minnesota, U.S.A. *Rock type*: diatexite migmatite; pelitic hornfels protolith, pyroxene hornfels facies, T 800–850°C, P 1.5–2 kbar. *Scale*: the long side of the photomicrograph corresponds to 9 mm. Plane-polarized light. *Image*: E.W. Sawyer.

Figure F69

Fig. F69. Partial melting and the formation of migmatites at Saint-Malo occurred largely through the breakdown of muscovite and was, consequently, mostly in the stability field of biotite. Therefore, a large proportion of the biotite in the migmatites is of residual origin and can be used as a marker for an investigation of the magma-flow history and the processes involved; it can also be used to identify sites of crystal accumulation in the migmatites. This photomicrograph shows the parallel alignment of large, bladed crystals of biotite and of subhedral, tabular plagioclase, respectively of residual origin and early products of melt crystallization, oriented when the magma that contained them was subject to a shear strain. Subsequent crystallization produced large, equant crystals of quartz and microcline and smaller, equant crystals of plagioclase and biotite between the oriented crystals of biotite and plagioclase. Because most of the crystallization occurs near the solidus in melts of granitic composition, the late-crystallizing minerals cannot grow freely and cannot be rotated into the foliation. Thus, they tend to develop interlocking shapes. In this way, magmatic flow may have ended abruptly in this migmatite owing to a sharp increase in viscosity as the magma neared solidification. As anhydrous phases progressively crystallized, the H_2O content of the remaining melt increased, and at the late stages of crystallization, an aqueous fluid was exsolved. It may have reacted with the minerals already crystallized. Both these effects are common at Saint-Malo, where muscovite is typically one of the last phases to crystallize from the melt (e.g., Fig. F64) and, as shown in this photomicrograph, biotite is partially replaced by muscovite.

Location: Saint-Malo migmatite terrane, Brittany, France. *Rock type*: diatexite migmatite; semipelitic protolith, anatexis at *T* ca. 750°C, *P* 4–7 kbar. *Scale*: the long side of the photomicrograph corresponds to 11 mm. Cross-polarized light. *Image*: E.W. Sawyer.

Further reading: Brown, M. (1979): The petrogenesis of the St-Malo migmatite belt, Armorican Massif, France, with particular reference to the diatexites. *Neues Jahrbuch für Mineralogie, Abhandlungen* **135**, 48–74.

Figure F70

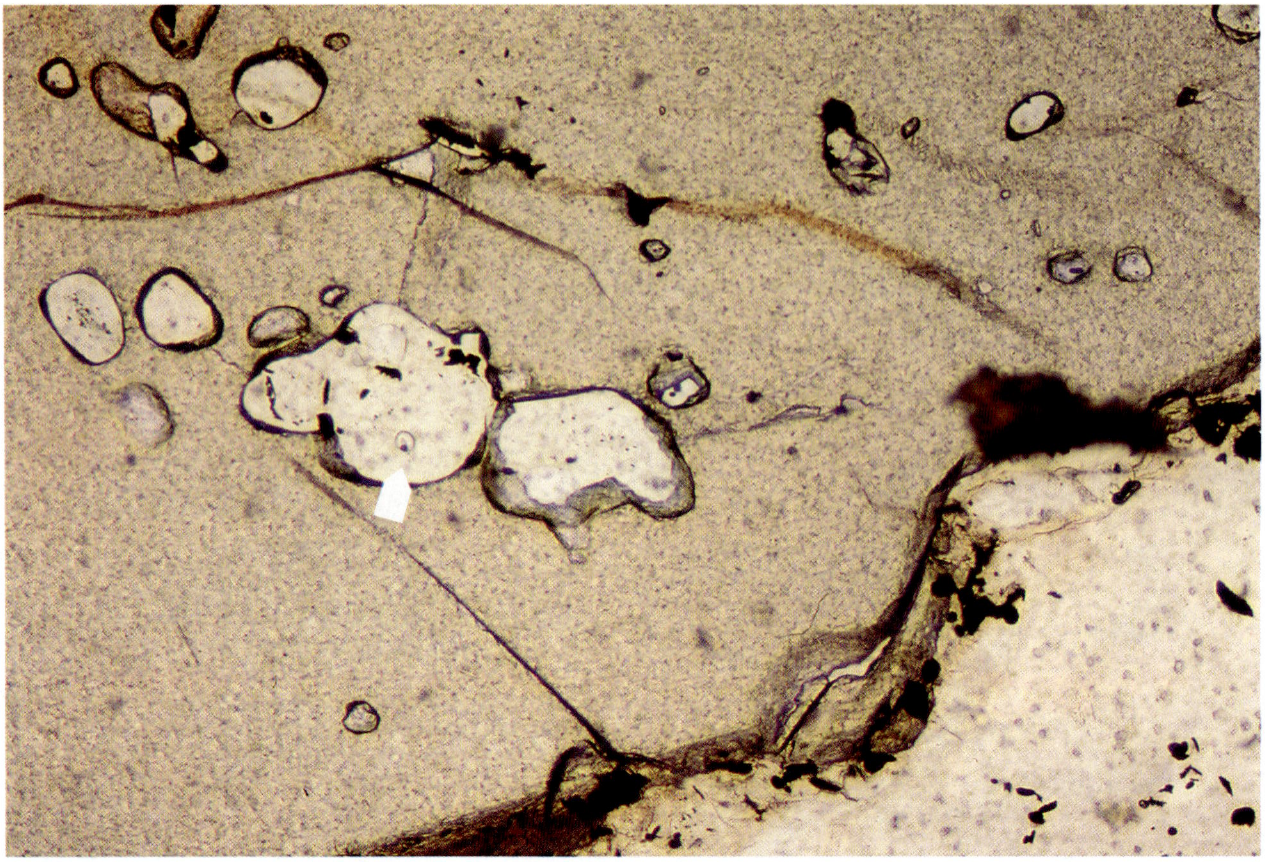

Fig. F70. Details of the mineral inclusions in a garnet porphyroblast from a metapelitic enclave that was rapidly quenched when erupted in a dacitic lava. Most of the inclusions in the garnet crystal are quartz, but one of the quartz inclusions contains a glass inclusion, and this indicates the former presence of anatectic melt in the rock. The metapelitic enclave contains the mineral assemblage biotite + garnet + sillimanite; quartz is virtually absent, and is only present as inclusions in the other minerals. This indicates that the quartz in the matrix of the pelite was totally consumed by the melting reaction, and hence that the mineralogy of the enclave is residual.

Location: El Joyazo, Spain. *Rock type*: melt inclusion in quartz in garnet; metapelite protolith melted at T 850 ± 50°C, $P > 5$ kbar, but quenched by eruption. *Scale*: the width of this plane-polarized light photograph is 1.35 mm. *Image and caption*: Bernardo Cesare.

Further reading: Cesare, B. (2000): Incongruent melting of biotite to spinel in a quartz-free restite at El Joyazo (SE Spain): textures and reaction characterization. *Contributions to Mineralogy and Petrology* **139**, 273–284.

Cesare, B., Salvioli Mariani, E. & Venturelli, G. (1997): Crustal anatexis and melt extraction in restitic xenoliths at El Joyazo (SE Spain). *Mineralogical Magazine* **61**, 15–27.

Figure F71

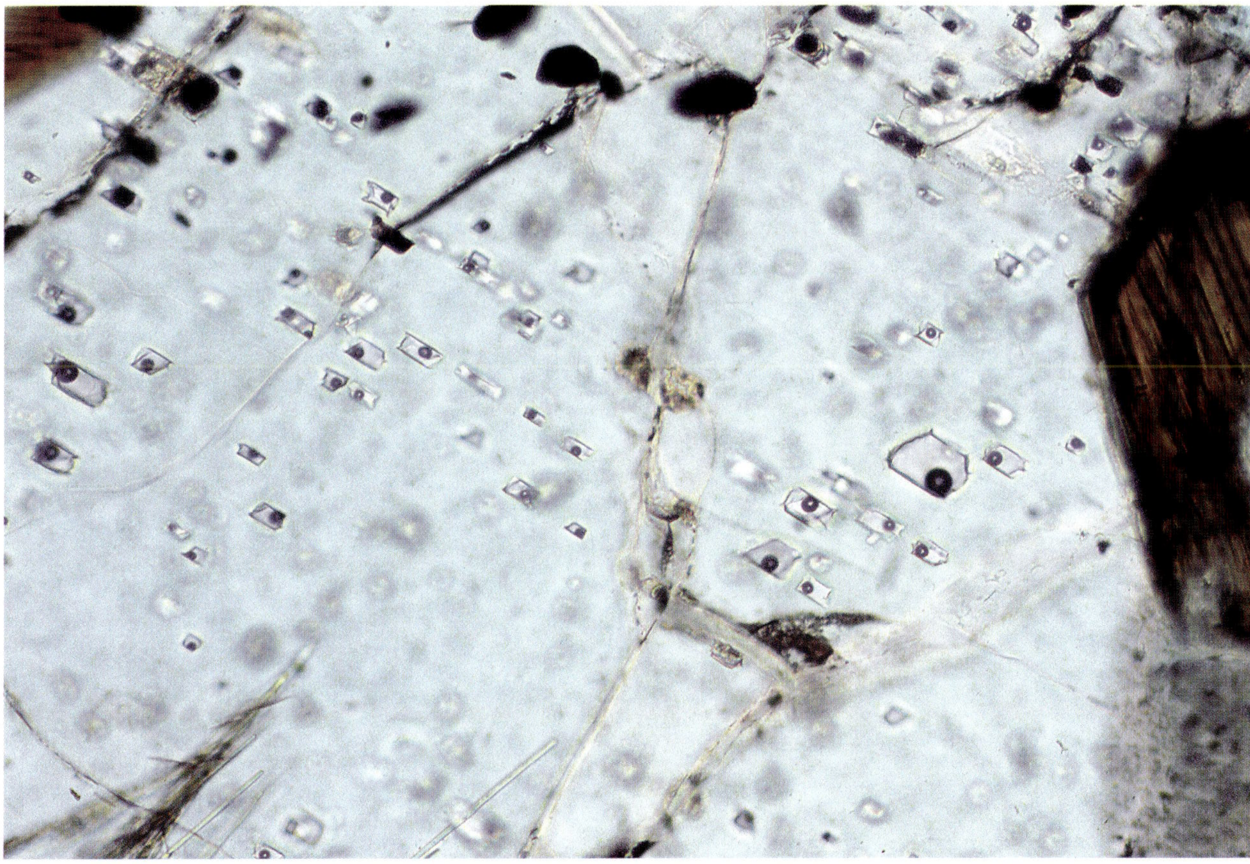

Fig. F71. Numerous slightly elongate and equant inclusions of glass form parallel arrays in a plagioclase crystal in a metapelitic enclave brought rapidly to the surface by an eruption of dacitic lava. In some places, the glass inclusions have a negative crystal shape, but most are outlined by two parallel faces, which correspond to specific crystallographic directions in the plagioclase host; the ends of many inclusions are somewhat irregular. The glass is clear and colorless, and shows no signs of having either crystallized or devitrified. Virtually all of the melt inclusions contain a conspicuous "shrinkage bubble," which formed as the glass contracted on cooling. The size and number (i.e., total volume) of the shrinkage bubbles in a melt inclusion depend upon cooling rate (Lowenstern 1995). Furthermore, the bubbles may either be empty or contain volatiles, such as H_2O, that were dissolved in the melt.

Location: El Joyazo, Spain. *Rock type*: glass inclusions in plagioclase; metapelite protolith, melted at *T* 850 ± 50°C, *P* > 5 kbar, quenched by eruption. *Scale*: the width of photomicrograph is 0.65 mm. Plane-polarized light. *Image and caption*: Bernardo Cesare.

Further reading: Lowenstern, J.B. (1995): Applications of silicate melt inclusions to the study of magmatic volatiles. *In* Magmas, Fluid and Ore Deposits (J.F.H. Thompson, ed.). Mineralogical Association of Canada Short Course **23**, 71–99. See Fig. F72 for more reading.

Figure **F72**

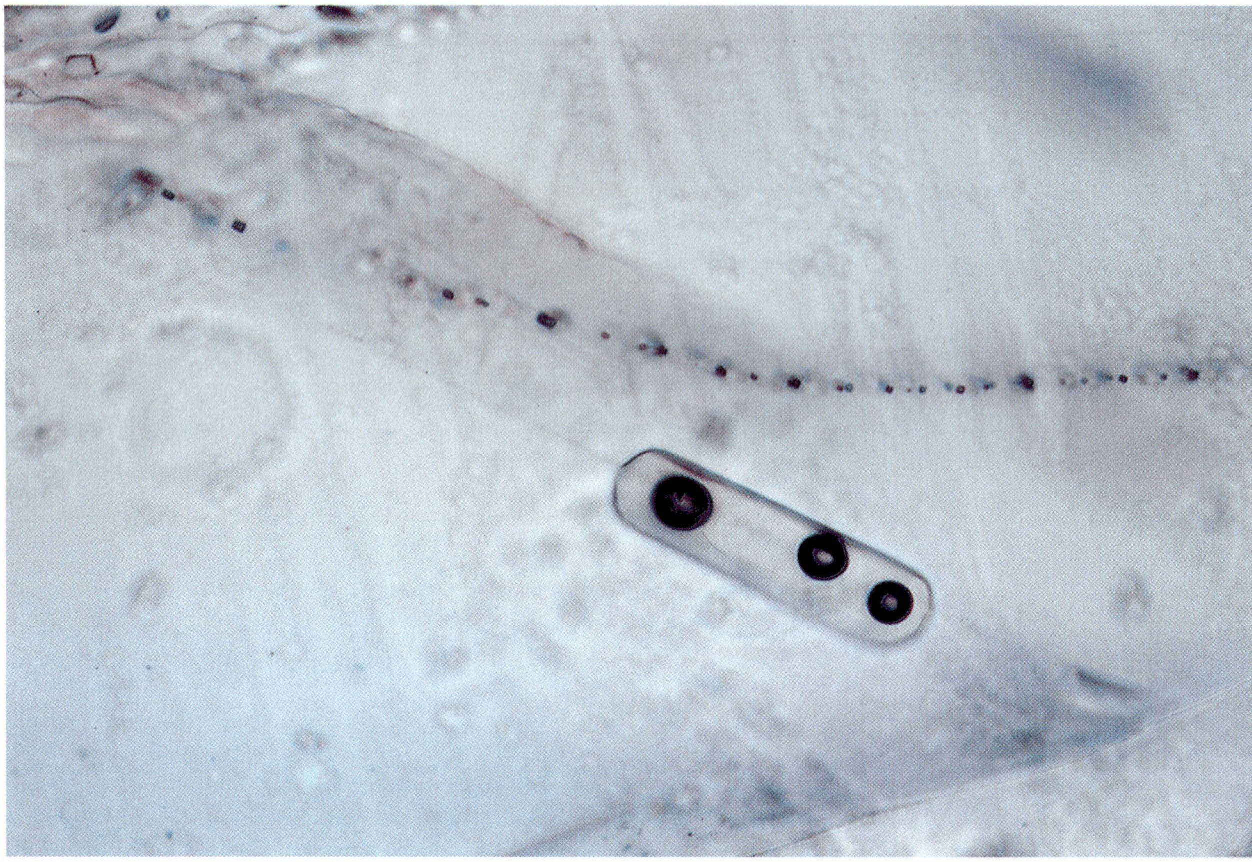

Fig. F72. The glass inclusion near the center of this photomicrograph is 60 μm long and located in a cordierite crystal from an enclave of residual pelitic material in a volcanic rock. The inclusion has the bulk composition of a peraluminous rhyolite, and is typical of pelite-derived leucogranites in general; it represents the anatectic melt that was trapped as the host cordierite crystal grew. The melt inclusion contains three conspicuous "shrinkage bubbles".

Location: Mazarrón, southeast Spain. *Rock type*: glass inclusion; pelite protolith, partially melted at T 850 ± 50°C, P 2–3 kbar. *Scale*: the field of view is 0.3 mm wide. Plane-polarized light. *Image and caption*: Bernardo Cesare.

Further reading: Cesare, B., Marchesi, C., Hermann, J. & Gómez-Pugnaire, M.T. (2003): Primary melt inclusions in andalusite from anatectic graphitic metapelites: implications for the position of the Al_2SiO_5 triple point. *Geology* **31**, 573–576.

Figure F73

Fig. F73. This photograph shows a large and irregularly shaped glass inclusion (55 μm long) in a plagioclase crystal from an enclave of residual metapelite in a volcanic rock. The glass inclusion has a rhyolitic composition, and contains a double shrinkage bubble that indicates the presence of a fluid, probably derived from the H_2O that was dissolved in the anatectic melt. The plagioclase crystal also contains two more large and irregularly shaped glass inclusions that have conspicuous shrinkage-induced bubbles, but these bubbles are too small to check if they also are double. Furthermore, the plagioclase contains innumerable much smaller inclusions, many of which are square or rectangular in outline, and almost all of them contain tiny bubbles.

Location: Mazarrón, southeast Spain. *Rock type*: glass inclusions in plagioclase; pelitic protolith, partially melted at T 850 ± 50°C, P 2–3 kbar. *Scale*: the field of view is 0.35 mm wide. Plane-polarized light. *Image and caption*: Bernardo Cesare.

Further reading: as for Fig. F72.

Figure F74

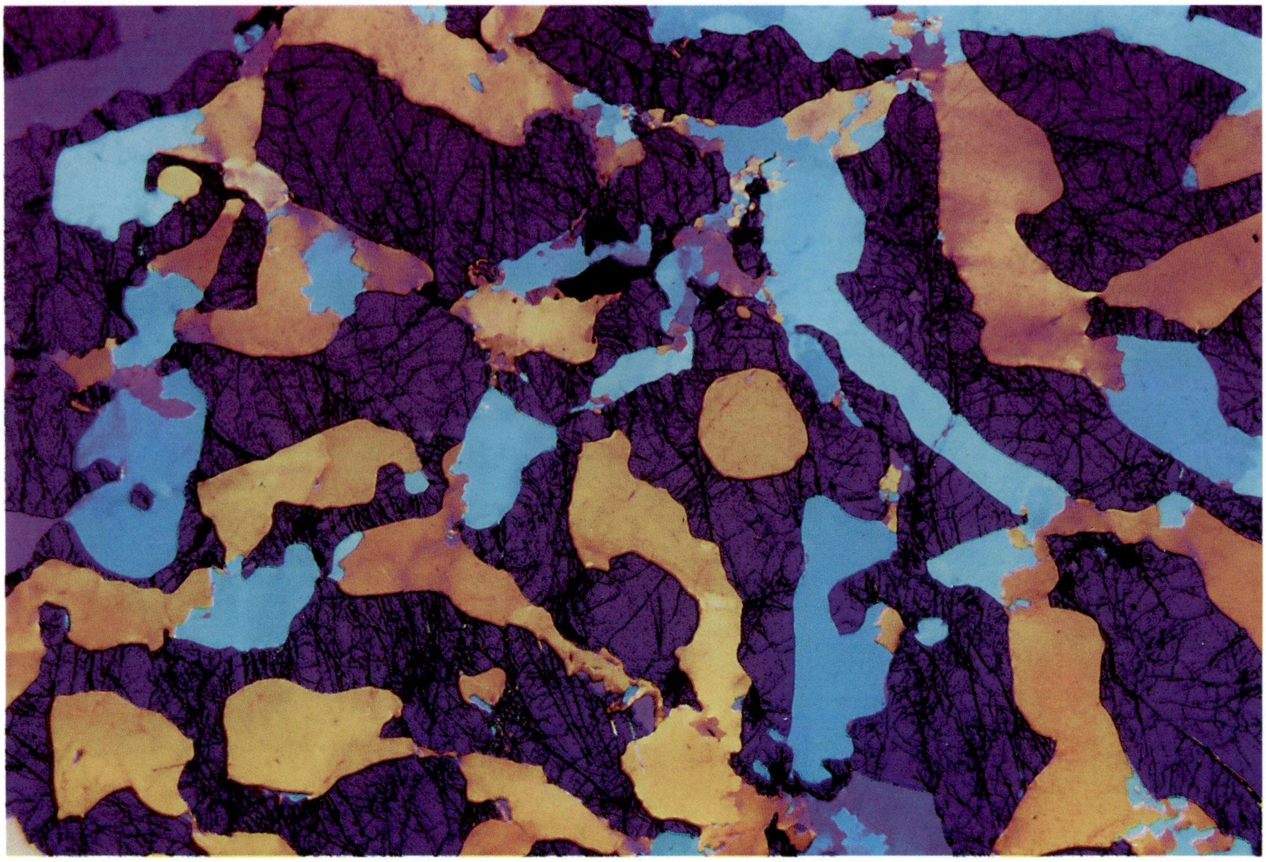

Fig. F74. Close-up of the intergrowth between quartz and garnet from the core of a patch of neosome in which a leucocratic (quartz + K-feldspar + plagioclase) rim and a melanocratic (garnet) core developed in a biotite + plagioclase + quartz semipelite host. Figure B30 is an outcrop photograph of this type of intergrowth between quartz and garnet. The mineral assemblage in the host corresponds to the reactants, and the neosome, to the products in a biotite-dehydration-melting reaction similar to biotite + quartz + plagioclase + sillimanite = garnet + melt + K-feldspar. The reaction may possibly have terminated because all the sillimanite in the semipelitic protolith was consumed. Note that the quartz appears to comprise just two main crystals (one yellow and one blue in this image) in the host garnet (purple). The partial melting is Paleoproterozoic in age, but the undulose extinction in the quartz is due to much younger deformation associated with the Paleozoic Alice Springs orogeny. Many of the patches of neosome occur in small shear zones; on the basis of their major- and trace-element compositions, they have lost some of their melt fraction.

Location: Wuluma Hills area, Arunta Inlier, Australia. *Rock type*: neosome in a metatexite migmatite; semipelitic protolith, anatexis at T 825–875°C, P ca. 5 kbar. *Scale*: the long side of the photograph equals 9 mm. Cross-polarized light and quartz accessory plate. *Image*: E.W. Sawyer.

Further reading: Barbey, P., Marignac, C., Montel, J.-M., Macaudière, J., Gasquet, D. & Jabbori, J. (1999): Cordierite growth textures and the conditions of genesis and emplacement of crustal granitic magmas: the Velay Granite Complex (Massif Central, France). *Journal of Petrology* **40**, 1425–1441.

Waters, D.J. (2001): The significance of prograde and retrograde quartz-bearing intergrowth microstructures in partially melted granulite-facies rocks. *Lithos* **56**, 97–110.

Figure F75

Fig. F75. This photomicrograph shows the internal part of another neosome in a biotite + plagioclase + quartz semipelitic host from the same locality as Fig. F74. In this example, the morphology of the garnet–quartz intergrowth is somewhat different. The quartz occurs as deep embayments in the garnet and as equant, or slightly elongate, bleb-like inclusions within it. Some of the larger inclusions of quartz also contain small roundish crystals of partially altered plagioclase. The garnet is surrounded by a rim of quartz, which is, in turn, surrounded by an annulus of K-feldspar that contains a few small grains of plagioclase and quartz. The K-feldspar and garnet in the neosome are interpreted to be the solid products of an incongruent melting reaction such as biotite + sillimanite + quartz + plagioclase = melt + garnet + K-feldspar. Sillimanite is no longer present in the host and is considered to have been entirely consumed by the melt-producing reaction; biotite is absent from the neosome and only found in the host.

Location: Wuluma Hills area, Arunta Inlier, Australia. *Rock type*: neosome from a metatexite migmatite; semipelitic protolith, partial melting at *T* 825–875°C, *P* ca. 5 kbar. *Scale*: the long side of the photograph equals 14 mm. Cross-polarized light. *Image*: E.W. Sawyer.

Figure **F76**

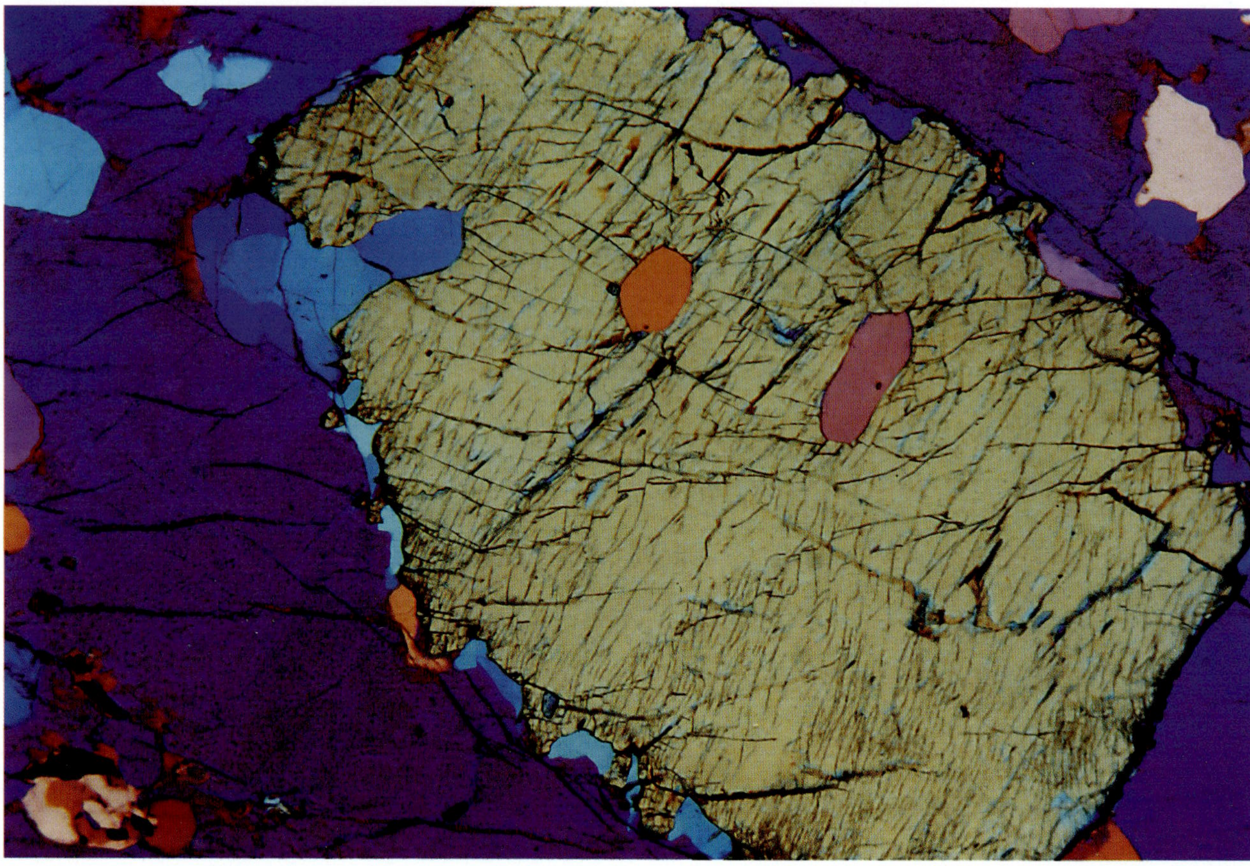

Fig. F76. Occurrences of a garnet–quartz intergrowth are very common in the neosome of the Wuluma Hills migmatites; intergrowths between orthopyroxene and quartz, and between cordierite and quartz, also occur, however. The orthopyroxene crystal (yellow) in this intergrowth contains two large, rounded inclusions of quartz (orange and pink) and is almost completely surrounded by a very narrow rim of small crystals of quartz (mostly blue), part of which also forms a deep embayment in the orthopyroxene. The quartz and orthopyroxene are surrounded by a single grain of perthitic K-feldspar (red and purple). Although not visible in this photograph, the K-feldspar is, in turn, surrounded by a rim of medium-grained quartz + plagioclase + K-feldspar that constitutes the leucosome part of the neosome. The orthopyroxene and the large K-feldspar grain surrounding it are interpreted as the solid products of the incongruent reaction: biotite + quartz + plagioclase = orthopyroxene + K-feldspar + melt. They constitute the residuum part of the neosome, and it is, of course, *in situ*. The outer granitic part of the neosome may have been largely derived from the crystallization of the anatectic melt produced by the reaction, and can therefore be considered as the *in situ* leucosome portion.

Location: Wuluma Hills area, Arunta Inlier, Australia. *Rock type*: neosome in a metatexite migmatite; Al-poor psammite protolith, anatexis at *T* 825–875°C, *P* ca. 5 kbar. *Scale*: the long side of the photograph corresponds to 9 mm. Cross-polarized light and the quartz accessory plate. *Image*: E.W. Sawyer.

Figure F77

Fig. F77. Although most of the intergrowth microstructures contain only one ferromagnesian phase, some contain several. The core of this neosome is a large (15 mm) crystal of cordierite (yellow, bottom left), which is enclosed by a thin (up to 2 mm thick), discontinuous rim of quartz (mostly blue and purple) that locally fills embayments in the cordierite. This cordierite + quartz core is enclosed by an annular zone containing garnet, K-feldspar, and idioblastic orthopyroxene with lesser amounts of biotite, quartz, and plagioclase. This zone is in turn enclosed by an outer, quartz + plagioclase + K-feldspar leucocratic rim (not present in the micrograph) that is of broadly granitic composition and is in contact with the host, a biotite–quartz–orthopyroxene–plagioclase–cordierite semipelitic schist. Locally, this leucocratic outer zone is highly strained and attenuated. The outer granitic zone of the neosome is interpreted to have been derived from the crystallization of anatectic melt and can be viewed as an *in situ* leucosome. In contrast, the inner zones with cordierite, garnet, orthopyroxene, and K-feldspar are interpreted to be the solid products of the partial-melting reaction, and are the residuum portion of the neosome. The biotite present in the zone that surrounds the cordierite crystal is mostly only on the margins of the garnet and the orthopyroxene. Hence, this biotite is interpreted to have formed from both garnet and orthopyroxene by reaction with the last remaining anatectic melt in the neosome. Thus, the biotite crystallized at a late stage of mineral growth in the neosome and formed during cooling and subsequent to the loss of some melt from the neosome.

Location: Wuluma Hills area, Arunta Inlier, Australia. *Rock type*: neosome in a metatexite migmatite; pelitic protolith, anatexis at *T* 825–875°C, *P* ca. 5 kbar. *Scale*: the long side of the photograph equals 9 mm. Cross-polarized light and quartz accessory plate. *Image*: E.W. Sawyer.

Further reading: Barbey, P., Marignac, C., Montel, J.-M., Macaudière, J., Gasquet, D. & Jabbori, J. (1999): Cordierite growth textures and the conditions of genesis and emplacement of crustal granitic magmas: the Velay Granite Complex (Massif Central, France). *Journal of Petrology* **40**, 1425–1441.

Waters, D.J. (2001): The significance of prograde and retrograde quartz-bearing intergrowth microstructures in partially melted granulite-facies rocks. *Lithos* **56**, 97–110.

White, R.W., Powell, R. & Halpin, J.A. (2004): Spatially focussed melt formation in aluminous metapelites from Broken Hill, Australia: the importance of mineral texture development in melt formation. *Journal of Metamorphic Geology* **22**, 825–845.

Figure **F78**

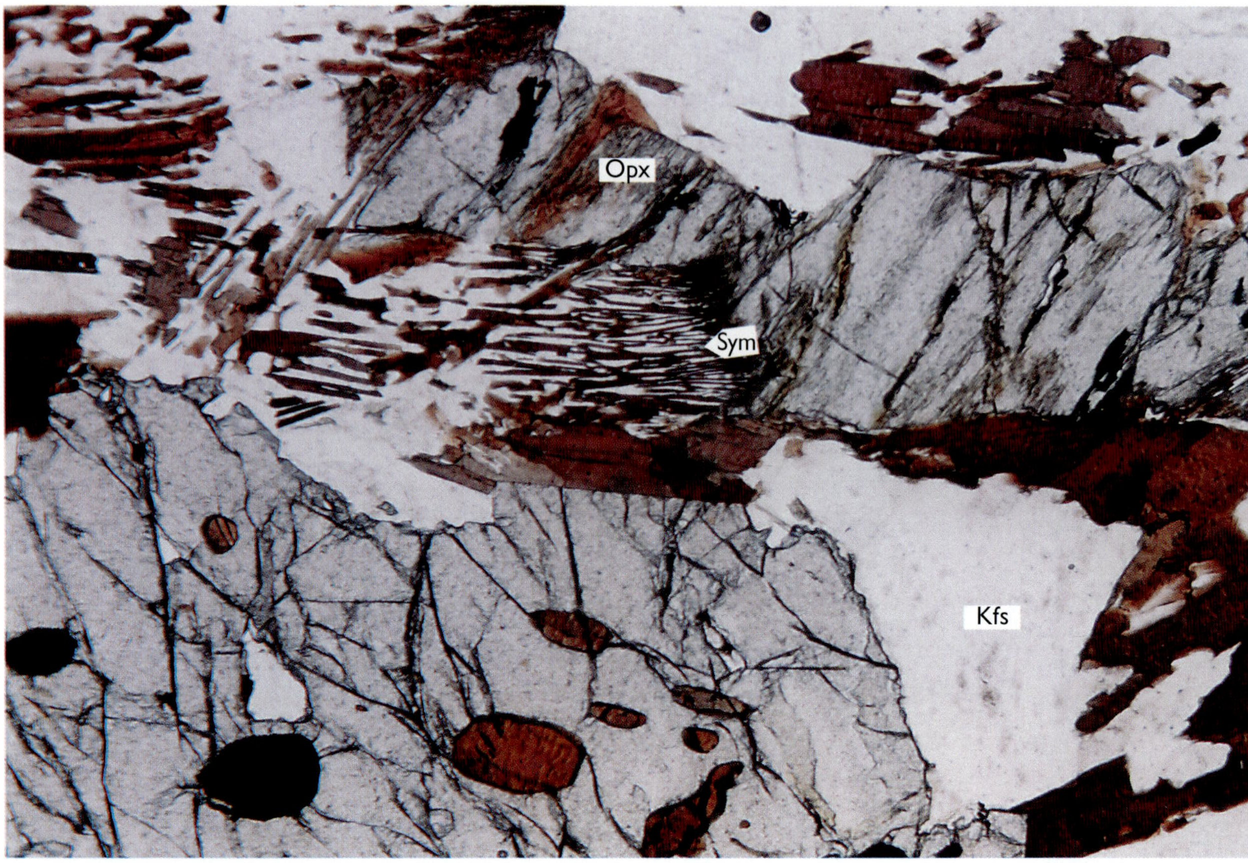

Fig. F78. This photomicrograph is from a melt-depleted, semipelitic layer in a granulite-facies metatexite migmatite. It shows a symplectitic intergrowth (Sym) of late-generation biotite + quartz replacing orthopyroxene (Opx). The garnet (bottom left) in the rock is not affected, and contains several rounded grains of biotite interpreted as belonging to the older, reactant assemblage. The colorless, low-relief matrix consists of perthitic K-feldspar (Kfs) + plagioclase + quartz. Garnet, orthopyroxene, and K-feldspar are interpreted to be the products of various incongruent melting reactions involving the breakdown of biotite and, initially, sillimanite. Most of the melt generated by the breakdown of the biotite has been removed, and a residual plagioclase + K-feldspar + garnet + quartz + orthopyroxene + biotite assemblage remains. The symplectite microstructure formed when the small amount of anatectic melt trapped in the matrix reacted with the residual orthopyroxene in a reaction such as melt + orthopyroxene = biotite + quartz. The reaction that formed the symplectite is essentially the reverse of the partial-melting reaction responsible for generating the orthopyroxene.

Location: Ashuanipi Subprovince, Quebec, Canada. *Rock type*: residual part of a metatexite migmatite; semipelite protolith, anatexis at *T* 825–875°C, *P* 6–7 kbar. *Scale*: the long side of the photograph equals 2.7 mm. Plane-polarized light. *Image*: E.W. Sawyer.

Further reading: Waters, D.J. (2001): The significance of prograde and retrograde quartz-bearing intergrowth microstructures in partially melted granulite-facies rocks. *Lithos* **56**, 97–110.

Figure F79

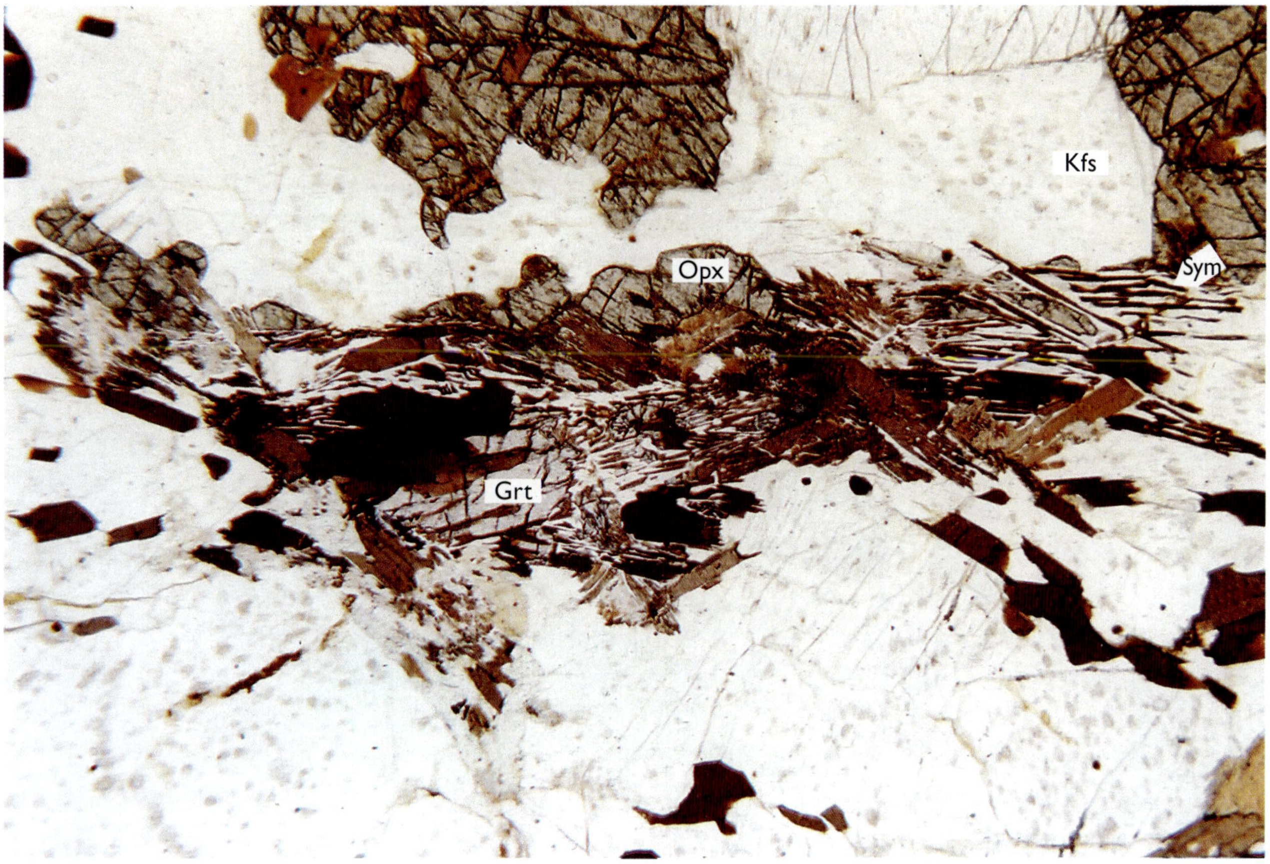

Fig. F79. A symplectite (Sym) of quartz + bladed biotite and larger, more equant crystals of biotite replace both garnet (Grt) and orthopyroxene (Opx) in this metapelite layer from a melt-depleted metatexite migmatite. The domain that contains the symplectite is bordered by a leucocratic region that consists principally of perthite (faintly mottled appearance) and minor amounts of plagioclase and quartz. The formation of the symplectite is interpreted to be the result of a reaction between the last-remaining anatectic melt in the residuum and the residual garnet, orthopyroxene, and K-feldspar (Kfs). The reaction responsible is inferred to have been melt + garnet + orthopyroxene + K-feldspar = biotite + quartz. It occurred after almost all of the melt generated during anatexis had migrated away.

Location: Ashuanipi Subprovince, Canada. *Rock type*: melt-depleted (residual) part of a metatexite migmatite; metapelite protolith, anatexis at T 825–875°C, P 6–7 kbar. *Scale*: the long side of the photograph equals 14 mm. Plane-polarized light. *Image*: E.W. Sawyer.

Figure **F80**

Fig. F80. This example, from a diatexite migmatite, shows a rectangular, biotite-rich domain replacing an idioblastic, rectangular crystal of orthopyroxene. The orthopyroxene is interpreted either as the product of incongruent melting of biotite, or as a product of crystallization of the anatectic melt. The replacement of the orthopyroxene crystal by biotite and quartz appears to have occurred in two stages. The biotite forms thin plates in a symplectite with quartz in the portion where corroded relics of the orthopyroxene still remain. However, the biotite and quartz are conspicuously coarser grained and generally not intergrown in the region where no relics of the orthopyroxene remain. The orthopyroxene crystal in the photomicrograph is part of a large-scale framework structure in the diatexite that consists principally of large subhedral crystals of plagioclase (top left) and K-feldspar (bottom right). Smaller anhedral crystals of K-feldspar, plagioclase, quartz, and biotite are interstitial to the framework structure, and are believed to have crystallized from the melt trapped within the crystal framework. The overall microstructure of the diatexite migmatite is, therefore, interpreted to be magmatic. On the basis of the abundance of K-feldspar, this diatexite is interpreted to have crystallized from a relatively evolved or fractionated magma. The symplectite microstructure is interpreted to be the result of a reaction between the remaining melt in the diatexite, or an H_2O-rich fluid exsolved from the melt as it crystallized, and the orthopyroxene crystals through a reaction such as melt + orthopyroxene = biotite + quartz + plagioclase. The replacement of orthopyroxene by a symplectite of biotite + quartz is rare in the plagioclase-rich, cumulate diatexite migmatites from the Ashuanipi Subprovince, but is very common in the more evolved, K-feldspar-rich ones. This may simply reflect a higher content of H_2O in the evolved magmas. This microstructure suggests that the large, rectangular aggregates of biotite present in some orthopyroxene-free diatexite migmatites in the Ashuanipi Subprovince may in fact be pseudomorphic after orthopyroxene.

Location: Ashuanipi Subprovince, Quebec, Canada. *Rock type*: diatexite migmatite; metagreywacke protolith, anatexis in the granulite facies, T 825–875°C, P 6–7 kbar. *Scale*: the long side of the photograph corresponds to 3.5 mm. Cross-polarized light. *Image*: E.W. Sawyer.

Figure F81

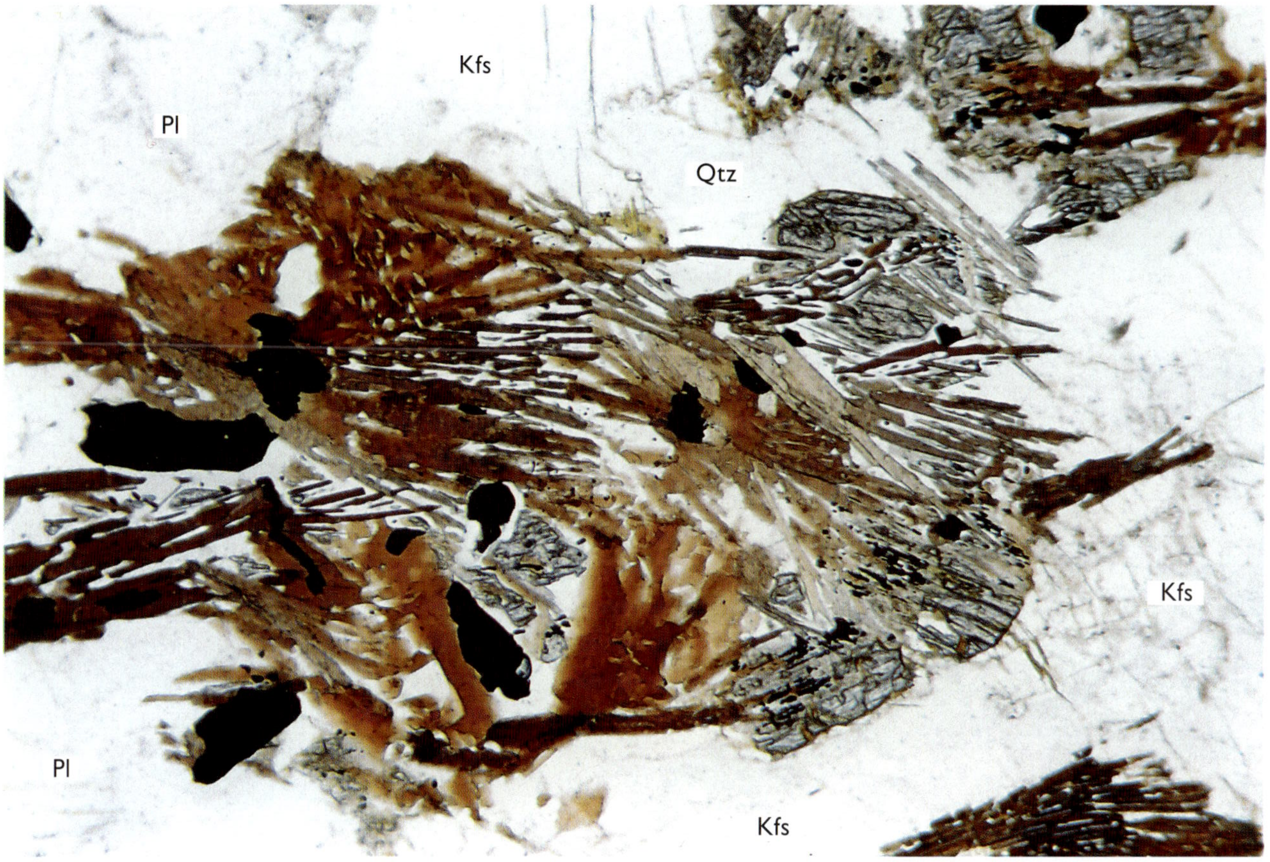

Fig. F81. The photomicrograph shows a cluster of orthopyroxene crystals in the perthite + plagioclase + quartz matrix of a diatexite migmatite. The origin of the orthopyroxene is uncertain; it could be the product of the incongruent melting of biotite and, hence, interpreted as residual crystals in the diatexite magma, or alternatively, it could have crystallized from the diatexite magma. The matrix around the orthopyroxene consists of large, subhedral crystals of K-feldspar (Kfs) and plagioclase (Pl) with small interstitial crystals of quartz and minor amounts of feldspar. On the basis of its bulk composition, the diatexite is interpreted to have crystallized from a moderately evolved anatectic melt. The orthopyroxene crystals have been partially replaced by a biotite + quartz symplectitic intergrowth. Symplectite is not generally developed at the contacts between the large subhedral crystals of K-feldspar and the orthopyroxene; rather, it appears to be concentrated in regions where orthopyroxene is in contact with the small grains of interstitial quartz and feldspar. This spatial relationship suggests that the symplectite results from reaction between the orthopyroxene crystals and the interstitial melt, or an H_2O-rich fluid exsolved from it.

Location: Ashuanipi Subprovince, Canada. *Rock type*: diatexite migmatite; metagreywacke protolith, anatexis in the granulite facies; *T* 825–875°C, *P* 6–7 kbar. *Scale*: the long side of the photograph corresponds to 2.7 mm. Plane-polarized light. *Image*: thin section provided by Robert Thériault.

Figure **F82**

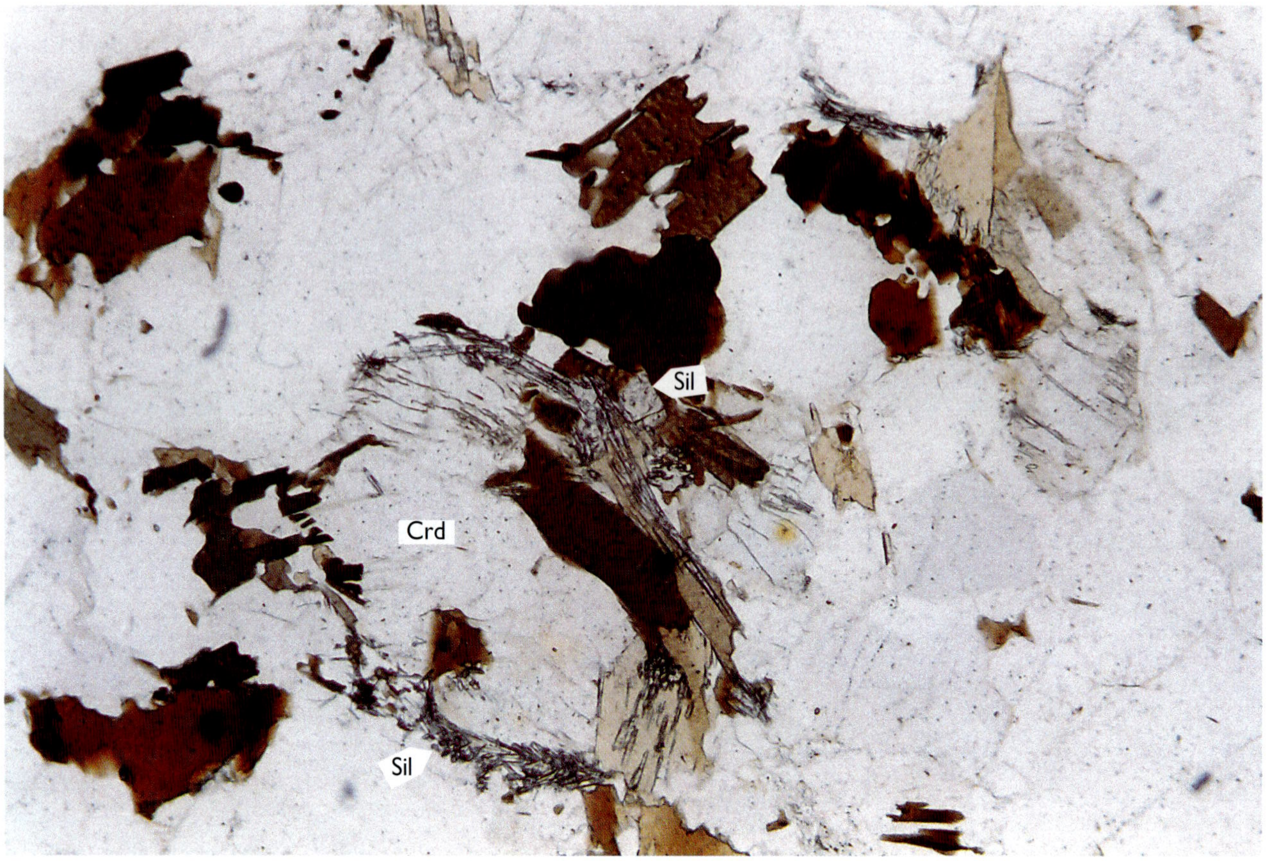

Fig. F82. The large crystal of cordierite (Crd) in this residual diatexite shows incipient replacement by biotite and sillimanite around the edges. Sillimanite crystals at the contact between cordierite and its quartz + plagioclase + K-feldspar matrix are relatively small, but some of the sillimanite crystals (Sil) in the biotite are larger and more equant in their form. The cordierite can be identified by the yellow pleochroic halo around small, high-relief inclusions, and by the characteristic fractures across the grains. The cordierite in this rock is a product of the incongruent breakdown of biotite and sillimanite and was replaced, following the loss of some of the melt fraction, as the diatexite crystallized. A reaction between the last increments of melt left in the rock and the surfaces of the crystals of residual cordierite generated the biotite + sillimanite fringes around the cordierite. See Waters (2001) for an alternative interpretation of the replacement of garnet or cordierite by aggregates of biotite and sillimanite.

Location: Lac Kenogami, Grenville Province, Quebec, Canada. *Rock type*: residual diatexite migmatite; metapelitic protolith, anatexis in the lower granulite facies. *Scale*: the long side of the photograph corresponds to 2.5 mm. Plane-polarized light. *Image*: E.W. Sawyer.

Further reading: Büttner, S.H., Glodny, J., Lucassen, F., Wemmer, K., Erdmann, S., Handler, R. & Franz, G. (2005): Ordovician metamorphism and plutonism in the Sierra de Quilmes metamorphic complex: implications for the tectonic setting of the northern Sierras Pampeanas (NW Argentina). *Lithos* **83**, 143–181.

Waters, D.J. (2001): The significance of prograde and retrograde quartz-bearing intergrowth microstructures in partially melted granulite-facies rocks. *Lithos* **56**, 97–110.

Figure F83

Fig. F83. This plagioclase crystal from the center of a large domain of leucosome exhibits an unusual, mottled zonation (An_{47-23}) in its core, and has concentric overgrowths of a more sodic plagioclase (An_{31-26}). The host migmatites (mostly diatexite at the location) are intruded by sheets of quartz gabbro and meladiorite, and contain partially hybridized enclaves of this mafic material. As such, plagioclase cores are not encountered in the leucosome remote from the mafic intrusive bodies; the complex zonation in this plagioclase is attributed to the presence of the mafic intrusive bodies in the diatexite. The complexly zoned plagioclase may (1) be inherited by contamination of the anatectic diatexite magma by the relatively calcic mafic magma or (2) result from the remelting of early-formed crystals of plagioclase in the anatectic magma as a consequence of heating by the intrusion of the mafic magmas into the diatexite migmatite, followed by thermal re-equilibration and precipitation of more sodic plagioclase from the diatexite magma as concentric zones on the core. In either case, the presence of plagioclase crystals with a complexly zoned and more calcic core provides evidence for the presence of mafic magma in the crust-derived anatectic one.

Location: Glenelg River Complex, Victoria, Australia. *Rock type*: plagioclase from leucosome. *Scale*: the long edge of the photograph represents 2.5 mm. Cross-polarized light. *Image and caption*: Tony I.S. Kemp.

Further reading: Castro, A. (2001): Plagioclase morphologies in assimilation experiments. Implications for disequilibrium melting in the generation of granodiorite rocks. *Mineralogy and Petrology* **71**, 31–49.

Kemp, A.I.S. (2004): Petrology of high-Mg, low-Ti igneous rocks of the Glenelg River Complex (SE Australia) and the nature of their interaction with crustal melts. *Lithos* **78**, 119–156.

Figure **F84**

Fig. F84. This photomicrograph shows the edge of an *in situ* leucosome developed at the start of partial melting in a quartz-rich layer of the Biwabik Formation in the footwall of the contact aureole around the Duluth Igneous Complex. Because the protolith is a white quartzite, there is residuum in this migmatite, but no melanosome. The melt-rich (left) and residuum-rich (right) parts of the neosome are better distinguished by their distinctive mineralogies and especially their microstructures. The part that formed from the anatectic melt consists mostly of a graphic intergrowth of feldspar and quartz in the leucosome and the larger intergranular domains in the residuum, but is a film of plagioclase in the smallest intergranular pores (A) and along grain boundaries (B) that extend into the residuum. The residual part consists mostly of large corroded grains of quartz and some grains of opaque Fe-oxide. The boundary between the melt-derived and the residual parts of the neosome is highly irregular at the grain scale; the melt-derived part can be traced along grain boundaries between the residual quartz grains for a distance of at least four grain diameters into the matrix of the quartzite. Grains of residual quartz become surrounded by anatectic melt and then detached from the residuum and incorporated into the melt-derived part; this pattern is very similar to that shown at the outcrop scale by some transitions from metatexite to diatexite migmatites (see Figs. D41–D43). Note that recrystallization has occurred along some of the former melt–quartz contacts, and that the quartz has developed small lobes, which result in a cuspate outline. Anatectic melt either did not develop, or did not penetrate along, all of the quartz–quartz grain boundaries (C), and several two- or three-grain aggregates of quartz can be seen.

Location: Duluth Igneous Complex, Minnesota, U.S.A. *Rock type*: metatexite migmatite; quartzite protolith, pyroxene hornfels facies, *T* 850–900°C, *P* 1.5–2 kbar. *Scale*: the long side of the photomicrograph corresponds to 7 mm. Cross-polarized light. *Image*: E.W. Sawyer.

Figure F85

Fig. F85. This photomicrograph shows part of a larger (about 15 mm across) diffuse leucosome generated in a pelitic layer that is much closer to the contact of the Duluth Igneous Complex than that shown in Figs. F86 and F87. Consequently, the partial melting in this rock was at a higher metamorphic temperature and consumed much of the reactant biotite. Thus, the fraction of melt that was produced was greater, and the residuum richer in cordierite and K-feldspar, the solid products of the incongruent breakdown of biotite. Much of the melt remained where it formed and enabled the residuum to recrystallize to a much coarser grain-size and to develop a relatively isotropic microstructure typical of contact metamorphic rocks. Some of the melt that did segregate formed this patch of *in situ* leucosome; it has an irregular and very diffuse outline, and in contrast to that in Fig. F86, has no biotite-rich melanosome around it. In this case, there is an enrichment of both cordierite and K-feldspar around the leucosome, but this rim of residual material is not melanocratic and, therefore, inconspicuous in both the outcrop and the hand sample. The fine-scale sedimentary layering visible in many rocks of the outer part of the migmatite aureole is no longer visible in this one, it was lost during grain growth and recrystallization. The complete absence of a foliation indicates that no flow of the melt with its crystals occurred. Thus, this is a nebulitic migmatite that contains scattered small patches of leucosome where some anatectic melt was able to segregate. The leucosome contains small polygonal, (commonly) hexagonal sector-twinned crystals of cordierite, and somewhat larger rectangular crystals of K-feldspar and plagioclase, set in a matrix of large anhedral crystals of quartz, K-feldspar, and untwinned plagioclase. This microstructure indicates that the leucosome crystallized from a melt. The euhedral habit of both cordierite and K-feldspar in the leucosome indicates that they crystallized in a melt, but their exact origin is uncertain. They could be either the solid products of the incongruent melting of biotite that happened to grow in the anatectic melt, or they could have crystallized later from the anatectic melt itself as it cooled. The large anhedral crystals of quartz, plagioclase, and K-feldspar were the last minerals to form in the leucosome, and certainly crystallized from the melt. Large, bladed crystals of biotite occur throughout all parts of the migmatite, and they are interpreted to be either a late product of crystallization of the melt, or to have formed when the final increments of melt that remained reacted with the residual cordierite.

Location: Duluth Igneous Complex, Minnesota, U.S.A. *Rock type*: nebulitic migmatite; metapelitic protolith, pyroxene hornfels facies, T 800–850°C, P 1.5–2 kbar. *Scale*: the long side of the photomicrograph corresponds to 11 mm. Cross-polarized light. *Image*: E.W. Sawyer.

Figure F86

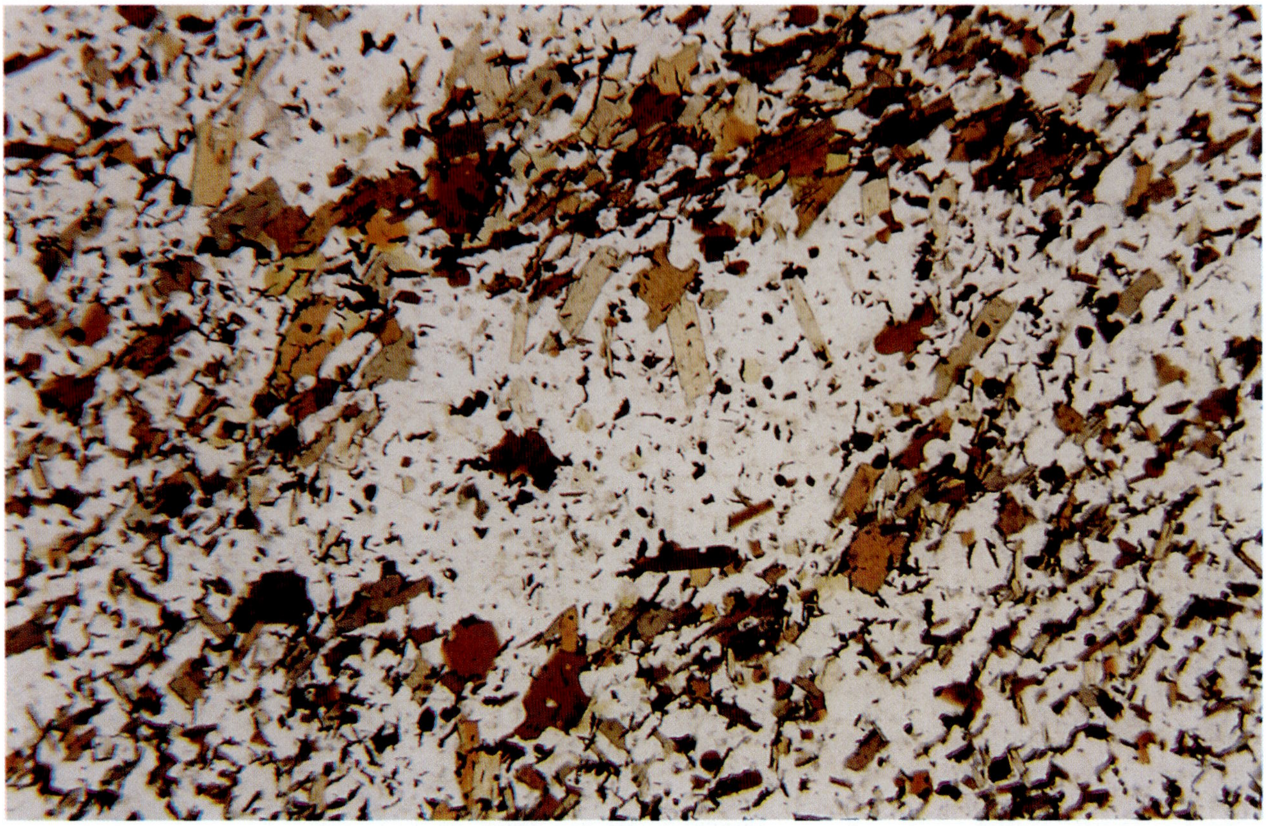

Fig. F86. This small, rounded patch of neosome has developed in a plagioclase-poor quartz–biotite–cordierite–K-feldspar metapelite close to the "melt-in" isograd, which marks the onset of partial melting in the aureole around the Duluth Igneous Complex. The whole-rock composition of this sample is the same as similar rocks just below the "melt-in" isograd, which suggests that no anatectic melt has been lost or gained and, therefore, that the patch of neosome formed *in situ*. The patches of neosome in this rock are too small (<5 mm across) for practical extraction and separation of the leucosome part from the melanosome. The separation of the partial-melt fraction from its residuum has formed a small domain of leucosome 1–3 mm across that contains the assemblage quartz + K-feldspar + plagioclase + biotite + cordierite. The photomicrograph shows the leucosome to be surrounded by melanosome (cf. the neosome in Fig. F85) that is enriched in biotite relative to the adjacent paleosome. Biotite may not have been involved in the reaction that caused partial melting, but has grown to a noticeably larger grain-size than biotite in the paleosome. The leucosome part contains two populations of biotite: small crystals that may be residual in origin and a few much larger, bladed crystals of biotite that may have crystallized from the melt or may have formed by reaction between the melt and the residual phases. The contacts between the leucosome and melanosome parts of this neosome are well defined, but not sharp on the grain scale.

Location: Duluth Igneous Complex, Minnesota, U.S.A. *Rock type*: patch metatexite migmatite; Virginia Formation pelite protolith, partial melting in the pyroxene hornfels facies, *T* ca. 750°C, *P* 1.5–2 kbar. *Scale*: the long side of the photomicrograph corresponds to 4.5 mm. Plane-polarized light. *Image*: E.W. Sawyer.

Figure F87

Fig. F87. The photomicrograph shows the same patch of neosome as in Fig. F86, but in cross-polarized light in order to show the microstructure in the leucosome portion. The quartz, K-feldspar, and plagioclase in the leucosome are much coarser-grained (0.25 mm) than the same minerals in the adjacent paleosome (>0.15 mm). The largest white and grey crystals in the leucosome are quartz, and the smaller, rectangular grains are euhedral crystals of plagioclase and K-feldspar. Tiny (<0.07 mm, too small to see in this photomicrograph) inclusions of cordierite are common in the quartz, and because of their euhedral form, they are interpreted to have crystallized from the melt. Cordierite and K-feldspar in the paleosome are the products of subsolidus metamorphic reactions and form larger (0.15 mm) xenoblastic crystals that contain rounded inclusions of quartz and plagioclase. Some of the leucosome material in the core of other patches of neosome in this and adjacent rocks contains small blades or radiating aggregates of muscovite at its center; the last melt to crystallize in these patches of leucosome presumably was enriched in H_2O and crystallized muscovite in addition to, or in place of, K-feldspar.

Location: Duluth Igneous Complex, Minnesota, U.S.A. *Rock type*: patch metatexite migmatite; Virginia Formation pelite protolith, partial melting in the pyroxene hornfels facies, *T* ca. 750°C, *P* 1.5–2 kbar. *Scale*: the long side of the photomicrograph corresponds to 4.5 mm. Cross-polarized light. *Image*: E.W. Sawyer.

Figure **F88**

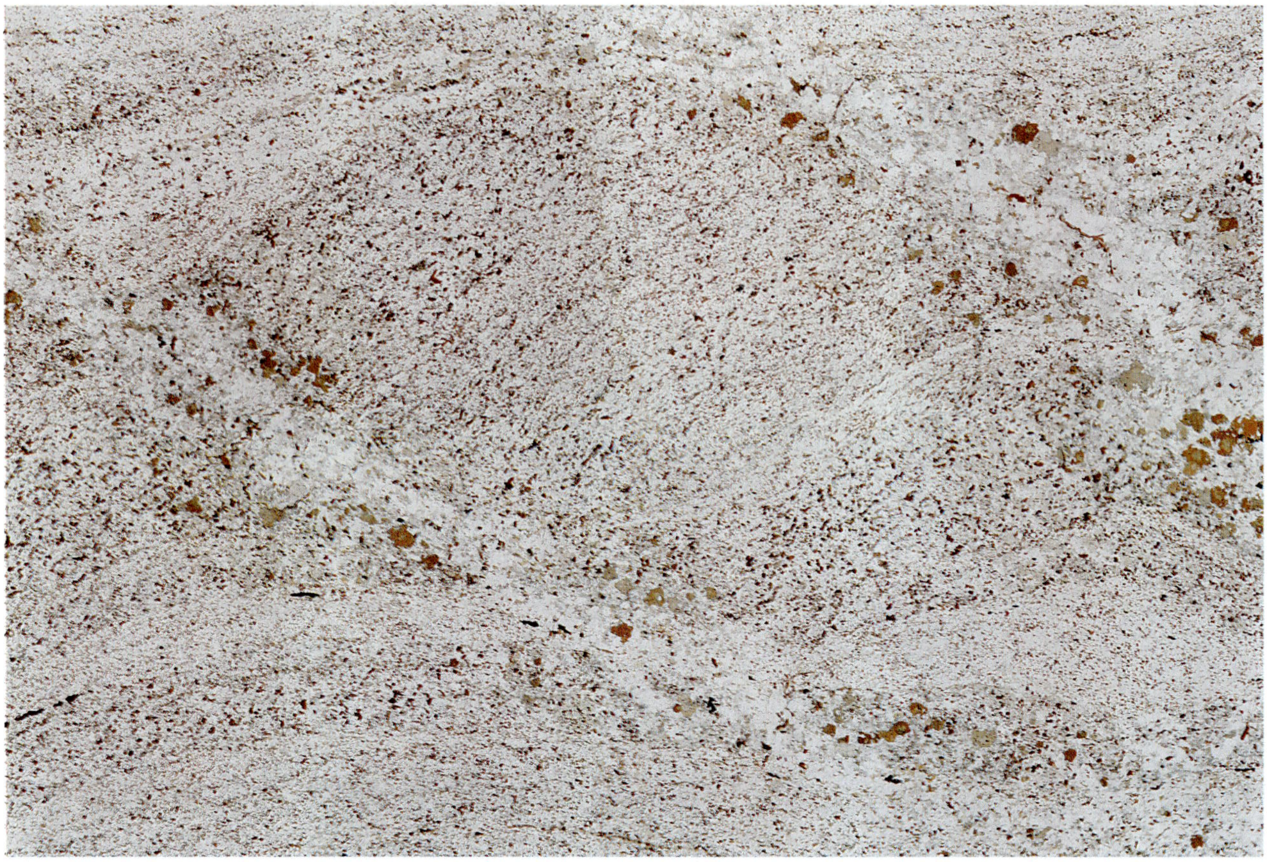

Fig. F88. This mosaic shows domains of leucosome from just inside the "melt-in" isograd of the contact aureole at the Duluth Igneous Complex. Some key structural and petrological features are well illustrated in this migmatite. The bedding in the metasedimentary hornfels is distinguished by the layers of different grain size and color; the darker layers are richest in biotite and Fe-oxide minerals. Most of the metasediment contains the assemblage cordierite + K-feldspar + quartz + plagioclase + biotite and shows no evidence for anatexis; it is, therefore, paleosome. The grain size of the leucosome (up to 0.4 mm) is considerably greater than that of the metasediment (0.1 mm). Leucosome is located in small extensional shear zones that formed when a relatively more competent, cordierite-rich, semipelitic bed was extended along its layering. A slight sigmoidal curvature of the foliation indicates dextral movement for the shear zones as the layer was extended. Most contacts between the leucosome and the metasediment are diffuse, and the leucosome thus seems to have formed *in situ*. There are several significant petrological features in this migmatite. Because there is no systematic difference in the modal proportion of cordierite in the host rock close to the leucosome compared to farther away, cordierite was probably not a product of the melt-producing reaction; it had formed, together with K-feldspar, by an earlier subsolidus reaction. The slightly darker color of the host rocks immediately around the leucosome is the result of alteration of the cordierite; cordierite farther (>2–3 mm) from the leucosome is unaltered. The domains of leucosome are composed predominantly of quartz, plagioclase, K-feldspar, and minor amounts of muscovite; the large brown grains in the leucosome and the adjacent host-rocks are tourmaline. Together, these petrological features suggest that the rocks did not undergo fluid-absent melting of biotite. If they did partially melt, then it was probably by the fluid-present melting of quartz, plagioclase, and K-feldspar; moreover, the fluid contained boron. If partial melting did occur in these rocks, then the leucosome is *in situ* or in-source, but if melting did not occur, then the term leucocratic vein applies. In either case, it is likely that the H_2O released as the melt crystallized migrated from the leucosome, hydrated the cordierite in the adjacent wallrocks, and enabled growth of tourmaline there.

Location: Duluth Igneous Complex, Minnesota, U.S.A. *Rock type*: net-structured metatexite migmatite; pelitic protolith, pyroxene hornfels facies, *T* ca. 750°C, *P* 1.5–2 kbar. *Scale*: the long side of the photomicrograph corresponds to 45 mm. Plane-polarized light. *Image*: E.W. Sawyer.

Figure F89

Fig. F89. This photomicrograph shows part of a small *in situ* patch of neosome formed in a mafic granulite where the degree of partial melting was very low. Therefore, this image shows a typical relationship between the leucosome and melanosome that can be seen in many examples of neosome formed at the onset of anatexis. The patch of neosome is about 3 cm long and oriented orthogonal to the weak layering and foliation in the mafic granulite paleosome. The part of the neosome shown in the center of this image is diffuse and is especially difficult to see in the cross-polarized light image (Fig. F90). The neosome comprises a plagioclase + quartz leucosome and a coarse-grained (0.5–1 mm), clinopyroxene-dominated melanosome, both of which have a granoblastic-polygonal microstructure owing to subsequent textural re-equilibration, which is typical of regional metamorphic terranes. The plagioclase + hornblende + clinopyroxene paleosome exterior to the neosome is noticeably finer-grained (ca. 0.3 mm), and has a more isotropic distribution of its constituent minerals than the neosome. The difference in grain size and distribution of minerals that results from anatexis is a key factor in distinguishing neosome from paleosome in migmatites where the degree of partial melting was low. The boundary between the leucosome and melanosome is highly irregular. The spatial relationship among the leucosome, the melanosome, and the paleosome is variable in this migmatite. Parts of the neosome conform to the general case for *in situ* formation and have the melanosome between the leucosome and the paleosome (see Fig. F86). However, in places, the melanosome assumes the contrary position and is in the center of the neosome, with the leucosome in contact with the paleosome. Melting may have begun at several places, and these subsequently coalesced to produce the irregular distribution.

Location: Mount Hay, Arunta Inlier, Australia. *Rock type*: patch metatexite migmatite; metamafic protolith, anatexis in the granulite facies, T 825–875°C, P 6–7 kbar. *Scale*: the long side of the photo is equal to 11 mm. Plane-polarized light. *Image*: E.W. Sawyer.

Figure **F90**

Fig. F90. Same as Fig. F89, but taken in cross-polarized light to better show the coarser grain-size of the neosome relative to the paleosome.

Location: Mount Hay, Arunta Inlier, Australia. *Rock type*: patch metatexite migmatite; metamafic protolith, anatexis in the granulite facies, *T* 825–875°C, *P* 6–7 kbar. *Scale*: the long side of the photo is equal to 11 mm. Cross-polarized light. *Image*: E.W. Sawyer.

Figure F91

Fig. F91. At the outcrop scale, the neosome in this migmatite consists of well-defined portions of leucosome and melanosome. However, this photomicrograph shows that the contact between the two is very irregular at the grain scale, a feature generally regarded as more consistent with *in situ* formation of neosome than the injection of an anatectic melt. A mass-balance calculation for this migmatite using the bulk composition of the leucosome, its adjacent melanosome, and the assumed protolith supports an *in situ* origin for the neosome, that is, segregation of melt from its residue occurred without loss of melt to an external sink. The protolith was a hornblende + plagioclase + clinopyroxene + garnet metamafic rock, and the neosome that resulted from partial melting comprises a clinopyroxene + garnet + plagioclase + hornblende melanosome and a plagioclase + quartz leucosome. The garnet and clinopyroxene have a greater modal abundance and form larger crystals in the melanosome compared to the protolith (not visible in the photomicrograph). The clinopyroxene in the melanosome has experienced minor retrogression and has, locally, developed a thin rim of dark green hornblende.

Location: Abitibi Subprovince, Superior Province, Quebec, Canada. *Rock type*: metatexite migmatite; metamafic protolith, anatexis in the uppermost amphibolite facies, *T* 800–850°C and *P* 8–10 kbar. *Scale*: the long side of the photomicrograph corresponds to 7 mm. Plane-polarized light. *Image*: E.W. Sawyer.

Figure **F92**

Fig. F92. The photomicrograph shows the same migmatite as in Fig. F91, but it is taken in cross-polarized light. This reveals the far larger grain-size in the leucosome relative to the melanosome. Furthermore, some lobate grain-boundaries on the quartz and undulose extinction are evident in both the quartz and the plagioclase in the leucosome. These features are the result of subsolidus, intracrystalline plastic deformation, which has removed all the microstructural criteria normally used to infer crystallization from a melt. Some small, irregularly shaped crystals of plagioclase in the melanosome are conspicuously zoned, and generally have a more calcic rim; these grains are interpreted to be residual plagioclase. Microstructures such as this, which have been highly modified by subsequent deformation and recrystallization so that they no longer retain evidence that they crystallized from a melt, are typical of the leucosome from regional metamorphic terranes.

Location: Abitibi Subprovince, Superior Province, Quebec, Canada. *Rock type*: metatexite migmatite; metamafic protolith, anatexis in the uppermost amphibolite facies, T 800–850°C and P 8–10 kbar. *Scale*: the long side of the photomicrograph corresponds to 7 mm. Cross-polarized light. *Image*: E.W. Sawyer.

Further reading: Sawyer, E.W. (1991): Disequilibrium melting and the rate of melt–residuum separation during migmatization of mafic rocks from the Grenville Front, Quebec. *Journal of Petrology* **32**, 701–738.

Figure F93

Fig. F93. This photomicrograph shows the contact between a stromatic leucosome and its melanosome in a metatexite migmatite that was derived from a metapelitic protolith. The contact between the leucosome and the melanosome is abrupt at both the outcrop scale and the hand-sample scale, but it is irregular on the grain scale. The melanosome consists of the assemblage biotite + garnet + sillimanite + K-feldspar + quartz + plagioclase, and is interesting because it exhibits an internal mineralogical and textural zonation. Most of the garnet occurs close to the contact with the leucosome, whereas the sillimanite is located farther away and is not actually in contact with the leucosome. The biotite nearest to the leucosome is considerably coarser-grained than biotite farther away and may, in part, replace the garnet. The leucosome is granitic in composition, and has the mineral assemblage quartz + plagioclase + K-feldspar + minor biotite.

Location: Central Metasedimentary Belt, Grenville Province, Quebec, Canada. *Rock type*: stromatic metatexite migmatite; metapelitic protolith, anatexis in the lower granulite facies. *Scale*: the long side of the photomicrograph corresponds to 9 mm. Plane-polarized light. *Image*: E.W. Sawyer.

Figure F94

Fig. F94. The photomicrograph shows the same rock as in Fig. F93, but with crossed polarizers to show that the leucosome has a much coarser grain-size than the melanosome, even after its grain size has been reduced by deformation. However, the grain size of the minerals in the melanosome decreases systematically away from the leucosome. Deformation, recovery, and recrystallization have affected the minerals in the leucosome and have eradicated all the microstructural evidence that might have indicated crystallization from an anatectic melt. The quartz in the leucosome has undergone some grain-boundary migration in response to deformation, and this has resulted in the curved, bulging grain-boundaries. Some plagioclase crystals show curved albite-law twins and undulose extinction.

Location: Central Metasedimentary Belt, Grenville Province, Quebec, Canada. *Rock type*: stromatic metatexite migmatite; metapelitic protolith, anatexis in the lower granulite facies. *Scale*: the long side of the photomicrograph corresponds to 9 mm. Cross-polarized light. *Image*: E.W. Sawyer.

Figure F95

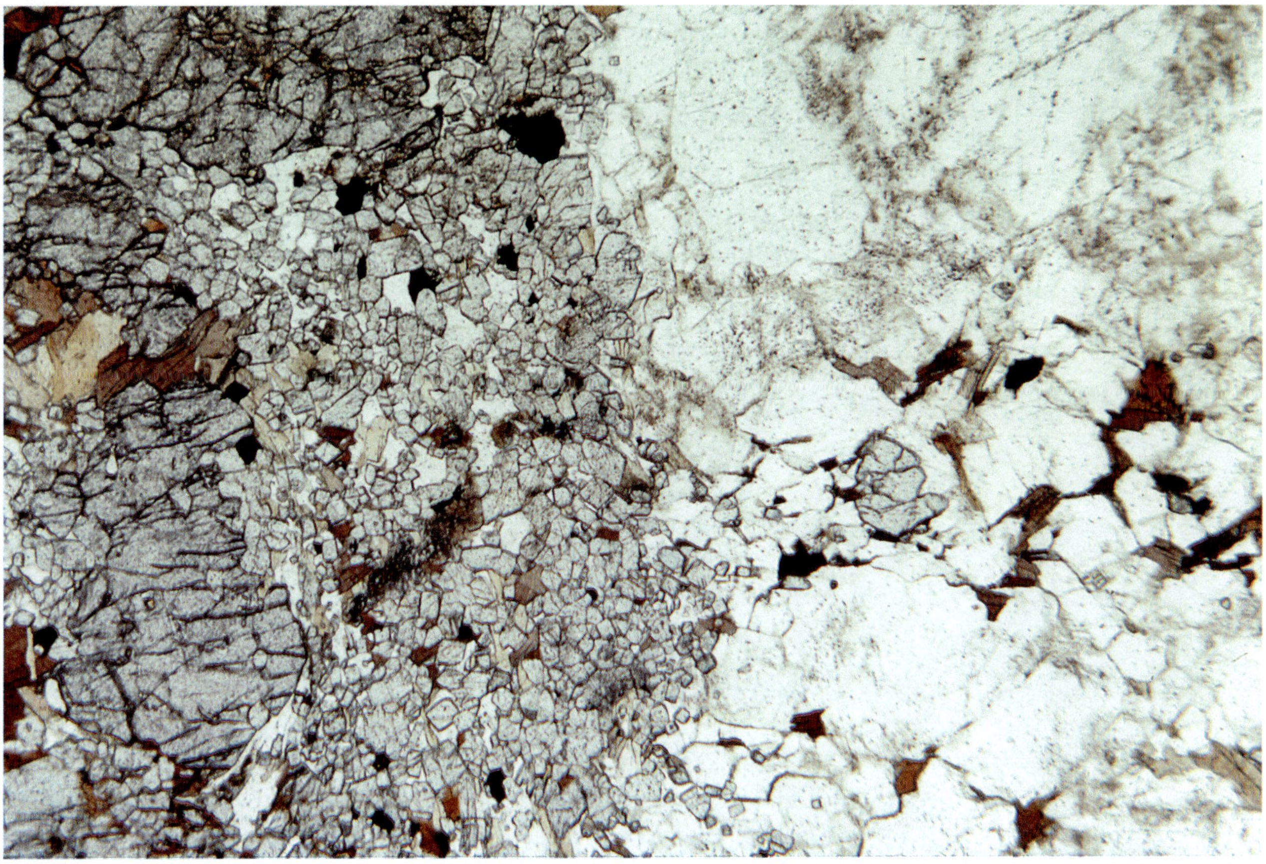

Fig. F95. This photomicrograph shows the contact between a stromatic leucosome and its residuum in a metatexite migmatite formed from a plagioclase-poor aluminous metapelite in a regional metamorphic terrane. The contact between the granitic leucosome, containing quartz + plagioclase + K-feldspar + biotite + garnet, and the adjacent residuum, containing sillimanite + garnet + K-feldspar + biotite + quartz + ilmenite, is sharp and planar in both the hand sample and the photomicrograph. The irregularity of the contact is on the scale of several grain-diameters, like those from other migmatites (e.g., Figs. F84, F91, and F93), but in this case, the grain size of the sillimanite is only 0.17 mm. Hence, the contact appears to be planar principally because of the fine grain-size of the sillimanite. The rock adjacent to the leucosome is dominated by sillimanite and is, therefore, not darker than nearby paleosome; it is residuum. Large crystals of garnet, in places embayed and partially replaced by aggregates of large crystals of biotite, occur within the sillimanite. The grains of garnet are generally not in direct contact with the leucosome, however. Hence, the mineralogical zoning in the residuum next to the leucosome is the converse of the situation in Figs. F93 and F94. Although not shown in the photomicrograph, the residuum farther away from the leucosome contains less sillimanite, and more biotite and K-feldspar; the latter, along with the garnet, represents the solid products of the incongruent breakdown of biotite through a reaction like biotite + sillimanite + quartz = melt + garnet + K-feldspar. Much of the biotite in the leucosome occurs as large grains along the boundaries between feldspar grains, or as irregularly shaped grains that appear to have grown interstitially with respect to quartz and feldspar.

Location: Mount Hay, Arunta Inlier, Australia. *Rock type*: stromatic metatexite migmatite; aluminous metapelite protolith, melted in the granulite facies, *T* 825–875°C, *P* 6–7 kbar. *Scale*: the long side of the photograph is equal to 9 mm. Plane-polarized light. *Image*: E.W. Sawyer.

Figure **F96**

Fig. F96. This photomicrograph, of the same field of view as Fig. F95, is taken in cross-polarized light to show that the leucosome is much coarser-grained than its associated residuum. Although the quartz and feldspar in the leucosome show undulose extinction and some evidence of migration of grain boundaries, some crystals of plagioclase and perthitic K-feldspar are subhedral and retain rational faces against quartz, a microstructure that is commonly used to infer growth from a melt.

Location: Mount Hay, Arunta Inlier, Australia. *Rock type*: stromatic metatexite migmatite; aluminous metapelite protolith, melted in the granulite facies, *T* 825–875°C, *P* 6–7 kbar. *Scale*: the long side of the photograph is equal to 9 mm. Cross-polarized light. *Image*: E.W. Sawyer.

Figure F97

Fig. F97. This photomicrograph shows part of a single, biotite-rich schliere 1–2 mm wide in a K-feldspar (now microcline) + quartz + plagioclase + biotite diatexite migmatite derived from an evolved melt. Schlieren in this migmatite occur in a band about 30 cm wide that can be traced several meters. Most individual schliere are thin (millimetric) biotite-rich foliae that extend for only 30–60 cm. Locally, there are wider (up to 10 mm) and longer schlieren, but they are less biotite-rich (they contain biotite + plagioclase). The microstructure of the thin schliere in the photomicrograph is typical; it is no more than a few biotite grains wide, and locally widens where it encloses crystals of plagioclase. It contains crystals of biotite that have a high aspect-ratio and that are arranged in a systematic, imbricate or tiled manner, with their long axes at a low angle to the schliere. This schliere may have been assembled as biotite flakes, forced to rotate by the shear imposed on the diatexite magma; they impinged on their neighbors and formed imbricate clusters of grains. There are small clusters of biotite that appear to be incipient schliere next to the principal one. Interestingly, these incipient schliere are separated from the principal schliere, and each other, by a layer of plagioclase that is just a single subhedral crystal (ca. 1 mm) of plagioclase wide. This microstructure is reminiscent of the 10-mm-wide biotite + plagioclase schlieren present elsewhere in the rock, composed of a net-like arrangement of biotite filaments that enclose subhedral crystals of plagioclase.

Location: Opatica Subprovince, Quebec, Canada. *Rock type*: biotite schliere in a diatexite migmatite; leucotrondhjemite protolith, *T* ca. 750°C, *P* 5–7 kbar. *Scale*: the width of the photomicrograph corresponds to 11 mm. Plane-polarized light. *Image*: E.W. Sawyer.

Further reading: Milord, I. & Sawyer, E.W. (2003): Schlieren formation in diatexite migmatite: examples from the St. Malo migmatite terrane, France. *Journal of Metamorphic Geology* **21**, 341–362.

Figure **F98**

Fig. F98. The main part of this schliere is composed of very elongate crystals of biotite that are arranged in an imbricate or tiled pattern. Consequently, the schliere is, in general, only a few grains wide. However, the biotite that forms the bridge between the two parallel schliere has a lower aspect-ratio and is not organized in an imbricate fashion. It also appears to be just one crystal of biotite wide. Crystals of biotite in the large subhedral crystals of plagioclase located between the two schlieren are oriented parallel to the schlieren, i.e., orthogonal to the bridge of biotite. Thus, the lower aspect-ratio of the biotite in the bridge could exist because the biotite there (1) crystallized from melt located between the adjacent subhedral crystals of plagioclase or (2) recrystallized from smaller grains originally oriented parallel to the schlieren. It is interesting to note that the tabular crystals of plagioclase in the host also are imbricated or tiled in the same sense as the biotite. This microstructure suggests that at the time the biotite schliere was forming, the magma also contained the tabular plagioclase crystals. Both the biotite and the plagioclase were aligned by the shear imposed on the diatexite magma.

Location: Opatica Subprovince, Quebec, Canada. *Rock type*: biotite schlieren in a diatexite migmatite; leucotrondhjemite protolith; upper-amphibolite-facies anatexis, T ca. 750°C, P 5–7 kbar. *Scale*: the width of the photomicrograph corresponds to 9 mm. Plane-polarized light. *Image*: E.W. Sawyer.

Figure F99

Fig. F99. Some of the biotite schlieren that occur in the melt-rich diatexite migmatites in the Opatica Subprovince have a conspicuous, narrow, white rim. This photomicrograph shows a biotite-rich schliere that has a plagioclase + quartz border 3 mm wide on each side that separates it from the coarse-grained K-feldspar + quartz + plagioclase leucocratic diatexite. The grain size of quartz and plagioclase decreases across the rim toward the schliere. The origin of the white, K-feldspar-free rim is problematical. The bulk composition of the diatexite can be modeled as an evolved melt, but the biotite schliere and its white plagioclase + quartz border do not resemble an anatectic melt in terms of composition. It can be modeled as an accumulation of these minerals, although how this could happen remains unknown. The change in microstructure and mineral assemblage could indicate that a local diffusion of components, such as K, may play a part in the formation of some schlieren.

Location: Opatica Subprovince, Quebec, Canada. *Rock type*: biotite schliere in a diatexite migmatite; leucotrondhjemite protolith, T ca. 750°C, P 5–7 kbar. *Scale*: the length of the photomicrograph corresponds to 11 mm. Cross-polarized light. *Image*: E.W. Sawyer.

Figure **F100**

Fig. F100. This disc-shaped schliere is about 20 mm across and 2 mm wide; it occurs in a diatexite migmatite that was derived from a pelitic protolith. It contains the mineral assemblage biotite + sillimanite + garnet + quartz + accessory phases and trace amounts of plagioclase. The central part of the schliere consists primarily of biotite that encloses prismatic sillimanite and idioblastic crystals of garnet; minor amounts of quartz occur as small, irregularly shaped grains between some of the biotite flakes. Domains of quartz + sillimanite + plagioclase occur with biotite at the border of the schliere. K-feldspar and cordierite are confined to the diatexite host, although they are locally in contact with the schliere. Most of the biotite crystals in the schliere have a high aspect-ratio. In the thicker parts of the schliere, however, the biotite microstructure is decussate, whereas in the thinner parts, it is imbricate or tiled. These relationships suggest that this schliere may be an attenuated piece of residual material, rather than the result of the accumulation of biotite crystals.

Location: Lac Kenogami, Grenville Province, Quebec, Canada. *Rock type*: biotite schliere in a diatexite migmatite; pelitic protolith, anatexis in the lower granulite facies; *T* ca. 800°C, *P* ca. 5 kbar. *Scale*: the long side of the photomicrograph corresponds to 11 mm. Plane-polarized light. *Image*: E.W. Sawyer.

Figure F101

Fig. F101. A thin (0.3 mm) mafic selvedge composed of biotite occurs between a 2.5-cm-wide K-feldspar–plagioclase–quartz leucocratic vein and its host biotite–quartz–plagioclase psammitic schist (metagreywacke). The leucocratic vein is discordant to the compositional layering, interpreted to be bedding, and the foliation in the psammitic host. The absence of K-feldspar, sillimanite, cordierite, and patches of neosome is evidence that the psammite probably did not undergo partial melting. However, thin pelitic layers elsewhere in the outcrop did partially melt, as they contain neosome, cordierite, and K-feldspar. Therefore, the psammite layers can be considered to be paleosome in the migmatite. Note that the crystals of biotite in the mafic selvedge are strongly aligned and oriented parallel to the contact with the leucocratic vein, and not parallel to the foliation in the host. The biotite crystals also are slightly longer (0.6 mm) than those in the host (0.4 mm). Two grains of plagioclase (indicated by arrows) at the edge of the leucocratic vein have grown after the mafic selvedge formed and deflected it. The feldspar, mostly K-feldspar, in the center of the vein shows more intense alteration to white mica than the plagioclase at the edges of the vein.

Location: Quetico Subprovince, Ontario, Canada. *Rock type*: mafic selvedge to a leucocratic vein in a metatexite migmatite; metagreywacke host, anatexis at *T* 700–800°C, *P* 3–4 kbar. *Scale*: the long side of the photomicrograph corresponds to 10 mm. Cross-polarized light. *Image*: E.W. Sawyer.

Figure **F102**

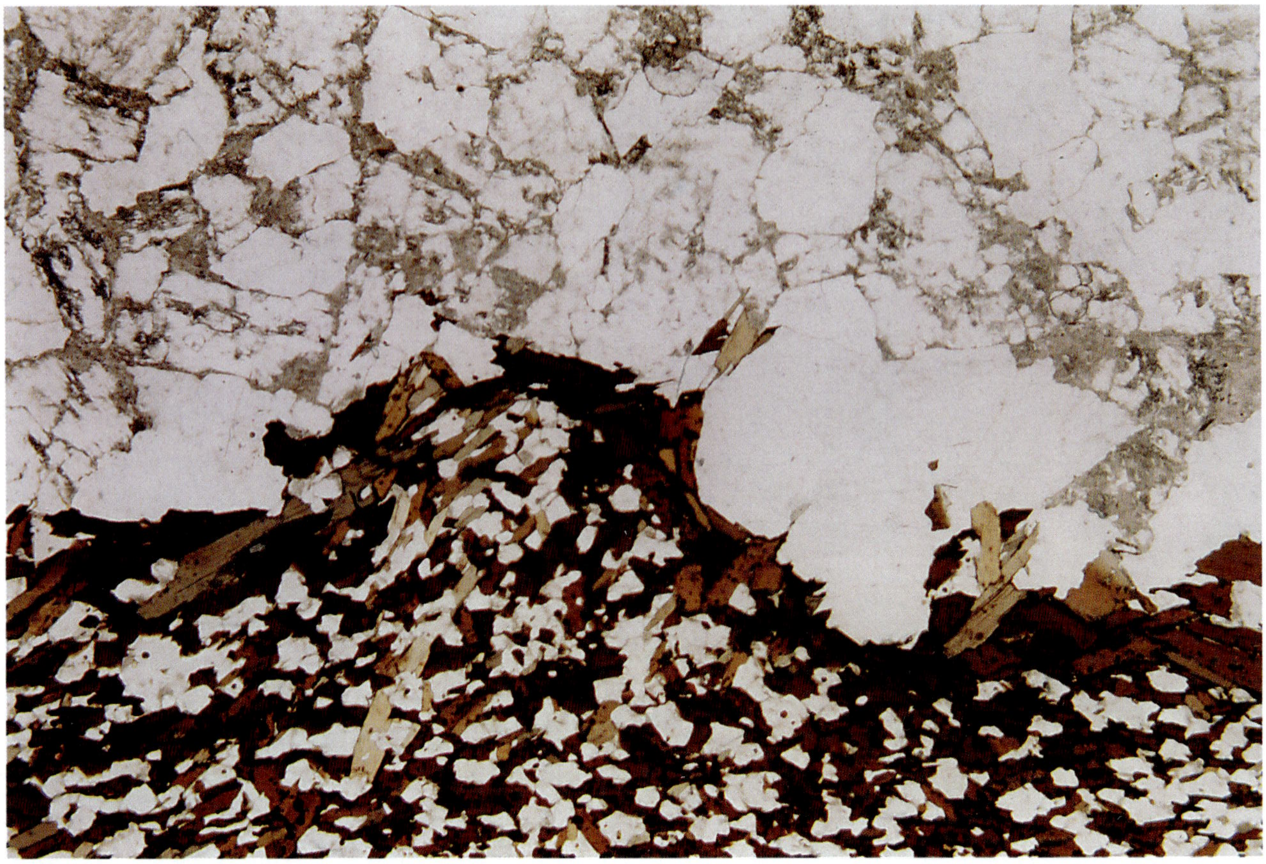

Fig. F102. This photomicrograph shows a mafic selvedge 1 mm wide developed between an apatite- and garnet-bearing quartz–plagioclase–K-feldspar leucocratic vein 15 mm wide and its quartz–biotite–plagioclase metagreywacke host. The mafic selvedge is only a few grains of biotite wide and devoid of quartz or plagioclase; quartz and plagioclase thus may have been removed as the mafic selvedge formed. Crystals of biotite in the selvedge are noticeably larger, and like those in the host, do not have much of a preferred orientation. Two large grains of plagioclase located at the edge of the leucocratic vein have undergone a late increment of growth that may have occurred after the mafic selvedge formed, as both crystals appear to have grown across the biotite that forms the selvedge. The material for this late episode of grain growth possibly was derived from the redistribution of the plagioclase component as the mafic selvedge formed.

Location: Quetico Subprovince, Ontario, Canada. *Rock type*: mafic selvedge to a leucocratic vein in a metatexite migmatite; metagreywacke host, anatexis at *T* 700–800°C, *P* 3–4 kbar. *Scale*: the long side of the photomicrograph corresponds to 14 mm. Plane-polarized light. *Image*: E.W. Sawyer.

Figure F103

Fig. F103. In this example, the host is a psammitic metagreywacke poor in both Al and K. Consequently, it contains the mineral assemblage plagioclase + biotite + quartz + hornblende. The discordant K-feldspar–plagioclase–quartz leucocratic vein that is intruded into the psammite has a markedly different microstructure from that in Fig. F101. Quartz was one of the last minerals to crystallize in the interstices of the framework structure formed by the early crystallization of euhedral and subhedral K-feldspar and plagioclase. Hence, the crystals of quartz have a blocky outline, as there are many straight contacts with the feldspar crystals, consistent with crystallization from a melt. The leucocratic vein is 11 mm wide. A narrow mafic selvedge 1.8 mm wide that contains both biotite and hornblende has formed between the vein and its host. In contrast to many mafic selvedges, this one shows a more gradual increase in the grain size of both biotite and hornblende (Hb) from the host toward the edge of the vein; both are coarser by a factor of about five relative to the host. There is also a more gradual depletion of quartz and plagioclase across the mafic selvedge.

Location: Quetico Subprovince, Ontario, Canada. *Rock type*: mafic selvedge to a leucocratic vein in a metatexite migmatite; metagreywacke host, anatexis at T 700–800°C, P 3–4 kbar. *Scale*: the long side of the photomicrograph corresponds to 11 mm. Cross-polarized light. *Image*: E.W. Sawyer.

Figure **F104**

Fig. F104. This photomicrograph shows the mafic selvedges at both sides of a discordant K-feldspar–quartz–plagioclase leucocratic vein 10 mm wide injected into a hornblende–biotite–quartz–plagioclase metagreywacke schist. Plagioclase (cleavage and cloudy alteration) was the first major mineral to crystallize in the vein and was typically euhedral or subhedral in shape. Quartz (white featureless mineral) was the last major phase to crystallize, and consequently, it has an irregular shape and blocky outline that was inherited from the shape of the melt-filled pore-space within the framework of feldspar crystals. Both mafic selvedges are narrow and show a five- to ten-fold increase in grain size over that of typical crystals of hornblende and biotite in the host. In general, the biotite in the mafic selvedge is oriented parallel to the contact of the vein, but locally there are grains, or aggregates of grains, that have other orientations. Most commonly, the differently oriented grains are spatially associated with subhedral grains of plagioclase that appear to have had a late, outwardly directed increment of growth (or perhaps overgrowth) into the mafic selvedge. The late increment of plagioclase growth from the edge of the vein may be related to the elimination of small crystals of plagioclase (and quartz) from the innermost part of the mafic selvedge. Inclusions of accessory phases and their surrounding pleochroic halos are much larger, and more conspicuous, in the biotite and hornblende of the selvedge, compared to those in the same minerals in the host rock.

Location: Quetico Subprovince, Ontario, Canada. *Rock type*: mafic selvedge to a leucocratic vein in a metatexite migmatite; metagreywacke host, anatexis at T 700–800°C, P 3–4 kbar. *Scale*: the long side of the photomicrograph corresponds to 14 mm. Plane-polarized light. *Image*: E.W. Sawyer.

G.

MIGMATITE-LIKE ROCKS

Many rocks that are not migmatites have been mistaken for them and shown on maps as migmatites. In this section, I show some of the rock types that are commonly mistaken for migmatites; I start with subsolidus segregations, and follow with syntectonic plutons, and end with arrays of felsic veins.

Layer-parallel or stromatic subsolidus segregations [Figs. G1, G2]

Segregations consisting of laterally persistent quartz or quartzofeldspathic layers flanked by biotite-rich selvedges are common in many low-, medium-, and high-grade metamorphic terranes. Typically, the segregations are located parallel to the compositional layering or to the foliation in the host; consequently, the ones in higher-grade hosts, in particular, can resemble stromatic metatexite migmatites. Numerous studies have shown that the segregations are the result of stress-induced chemical transfer of the more mobile quartzofeldspathic components, leaving the insoluble minerals, such as biotite, behind to form the dark selvedge.

Fleck segregations [Figs. G3, G4]

The term *fleck* is reserved for equant, centimeter-sized segregations, typically formed under hydrostatic conditions by subsolidus, diffusion-controlled growth; partial melting is not involved. Typically, the segregations have a globular form, although more complex shapes can arise from the coalescence of adjacent flecks. Characteristically, flecks have a melanocratic core surrounded by a leucocratic, quartzofeldspathic mantle.

Syntectonic plutons: rocks that resemble metatexite migmatites [Figs. G5–G10]

Magmas pass from a suspension of crystals in melt to a framework of crystals containing interstitial melt and, finally, to a solid as they crystallize. When a framework of touching crystals with interstitial melt (a stage called the rigid percolation threshold) forms, the crystallizing magma is in the same state as rocks in the early stages of anatexis when a framework of residual crystals contains anatectic melt. Therefore, deformation of a pluton at this stage of crystallization can produce rocks with a morphology that is remarkably similar to some types of migmatite. Furthermore, if parts of a pluton that contain magmas of different composition are deformed to high, synmagmatic shear strains, then these regions can develop a strongly layered appearance with bands of distinct composition, grain size, and microstructure, which can resemble some stromatic migmatites.

Syntectonic plutons: rocks that resemble diatexite migmatites [Figs. G11–G14]

Diatexite migmatites contain a high fraction of melt relative to paleosome; therefore, similar-looking rocks might be expected to form in environments where the ratio of magma to solid rock was high also. The margin of felsic intrusive bodies where xenoliths are entrained in magma is one such setting.

Arrays of felsic veins that look like migmatites [Figs. G15–G20]

The injection of felsic veins into mesocratic, high-grade hosts can produce rocks that resemble many migmatites, particularly if a melanocratic rim is developed around the veins. In these figures, I show some examples of veined rocks that could be mistaken for migmatites.

Figure G1

Fig. G1. This photograph shows folded subsolidus segregations in a lower-amphibolite-facies metapelite and metagreywacke host. The segregations are located parallel to the compositional layering, which is interpreted to be bedding, and have a prominent internal structure. The interior of the segregation is leucocratic and consists principally of quartz and plagioclase; however, it can be subdivided into plagioclase-rich and quartz-rich layers plus a few thin seams of biotite. There are biotite-rich selvedges on either side of the segregation, and petrographic and geochemical mass-balance estimates show that they formed by a closed-system segregation of more mobile components (quartz and plagioclase) from the less mobile (biotite). Subsolidus segregations typically develop a quartz-rich core and plagioclase-rich rim as they grow, as first quartz and then plagioclase become mobile. Consequently, the internal layering of the thick part of the segregation in the photograph could mean that it is in fact made up of several closely spaced, thinner layer-parallel segregations. The compositional layering visible in low-grade subsolidus segregations means that they are less likely to be mistaken for leucosome in stromatic migmatites. However, if the zoned structure did not develop, or has been subsequently lost through recrystallization at higher metamorphic temperatures, then subsolidus segregations can resemble plagioclase-rich leucosome.

Location: Quetico Subprovince, Ontario Canada. *Rock type*: plagioclase–quartz subsolidus segregation, lower amphibolite facies, T 600°C, P 3–4 kbar. *Scale*: the ruler is 15 cm long. *Image*: E.W. Sawyer.

Further reading: Sawyer, E.W. & Robin, P.-Y. F. (1986): The subsolidus segregation of layer-parallel quartz–feldspar veins in greenschist to upper amphibolite facies metasediments. *Journal of Metamorphic Geology* **4**, 237–260.

Figure G2

Fig. G2. The outcrop in this photograph contains interbedded metagreywacke and metapelite on the high-temperature side of the "melt-in" isograd mapped in metapelites. It is unlikely that thicker, light-colored beds of metagreywacke in the photograph have undergone partial melting; they are not sufficiently aluminous, and contain the paragenesis plagioclase + biotite + quartz. The ruler lies close to a vein of coarse-grained granite (K-feldspar + plagioclase + quartz + garnet + muscovite assemblage) that has intruded the metasedimentary rocks; in places, it has developed a very narrow, biotite-rich mafic selvedge. An offshoot from this vein of granite occupies a dextral shear zone and truncates a group of four or five layer-parallel subsolidus segregation veins (A); another subsolidus segregation (B) occurs on the other side of the shear zone. Each has a narrow, mafic selvedge that contains the assemblage biotite + plagioclase + accessory minerals. These veins have about the same plagioclase:quartz ratio (ca. 0.4), but a coarser grain-size and a more homogeneous distribution of the quartz and plagioclase than the subsolidus segregation veins in Fig. G1. These differences are attributed to recrystallization at the higher temperatures experienced by the rocks of this outcrop, rather than growth at the higher temperatures. Although the subsolidus segregations have a leucocratic core and a melanocratic rim like the neosome in many migmatites, there are some significant differences indicating that they did not form by anatexis. (1) The leucocratic parts do not contain K-feldspar; K-feldspar is expected if a melt is derived from the metapelite. (2) The mafic selvedge portions do not contain a mineral assemblage compatible with incongruent melting involving the breakdown of biotite.

Location: Quetico Subprovince, Ontario Canada. *Rock type*: metatexite migmatite containing granite veins and subsolidus segregations; metapelite and metagreywacke protolith, upper amphibolite facies, *T* 700–750°C, *P* 3–4 kbar. *Scale*: the ruler is 15 cm long. *Image*: E.W. Sawyer.

Further reading: Blom, K.A. (1988): Subsolidus migmatization in high-grade meta-tuffs (Kurkijärvi, southwest Finland). *Lithos* **21**, 263–278.

Sawyer, E.W. & Barnes, S.-J. (1988): Temporal and compositional differences between solidus and anatectic migmatite leucosomes from the Quetico metasedimentary belt, Canada. *Journal of Metamorphic Geology* **6**, 437–450.

Figure **G3**

Fig. G3. Photograph of a polished slab of biotite–plagioclase–quartz–K-feldspar gneiss (or schist) that contains well-developed fleck structures. All the flecks on this slab have a magnetite core and a quartzofeldspathic mantle. Most of the flecks are about 8 mm long and slightly elongate in shape, with the long axis oriented parallel to the foliation in the host rock. In some nearby rocks, the flecks are conspicuously elongate ellipsoids, with their long axes oriented parallel to the mineral lineation in their host. Therefore, the formation of flecks either predates, or was synchronous with, the penetrative deformation in the rocks. Flecks are not randomly distributed in this rock; some foliation-parallel domains contain more, or larger, flecks than others, and this is attributed to minor variations in composition from one layer to the next that favor the development of flecks in some layers over others. Note the close macroscopic resemblance of these fleck structures to some types of patch migmatites, for example, those shown in Figs. D9 and D10. Flecks with cores consisting of hornblende and of tourmaline also have been reported.

Location: Wet Mountains, Colorado, U.S.A. *Rock type*: flecked gneiss; biotite–gneiss protolith, metamorphic conditions *T* 600–700°C, *P* 3–5 kbar. *Scale*: tape in centimeters. *Image*: Bob Trumbull.

Further reading: Trumbull, R.B. (1988): Petrology of flecked gneisses in the northern Wet Mountains, Fremont County, Colorado. *Geological Society of America, Bulletin* **100**, 247–256.

Figure G4

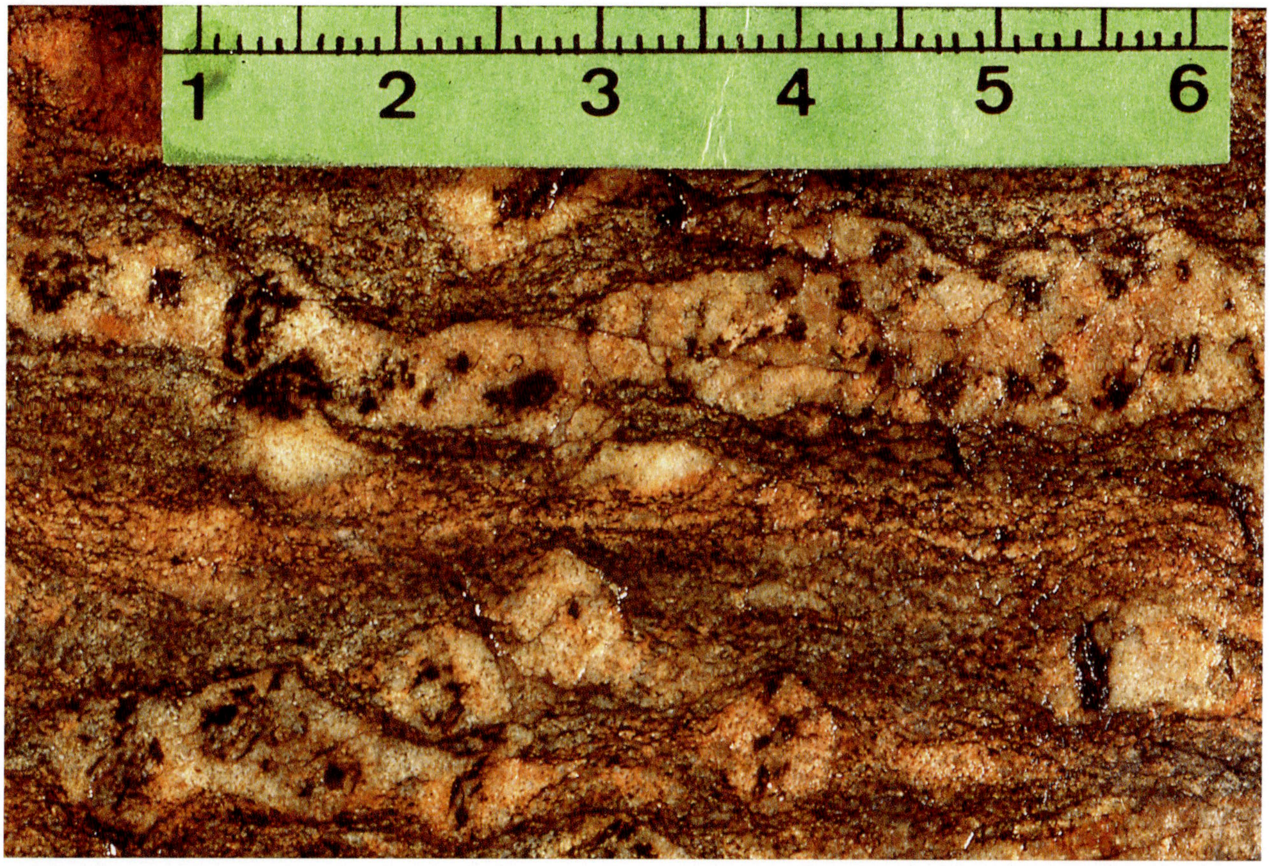

Fig. G4. The flecks in this biotite schist have hornblende in the core and are surrounded by a mantle of quartz, plagioclase, and K-feldspar. Although the rock contains some isolated flecks, it also contains discontinuous bands formed by the coalescence of many closely spaced flecks. Overall, the bands are oriented parallel to the foliation in the host, but in detail the bands are locally discordant to the biotite foliae (e.g., just below the start of the scale). It is interesting to note that a proportion of the hornblende crystals are elongate in the direction perpendicular to the length of the fleck bands and to the foliation in the host. Some of the flecks in this rock are equant in shape, but others are lenticular and weakly sigmoidal in shape. Fleck formation thus seems to have continued after the deformation ended in this rock. In contrast to the magnetite flecks in Fig. G3, the hornblende in these flecks is (1) intergrown with quartz and feldspar, (2) commonly irregular in shape, (3) not invariably centered in the fleck, and (4) commonly present in clusters, rather than as single crystals. The outcrop appearance of the hornblende flecks, and especially the band of coalesced flecks, closely resembles neosome in some migmatites, most notably those formed by the incongruent melting of biotite in which the melt and solid reactant fractions have not separated (cf. Figs. B19, D9, and D10). The most noticeable difference is the fine grain-size of the quartzofeldspathic mantle of the flecks.

Location: Wet Mountains, Colorado, U.S.A. *Rock type*: flecked gneiss; biotite gneiss protolith, metamorphic conditions, *T* 600–700°C, *P* 3–5 kbar. *Scale*: the tape is in centimeters. *Image*: Bob Trumbull.

Further reading: see Fig. G3.

Figure G5

Fig. G5. The outcrop is part of a syntectonic pluton that is about 20 km across. The most abundant lithology is grey tonalite, most of which has a foliation obtained in the magmatic or submagmatic state. At some places, the tonalite still was a magma when it was intruded by mafic magma, and this resulted in the formation of trains of rounded, pillow-shaped mafic enclaves. Elsewhere, the tonalite had crystallized to a greater extent, and the injection of mafic magma into it formed dikes, some of which, as in this outcrop, were later disrupted to form scattered, subangular mafic blocks in a foliated tonalite host. A leucocratic tonalitic to granodioritic magma also was present, and formed conspicuous white layers parallel to the foliation in the tonalite, and around the blocks of mafic material. Close inspection of the outcrop reveals the presence of an array of interconnected grey veins of nonfoliated tonalite that enclose blocks of the foliated tonalite; the contacts between the veins and the host are gradational. In addition, there are (e.g., at the left end of the mafic enclave in the center) transitional contacts from foliated tonalite through nonfoliated tonalite to the leucotonalite–granodiorite, indicating that all must have contained some melt at the same time. This array of diffuse veins enclosing a foliated host closely resembles some net-structured metatexite migmatites, and is the principal reason why this outcrop could be mistaken for a migmatite. However, there is no petrological evidence that the rocks in this outcrop underwent partial melting. Locally, metamafic rocks and siliciclastic metasediments form the wallrocks to the pluton, and they contain lower-amphibolite-facies assemblages of minerals, which also suggests that the pluton did not undergo anatexis. Furthermore, the regional "melt-in" isograd is located at least 20 km away. Consequently, the transitional contacts are interpreted to indicate that the network of tonalite and leucotonalite–granodiorite veins were derived from a residual (i.e., a fractionated or evolved) melt that was expelled from the foliated tonalite part of the outcrop as it was deformed during the final stages of its crystallization. The youngest veins in the outcrop are granitic in composition, and are distinguished by their pink color and pegmatitic microstructure (comb-structure of feldspar crystals grown in from the vein walls and the quartz-rich center); these veins are 17 Ma younger than the ages of crystallization of the tonalite pluton.

Location: Ouescapis Pluton, Opatica Subprovince, Quebec, Canada. *Rock type*: syntectonic tonalite pluton, lower amphibolite facies. *Scale*: the ruler is 15 cm long. *Image*: E.W. Sawyer.

Figure G6

Fig. G6. This photograph shows part of a syntectonic, quartz monzodiorite pluton that has been intruded by a mafic magma as it crystallized. In the lower part of the outcrop, the mafic magma, which was less competent than its monzodiorite host when it was deformed, has been attenuated into thin, dark-colored wisps that resemble schlieren in diatexite migmatites. In contrast, the mafic dike in the center of the photograph was more competent than its host, and was disrupted into a train of boudins. The monzodiorite was in a submagmatic state (i.e., a crystal framework with as much as 60 vol.% of interstitial melt) when the boudins formed, and since the regions between the boudins are domains of low pressure during the boudinage process, they sucked melt out of the interstices in the framework of feldspar and amphibole crystals. The melt that segregated from the monzodiorite matrix had an evolved composition and produced small veins of leucogranodiorite between the boudins when it crystallized. The morphology of this part of the syntectonic pluton consists of leucocratic veins located between mafic boudins in a mesocratic host and, therefore, resembles some metatexite migmatites (e.g., Figs. B39, D18, D19, and D25). However, in this case, there is no petrographic evidence for partial melting in either the mafic layers or in the monzodiorite. The composition of the leucocratic rocks is consistent with a fractionated melt remaining after the host monzodioritic magma had largely crystallized.

Location: Canet pluton, Opatica Subprovince, Quebec, Canada. *Rock type*: syntectonic monzodiorite pluton, greenschist facies. *Scale*: the ruler is 15 cm long. *Image*: E.W. Sawyer.

Figure G7

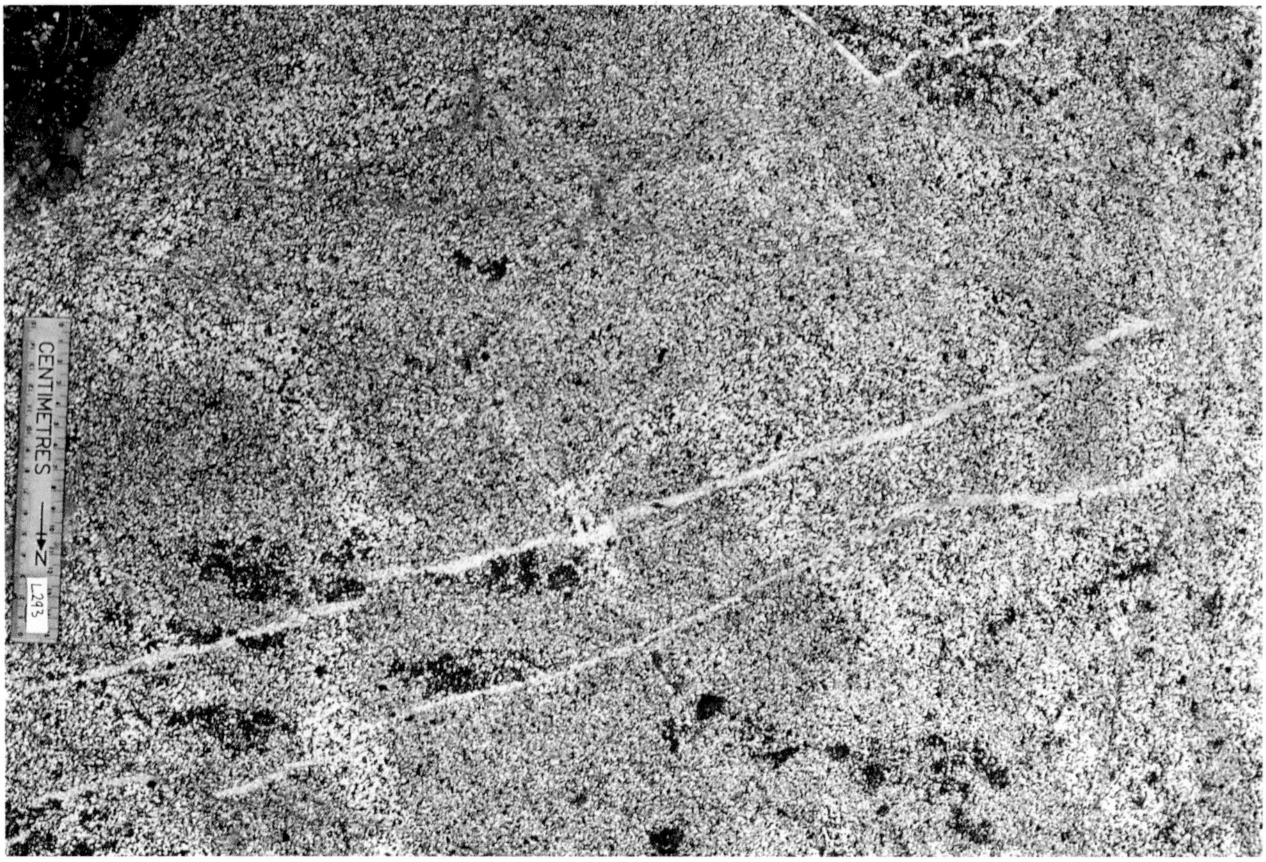

Fig. G7. This photograph shows an outcrop in the 2824 ± 3 Ma Lac Rodayer pluton, a typical Archean pluton of the tonalite–trondhjemite–granodiorite suite. Three types of tonalite can be distinguished in the outcrop by subtle differences in color, grain size, and microstructure. The bottom left corner of the photograph has the darkest rock, and it is a biotite–hornblende–quartz–plagioclase tonalite. This grades to a slightly lighter-colored biotite–hornblende–quartz–plagioclase tonalite that is also slightly coarser-grained, and occupies the top half of the outcrop. Both of the grey tonalites form patches with somewhat rounded shapes, and are enveloped by a leucocratic K-feldspar–hornblende–biotite–quartz–plagioclase tonalite that is coarse-grained; this is prevalent at the bottom right. The region of interest is to the right of the scale. There, the leucocratic tonalite forms an array of diffuse veins that divides the darkest tonalite into several blocks. Overall, therefore, this outcrop contains leucocratic, mesocratic, and melanocratic parts, and moreover, the leucocratic parts are diffuse veins in the melanocratic part, a relationship not unlike that in some migmatites. The best evidence that the relationships in this outcrop are not due to anatexis comes from nearby outcrops, where magma-mingling features are obvious. However, all the rocks in this outcrop contain euhedral crystals of plagioclase that have a complexly zoned core, a feature that appears to be characteristic of magma mingling, not anatexis. The presence of K-feldspar and a high content of incompatible elements in the leucocratic parts suggest that the magma from which they formed was derived by the fractional crystallization of a tonalitic magma. The orientation of the leucocratic veins suggests that the fractionated melt may have been expelled from the host tonalites during minor deformation at a point when this part of the pluton was almost solid.

Location: Lac Rodayer pluton, Opatica Subprovince, Quebec, Canada. *Rock type*: syntectonic tonalite–trondhjemite–granodiorite pluton, amphibolite facies. *Scale*: the ruler is 15 cm long. *Image*: E.W. Sawyer.

Figure G8

Fig. G8. This photograph shows a granite with the mineral paragenesis K-feldspar + plagioclase + quartz + biotite + amphibole that was emplaced into a dextral, strike-slip shear zone and was strongly deformed as it crystallized. Abundant tabular plagioclase and K-feldspar crystals in the magma created a framework structure containing interstitial melt. Deformation of this framework of crystals as the pluton cooled first oriented its crystals into a strong foliation, and then created shear bands, asymmetrical boudins, and other dilatant structures in it. During this process, the loss of interstitial pore-space by compaction and the creation of low-pressure sites at the dilatant structures generated local pressure-gradients, which caused the melt to move from the pores to the dilatant structures, in a process identical to the formation of leucosome in dilatant structures in migmatites. The photograph shows a segregation of the residual melt in an asymmetrical boudin structure. The resemblance to leucosome in some migmatites is clearly evident (see Figs. B6, D18, D19, D22, and D23); the segregation is coarse-grained, leucogranitic in composition, has diffuse borders and a granitic microstructure. Although it has no melanocratic border, principally because the granite has a low modal content of ferromagnesian minerals, there is a depletion halo three centimeters wide around the segregation. This halo is depleted in material of the same composition as the segregation compared to the matrix 10 and 30 cm away from the segregation, and is correspondingly enriched in the proportion of plagioclase and K-feldspar that it contains. The halo indicates a local origin for the melt in the segregation. Factors that distinguish this segregation from leucosome in a migmatite are the absence of any evidence for a partial-melt-producing reaction in the host and the highly evolved bulk composition of the leucocratic segregation, which is more compatible with the final product of a long sequence of fractional crystallization in the pluton rather than anatexis of the granite.

Location: Near Tadoussac, Grenville Province, Quebec, Canada. *Rock type*: leucogranite segregation in a syntectonic hornblende granite pluton. *Scale*: the ruler is 15 cm long. *Image*: E.W. Sawyer.

Figure G9

Fig. G9. Many syntectonic plutons contain parts that are strongly banded on a centimeter scale and that resemble migmatites with a stromatic morphology. In this example, thin bands of leucotonalite and various bands of melatonalite are interlayered with mesocratic grey tonalite; two aspects are particularly suggestive of migmatites. (1) Some of the most leucocratic layers are in contact with very melanocratic ones (e.g., near the lens cap); this is reminiscent of the leucosome–melanosome pairs in migmatites in which movement of anatectic melt to the leucosome created the adjacent melanosome. (2) The horizontal, discordant leucocratic vein in the center of the photograph occupies a small shear zone, and appears to have gradational contacts with its host rocks at both ends, a feature that could indicate local derivation of the vein. However, other evidence against this rock being a migmatite is compelling. (1) Some of the leucocratic veins, including the coarse-grained ones, are discordant to the layering, and many have no melanocratic rim at all, whereas others have but a very narrow one. The leucocratic veins thus seem to have been intruded into a tonalite that already had compositional layering, and reaction between the veins and their wallrocks seems responsible for the mafic selvedges. (2) There is no evidence for partial-melting reactions in the rocks that would establish the required petrogenetic link between the melanocratic layers and the leucocratic ones, and between both these and the mesocratic grey tonalite. The features of this outcrop could be synmagmatic and principally due to the injection of leucocratic and melanocratic material in veins into a crystallizing grey tonalite, with some reaction between the veins and the host to create melanocratic rims. The vein complex underwent high shear strain during crystallization; this rotated and attenuated the veins into near-parallelism, folded some and, in the later stages of crystallization, led to the expulsion of residual melt into dilatant structures to form leucosome-like veins. Some veins of coarse-grained material were more competent than their hosts and underwent boudinage; they could have been solid, but they may simply have had a smaller fraction of melt.

Location: Pyhäsalmi area, Finland. *Rock type*: Svecokarelian syntectonic tonalite pluton. *Scale*: the lens cap is 6 cm in diameter. *Image*: S.-J. Barnes.

Figure G10

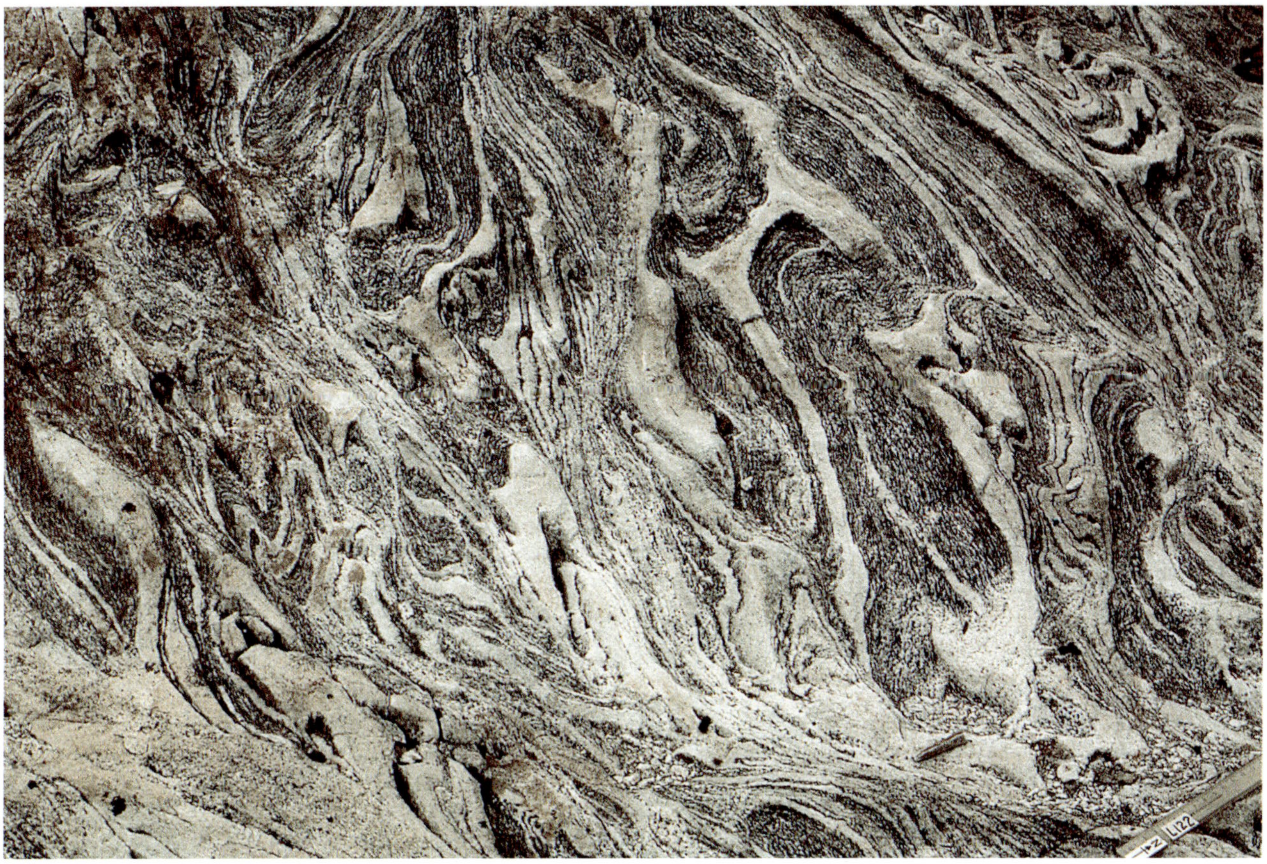

Fig. G10. The outcrop in this photograph contains three rock types: (1) fine-grained leucogranodiorite layers, (2) medium-grained, mesocratic plagioclase + quartz + biotite tonalite gneiss layers, and (3) rare, discontinuous, biotite-rich melanocratic layers located between some of the leucocratic and mesocratic layers. This outcrop has been interpreted as a migmatite in which partial melting of the mesocratic tonalite gneiss formed the leucocratic and melanocratic layers. However, three factors argue against such an interpretation. First, there appears to be far too much of the leucocratic material relative to the melanocratic fraction; in other words, there is too much "melt" and not enough "residuum" in the outcrop. Secondly, in order to produce a K-feldspar-bearing melt from the tonalite gneiss, a melting reaction involving the incongruent breakdown of biotite seems necessary, and there is no evidence for one. Thirdly, the whole-rock composition of the leucocratic layers resembles leucogranodiorite veins in a nearby hornblende monzodiorite to granodiorite pluton, and these leucogranodiorite veins formed by fractional crystallization of the major magmatic phase of the pluton. In addition, the tonalite gneiss layers have the same major- and trace-element compositions as a nearby pluton of tonalite. Mapping indicates that this migmatite-like rock formed where abundant dikes of leucogranodiorite emanating from a syntectonic monzodiorite to granodiorite pluton have intruded the margins of a tonalite pluton. Both the host and the dikes were then rotated into parallelism and attenuated by a synmagmatic shear zone, and then folded. The biotite-rich selvedges formed as a result of reaction between the two different lithologies. The high degree of recrystallization present in these rocks indicates that deformation in the shear zone continued until very close to the solidus. However, there are some coarser-grained diffuse regions that contain euhedral crystals of feldspar located within small subsidiary shear zones, and these are the sites where the last melt accumulated.

Location: Border between the Abitibi and Opatica subprovinces, Quebec, Canada. *Rock type*: sheared vein complex. *Scale*: the ruler is 15 cm long. *Image*: E.W. Sawyer.

Figure G11

Fig. G11. Many Archean and Proterozoic syntectonic tonalite–granodiorite plutons have portions that resemble migmatites. This photograph shows the termination of one such domain, and Fig. G12 shows its interior. The lower and upper parts of this photograph consist of tonalite; it is weathered brownish grey at the bottom, but is fresh at the top. The tonalite is foliated, and has a centimeter-scale compositional layering due to heterogeneities in the original magma, later attenuated by shearing in the magmatic or submagmatic state. The mafic enclaves are considered to be the remains of a layer of mafic magma intruded contemporaneously with the tonalite. However, they could be xenoliths. The migmatite-like appearance is due to the presence of an elongate, tapered body of leucogranodiorite that crosses the layering and partially engulfs the mafic enclaves. The compositional layering in the tonalite curves on both sides of the leucogranodiorite, indicating that it occupies a dextral shear zone. The compositional layering of the tonalite fades into the leucogranodiorite, which gives it a diffuse aspect that resembles some nebulitic and diatexite migmatites (e.g., Fig. B15). The microstructure of the leucogranodiorite is magmatic and essentially isotropic, and indicates that it crystallized after the shearing. Lens-shaped bodies of coarse-grained granite occur in the leucogranodiorite and the tonalite, and postdate the layering in the latter, but are extended by the dextral shearing; hence, they are older than the microstructure in the leucogranodiorite. Small, dark, irregular patches in the leucogranodiorite are fragments from the mafic enclaves that have been altered to biotite. Pelitic schists in the country rocks are of staurolite + andalusite grade; hence, partial melting of the tonalite is unlikely to have occurred, and thus this is not a migmatite. Gradational contacts between the leucogranodiorite and its host indicate that both contained melt at the same time. Consequently, the leucogranite is interpreted to be derived from the fractionated residual melt drawn in from the host tonalite as the dextral shear zone formed. The zone of dilatancy likely formed once the tonalite had crystallized sufficiently (about 55%) to pass the rigid percolation threshold.

Location: La Grande Subprovince, northern Quebec, Canada. *Rock type*: syntectonic segregation of residual melt in a tonalite pluton, lower amphibolite facies. *Scale*: the ruler is 15 cm long. *Image*: E.W. Sawyer.

Figure G12

Fig. G12. The tonalite country-rock is visible in the top left and bottom right corners of this photograph. The compositional layering in the bottom right can be traced to the left, but becomes progressively more indistinct toward the ruler, and this part resembles a nebulitic migmatite. The center of the picture consists of leucogranodiorite, which contains numerous mafic enclaves. Many of the enclaves were frozen in the process of disaggregation. The morphology that results from the mafic fragments scattered within leucogranodiorite resembles that of some diatexite migmatites (see Figs. D42, D44, D47, and D50). Mafic enclaves and fragments have a melanocratic rim along contacts with the leucogranodiorite. Melanocratic rims are also found in migmatites, but in this rock, biotite replaces hornblende, which suggests a reaction between the leucogranodiorite and the enclave, rather than the extraction of melt from the latter. The leucogranodiorite between the largest enclaves is isotropic and has no trace of the layering in the tonalite; therefore, this is newer material. Consequently, this portion of the pluton is interpreted to have become dilatant during the late stages of its crystallization. The evolved melt residing in the pores of the tonalite then migrated into this dilatant region and intruded both the mafic enclaves and the tonalite. The influx of melt into this area increased the local fluid pressure and partially disaggregated the mafic layer and the surrounding tonalite, which created a diatexite-like rock. Other effects were reaction of the mafic rocks with H_2O-rich leucogranodioritic melt (biotite rim replacing hornblende) and the partial destruction of the old structures in the tonalite (e.g., near the ruler).

Location: La Grande Subprovince, northern Quebec, Canada. *Rock type*: syntectonic segregation of residual melt in a tonalite pluton, lower amphibolite facies. *Scale*: the ruler is 15 cm long. *Image*: E.W. Sawyer.

Figure **G13**

Fig. G13. This photograph shows a border phase of a felsic intrusive body. The wallrocks are at the bottom of the photograph, and have been veined and broken into angular blocks. Within a short distance into the felsic intrusive body, the xenoliths are smaller, and many are noticeably rounded, whereas others have become fragmented and partially disaggregated. In other words, they show signs of magmatic erosion. Most of the fragments of wallrock have developed a dark, coarse-grained reaction rim or rind; the larger enclaves still retain the original mineralogy and microstructure of the wallrock in the core. However, the smaller enclaves are coarsely recrystallized throughout. The felsic magma around the enclaves has become strongly contaminated with crystals and larger fragments from the disaggregated enclaves. The morphological similarities with diatexite migmatites are strong (cf. Figs. D42–D44); in particular, the dark-colored enclaves and the rim around the enclaves and along felsic veins in the wallrocks resemble the melanosome, i.e., the melt-depleted residuum in migmatites. However, there is no evidence for a felsic fraction derived from a melt-producing reaction in these enclaves, and the dark rim can be explained by reaction of the felsic melt with the wallrock. Potassium from the melt diffuses into the enclave and converts amphibole there to biotite, and the Ca released by the conversion of the amphibole to biotite diffuses out into the felsic magma and generates plagioclase there.

Location: near Biella, northwestern Italy. *Rock type*: border phase of a felsic intrusive body. *Scale*: The width of the image corresponds to 1.4 m. *Image*: E.W. Sawyer.

Figure G14

Fig. G14. Small plutons, dikes, or sills of evolved leucogranite or aplite commonly develop a layered structure. In some cases, the repeated switching back and forth, or oscillation, of conditions, such as the activity of H_2O, during crystallization produces repeated differences in grain size, microstructure, or mineralogy that result in the development of a fine-scale layering. However, in other examples, very light-colored quartzofeldspathic layers alternate with somewhat darker layers that are richer in biotite; this type of layering may be due to the presence of two magmas. This photograph shows the interior of a sill of leucogranite that is hosted by a sequence of pelitic to psammitic metasediments. The leucogranite consists of alternating very light-colored layers that are granitic in composition and contain <1 modal % biotite, and darker granitic to granodioritic layers that are slightly richer in plagioclase and contain up to 10 modal % biotite. The layering is interpreted to result from the presence of two magmas when the sill was emplaced. One was very felsic and had a fractionated composition, whereas the other was less evolved and had a higher biotite content. The two magmas were attenuated together, either as the sill was emplaced, or afterward during shearing, but before it had solidified. The result is a prominent but wispy layering that contains flow structures and truncated layers and that, overall, strongly resembles some schollen and schlieric diatexite migmatites.

Location: Eastmain, Nemiscau Subprovince, Quebec, Canada. *Rock type*: layered leucogranite; host metasedimentary rocks in the middle to upper amphibolite facies. *Scale*: the ruler is 15 cm long. *Image*: E.W. Sawyer.

Figure G15

Fig. G15. The principal rock-type in this outcrop is a grey biotite granodiorite in which the foliation is parallel to a weak compositional layering. The granodiorite contains numerous veins of pink granite, most of which are parallel to the foliation and compositional layering, but some are discordant. The granitic veins are folded, and by using the vergence of adjacent folds, a second-order fold closure can be identified in the center of the outcrop. Both parallel and similar fold forms are present. There are also examples of "eye" structures (i.e., closed loop structures), but these are not interpreted as sheath folds, because the original attitude of the veins cannot be deciphered. The granite veins on the far right are much thinner than elsewhere, and have a layered, or stromatic, morphology in which individual veins exhibit a pinch-and-swell structure, which indicates that they have been attenuated. Many of the veins in the outcrop have a narrow mafic selvedge; the mafic selvedge is narrowest where the veins are the most attenuated. The presence of boudins, pinch-and-swell and parallel folds all suggest that the granitic veins were more competent than the granodiorite host during the deformation. As one explanation for this scenario, the veins were completely crystalline when the folding occurred. The presence in this outcrop of a mesocratic host (the biotite granodiorite) that contains younger and spatially related leucocratic and melanocratic portions is reminiscent of the relationships seen between the constituent parts of many metatexite migmatites. If the grey granodiorite had undergone partial melting, then the melanocratic border around the granitic veins might be interpreted as the residuum left after the extraction of an anatectic melt that became the granitic veins. However, the proportion of melanocratic rock relative to the amount of pink granite present in the outcrop seems to be too small to account for the *in situ* formation of all the granite present (see also Fig. B34). Therefore, it seems likely that most, and possibly all, of the granite veins have been injected. If the granite in the veins was not locally derived (i.e., the melanocratic rims are not residuum) but rather injected, then this is not a migmatite. The felsic material should be interpreted as a complex of granitic veins, and the mafic selvedges, as the product of the local alteration of the granodiorite by the granite. However, if some of the granitic material in the veins was produced *in situ* by melting, then this is a migmatite, but an open-system one in which the main mechanism responsible for creating this morphology is the injection of a granitic magma in veins.

Location: Saturday Peak, Monashee Complex, British Columbia, Canada. *Rock type*: granite veins in a biotite granodiorite. *Scale*: the hammer in the center is 40 cm long. *Image*: Paul McNeill.

Figure G16

Fig. G16. This photograph shows an outcrop of foliated granodiorite that contains numerous parallel granitic veins. The granite veins have a very narrow, melanocratic rim. Most of the granitic veins are thin and oriented parallel to each other and to the main compositional layering in the host granodiorite; because most of the veins are laterally persistent, they give the outcrop a layered morphology. Many of the granitic veins exhibit a pinch-and-swell structure, or have wider portions that correspond to partitions between asymmetrical foliation boudins in the host granodiorite. However, some granitic veins are discordant to the layering, which suggests that they are dikes, whereas a few others occur as isolated, lenticular bodies in asymmetrical boudinage structures. Locally, granitic veins outline isoclinal fold hinges. The ratio of the melanocratic border to granitic veins is low in this outcrop, which, taken with the discordant nature of some veins, suggests that much of the granitic material may have been injected as dikes. Thus, the situation is the same as with the rocks shown in the previous figure (G15); in order for the rocks in this outcrop to be called migmatite, there has to be some evidence that the granodiorite, and the granitic veins plus their associated melanocratic rims, are genetically linked by partial melting. If this cannot be shown, then the outcrop is a granitic vein complex hosted by the granodiorite. The macroscopic characteristics of the granitic veins in this outcrop indicate that when they were deformed, they were more competent than the host granodiorite. Although the rocks shown in this figure and the previous one have formed in essentially identical ways, their final morphologies are strikingly different. This is because the rocks shown in this figure were subject to higher strains, which rotated and attenuated (transposed) the granitic veins far more, and consequently generated their strongly layered appearance.

Location: Mount Odin area, Monashee complex, British Columbia, Canada. *Rock type*: sheared dike complex. *Scale*: the head of the ice axe is 35 cm wide. *Image*: Paul McNeill.

Figure **G17**

Fig. G17. Melanocratic layers of garnet–sillimanite–biotite schist alternate with leucocratic quartzofeldspathic layers in this outcrop to produce a morphology that closely resembles a stromatic metatexite migmatite. The composition of the melanocratic layers is very uniform over the entire outcrop, and is consistent with the interpretation that it is residuum left after the extraction of an anatectic melt. The leucocratic layers, however, are problematic, as they exhibit a wide range of microstructures and modal composition. Some parts are fine-grained, and contain the mineral assemblage plagioclase + quartz + biotite ± K-feldspar, whereas the most leucocratic parts are richer in K-feldspar and, in places, display a feldspar–quartz intergrowth microstructure. The absence of garnet and sillimanite from all the leucocratic layers is not what one would expect if the leucocratic layers had resulted from the anatectic melt generated in making the biotite + sillimanite + garnet residuum from a metapelitic protolith. Some of the leucocratic layers are folded and others are boudinaged, which is the result of deformation during, or after, their crystallization.

Location: Val Strona, Italy. *Rock type*: deformed felsic dikes. *Scale*: the ruler is 15 cm long. *Image*: E.W. Sawyer.

Figure G18

Fig. G18. This photograph of another part of the outcrop shown in Fig. G17 shows evidence indicating that this rock is in fact not a migmatite. Here, some of the leucocratic layers are as much as 30 cm wide (e.g., bottom right), and show clearly that the leucocratic layers were derived from two different magmas. One gave rise to a grey biotite–K-feldspar–quartz–plagioclase granodiorite or tonalite with a fine to medium grain-size, whereas the other produced white and very leucocratic rocks that have a coarser grain-size and, locally, a pegmatitic microstructure. The leucocratic body in the bottom right of the photograph shows that the more leucocratic magma intruded the other and that both enclose fragments of the melanocratic biotite–sillimanite–garnet schist as xenoliths. Because no genetic link due to partial melting can be established between the leucocratic and the melanocratic rocks, the rock is not a migmatite. The relationships observed in some of the leucocratic rocks suggest that the morphology of this rock is the result of diking of a melt-depleted garnet–sillimanite–biotite schist by two felsic magmas and their subsequent deformation.

Location: Val Strona, Italy. *Rock type*: deformed felsic dikes. *Scale*: the ruler is 15 cm long. *Image*: E.W. Sawyer.

Figure G19

Fig. G19. Metasedimentary schists derived from siliciclastic turbidite sequences are a common component of cratons around the world. In most Precambrian terranes, Al-poor rocks containing the assemblage plagioclase + quartz + biotite, derived from greywacke protoliths, dominate in the metasedimentary belts. For example, the Archean metasedimentary subprovinces in Quebec and the Proterozoic ones in the Narvik area of Norway each contain >80% greywacke, whereas metapelites are a minor constituent. Partial melting in these Al-poor rocks does not normally begin until the assemblage biotite + plagioclase + quartz becomes unstable, at temperatures of about 800°C. However, in many terranes, these rocks take on the appearance of stromatic migmatites in the middle amphibolite facies because of the presence of thin granitic veins parallel or subparallel to the bedding or foliation. This photograph shows a single granitic vein that contains the mineral assemblage plagioclase + quartz + K-feldspar + biotite + garnet. Similar granitic layers and larger bodies (dikes and small plutons) nearby do not contain garnet, but some have muscovite, whereas others contain spodumene. Some of the feldspar crystals in the granite layers have rational faces against quartz, and this is taken as evidence for crystallization from a melt. The edges of the granitic vein are not smooth; they have millimeter- to centimeter-scale indentations and irregularities that are not necessarily the result of boudinage or pinch-and-swell structures (see Figs. F102 and F104). A very narrow, mafic selvedge around the granite veins is ubiquitous, and it consists mainly of biotite; however, the selvedge in this photograph contains a few crystals of garnet that are identical to that in the granite vein. In general, the mafic selvedges in this type of migmatite-like rock do not contain sillimanite, cordierite, orthopyroxene, garnet, or K-feldspar, but muscovite may be present. The major- and trace-element composition (e.g., high Li, Cs, Th, U, and K_2O) of the granite vein in the photograph indicates that it is the product of a highly fractionated anatectic melt derived from a metapelitic or metagreywacke protolith. The absence of mineral assemblages or microstructures indicating that a melt-producing reaction occurred in any of the surrounding rocks, or in the mafic selvedges, indicates that these rocks do not consist of migmatite. They are interpreted as veined rocks, albeit veined by a felsic melt that was derived from a protolith similar in bulk composition to the host. However, the partial melting occurred at greater depth, where temperatures were sufficient. As the anatectic melt migrated upward, it crystallized and became progressively more fractionated, and was finally injected into and crystallized in an amphibolite-facies subsolidus host-rock.

Location: Nemiscau Subprovince, Quebec, Canada. *Rock type*: granite vein in a metagreywacke, middle amphibolite facies, *T* ca. 600°C, *P* 3–5 kbar. *Scale*: the scale is 15 cm long. *Image*: E.W. Sawyer.

Figure G20

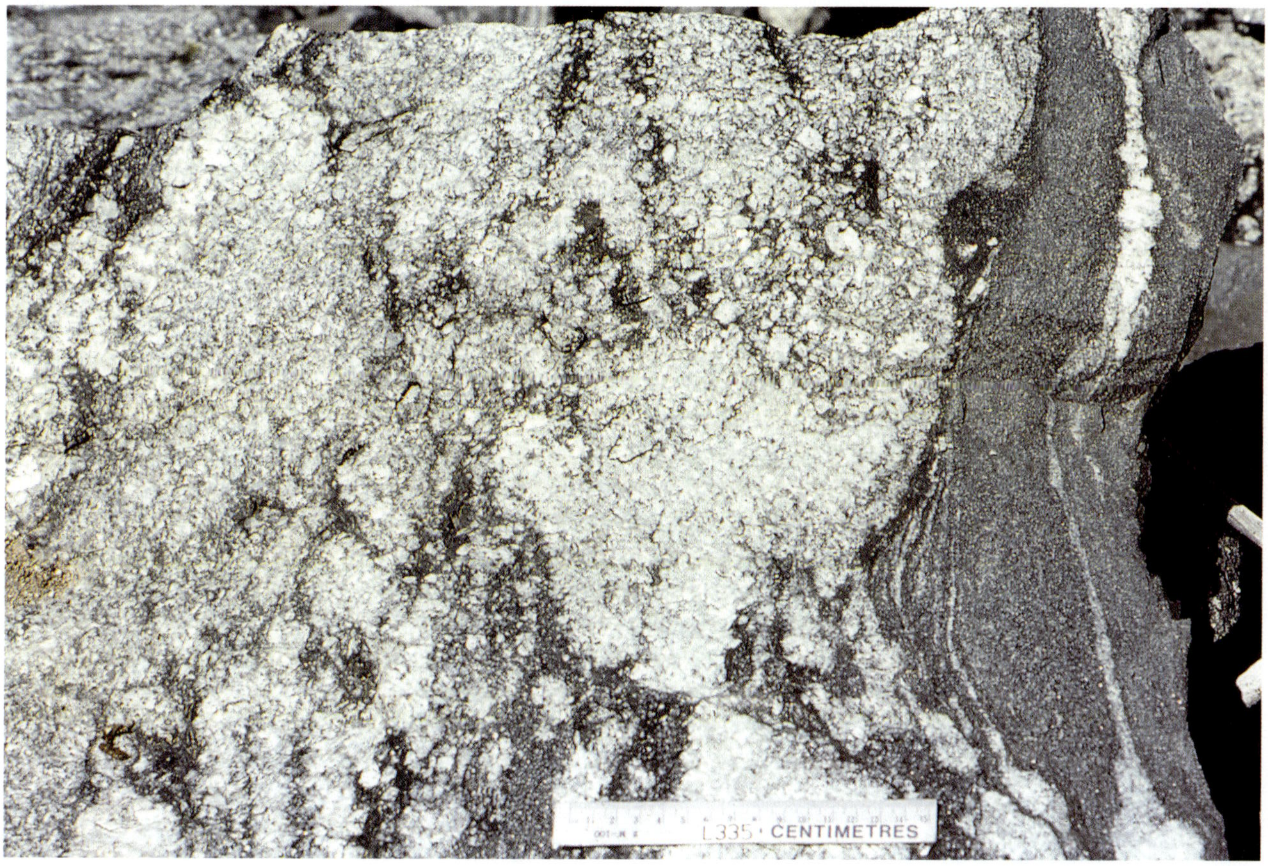

Fig. G20. The higher proportion of granitic veins makes the intrusive origin of this migmatite-like rock clear. Nevertheless, the outcrop resembles some diatexite migmatites (cf. Figs. D43, D45, D49, and D54). A layer of the biotite–quartz–plagioclase schist occurs at the right of the photograph. The schist does not contain orthopyroxene, nor does it have any microstructures indicative that it partially melted. However, it had reached the middle amphibolite facies when the granite sheets were emplaced into it. A narrow mafic selvedge consisting mostly of biotite occurs between the granite and its host. Other less continuous melanocratic patches occur within the granite, and these may mark the position of other granite–host contacts that have been subsequently overprinted by granite veins. These relics of mafic selvedges consist almost entirely of biotite (no garnet, cordierite, orthopyroxene, or K-feldspar is present). The mafic selvedge is interpreted to be the result of a reaction between the wallrock and the granitic magma, or more likely of an aqueous fluid exsolved from it. The whole-rock composition of the granite sheets in this outcrop indicates that they formed from an evolved melt derived from a pelitic or greywacke source. The U–Pb zircon ages of the granite sheets is 2672 ± 2 Ma, about 4 Ma younger than U–Pb monazite ages from nearby migmatites. Consequently, these are interpreted as rocks from below the "melt-in" isograd that have been veined by an evolved melt that escaped from a deeper anatectic zone.

Location: Nemiscau Subprovince, Quebec, Canada. *Rock type*: granite veins or dikes in metagreywacke, middle amphibolite facies. *Scale*: the ruler is 15 cm long. *Image*: E.W. Sawyer.

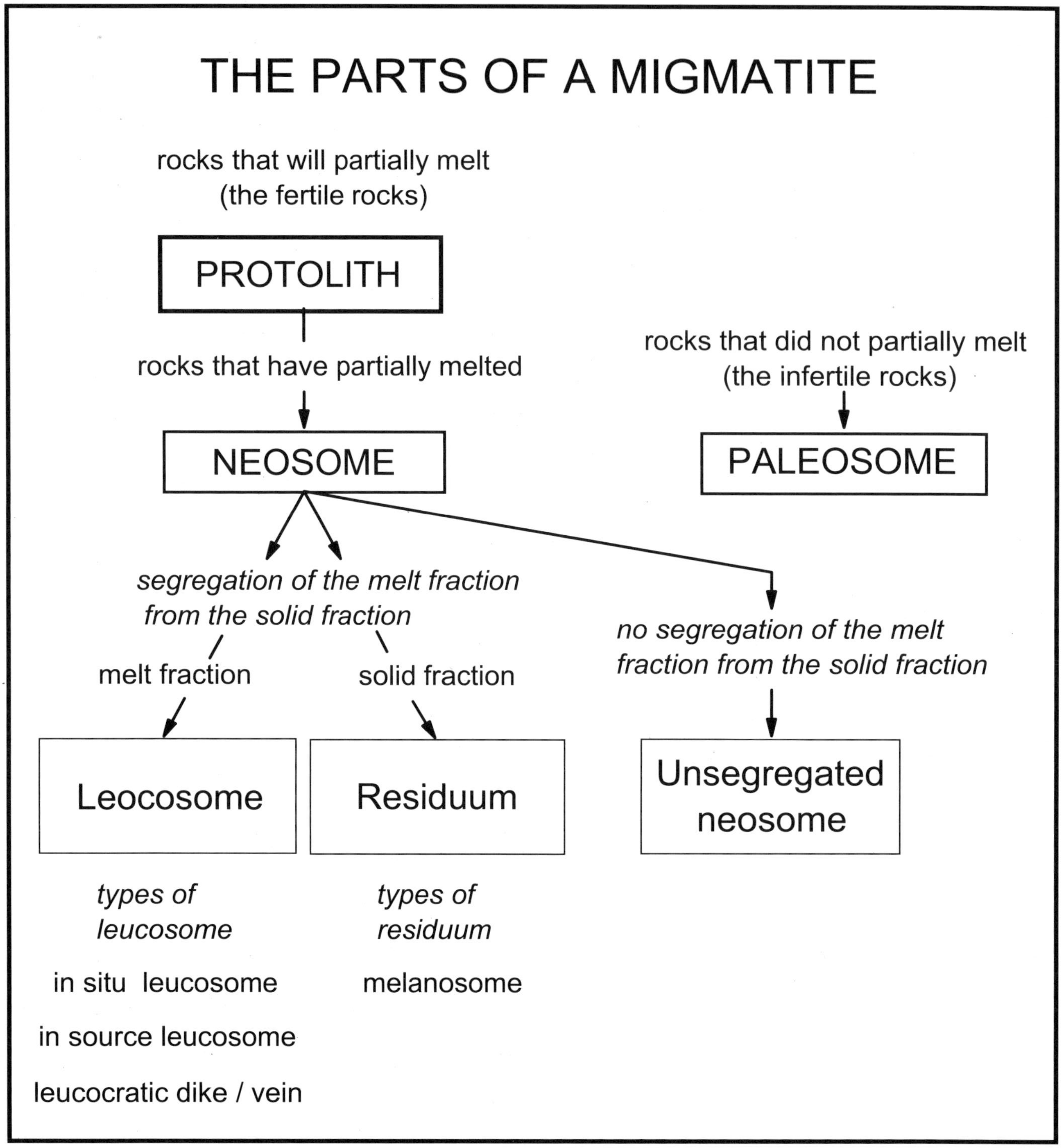
THE PARTS OF A MIGMATITE
rocks that will partially melt
(the fertile rocks)
PROTOLITH
rocks that have partially melted
NEOSOME
rocks that did not partially melt
(the infertile rocks)
PALEOSOME
segregation of the melt fraction
from the solid fraction
melt fraction
solid fraction
no segregation of the melt
fraction from the solid fraction
Leocosome
Residuum
Unsegregated
neosome
types of
leucosome
in situ leucosome
in source leucosome
leucocratic dike / vein
types of
residuum
melanosome